Science Networks. Historical Studies
Volume 23

Edited by Erwin Hiebert, Eberhard Knobloch
and Erhard Scholz

José Ferreirós

Labyrinth of Thought

A History of Set Theory
and Its Role in Modern Mathematics

Birkhäuser Verlag
Basel · Boston · Berlin

Author:

José Ferreirós
Dpto. de Filosofía y Lógica
Universidad de Sevilla
Avda. San Francisco Javier, s/n
E-41005 Sevilla
e-mail: jmferre@cica.es

Library of Congress Cataloging-in-Publication Data
Ferreirós Domínguez, José.
    Labyrinth of thought : a history of set theory and its role in
modern mathematics / José Ferreirós.
        p. cm. -- (Science networks historical studies ; v. 23)
    Includes bibliographical references and index.
    ISBN 0-8176-5749-5 (alk. paper). – ISBN 3-7643-5749-5 (alk. paper)
    1. Set theory--History.   I. Title.   II. Series.
    QA248.F467 1999
    511.3'22'09 – dc21

Deutsche Bibliothek Cataloging-in-Publication Data
Ferreirós Domínguez, José:
Labyrinth of thought : a history of set theory and its role in modern
mathematics / José Ferreirós. - Basel ; Boston ; Berlin : Birkhäuser, 1999
    (Science networks ; Vol. 23)
    ISBN 3-7643-5749-5 (Basel...)
    ISBN 0-8176-5749-5 (Boston)

ISBN 3-7643-5749-5 Birkhäuser Verlag, Basel – Boston – Berlin

©1999 Birkhäuser Verlag, P.O. Box 133, CH-4010 Basel, Switzerland
Camera-ready copy prepared by the author.
Printed on acid-free paper produced from chlorine-free pulp. TCF ∞
Cover design: Micha Lotrovsky, Therwil, Switzerland
Cover illustration: Riemann and Dedekind (left), Cantor (center), Zermelo and Russell (right)
Printed in Germany
3-7643-5749-5

9 8 7 6 5 4 3 2

A Dolores, por todos estos años.

Contents

Introduction xi

  1. Aims and Scope xiii

  2. General Historiographical Remarks xvii

Part One: The Emergence of Sets within Mathematics 1

I   **Institutional and Intellectual Contexts in
German Mathematics, 1800–1870** 3

  1. Mathematics at the Reformed German Universities 4

  2. Traditional and 'Modern' Foundational Viewpoints 10

  3. The Issue of the Infinite 18

  4. The Göttingen Group, 1855–1859 24

  5. The Berlin School, 1855–1870 32

II   **A New Fundamental Notion: Riemann's Manifolds** 39

  1. The Historical Context: Grössenlehre, Gauss, and Herbart 41

  2. Logical Prerequisites 47

  3. The Mathematical Context of Riemann's Innovation 53

  4. Riemann's General Definition 62

  5. Manifolds, Arithmetic, and Topology 67

  6. Riemann's Influence on the Development of Set Theory 70

  Appendix: Riemann and Dedekind 77

III **Dedekind and the Set-theoretical Approach to Algebra** 81

  1. The Algebraic Origins of Dedekind's Set Theory, 1856–58 82

  2. A New Fundamental Notion for Algebra: Fields 90

  3. The Emergence of Algebraic Number Theory 94

4. Ideals and Methodology 99

5. Dedekind's Infinitism 107

6. The Diffusion of Dedekind's Views 111

IV   The Real Number System 117

1. 'Construction' vs. Axiomatization 119

2. The Definitions of the Real Numbers 124

3. The Influence of Riemann: Continuity in Arithmetic and Geometry 135

4. Elements of the Topology of ℝ 137

V   Origins of the Theory of Point-Sets 145

1. Dirichlet and Riemann:
   Transformations in the Theory of Real Functions 147

2. Lipschitz and Hankel on Nowhere Dense Sets and Integration 154

3. Cantor on Sets of the First Species 157

4. Nowhere Dense Sets of the Second Species 161

5. Crystallization of the Notion of Content 165

**Part Two: Entering the Labyrinth –
Toward Abstract Set Theory** 169

VI   The Notion of Cardinality and the Continuum Hypothesis 171

1. The Relations and Correspondence Between Cantor and Dedekind 172

2. Non-denumerability of ℝ 176

3. Cantor's Exposition and the 'Berlin Circumstances' 183

4. Equipollence of Continua ℝ and ℝ^n 187

5. Cantor's Difficulties 197

6. Derived Sets and Cardinalities 202

7. Cantor's Definition of the Continuum 208

8. Further Efforts on the Continuum Hypothesis 210

VII   Sets and Maps as a Foundation for Mathematics 215

1. Origins of Dedekind's Program for the Foundations of Arithmetic 218

2. Theory of Sets, Mappings, and Chains 224

3. Through the Natural Numbers to Pure Mathematics 232

4. Dedekind and the Cantor–Bernstein Theorem 239

5. Dedekind's Theorem of Infinity, and Epistemology 241

6. Reception of Dedekind's Ideas 248

VIII The Transfinite Ordinals and Cantor's Mature Theory 257

1. "Free Mathematics" 259

2. Cantor's Notion of Set in the Early 1880s 263

3. The Transfinite (Ordinal) Numbers 267

4. Ordered Sets 274

5. The Reception in the Early 1880s 282

6. Cantor's Theorem 286

7. The *Beiträge zur Begründung der transfiniten Mengenlehre* 288

8. Cantor and the Paradoxes 290

Part Three: In Search of an Axiom System 297

IX   Diffusion, Crisis, and Bifurcation: 1890 to 1914 299

1. Spreading Set Theory 300

2. The Complex Emergence of the Paradoxes 306

3. The Axiom of Choice and the Early Foundational Debate 311

4. The Early Work of Zermelo 317

5. Russell's Theory of Types 325

6. Other Developments in Set Theory 333

X   Logic and Type Theory in the Interwar Period 337

1. An Atmosphere of Insecurity: Weyl, Brouwer, Hilbert 338

2. Diverging Conceptions of Logic 345

3. The Road to the Simple Theory of Types 348

4. Type Theory at its Zenith 353

5. A Radical Proposal: Weyl and Skolem on First-Order Logic 357

XI   Consolidation of Axiomatic Set Theory 365

1. The Contributions of Fraenkel 366

2. Toward the Modern Axiom System: von Neumann and Zermelo 370

3. The System von Neumann–Bernays–Gödel 378

4. Gödel's Relative Consistency Results 382

5. First-Order Axiomatic Set Theory 386

6. A Glance Ahead: Mathematicians and Foundations
after World War II 388

Bibliographical References 393

Index of Illustrations 422

Name Index 423

Subject Index 430

Introduction

> I descried, through the pages of Russell, the doctrine of sets, the *Mengenlehre*, which postulates and explores the vast numbers that an immortal man would not reach even if he exhausted his eternities counting, and whose imaginary dynasties have the letters of the Hebrew alphabet as ciphers. It was not given to me to enter that delicate labyrinth. (J. L. Borges, *La cifra*, 1981)

The story of set theory, a 'doctrine' that deals with the labyrinth of infinity and the continuum, is sometimes told in a way that resembles beautiful myths. The Greek goddess Athena sprang full-grown and armored from the forehead of Zeus, and was his favorite child.[1] Set theory is generally taken to have been the work of a single man, Georg Cantor, who developed single-handedly a basic discipline that has deeply affected the shape of modern mathematics. He loved his creature so much that his life became deeply intertwined with it, even suffering mental illness for its sake. The comparison between theory and goddess is interesting in other ways, too. Athena was a virgin goddess, whereas set theory is comparable to number theory in its purity and abstractness. She was the goddess of wisdom and of the polis, while set theory plays an organizing role in the polis of mainstream modern mathematics and represents one of the highest achievements of mathematical wisdom. She was also the goddess of war, which brings to mind the polemics and disputes brought forward by the discipline that will occupy our attention in the present study.

There certainly is wisdom in the search for founding fathers or founding myths, especially when a new approach or a new discipline is fighting for recognition. Moreover, mathematicians tend to concentrate on advanced results and open problems, which often leads them to forget the ways in which the orientation that made their research possible actually emerged.[2] Both reasons help us understand why the tradition of ascribing the origins of set theory to Cantor alone goes back to the early 20th century. In [1914], Hausdorff dedicated his handbook, the first great manual of set theory, to its "creator" Cantor. One year later, the *Deutsche Mathematiker-Vereinigung* sent a letter to the great mathematician, on the occasion of his 70th

---

[1] Athena, goddess of the Parthenon in Athens, was identified with the Roman Minerva.

[2] On founding fathers, see [Bensaude-Vincent 1983]. Kuhn [1962, chap. 11] analyzed how scientists, in their systematic work, are continuously rewriting and hiding the real history.

birthday, calling him "the creator of set theory."[1] Later on, Hilbert chose set theory as a key example of the abstract mathematics he so strongly advocated, and he constantly associated it with the name of Cantor (see, e.g., [Hilbert 1926]). This is natural, for Cantor turned the set-theoretical approach to mathematics into a true branch of the discipline, proving the earliest results in transfinite set theory and formulating its most famous problems.

But this traditional view has also been contested from time to time. Some twenty years ago, Dugac wrote that the birthplace of set theory can be found in Dedekind's work on ideal theory [Dugac 1976, 29]. Presented this way, without further explanation, this assertion may seem confusing.[2] In the introduction to Dugac's book, Jean Dieudonné wrote opinionatedly:

> the 'paradise of Cantor,' that Hilbert believed to be entering, was in the end but an artificial paradise. Until further notice, what remains alive and fundamental in Cantor's work is his first treatises on the denumerable, the real numbers, and topology. But in these domains it is of justice to associate Dedekind to him, and to consider that both share equally the merit of having founded the set-theoretical basis of present mathematics.[3]

This viewpoint is also in agreement with some older views, for instance with Zermelo's in his famous paper on the axiomatization of set theory, which he called the "theory created by Cantor and Dedekind" [1908, 200]. If one takes into account that Dedekind's set-theoretical conceptions were very advanced by 1872, and that he and Cantor became – so we are told – good friends from that time (see chapter VI), the customary story appears problematic. This uneasy state of affairs was actually the starting point for my own work, though its scope grew and changed substantially in time.

The thrust of the argument most frequently used by those who attributed the authorship of set theory to Cantor is just the following. Cantor was the man who, in the latter half of the 19th century, introduced the infinite into mathematics; this, in turn, became one of the main nutrients in the spectacular flowering of modern mathematics.[4] If this is the whole argument, one can simply say that its premise is historically inaccurate, so the conclusion does not follow. For at least another author, Dedekind, introduced the actual infinite unambiguously and influentially even before Cantor. On the other hand, Cantor did inaugurate transfinite set theory, after others had started to rely on actual infinity and while the theory of point-sets was being studied by several mathematicians.

---

[1] [Purkert & Ilgauds 1987, 165-66]: "Schöpfer der Mengenlehre."

[2] A similar but subtler pronouncement can be found in [Edwards 1980, 346].

[3] [Dugac 1976, 11]: "le 'paradis de Cantor' où Hilbert croyait entrer n'etait au fond qu'un paradis artificiel. Jusqu'à nouvel ordre, ce qui reste vivant et fondamental dans l'œuvre de Cantor, ce sont ses premiers travaux sur le dénombrable, les nombres réels et la Topologie. Mais dans ces domaines, il n'est que juste de lui associer Dedekind, et de considérer qu'ils partagent à titre égal le mérite d'avoir fondé les bases 'ensemblistes' de la mathématique d'aujourd'hui."

[4] I have paraphrased [Lavine 1996, 1], but see [Hausdorff 1914], [Fraenkel 1923; 1928].

The above shows that, despite the large number of historical works which have dealt with set theory in one way or another, we still lack an adequate and balanced historical description of its emergence. The present study attempts to fill the gap with a general overview that synthesizes much previous work and at the same time tries to provide new insights.[1]

1. Aims and Scope

The traditional historiography of set theory has reinforced several misconceptions as regards the development of modern mathematics. Excessive concentration upon the work of Cantor has led to the conclusion that set theory originated in the needs of analysis, a conclusion embodied in the very title of [Grattan-Guinness 1980]. From this standpoint, it would seem that the successful application of the set-theoretical approach in algebra, geometry, and all other branches of mathematics came afterwards, in the early 20th century. These novel developments thus appear as unforeseen successes of Cantor's brainchild, whose most explicit expression would be found in Bourbaki [Meschkowski 1967, 232–33].

The present work will show, on the contrary, that during the second half of the 19th century the notion of sets was crucial for emerging new conceptions of algebra, the foundations of arithmetic, and even geometry. Moreover, all of these developments antedate Cantor's earliest investigations in set theory, and it is likely that some may have motivated his work. The set-theoretical conception of different branches of mathematics is thus inscribed in the very origins of set theory. It is the purpose of Part One of the present work to describe the corresponding process. That will lead us to consider whether there was a flux of ideas between the different domains, trespassing disciplinary boundaries.

Part Two analyzes the crucial contributions to abstract set theory made in the last quarter of the 19th century. This means, above all, Cantor's exploration of the transfinite realm – the labyrinth of infinity and the continuum – which started with his radical discovery of the non-denumerability of $\mathbb{R}$ in December 1873. It also means Dedekind's work on sets and mappings as a basis for pure mathematics, elaborated from 1872. And, as a natural consequence, the interaction between both authors, who maintained some personal contacts and an episodic but extremely interesting correspondence. Particular attention will be given to the reception of their new ideas among mathematicians and logicians, ranging from the well-known opposition of Kronecker and his followers, to the employment of transfinite numbers in function theory, the rise of modern algebra and topology, and the expansion of logicism.

---

[1] Among previous historical contributions, the following, at least, stand out: [Jourdain 1906/14], [Cavaillès 1962] (written in 1938), [Medvedev 1965], a number of books centering on Cantor – above all [Dauben 1979] and [Hallett 1984], but also [Meschkowski 1967], [Purkert & Ilgauds 1987] – , [Dugac 1976], and finally [Moore 1982].

In Part Three, a synthetic account of the further evolution of set theory up to 1950 is offered, concentrating on foundational questions and the gradual emergence of a modern axiomatization. Apart from the attempt to offer, perhaps for the first time, a comprehensive overview, a novel feature that deserves special mention is the attention given to a frequently forgotten aspect of this period – the bifurcation of two alternative systems, Russellian type theory and Zermelian set theory, and their subsequent convergence. Along the way we shall review a wide range of topics, including aspects of the so-called 'foundational crisis,' constructivist alternatives to set theory, the main axiomatic systems for set theory, metatheoretical work on them (in particular, Gödel's results), and the formation of modern first-order logic in interaction with formal systems of set theory.

One can see that the topics dealt with concentrate gradually from part to part. Part One studies the emergence of the set-theoretical approach against the background of more traditional viewpoints. Without abandoning the question of how the language of sets became dominant, Part Two concentrates on abstract set theory, and Part Three has an even more restricted focus on the *foundations* of abstract set theory.[1]

The reader must have noticed that we shall not just focus on set theory in the strictest sense of the word. If by those words the reader understands abstract set theory as it is presently studied by authors who are classified as mathematical logicians, he or she should turn to Parts Two and Three.[2] Such a conception overlooks the question of how the set-theoretical approach to mathematics arose and why it came to play a central role in modern mathematics. These questions are certainly of much wider interest, and they seem no less important for the historian. Thus I have decided not to deal exclusively with (transfinite) set theory, but to pay careful attention to the set-theoretical approach too, asking questions like the following: How did mathematicians arrive at the notion of set? How did they become convinced that it offered an adequate basis and language for their discipline? How could a mathematician (like Cantor) convince himself that it is important to develop a theory of sets?

The whole historical development that we shall analyze is best described as a *progressive differentiation* of subdisciplines within the historical context of research programs that originally were unitary. We shall observe that, at first, the notion of set was employed in several different domains and ways, with no recognition of subdisciplinary boundaries. Gradually, mathematicians recognized differences between several aspects that originally appeared intertwined in the traditional objects of mathematics – metric properties, topological features, algebraic structures, measure-theoretic properties, and finally abstract set-theoretical aspects. In the process, new subdisciplines emerged.

I realize that, in following this path, I risk the danger of being sharply criticized by lovers of neat conceptual distinctions. In general, it is not commendable to proj-

---

[1] An attempt to treat the development of abstract and descriptive set theory in full until about 1940 would have required another volume, and of course a long period of preparation.

[2] And complement them with other works, like [Moore 1982], [Kanamori 1996].

ect present-day disciplinary boundaries on the past. In this connection, it is interesting to consider the rather negative review that K. O. May wrote [1969] of a book by the late Medvedev [1965] on the development of set theory (a book that I have not read, since it has never been translated from Russian, but which may be similar to the present work in some respects). May argued that, in order to counter the opinion that Cantor was the creator of set theory, Medvedev confused abstract set theory with the topological theory of point-sets. But Cantor himself did not differentiate the theory of point-sets from abstract set theory until as late as 1885, fifteen years after he had started to do original work involving sets (see chapter VIII).[1] May also criticized Medvedev's search for 19th-century precedents of the notion of set, like Gauss, since they never went beyond an *implicit* use of actual infinity. This is a more serious criticism, yet one may reflect on the fact that Dedekind, who indulged in a very explicit use of sets and actual infinity, mentioned Gauss's work in order to justify his viewpoint (see §III.5).[2]

One has to say that the theory of sets only becomes mathematically interesting, and controversial, with the acceptance of infinite sets. Here one should share May's reservations, and I have decided to establish the acceptance of actual infinity as a *criterion* for 'serious' involvement with the notion of set.[3] In this connection, one should distinguish between the acceptance of the *actual infinite* and the elaboration of a theory of the *transfinite*. Cantor's work was very important on both accounts, but it was only in the second domain that he struck out on his own. It is commonly said that the mathematical tradition rejected the actual infinite, but that position was not universal in 19th century Germany (see §I.3).

Distinctions like the preceeding one are very useful in any attempt to understand the development of set theory, and they are important for clarifying my approach and preventing misunderstandings. A second distinction, already indicated and intimately related to the first, is that between set theory as an autonomous branch of mathematics – as in *transfinite* set theory or *abstract* set theory – and set theory as a basic tool or language for mathematics: the set-theoretical *approach* or the *language* of sets. As indicated above, abstract set theory came about after the set-theoretical approach began to develop, not the other way around.

A third important distinction is that between set theory as an approach to or a branch of mathematics and set theory as a *foundation* for mathematics. In our present picture of set theory, its three aspects, as a language or approach to mathematics, a sophisticated theory, and a (purported) foundation for mathematics, are so intertwined that it may seem artificial to distinguish them. Nevertheless, within the context of the early history of sets it is essential to take these distinctions into account.

---

[1] The mathematical community as a whole began to assimilate the distinction only in the 1900s; abstract set theory appears in full clarity with the work of Zermelo in the same decade.

[2] The style of internal references is as follows: '§2.1' indicates section 2.1 within the same chapter where the reference is found; '§VI.4' indicates section 4 in chapter VI.

[3] The criterion of actual infinity justifies to some extent the exclusion of Weierstrass (§IV.2); also important in this connection are methodological questions (§I.4 and §IV.2).

Otherwise it becomes impossible to produce a clear picture of the development, which includes a rather complex interaction between the three aspects. The idea that sets constitute the foundation of mathematics emerged very early, and Cantor was by no means the leading exponent of this view. Its strongest proponent as of 1890 was Dedekind, but I shall argue that Riemann was also an important voice advocating this position.[1] Riemann took a significant step in the direction of introducing the language of sets, coupling this with the conception that sets are the basic objects of mathematics. There are good reasons to regard his early contribution as a significant influence on the early set-theoretical attempts of both Dedekind and Cantor.

Thus, Parts One and Two deal mostly with a small group of mathematicians – above all Riemann, Dedekind and Cantor – whose work was closely interconnected. This made it possible to present quite a clear picture of the evolution from some initial set-theoretical glimpses, to what we can presently recognize as abstract set theory. To judge from May's review, this may be what Medvedev's work missed, making his research on the immediate precedents of set theory seem irrelevant.

The notion of set seems to be, to some extent, natural for the human mind. After all, we employ common names (like 'book' or 'mathematician') and one is easily led to consider sets of objects as underlying that linguistic practice. For this simple reason, it is too easy to find historical precedents for the notion of set, and an incursion into such marshy terrain can easily become arbitrary or irrelevant. It is precisely to avoid this risk that I have concentrated on contributions which can be shown to have been directly linked to the work of the early set theorists. Since all of these men were German mathematicians, the first two parts of the book concentrate on mathematical work written in the German language.

In this connection, I would like to warn the reader that a study of the origins of the notion of set in Germany is immediately confronted with terminological difficulties. In contrast to Romance languages and English, where there were rather clear candidates for denoting the concept (*ensemble, insieme, conjunto, set*), the German language did not suggest a best choice. To give an example, both Dedekind and Cantor accepted that 'ensemble' was an ideal translation into French,[2] but they used different German terms. Dedekind chose the word 'System,' Cantor changed his choice several times, but mostly used the words 'Mannigfaltigkeit' and 'Menge.'[1] In the early period, each mathematician made his own selection, and one must carefully establish whether they were talking about sets, as commonly understood, or something else. The terminological question becomes particularly critical in the case of Riemann – the whole issue of his role in the early history of sets depends on how we interpret his notion of 'Mannigfaltigkeit' (see chapter II). I have decided to translate the relevant terms rigidly throughout the text, but it is important that the reader keep the problem of terminological ambiguity in mind. The mere fact that one finds the word 'Menge' [set] in a text does not in itself mean that the author is employing the right notion. And the mere fact that an author uses another word does

---

[1] See chapter II and also [Ferreirós 1996].

[2] See Dedekind [1877] and Cantor's 1885 paper in [Grattan-Guinness 1970].

not mean that he is *not* employing the notion of set.[1]

To end this section, I would like to comment on the opinion of some historians who have emphasized the difference between early notions resembling that of set (e.g., in Bolzano or in Riemann) and the modern notion, which has commonly been attributed to Cantor.[2] As a result of taking sets to be determined by concepts, Riemann and Bolzano regard them as endowed with an antecedently given structure. Today one considers an abstract set and then freely imposes structures on it. This contrast might be turned into an argument against the thesis that any of these 19th century authors were important in the development of set theory. My answer is that, if such a criterion were rigidly applied, even Cantor's work would not belong to the history of set theory. All earlier authors, including Cantor and Dedekind, started with conceptions akin to those of Riemann and Bolzano, and it was only gradually that they (eventually) arrived at an abstract approach. They began with the concrete, complex objects of 19th-century mathematics, and in the course of their work they realized the possibility of distinguishing several different kinds of features or properties (that we would label as metric, or topological, or algebraic, or abstract properties). The abstract, extensional notion of set developed gradually out of the older idea of concept-extension (see §II.2). Therefore, it is historically inadequate to contrast a 'concrete' with an abstract approach as exclusive alternatives; rather they should be regarded as initial and final stage in a complex historical process.[3]

2. General Historiographical Remarks

Many of the recent historical works dealing with the emergence of set theory are of a biographical character. Certainly, the biographical approach to history has its strengths, but it also has weaknesses. It becomes quite difficult to avoid a certain partiality as an effect of excessive concentration on a single author, or simply due to empathic identification.[4] From what precedes, it should be clear that the present writer has made an option for a less narrowly focused historical treatment – a collective approach. It is almost self-evident that great scientific contributions are collective work, and the emergence of set theory and the set-theoretical approach to mathematics is no exception.

In Parts One and Two, the work of a small group of authors is studied through a 'micro' approach, and the peculiarities of their orientation are analyzed by comparison with competing schools or approaches. As much attention is paid to informal

---

[1] There were many more linguistic variants: 'Klasse,' 'Inbegriff,' 'Gebiet,' 'Complex,' 'Vielheit,' 'Gesamtheit,' 'Schaar,' and so on. We shall encounter them along the way.

[2] See, for example, [Scholz 1990a, 2] and [Spalt 1990, 192-93].

[3] It is quite obvious that, in general, the process of invention/discovery will go from the familiar and concrete to the abstract.

[4] This danger is present, for instance, in the most comprehensive historical account of Cantor's life and work [Dauben 1979]. It is instructive to compare the partially overlapping contributions of Dauben and Hawkins to [Grattan-Guinness 1980].

communication (personal contacts, correspondence) as publications. Questions of influence thus become central, although I admit that a study of such questions must enter into the domain of the hypothetical much more than a more classical study of published work. Hopefully the reader will acknowledge the value of an inquiry of this kind.

I have already mentioned that the reader will notice a gradual narrowing of the conceptual issues treated as the narrative develops from the emergence of the set-theoretical approach (Part One) to the foundations of abstract set theory (Part Three). This step-by-step concentration goes in counterpoint with a progressive widening of the group of actors. Set theory began as the brainchild of just a few original thinkers, and it gradually became a community enterprise; as a natural side-effect, the amount of work that the historian has to cover increases exponentially.

The early history of set theory can only be adequately written when one abandons present disciplinary boundaries and makes at least some attempts to consider cross-disciplinary interactions. One has to pay attention to ideas and results proposed in several different branches of mathematics, if only because mathematicians of the late 19th century were by no means as narrowly specialized as they tend to be today.[1] It must be taken into account that clear boundaries between various branches of mathematics were not yet institutionalized before 1900. The careers of the figures that we shall study are clear examples that specialization was only beginning, and that 19th century mathematicians enjoyed a great freedom to move from branch to branch.

But even granting these premises, there are still different ways to approach the history of mathematics. As I see it, my own approach tends to concentrate on the development of mathematical knowledge – the processes of invention/discovery, the evolution of views held by mathematicians (both single individuals and communities), the research programs that the historical actors tried to advance, the schools and traditions that influenced their work. In this connection, it is convenient to clarify a few general historiographical notions that will be used in the sequel.

Already in the 19th-century it was common to speak of scientific and mathematical 'schools,' although the meaning of the term differed somewhat from present usage in the history of science. Frequently the term carried a pejorative connotation, suggesting a one-sided orientation with excessive attention to some specialty, as happened when some authors referred to the Berlin school. Here we shall employ the word *school* exclusively in the customary sense of recent historiography, which started over two decades ago with a well-known paper of J. B. Morrell.[2] In the present context, a research school is a group led normally by only one mathematician, localized within a single institutional setting, and counting on a significant supply of advanced students. As a result of continuous social interaction and intellectual col-

---

[1] The danger of excessive concentration on a single discipline is present in the excellent collection [Grattan-Guinness 1980], the multi-disciplinary approach can be found, e.g., in [Moore 1982], which is also the best example of collective historiography in connection to our topic.

[2] On the historiographical issue see the recent overview [Servos 1993], on the example of Berlin [Rowe 1989].

laboration, members of a school come to share conceptual viewpoints and research orientations: philosophical or methodological ideas concerning how to do their research, heuristic views regarding what problems are worth being pursued, which paths are dead-ends, and so on. Research schools are not governed by written regulations, they emerge spontaneously as an implicit reciprocal agreement between professor and students "to form a symbiotic learning and working environment based on the research interests of the professor" [Rowe 2002]. Schools are natural units within larger institutions, such as universities and faculties. Their great importance comes from the fact that they seem to constitute the crucial link between the social and the cognitive vectors of mathematical (or scientific) work and research.

One can mention several prominent examples of mathematical schools in 19th-century Germany, like Jacobi's Königsberg school, the school of Clebsch, and Klein's Leipzig school, but the most famous one is the so-called Berlin school. Some characteristics of this famous school will be studied in chapter I and contrasted with the views held by a group of mathematicians associated with mid-century Göttingen. Although I shall keep the traditional denomination 'Berlin school,' at this point I would like to mention a related issue that has been raised recently. David Rowe has suggested that one should differentiate schools from centers, linking the first exclusively with the name of their (single) leaders. According to this, we should speak of the Weierstrass school, not the Berlin school.[1] Certainly, in the 1880s Weierstrass and Kronecker entertained deep differences, so one should distinguish two schools at that time. But it is still open whether the kind of collaboration established by the Berlin mathematicians (above all Kummer and Weierstrass) in the 1860s and 1870s warrants talk of a single school. Here, I must leave the problem open.

In chapter I we shall speak of a Göttingen 'group' formed by Dirichlet, Riemann and Dedekind, which is not called a school because, though in many respects it had similar characteristics, it lacked a noteworthy output of researchers (probably because its temporal duration was very short). In other cases, I shall rename as a *tradition* what 19th-century authors called a 'school,' for instance the 'combinatorial tradition.'[2] Talk of a tradition implies that one can find a common research orientation in different actors that do *not* share a common institutional site, but are linked by traceable influences on each other. One should find a significant amount of shared conceptual elements that may have to do with preference for some basic mathematical notions, judgements concerning significant problems to be studied, methodological views affecting the approach to mathematics, and the like.[3]

---

[1] See [Rowe 2002]. An extreme view of schools as linked with single personalities was given by Hilbert in a 1922 address published by Rowe as an appendix.

[2] Another example would be the synthetic and analytic 'traditions' in geometry (formerly called 'schools').

[3] In the case of Göttingen, I could also have talked of a Göttingen tradition, since one can make the case of common methodological orientations, of an essentially abstract and modernizing kind, in a whole series of actors related to that center, from Gauss to Hilbert and Noether.

On the other hand, as I have used the term above, a 'research program' concerns a single individual's research projects, expectations, and preferences. Thus, I depart from the meaning given to the term by Lakatos [1970], which seems harmless for at least two reasons. Lakatos's term has never been frequently used, in his sense, by historians of science, and the term is employed here only rarely. Which is not to say that I have not tried to analyze carefully the research programs (in the specified sense) of men like Riemann, Dedekind, Cantor, Zermelo, Russell, and so on. Of course, the projects of an individual are deeply influenced by the mathematical situation in which he or she has become a mathematician, that is, by affiliation with a school or the influence of traditions. To give an example, I shall try to show that Cantor deviated from the orientation of the Berlin school due, in good measure, to the influence of the Göttingen tradition, i.e., the orientations embodied in the writings of members of the Göttingen group.

There are other historiographical terms that can be found more or less frequently, but which I shall not employ. It may be useful, however, to mention a couple of them in order to clarify the notions above. 'Invisible colleges' are groups of members of a single community or discipline, joined together by formal links (e.g., co-citation) and informal communication. One should reflect that normally we find members of *different*, and even opposing, research schools in a single invisible college (e.g., Riemann and Weierstrass, or Dedekind and Kronecker). It has been suggested that one might focus on a specific kind of invisible college, the 'correspondence network,' that would be particularly important in the case of mathematics [Kushner 1993]. Most of the actors studied here were, in fact, linked to each other through several correspondence networks, and it may be useful to read the following pages with that in mind. But the notion in question will not be explicitly used.

I stated above that my approach stresses the development of mathematical knowledge – how it was obtained, refined, and generally adopted. It is my hope that, by paying attention to these issues, it will become possible to unearth the multiple connections between set theory and the broader context of modern mathematics. Set theory emerged as part of an evolving new picture of the discipline, incorporating a more conceptual and decidedly abstract approach to traditional problems. It has been one of my goals to afford an understanding of the novelties of that approach and the opposition and difficulties it had to face. In this way I also hope to contribute to a richer understanding of the 'classical' world of 19th century mathematics. Standard historiography, with its tendency to project present-day conceptions (and myths), has frequently acted as a barrier cutting off a satisfactory interpretation both of the past and of the road to our present.

Another key goal has been to delineate the conceptual shifts affecting disciplines and notions that are frequently taken for granted – e.g., the very notion of logic and the relations between logic and set theory. Such conceptual changes become a fascinating topic and stimulate reflection on the basic ideas that we are familiar with, calling attention to alternative possibilities for theoretical development. I believe the work may turn out interesting for the working mathematician in this particular way.

In writing the present book I have worked as a historian of mathematics, but I would like to mention that, by training, my background is in logic and history & philosophy of science. That may still be visible in the particular orientation I have given to the selection of material, and in the approach that I have preferred to take. For this reason the volume should be of interest to philosophers of mathematics, who will find detailed case studies that offer much valuable material for philosophical reflection.

This work is a fully revised and expanded edition of a book originally published in Spanish, *El nacimiento de la teoría de conjuntos* [Ferreirós 1993a], which in turn was an outgrowth of my doctoral dissertation, presented two years earlier. Improved versions of part of the material, contained in the papers [Ferreirós 1993; 1995; 1996], have been incorporated into the new edition. I thank *Historia Mathematica* and the *Archive for History of Exact Sciences* for their permission to include that material.

The initial impulse came through Javier Ordóñez (Madrid), who introduced me to the history of science and directed my doctoral work; above everything else, I wish to thank his constant support and friendship. During a fruitful stay at the University of California at Berkeley, and later, I've had the opportunity to discuss diverse aspects of the history of set theory with Gregory H. Moore (Hamilton, Ontario). For many years I have also profited from interaction with a leading expert on Dedekind, Ralf Haubrich (Göttingen); he helped me in the most diverse ways during the production of this book. David E. Rowe (Mainz) has been kind enough to revise the present edition and give his expert advice on many issues, big and small. Even though he and others have helped me correct the English, I fear the final version will still show too clearly that the author is not a native speaker. I can only ask the reader for indulgence.

Several other people have been helpful at different stages in the preparation of this volume: Leo Alonso (Santiago de Compostela), Leo Corry (Tel-Aviv), John W. Dawson, Jr. and Cheryl A. Dawson (Pennsylvania), Antonio Durán (Sevilla), Solomon Feferman (Stanford), Alejandro Garciadiego (México), Ivor Grattan-Guinness (Bengeo, Herts), Jeremy Gray (Milton Keynes), Hans Niels Jahnke (Bielefeld), Ignacio Jané (Barcelona), Detlef Laugwitz (Darmstadt), Herbert Mehrtens (Berlin), Volker Peckhaus (Erlangen), José F. Ruiz (Madrid), Erhard Scholz (Wuppertal). Thanks are also due to the Niedersächsische Staats- und Universitätsbibliothek Göttingen, and particularly to Helmut Rohlfing, director of its Handschriftenabteilung, for their permission to quote some unpublished material and reproduce an illustration. Another illustration comes from the Bancroft Library, University of California at Berkeley.

And finally, how could I forget Dolores and Inés, who provided personal support and also frequent distraction from the work!

Sevilla, July 1999 *José Ferreirós*

Part One: The Emergence of Sets within Mathematics

Before coming to constitute the subject of a particular branch of mathematics, the notion of set emerged as a useful tool for the study of diverse problems in function theory, analysis, algebra, and even geometry. Thus, around 1870 one can see the notion employed in a number of different contributions published by German authors. It is the purpose of Part One to discuss carefully those contributions, how they began to configure a set-theoretical approach to mathematics, and the way they linked with the later development of abstract set theory.

Disregarding glimpses that may be found in the work of earlier authors (Gauss, Dirichlet, Steiner), the first systematic and influential proposal of the notion of set as basic for mathematics can be found in Riemann (chap. II). In connection with his function-theoretic work and his deep analysis of geometrical notions, Riemann proposed the notion of manifold – which, for him, meant something akin to class or set – as the basis of pure mathematics. One can find evidence that his proposals had a notable impact on the work of Cantor and Dedekind.

Riemann's friend, Dedekind, elaborated a set-theoretical approach to algebra and number theory in his study of factorization in algebraic numbers, the ideal theory of 1871 (chap. III). One year later, three different definitions of the real numbers became known, all of them relying more or less consciously on the notion of set. Thus, with Weierstrass, Dedekind, and Cantor the problem of the foundations of arithmetic merged into the set-theoretical stream (chap. IV). At the same time, stimulated by Riemann the theory of point-sets began to appear in connection with the study of discontinuous functions in real analysis. This was quite an active focus of research, which in the 1870s saw contributions by Heine, Hankel, du Bois-Reymond, Smith, Dini, and notably Cantor (chap. V).

More striking still – all this happens around 1868–72, while Cantor opened the realm of transfinite set theory with a surprising discovery made in December 1873. It is likely that, as a perceptive observer, Cantor was able to realize the great possibilities inherent in the notion of set judging from the role it was beginning to play in advanced contemporary work. This helps to explain how the project of an autonomous theory of sets came into being. It also makes clear that the set-theoretical orientation of mathematics was chronologically prior to the development of abstract set theory.

Figure 1. *Gustav Lejeune Dirichlet (1805–1859)*

I Institutional and Intellectual Contexts in German Mathematics, 1800–1870

It is a common characteristic of the various attempts to integrate the totality of mathematics into a coherent whole – whether we think of Plato, of Descartes, or of Leibniz, of arithmetization, or of the logicists of the nineteenth-century – that they have all been made in connection with a philosophical system, more or less wide in scope; always starting from *a priori* views concerning the relations of mathematics with the twofold universe of the external world and the world of thought.[1]

In order to understand the growth of mathematical knowledge it is sometimes important to identify and consider the role played by schools of mathematical thought. Such a school usually possesses an underlying philosophy by which I mean a set of attitudes towards mathematics. The members of a school tend to share common views on what kinds of mathematics is worth pursuing or, more generally, on the manner in which, or the spirit in which, one should investigate mathematical problems.[2]

As will become clear in the body of the present work, a certain trend within 19th-century German mathematics, the so-called *conceptual* approach, seems to have been strongly associated with the rise of set theory. Therefore, it seems convenient to start by analyzing two different 'mathematical styles,' those that reigned in Göttingen and Berlin immediately after 1855. Such will be the topic of §§4 and 5. The reason for that particular selection of institutions is simple: the main figures in the first two parts of the book are Riemann, Dedekind and Cantor. Riemann and Dedekind studied at Göttingen, where they began their teaching career, while Cantor took his mathematical training from Berlin. It will turn out that the conceptual approach was present at both universities, but in different varieties, that we will identify as an *abstract* and a *formal* variety. The *abstract conceptual* approach that could be found at Göttingen promoted the set-theoretical orientation strongly.

---

[1] Nicolas Bourbaki [1950, §1].
[2] Hawkins [1981, 234].

In order to better understand what the conceptual viewpoint meant, it is necessary to overview, however briefly, a number of trends in the foundations of mathematics that were influential in Germany during the first half of the century. As we have seen, an important part of the rise of set theory was the emergence of new language that expressed a novel understanding of mathematical objects – the language of sets and the idea that sets constitute the foundation of mathematics. This raised some questions that had already been discussed by the Greeks: what kind of existence do mathematical objects have, what methods are admissible in mathematics, and the famous issue of the actual infinite. These were philosophical questions, that the German mathematicians understood as such.[1] In fact, those questions are not the exclusive domain of mathematics, and there is some evidence that the wider intellectual and philosophical atmosphere in Germany had some impact upon the mathematical discussion. Our discussion of these topics in §§1 to 3 will not aim to be complete, but just to clarify and make plausible this last thesis. We will observe that there has been some misunderstanding concerning such matters, and particularly the question of the infinite, since the actual infinite was not rejected by all German mathematicians as of 1800 or 1850.

1. Mathematics at the Reformed German Universities

The German scientific community seems to have been somewhat peculiar, within the context of the international panorama in the nineteenth-century. One has the impression that a certain intellectual atmosphere marked or conditioned German approaches to the sciences, including mathematics.[2] That can be better understood by taking into account the institutional context of mathematical research, which emerged from the educational reform that was undertaken after the Napoleonic invasion. Such research as we will analyze throughout this book was clearly placed within the university context, but the 19th-century German universities were the result of a complex and radical transformation that began some time around 1810 (see [McClelland 1980]).

The traumatic Napoleonic invasion had the effect of making clear the need for important transformations in order to elevate the political, economical, military, and scientific situation in Germany to the level of France. To some extent, the Germans

---

[1] Riemann [1854, 255] regarded his discussion of the notion of manifold as "philosophical," just like Kronecker [1887, 251] his analysis of number. Cantor [1883] entered into the philosophical arena in order to justify his introduction of transfinite numbers, while Dedekind [1888, 336] felt the need to make clear that his "logical" analysis of number did not pressupose any particular "philosophical or mathematical" knowledge.

[2] As is well known, the meaning of the German word *Wissenschaft* is different from that of its English counterpart 'science:' it refers to any academic discipline, including history, philosophy, etc. When I write about the sciences, I mean the *Naturwissenschaften* – physics, chemistry, biology, etc. It is important to observe that mathematics was sometimes treated as being closer to the humanities in early 19th-century Germany.

explained the defeat to themselves by pointing to the high level of scientific education enjoyed by the French officers. That high level was a consequence of the educational reform undertaken in France after the Revolution, particularly the creation of the Parisian *École Polytechnique* in 1794. Here, a higher scientific education, including the calculus, was for the first time regularly available to students. The German states undertook parallel reforms, but some of them, especially in Prussia, did not simply copy the French model. On the contrary, they tried to forge their own peculiar model.[1] While in France the universities had been abolished as an outmoded medieval institution, in Germany there were some precedents for a reformed university, adapted to the cultural standards of the Enlightment, as was the case for Göttingen University in Hannover [McClelland 1980].

The reforming impulse which came out of the Napoleonic era merged in a natural way with the educational aspirations of the German Enlightment. In the second half of the eighteenth-century, this had given rise to so-called neohumanism, a movement that reacted against rationalistic viewpoints, aspiring to an integral formation [Bildung] of the individual, a kind of broad, harmonic education not guided by utilitarian aims. Such ideals were strongly fostered by historians and philologists like F. A. Wolf at the Prussian University in Halle, who is regarded as founder of the new 'scientific' philology and of the university seminar. Wolf devoted his efforts to turning history and philology, what he called the "science of antiquity" [Altertumswissenschaft], into a true "system," to "unify it into an organic whole and elevate it to the dignity of a well-ordered ... science."[2] This was not done simply for professional or academic reasons, but embodied a characteristic educational ideal. As Goethe said to Eckermann in 1827,

A noble man in whose soul God has put the capacity for future greatness of character and high intellect will develop most splendidly through the acquaintance and the intimate intercourse with the lofty characters of Greek and Roman antiquity.[3]

Neohumanist professors went beyond the traditional role of a university teacher, namely the transmission of well-established knowledge, to its expansion by means of criticism and research. Through the institution of the seminar, selected groups of students were taught how to do research by themselves, and research came to be seen as an indispensable ingredient of teaching. One further important aspect of this movement was its association with the late eighteenth-century struggle of the *Wissenschaften* (philosophy, history, philology, mathematics, etc.) for a recognition of

---

[1] Prussian university reform was undertaken during French occupacion, and there was a conscious attempt to establish clear counterparts of French cultural orientations: the university would be an exponent of German *Kultur* as opposed to French *civilisation*, and in particular the notion of *Bildung* is opposed to a more utilitarian *Ausbildung* [formation or instruction]. See [Ringer 1969].

[2] Quoted in [Paulsen 1896/97, vol. 2, 209]: "zu einem organischen Ganzen zu vereinigen und zu der Würde einer wohlgeordneten philosophisch-historischen Wissenschaft emporzuheben."

[3] As translated in [Jungnickel & McCormmach 1986, 4].

equality with, or even superiority over, law, medicine and theology – the traditional 'higher' university faculties.

The German reform of higher education is normally traced back to the founding of Berlin university in 1810. The Philosophical Faculty, housing the sciences, was given the task of preparing *Gymnasium* (secondary-school) professors, who were required by the Prussian state to pass an examination where their knowledge of philology, history and mathematics would be tested. Thus, the Philosophical Faculty was established as an equal to the professional faculties, where mathematics found a sounder institutional frame.[1] One should pause to consider what adaptation to the milieu of the Philosophical Faculty may have meant. Far from being mere practitioners, as frequently happened in the past, the mathematicians now became part of a small élite of university professors, and precisely within the context of the humanities, where the ideals of "living the sciences" (W. von Humboldt) and pure knowledge were strong. The implications of this move had been beautifully expressed by Schiller in a short poem that was well known at the time, *Archimedes and the Apprentice*:

> To Archimedes came an eager-to-learn youngster;
> > Initiate me, he said to him, into the divine art,
> That such magnificent fruits gave the Fatherland,
> > And the city walls protected from the sambuca.
> Divine you call the art! She is, the sage replied,
> > But so she was, my son, before she served the State.
> If you want fruits, those a mortal can also beget,
> > He who woos the *Goddess*, seek in her not the *maid*.[2]

This cultural orientation helps to explain the 19th-century tendency toward pure mathematics, which was particularly noticeable in Germany.[3]

By the early century, the Philosophical Faculty was dominated by history, philology and philosophy, then living its golden age in Germany. The sciences did not enjoy strong support, a situation that was badly felt by physicists (see [Jungnickel & McCormmach 1986]). Mathematics was in a better situation, since it was regarded as central for educational purposes, given its assumed relation with the training of logical and reasoning abilities. But, even so, the mathematics curriculum

---

[1] The import of such institutional changes for mathematical research has been carefully studied by Schubring [1983].

[2] Schiller, *Die Horen* (1795), in [Schiller 1980, 280]: *"Archimedes und der Schüler./* Zu Archimedes kam ein wissbegieriger Jüngling;/ Weyhe mich, sprach er zu ihm, ein in die göttliche Kunst,/ Die so herrliche Früchte dem Vaterlande getragen,/ Und die Mauern der Stadt vor der Sambuca beschützt./ Göttlich nennst du die Kunst! Sie ist's, versetzte der Weise,/ Aber das war sie, mein Sohn, eh sie dem Staat noch gedient./ Willst du nur Früchte, die kann auch eine Sterbliche zeugen,/ Wer um die *Göttin* freyt, suche in ihr nicht das *Weib*." Quoted in [Hoffmann, vol. 5, 624]. Jacobi wrote a parody of this poem, that is quoted in [Kronecker 1887, 252]. The 'sambuca' was a war machine used by the Romans against Syracuse, Archimedes' fatherland.

[3] Compare [Scharlau 1981]. I hardly need to make explicit that I am not making a case for a reductionistic explanation: the contextual factor will probably be only one among several others.

was elementary and rarely included the calculus [*op.cit.*, 6–7]. It was only gradually that university professors raised their standards, partly in response to the French model, partly in imitation of the philologists, and partly as a result of the better mathematical level of students coming out of the reformed *Gymnasia*. The teaching of research topics started in the late 1820s, particularly with the courses and seminars offered by Jacobi at Königsberg and, somewhat later, by Dirichlet at Berlin, both Prussian universities.

From about 1830 there were periods of tension between the interests of the scientists and those of the humanists, that can be followed up to the early twentieth-century (see [Pyenson 1983]). But, even so, the context of the Philosophical Faculty, and the neohumanist ideal of *Bildung*, promoted an atmosphere in which the 'two cultures' were not separated. Throughout the century, we find scientists interested in philosophy and philosophers interested in the sciences. In fact, the unity between philosophy and the sciences was repeatedly promulgated, for instance by the idealist philosopher Schelling, who originated the tradition of *Naturphilosophie*. Much has been written about the influence of this movement on German science, but here I would like to warn against simplifying assumptions. Many historians tend to identify neohumanism with idealism, and this with philosophy generally. But philosophy in Germany was not, by any means, identical with idealism, and many neohumanists opposed the Romantic trends, including idealistic philosophy. For instance, the above-mentioned Wolf was much closer to Kant than to the idealists in philosophical matters [Paulsen 1896/97, vol. 2, 212–214]. Among philosophers, early 19th-century followers of Kant who remained close to the sciences and opposed the idealism of Schelling and Hegel included Fries and Herbart, whom Riemann regarded as his master in philosophy.[1]

Thus, rather than paying attention to peculiarities of the idealists, it would be more useful to analyze the ideas they shared even with their detractors, since these marked the development of German science after the anti-idealist reaction of the 1830s and 40s. It is my contention that much of the special ways of German scientists can be explained by their adaptation to the context of the Philosophical Faculty, and by the fluid intellectual contact they established with the philosophers. Among the particular orientations that were promoted in the process are the preference for a strictly theoretical orientation, the concentration on narrowly defined specialties or branches of mathematics, and in many cases a close attention to the philosophical presuppositions of the advocated theories.[2] The main authors studied in the present work afford good examples of such traits, and their philosophical preferences will be analyzed briefly (see especially §§II.1–2, VII.5, VIII.1–2 and 8).

From the 1820s, there was a clear scientific renaissance in Germany [Klein 1926, vol. 1, 17]. Peculiar idiosyncratic approaches began to be abandoned, and

---

[1] On Hegel, Fries, Herbart and mathematics, see [König 1990].

[2] This had negative effects insofar as it implied lack of attention to applied topics and to interconnections between branches, etc. See [Rowe 1989], where the fight against some implications of neohumanism around 1900 is discussed; see especially [186–87, 190].

closer attention was paid to foreign ideas. A professional scientific community began to emerge and fight for new standards in higher education. In 1822 the *Naturphilosoph* Lorenz Oken founded the *Gesellschaft Deutscher Naturforscher und Ärzte* [German Association of Natural Scientists and Physicians], which from the 1828 meeting at Berlin would become dominated by scientists opposed to the ideas of *Naturphilosophie*. This move was related to the influence of Alexander von Humboldt, the famous traveler and naturalist, of neohumanist allegiances, who played an important role in promoting the sciences in his position as a court counselor in Prussia. In connection with mathematics, another important event was the founding in 1826 of the *Journal für die reine und angewandte Mathematik* [Journal for pure and applied mathematics], directed by the engineer and high-level civil servant A. L. Crelle, who had always been keenly interested in mathematical research. Thanks to the availability of original material by such men as Abel, Jacobi, Dirichlet, and Steiner, Crelle's undertaking was a great success, and acted as a binding agent for the emerging community of mathematicians.[1]

The careers and ideas of Jacobi, Kummer and others might be used in order to show the impact neohumanism and philosophy had upon German mathematicians.[2] In discussing the role of Jacobi in the German scientific renaissance and the establishment of the first important research school in mathematics, the Königsberg school, Klein writes:

If we now ask about the spirit that characterizes this whole development, we can in short say: it is a scientifically-oriented neohumanism, which regards as its aim the inexorably strict cultivation of pure science, and in search of that aim establishes a specialized higher culture, with a splendor never seen before, through a concentrated effort of all its powers.[3]

The cases of Grassmann, Riemann and Cantor underscore the freedom and new possibilities of thought that the fluid exchange of ideas between philosophers and mathematicians could sometimes promote. But, of course, any institutional arrangement has its pros and cons. Dirichlet's career reminds us of the negative effects that neohumanist standards sometimes had; it will also serve to clarify several aspects of the situation in Germany at the time.[4]

---

[1] One should indicate, as David Rowe has urged me to do, that the emergence of a mathematical community was not easy, due to the dispersion of professors, their adscription to different universities (with remnants of a guild mentality) and states, etc. The *Deutsche Mathematiker Vereinigung* was not easy to launch even in 1890.

[2] See [Jahnke 1991], and on Kummer [Bekemeier 1987, 196–203].

[3] [Klein 1926, vol. 1, 114]: "Fragen wir nun nach dem Geist, der diese ganze Entwicklung trägt, so können wir kurz sagen: es ist der naturwissenschaftlich gerichtete Neuhumanismus, der in der unerbittlich strengen Pflege der reinen Wissenschaft sein Ziel sieht und durch einseitige Anspannung aller Kräfte auf dies Ziel hin eine spezialfachliche Hochkultur von zuvor nicht gekannter Blüte erreicht."

[4] No full biography of Dirichlet has yet been written, the best is still Kummer's obituary (in [Dirichlet 1897, 311–44]). Important archival material can be found in [Biermann 1959].

In 1822, Gustav Lejeune Dirichlet, son of the town postmaster at Düren, made the wise decision to take advantage from his family's connections and study mathematics at Paris, not at any German university. There he attended free lectures at the Collège de France and the Faculté des Sciences, and subsequently he became a key figure in the transmission of the French tradition of analysis and mathematical physics to Germany. Early success with a paper on indeterminate equations of degree five, sent to the Académie des Sciences in 1825 and printed in the *Recueil des Mémoires des Savans étrangers*,[1] made him known in Parisian scientific circles. Dirichlet entered the circle around Fourier, then "secrétaire perpétuel" of the Académie, and established contact with A. von Humboldt, who would promote his career in Prussia. Thanks to the support of Humboldt, Dirichlet was awarded an honorary Ph.D. from Bonn University in 1827, became *Privatdozent* at Breslau against faculty opposition, and was named extraordinary professor in 1828. Humboldt was again instrumental in bringing his young friend to Berlin, where he was forming plans for the creation of an important scientific center in imitation of the École Polytechnique. Dirichlet became teacher of mathematics at the military academy, shortly thereafter *Privatdozent* at the University, and in 1831 he was appointed extraordinary professor. His brilliant career continued with further papers on number theory, and with a famous 1829 article on the convergence of Fourier series (see chap. V). In 1832, when only twenty seven, he became a member of the Berlin Academy of Sciences, and in 1839 he was promoted to an ordinary professorship.

But such a quick career had not conformed to all the rules then in force, and Dirichlet subsequently faced some difficulties. Since he had not studied at a German university, nor even completed his education at the *Gymnasium* [Schubring 1984, 56–57], Dirichlet lacked some of the knowledge required by neohumanist curricula. Although it had been possible to avoid most of the consequences by means of an expeditious *honoris causa*, he had not satisfied a formality on the occasion of his qualification [Habilitation] as a *Privatdozent* at Breslau. Aspirants were required to submit a second thesis, written in Latin, and to defend it in an open discussion [Disputation] with faculty members, also to be conducted in Latin.[2] Dirichlet did not master the spoken language and was relieved from the Disputation, but this had the effect that in 1839 the Berlin Faculty would not grant him full rights as a professor until he complied with that formality. That only happened in 1851; in the meantime, Berlin's most influential mathematics professor could not take part, as a voting member, in Ph.D. and Habilitation proceedings.[3]

Dirichlet did not create a school in the strict sense, but among those who were his students and felt strongly influenced by him we find such important names as Heine, Eisenstein, Kronecker, Christoffel, Lipschitz, Riemann and Dedekind. His

---

[1] Legendre, who acted as a reviewer of the paper, was able to use Dirichlet's results for a proof of Fermat's Last Theorem for the exponent 5.

[2] This feature of the *Habilitation* varied from place to place, but was similar in Breslau and Berlin.

[3] On this issue, see [Biermann 1959, 21–29].

contributions to number theory, following and making available Gauss's work, on Fourier analysis, multiple integrals, potential theory and mathematical physics, were all of fundamental importance. Moreover, in all of these fields he carried further the rigorization of mathematics. A famous passage in a letter from Jacobi to Humboldt says that it is only Dirichlet, not Gauss, Cauchy or Jacobi, who knows "what a completely rigorous mathematical proof is." Jacobi goes on to mention the other specialty of Dirichlet, analytic number theory, and writes that he "has chosen to devote himself mainly to those subjects which offer the greatest difficulties."[1] According to Eisenstein, Gauss, Jacobi and Dirichlet started a new style of argument in mathematics, which avoids "long and involved calculation and deductions" in favor of the following "brilliant expedient:" "it comprehends a whole area [of mathematical truths] in a single main idea, and in one stroke presents the final result with utmost elegance," in such a way that "one can see the true nature of the whole theory, the essential inner machinery and wheel-work."[2] This has frequently been called the conceptual approach to mathematics, and will be of our concern in §§4 and 5. Some aspects of Dirichlet's work, that are of consequence for the present study, will be mentioned in the following chapters.

2. Traditional and 'Modern' Foundational Viewpoints

There have always been many possible approaches to the philosophical problems raised by mathematics – empiricism, Platonic realism, intellectualism, intuitionism, formalism, and many other intermediate possibilities. From what we have seen, one may expect to find, among 19th-century German mathematicians, an influence of philosophical ideas and a greater speculative tendency than among their foreign colleagues. Interesting early examples that will not be discussed in detail here are those of Bolzano and Kummer. But more relevant for our purposes is the fact that the influence of philosophy seems to have led to an increase of intellectualist viewpoints in 19th-century Germany.

Kantian philosophy has always been more congenial to scientists than idealism, so it is not surprising that during and after the acme of idealism it retained an important status among them. A characteristic Kantian idea is that the subject (the philosophical I) enjoys a central position in the world, or at least in our knowledge of the world. The world is regarded as a representation, a set of phenomena that unfold in the screen of consciousness. Such phenomena are not simply determined

---

[1] Letter of 1846, in [Pieper 1897, 99]: "Er allein, nicht ich, nicht Cauchy, nicht Gauss weiss, was ein vollkommen strenger mathematischer Beweis ist, sondern wir kennen es erst von ihm. ... D[irichlet] hat es vorgezogen, sich hauptsächlich mit solchen Gegenständen zu beschäftigen, welche die grössten Schwierigkeiten darbieten; darum liegen seine Arbeiten nicht so auf der breiten Heerstrasse der Wissenschaft und haben daher, wenn auch grosse Anerkennung, doch nicht alle die gefunden, welche sie verdienen."

[2] As translated in [Wussing 1969, 270], where this passage from Eisenstein's autobiography, written when he was 20, is quoted in full.

by the stimuli coming from external objects, but also by some intrinsic characteristics of the subject's mind. This situation of the representational world as codetermined by subject and objects may be seen as parallel to that of scientific theories, when viewed from the hypothetico-deductive standpoint, since theories are codetermined by mathematically stated hypotheses and laboratory experiences. If we accept this parallelism, and take the conclusion that arises from it, we should come to think that mathematics is on the side of the subject. It should be related to the intrinsic characteristics of the subject or, to use Kant's phrase, to the *a priori* in the subject's mind, and we thus arrive at an intellectualist conception.

Intellectualism can be found in association with several different conceptions of the foundations of mathematics in early 19th-century Germany. Two examples are Kantian intuitionism, and also a brand of formalism linked to the so-called combinatorial school. Let us begin with the latter.

2.1. Formalist approaches. Around 1800, the combinatorial tradition was very influential among German mathematicians.[1] This trend was headed by the Leipzig professor of physics and mathematics Carl F. Hindenburg, who from 1794 to 1800 edited the first periodical devoted to mathematics in Germany, *Archiv für die reine und angewandte Mathematik*, a journal that he used for the promotion of his conception of mathematics. Combinatorialists saw themselves as heirs to Leibniz, who had written about the *ars combinatoria* as a "general science of formulas," providing general combinatorial laws, that would embrace algebra as a subdiscipline [Leibniz 1976, 54–56]. Hindenburg and his followers regarded combinatorial theory as the core of pure mathematics and the basis for the theory of series, which they saw in turn as the foundation of analysis. Thus, the central themes for this tradition were issues in pure mathematics, including the much debated problem of the foundations of the calculus. Analysis became, for them, a theory concerned with the transformations of finite or infinite series of symbols, transformations that could be analyzed combinatorially.[2]

The combinatorial approach was not far from contemporary viewpoints, such as Lagrange's formal conception of the calculus in his 1797 *Théorie des fonctions analytiques*, and the related development of the calculus of operations [Koppelmann 1971]. One may say that combinatorialism developed some trends that were clearly present in 18th-century mathematics, trends which would also lead to the British tradition of symbolical algebra [Knobloch 1981; Pycior 1987]. An influential formulation of ideas related to those of the combinatorialists was given by the

---

[1] Although it has been customary to refer to this trend as the combinatorial school, following 19th-century usage, I shall prefer the word 'tradition.' We reserve 'school' for those institutional arrangements in which small groups of mature mathematicians pursued more or less coherent research programs, joined by a certain style or 'philosophy' of research (in the sense of Hawkins), training advanced students with which they worked side-by-side [see Geison 1981, Servos 1993]. On the other hand, 'tradition' seems apt to convey the idea of influence and community of interests and 'philosophy,' but on a looser institutional, geographical and/or temporal basis (see the Introduction).

[2] On this topic, see [Netto 1908], [Jahnke 1987; 1990].

Berlin professor Martin Ohm, brother of the famous physicist and a figure of some importance for what follows. His work was particularly successful among *Gymnasium* teachers and also, one may conjecture, among self-taught mathematicians. Ohm clearly formulated the program of basing all of mathematics upon the notion of natural number, a program that can also be found in his colleague Dirichlet, in Kronecker, Weierstrass and Dedekind. According to Ohm [1822, vol. 1, xi–xiii] only natural numbers have a real existence, while the rest of mathematics can be seen as a theory of numerical signs. Ohm's reconstruction of pure mathematics was largely based upon the manipulation of formulas in accordance with the algebraic rules, but on the basis of a purely analogical justification – an inheritance of the combinatorial tradition, and a trait he has in common with British symbolical algebra. That viewpoint made it possible to establish the use of divergent series on a sound basis, thus rescuing a peculiar characteristic of 18th-century analysis that would be severely criticized by Ohm's great contemporaries Abel, Cauchy and Gauss.[1]

The combinatorial approach can be labeled a purely formalistic viewpoint, since it regarded mathematics as a symbolical or syntactic construction. Ohm's approach seems to have been only partly formalistic, since he accepted the natural numbers as given objects with their characteristic properties. But one might think that even this partly formalistic standpoint ought to be radically opposite to intellectualist tendencies. Symbols, however, may be taken to have primarily a mental existence, and this move changes the picture completely. In defining a formal power series, Ohm says that it is a function of indefinitely great degree, and, "therefore, an entire function that is never really representable, but only lives in the idea within ourselves" [Ohm 1855, 239]. Moreover, what is essential in the calculus, according to Ohm, is not numbers but operations, i.e., "actions of the understanding" – where understanding [Verstand] is a characteristically philosophical, and more specifically Kantian, term:

In the most diverse phenomena of the calculus (of arithmetic, algebra, analysis, etc.) the author sees, not properties of quantities, but properties of the operations, that is to say, actions of the understanding ... It turns out that one only calculates with "forms," that is, with symbolized operations, actions of the understanding that have been suggested ... by the consideration of the abstract whole numbers.[2]

The general symbolical rules, therefore, represent mental actions performed on mentally existing forms or symbols. Ohm's mention of operations that are sug-

---

[1] The fact that Ohm's approach was made rigorous by the differentiation between symbolical and numerical equalities, and the rules of interpretation of the symbolical calculus, has been emphasized by Jahnke [1987]. See also [Bekemeier 1987].

[2] [Ohm 1853, vii]: "In den verschiedensten Erscheinungen des Kalkuls (der Arithmetik, Algebra, Analysis, u.u.) erblickt der Verf. nicht Eigenschaften der Grössen, sondern Eigenschaften der Operationen, d.h. Akten des Verstandes ..."

gested by the consideration of whole numbers is a reference to the so-called "principle of permanence of formal laws," which is also characteristic of British symbolical algebra, and can be found much later in the work of Hermann Hankel [1867].

Combinatorialism was very successful only until about 1810, but textbooks of that orientation continued to be published up to the mid-century. Ohm's approach, on the other hand, seems to have been widely influential among *Gymnasium* teachers and those who were more or less self-educated in mathematics.[1] As late as 1860, the Göttingen professor Moritz A. Stern published a textbook in which formalistic conceptions akin to Ohm's were central (see [Jahnke 1991]). Several ideas strongly reminiscent of Ohm's can be found in early writings of Dedekind, perhaps coming through his teacher Stern, and in Weierstrass, apparently through his teacher C. Gudermann [Manning 1975, 329–40].

An intellectualist conception of mathematics, couched in the language of 'forms,' can also be found in Hermann Grassmann. Grassmann [1844, 33] began his work by elaborating a general theory of forms [Formenlehre] as a frame for mathematics. The transition from one formula to another he regarded as strictly parallel to a conceptual process that should happen simultaneously [*op.cit.*, 9]. A form, or "form of thought," was simply an object posited by thought as satisfying a certain definition, a "*specialized being* generated by thought." And "pure mathematics is the doctrine of forms" [*op.cit.*, 24]. This new, abstract conception of mathematics is related with the influence of several philosophers, from Leibniz to Schleiermacher, on Grassmann.[2] Although he accepted the existence of an spatial intuition, Grassmann's *Ausdehnungslehre* was not dependent upon intuition, since it constituted the abstract, purely mathematical foundation for geometry, which is empirical. Despite similarities with the combinatorialists and Ohm, Grassmann abandoned excessive reliance upon the symbolical, and more specifically he abandoned reasoning founded upon analogy, thus going in the modern direction. Incidentally, it is worth noting that his approach promoted, like no other in early 19th-century Germany, a formal axiomatic structuring of mathematical theories.

2.2. Intuitiveness and logicism. The great British empiricist Hume, in his epistemological works, emphasized the fallibility of empirical knowledge. Perhaps for this reason, one of Kant's basic presuppositions is that anything that is absolutely certain must not be empirical, but based solely upon the mental constitution of the subject. Necessary truth, including mathematical truth, must be *a priori*:

---

[1] A third edition of his *Attempt at a completely consistent* [consequenten] *System of Mathematics* (1822) came out in 1853/55.

[2] See [Lewis 1977] and [Otte 1989]. An interesting, short analysis of Grassmann's mathematical method can be found in [Nagel 1939, 215–19].

truly mathematical propositions are always *a priori*, not empirical, judgements, since they involve necessity, which cannot be gained from experience.[1]

As we have seen, Kant regarded the world as a representation in the subject, partly determined by the impressions received from the "things in themselves," partly by *a priori* characteristics of the subject's sensitive and conceptual abilities. As is well known, space and time were not to be seen as traits of the external world, but as *a priori* forms of our sensitivity or intuition [Anschauung], that determine our representation of the world with absolute necessity. Mathematics develops the consequences that can be extracted *a priori* from our pure intuition, it develops the theory of the possible constructions within our forms of intuition, space and time. Having its origin in the subject, it is now possible to understand the necessity that accompanies mathematical knowledge.

Kant's explanation was satisfactory as far as geometry was concerned, but his discussion of arithmetic and algebra was unsystematic and vague. An attempt to correct this situation can be found in William Rowan Hamilton, the best example of a Kantian mathematician, which is not to be found in Germany, but in Great Britain. In 1837 he published a paper, that went almost unnoticed, in which he regarded "algebra" as "the science of pure time" [Hamilton 1837]. This was a time when terminology was in flux: many mathematicians saw arithmetic, algebra and analysis as a unity, but the name used for that discipline varied greatly; some called it "analysis," Hamilton preferred "algebra," and, as we will see, several German mathematicians employed the term "arithmetic." Within the Kantian frame, Hamilton's conception gives rise to a quite satisfactory symmetric schema, in which geometry is the science of spatial intuition, and algebra the science of temporal intuition. Fifty years later, Helmholtz [1887] defended a version of that thesis when he proposed that the origins of natural numbers are related to the perception of time. Besides that speculative vision, Hamilton presented very interesting mathematical ideas, including the notion that what is essential in $\mathbb{R}$ is the presence of a continuous ordering, a detailed and quite rigorous treatment of arithmetic, and the brilliant idea of introducing the complex numbers as pairs of real numbers. It seems that, this time, the philosophical garb was an obstacle for the diffusion of interesting thoughts. Nevertheless, his mathematical and philosophical ideas become well known later, thanks to the preface to his *Lectures on Quaternions* [Hamilton 1853].[2]

Around this time, the Kantian conception was quite influential in Germany. It was common to defend that all mathematical knowledge bears the mark of its intuitiveness [Anschaulichkeit]. A good example is given by the correspondence between Apelt, a philosopher follower of Fries, and thus a neokantian, and the mathe-

---

[1] [Kant 1787, 14]: "eigentliche mathematische Sätze jederzeit Urtheile *a priori* und nicht empirisch sind, weil sie Nothwendigkeit bei sich führen, welche aus der Erfahrung nicht abgenommen werden kann."

[2] Both Cantor [1883, 191–92] and Dedekind [1888, 335] criticized this conception.

matician Möbius. Apelt commented upon Grassmann's *Ausdehnungslehre* that it was remarkable, but based on a false philosophy of mathematics:

> An *abstract* theory of extension [Ausdehnungslehre] as he is looking for, can only be developed out of concepts. But the source of mathematical knowledge is not in concepts, but in intuition.

Möbius replied that he had not been able to read much in Grassmann's book because of its abstract character, and granted that intuition is the "essential trait" of mathematics.[1]

Once again, this is a version of intellectualism, but quite different from the former brand. Here, the symbols and formalism are taken simply to denote those figures, numbers, etc. that are constructed in intuition. Such a viewpoint, that avoids enthroning the symbolical plane, and prefers to emphasize contentual aspects, is actually closer to the conception that won the day. Against the formalists, Gauss, Cauchy and their followers would have contended that mathematics does not study symbols, but numbers, functions and the like. Symbols and operations only have a sense insofar as they represent objects and processes that seem to stand on a different plane. This is one, though only one, of the reasons why the names of Hindenburg and Ohm acquired a ludicrous ring late in the century.[2]

The Kantian epistemological conception seems to have been widely influential in German scientific circles, even becoming a kind of basic 'common sense.' The emergence of logicist viewpoints late in the century, which is quite difficult to understand for present-day historians, can actually be better understood when seen in this context. The development of mathematics in the nineteenth-century showed a clear tendency away from the intuitive, and toward the abstract. The most common example is non-Euclidean geometry, the emergence of a variety of consistent, alternative theories, which had the effect of denying the character of being basic or intuitive to any one of them. One of Gauss's comments on this issue, in a letter to Bessel of 1830, employs a Kantian language for correcting Kant's opinion:

> According to my most intimate conviction, the theory of space has a completely different position with regards to our *a priori* knowledge, as the pure theory of magnitudes. Our knowledge of the former lacks completely *that* absolute conviction of its necessity (and therefore of its absolute truth) which is characteristic of the latter. We must humbly acknowledge that, if number is *only* a product of our minds, space also has a reality outside our minds, and that we cannot *a priori* prescribe its laws completely.[3]

---

[1] Quoted in [Grassmann 1894, vol. 3, part 2, 101–02].

[2] Mittag-Leffler wrote in 1886: "Kronecker emploie toutes les occasions à dire du mal de Weierstrass et de ses recherches. Il disait même l'autre jour en parlant de lui et Weierstrass que Gauss était peu connu et peu estimé de ses contemporaines, tandis que Hindenburg était le grand géomètre populaire de ce temps en Allemagne" [Dugac 1973, 162]. On Ohm, see [Bekemeier 1987, 77–82].

[3] [Gauss 1863/1929, vol. 8, 201]: "Nach meiner innigsten Überzeugung hat die Raumlehre zu unserm Wissen a priori eine ganz andere Stellung, wie die reine Grössenlehre; es geht unserer

Besides the example of non-Euclidean geometry, which only after 1860 began to affect the mathematical community as a whole, many more could be given. It suffices to recall the theory of real functions, the many anomalous examples that began to proliferate after the mid-century. Or algebraic examples such as the quaternions and other new kinds of numbers, and the abstract notion of group finding its place as an extremely useful tool for diverse applications. Or the evolution of number theory in the modern direction of algebraic numbers.

Now, suppose that we are thinking within a Kantian overall frame, and we are not willing to abandon the fundamental thesis that mathematics has its origins in the human mind, not in experience or the outer things. How can we possibly make sense of those changes? We need to take into account one more aspect of Kant's philosophy. According to the Königsberg philosopher, the *a priori* material of the human understanding does not simply consist in space and time as forms of intuition, it also includes a whole set of concepts or categories, that we systematically apply in categorizing the phenomena of the world. It suffices to take a look at the index of *Kritik der reinen Vernunft*, to see that Kant calls the doctrine of the forms of intuition "aesthetics," and the doctrine of the categories or concepts of the understanding "[transcendental] logic." Gauss's thoughts on geometry cast doubt on the real existence of an inborn form of spatial intuition, since the problem of the geometry of real space is coming to be seen, to some extent at least, as an empirical issue. The post-Kantian philosopher Herbart, who counted some mathematicians among his followers, had already abandoned Kant's postulate of the forms of intuition for purely philosophical motives, and criticized it sharply (see his 1824 *Psychologie als Wissenschaft* in [Herbart 1964, vol. 5, 428–29]).

Likewise, even Hamilton's supposedly intuitive foundation of algebra in pure time involves many abstract or conceptual elements that cannot possibly be related to any simple intuition. Among them are his consideration of "steps" in time, of ratios between such steps, and of pairs of previous elements [Hamilton 1837, 1853]. All this suggests that the basic thesis of the intuitiveness [Anschaulichkeit] of mathematics ought to be abandoned. On the other hand, the 19th-century development from the intuitive to the abstract confirms that mathematics has much more to do with pure concepts than was previously thought. To abandon the reference to intuition, while sticking to the idea that mathematics is *a priori*, is to consider mathematics as a theoretical development based solely upon concepts of the understanding. That is to say, in Kantian terminology, as a development of 'logic.'

I should make it explicit that the previous exposition has simplified some important aspects, like the complex issue of the relations between "formal logic" (Kant's term) and "transcendental logic." But it is not my purpose to produce a philosophical analysis of Kantianism. Although Frege considered himself a Kan-

---

Kenntniss von jener durchaus *diejenige* vollständige Überzeugung von ihrer Nothwendigkeit (also auch von ihrer absolute Wahrheit) ab, die der letztern eigen ist; wir müssen in Demuth zugeben, dass, wenn die Zahl *bloss* unseres Geistes Product ist, der Raum auch ausser unserm Geiste eine Realität hat, der wir a priori ihre Gesetze nicht vollständig vorschreiben können."

tian, I am not trying to defend that any one of the logicists was an orthodox Kantian. More interesting for present purposes is to consider the epistemological frame that was implicitly accepted by German scientists during the nineteenth-century. This seems to have been closely related to Kantianism, but not in an orthodox reading. Rather, scientists freely mixed some Kantian elements with several ideas taken from the sciences themselves. To give just an example, in the nineteenth-century it was quite common to think about human reason, instead of trying to be faithful to Kant's *pure* reason – following Fries, it was common to adopt a psychologistic reading of Kantian philosophy. Actually, in most cases the Kantian elements that can be found in a scientist or mathematician may be explained as coming through other scientists' work, rather than from a careful reading of the philosopher.

What is important, then, is that several mathematicians understood the abstract turn, and quite specifically, as we will see, the set-theoretical reformulation, as implying that mathematics is a development of logic [Frege 1884; Dedekind 1888].[1] Of course, to become a logicist was not simply to apply some Kantian thesis. By the late nineteenth-century, a serious occupation with logicist ideas meant to give a clear formulation of formal logic that might be seen as sufficient for founding mathematics, and this implied the need to go beyond received logical theory. We will pay attention to this issue later, particularly in chapters VII and X. At this point, however, it is important to emphasize that our present image of logicism is taken from the writings of Russell and his followers, which makes us loose historical perspective, since the epistemological frame and quite specifically the notion of logic changed decisively in the period 1850–1940. Initially, logicism was typically a German trend that makes full sense against the background of a 19th-century epistemology permeated by Kantian presuppositions. Logicism was a reaction against the specific Kantian theory of the origins of mathematics, a reaction based upon, and favoring, the abstract tendencies that became quite evident after the mid-century.[2]

As will be seen throughout this book, beginning with chapter II, most mathematicians and logicians in the second half of the nineteenth-century took the notion of set to be simply a logical notion. Actually, the logicist program would have been most implausible, for technical reasons, unless logic embraced some kind of set theory. During the early decades of our century, as a consequence of the set-theoretical paradoxes, the panorama changed radically. The paradoxes meant a revolution in the conception of logic; part, if not all, of set theory 'divorced' from logic, and the logicist program suddenly lost its plausibility. Subsequent changes in logical theory, which led to the wide acceptance of first-order logic as the main example of a logical system, even deepened the gap that separates us from 19th-

---

[1] The case of Riemann is different. He was a perfect example of the abstract trend, and some of his statements could incite logicist conclusions, but he was a careful philosopher and a follower of Herbart. Now, Herbartianism avoids all apriorism, and therefore it is quite incompatible with logicism (see §II.1–2).

[2] This topic will be taken up in chapter VII.

century authors. The concept of 'logic' is also a historical one, and any attempt at understanding the emergence of logicism – directly related to the history of set theory – must be quite aware of the radical transformations which that concept underwent from 1850 to 1940.

3. The Issue of the Infinite

It has frequently been written that the Aristotelian *horror infiniti* reigned among scientists and mathematicians until Cantor's vigorous defense of the possibility and necessity of accepting it. Like other extreme statements, this one does not bear a historical test. At least in Germany, and perhaps here it makes much sense to speak of national differences, there was a noticeable tendency to accept the actual infinite. The philosophical atmosphere could hardly have been more favorable, and there were several attempts to develop mathematical theories of the infinite. We shall begin with philosophy, and then consider the views of some mathematicians.

The history of philosophical attitudes toward actual infinity in 19th-century Germany would be a long one. By the beginning of the century, during the time of idealism, the potential infinity of mathematics was called the "bad infinite" by Hegel and his followers. The implication was clear: there is a 'good' infinite that is actual in the highest sense, the philosophical infinite, the Absolute.[1] It is well known that idealism affected an important sector of German scientists early in the century; among mathematicians, the best examples seem to be Steiner and Kummer. The philosophy of nature was also full of implications for the problem of infinity. Here the question normally took the form of deciding whether space and time are bounded or infinite, or whether the physical universe is or not made up of simple elements. These are the first and second "antinomies" discussed by Kant in his *Kritik der reinen Vernunft* [1787, 455–471]. The issue was taken up by later philosophers, for instance by the anti-idealist Herbart, whom Riemann took as his mentor in philosophy (see §II.4.2 for his views on infinity).

The work of Herbart shows a trait that seems to have been rather characteristic of German philosophy at the time – incorporation of elements taken from Leibniz's philosophy. In the early 19th-century there seems to have been a revival of Leibnizian ideas, also promoted by the idealists. And of course, Leibniz favored the actual infinite:

I am so much for the actual infinite that, instead of admitting that nature abhors it, as is vulgarly said, I sustain that it affects her everywhere, in order to better mark the perfections

---

[1] [Bolzano 1851, 7]: "Hegel und seine Anhänger ... nennen es verächtlich das schlechte Unendliche und wollen noch ein viel höheres, das wahre, das *qualitative Unendliche* kennen, welches sie namentlich in Gott und überhaupt im *Absoluten* nur finden." See Bolzano's criticisms of Cauchy, Grunert, Fries, etc. in [*op.cit.*, 9–13].

of its Author. Thus I believe there is no part of matter which is not, I do not say divisible, but actually divided; and consequently the least particle of matter must be regarded as a world full of an infinity of different creatures.[1]

Certainly, this is not Leibniz's position in all his writings,[2] but it is the one that characterizes the whole spirit and the statements in his *Monadologie*, a work that advanced a conception of the universe as made up of simple metaphysical units [Leibniz 1714, §§57, 64–67]. In the wake of the Leibnizian revival, the ideas of the *Monadologie* were put into new life. Herbart elaborated an ontology of simple units called the "Reale," which is reminiscent of Leibniz's monads; the physicist and philosopher Fechner defended a Leibnizian *Atomenlehre* [Theory of atoms; 1864], somewhat like the physiologist and philosopher Lotze in his famous work *Mikrokosmus* [1856/64]. As the reader can see, we are now talking about authors who were quite influential among the scientific community. In fact, even Cantor defended viewpoints similar to those of Fechner and Lotze [1932, 275–76]; he quoted Faraday, Ampère and Wilhelm Weber as his forerunners.

Interestingly, Leibnizian ideas can be found in all the main authors that we shall deal with in Part One. Riemann was strongly influenced by Herbart, and his fragments on psychology and physics present us with viewpoints that are quite close to the author of the *Monadologie*. Riemann (and Herbart) seem to favor the Leibnizian conception of space as an 'order of coexistence' of things (see §II.1.2). Also noteworthy is Riemann's preference for the hypothesis of a material *plenum* and contact action, instead of Newtonian action-at-a-distance.[3] His friend Dedekind was also in favor of this hypothesis, as he made clear in a noteworthy passage of his correspondence with Heinrich Weber:

As far as I'm concerned, I very much favor the continuous material filling of space and [the Riemannian] explanation of gravitation and light phenomena ... Riemann adopted these ideas quite early, not in his late years ... Without doubt, his efforts went toward basing the most general principles of mechanics, which he did not want to revoke at all, upon a new conception, more natural for the explanation of Nature. The effort of *self*-preservation and the dependence – expressed in the partial differential equations – of state-variations from states immediately surrounding in space and time, should be regarded as the original, not as the

---

[1] The passage is quoted by Bolzano [1851, iii] and Cantor [1932, 179]: "Je suis tellement pour l'infini actuel, qu'au lieu d'admettre que la nature l'abhorre, comme l'on dit vulgairement, je tiens qu'elle l'affecte partout, pour mieux marquer les perfections de son Auteur. Ainsi je crois qu'il n'y a aucune partie de la matière qui ne soit, je ne dis pas divisible, mais actuellement divisée; et par conséquent la moindre particelle doit être considerée comme un monde plein d'une infinité de créatures différentes."

[2] According to Laugwitz [König 1990, 9–12], Leibniz spoke about infinity on three different levels: a popular one, a second for mathematicians, and the third for philosophers. At the second level, which is that of the *Nouveaux essays*, he favors the potential conception of limits, but at the third he presents the kind of approach that is typical of his *Monadologie*.

[3] See [Riemann 1892, 534–38]. The interrelation that Riemann seeks to establish between psychology and physics is also reminiscent of Leibniz.

derived. At least, this is how I think that his plan was ... Unfortunately, it is all so fragmentary![1]

Dedekind's early approach to the natural numbers, traits of his elementary theory of sets, and even his logicist standpoint would seem to link back to Leibniz (see §VII.5). As regards Cantor, he frequently made explicit his strong interest in Leibniz's philosophy, and particularly in the attempt to elaborate a new, organicistic theory of Nature.[2]

This common trait in the otherwise divergent views of Riemann, Dedekind and Cantor could help explain their attitude toward infinity. The influence of Leibniz's *Monadologie* might be one of the key factors that impelled them to accept the actual infinite. Another key factor, of course, was the development of mathematical ideas themselves, for instance (though not exclusively) in the foundations of the calculus. This brings us to the issue of attitudes toward actual infinity among German mathematicians.

Ever since it was given by Cantor [1886, 371], the paradigmatic example of rejection of actual infinity has been a passage of an 1831 letter from the "prince of mathematicians" to Schumacher. Schumacher had sent an attempted proof of the Euclidean parallel postulate, and Gauss replied:

But, as concerns your proof of 1), I object above all the use of an infinite magnitude as if it were *complete*, which is never permitted in mathematics. The infinite is only a *façon de parler*, when we are properly speaking about limits that certain relations approach as much as one wishes, while others are allowed to increase without limit.[3]

It has been argued that these statements had a very particular aim, and cannot be used against the set-theoretical infinite [Waterhouse 1979]. Schumacher made some

---

[1] Dedekind to Weber, March 1875 [Cod. Ms. Riemann 1, 2, 24]: "Was mich betrifft, so bin ich für die stetige materielle Erfüllung des Raumes und die Erklärung der Gravitations- und Lichterscheinigungen im höchsten Grade eingenommen ... Diese Gedanken hat Riemann sehr früh, nicht erst in seiner letzten Zeit, ergriffen ... Sein Streben ging ohne Zweifel dahin, den allgemeinsten Principien der Mechanik, die er keineswegs umstossen wollte, bei der Naturerklärung eine neue, natürlichere Auffassung unterzulegen; das Bestreben der *Selbst*erhaltung und die in den partiellen Differentialgleichungen ausgesprochene Abhängigkeit der Zustandsveränderungen von den nach Zeit und Raum unmittelbar benachbarten Zuständen sollte er als 130 das Ursprüngliche, nicht Abgeleitete angesehen werden. So denke ich mir wenigstens seinen Plan. ... Leider ist Alles so lückenhaft!"

[2] See [Cantor 1883, especially 177, 206–07], [Cantor 1932, 275–76], [Schoenflies 1927, 20], [Meschkowski 1967, 258–59]. It is worth mentioning that in the 1870s there was a group of theologians who accepted the actual infinite and were important for Cantor – above all Gutberlet and cardinal Franzelin; see [Meschkowski 1967], [Dauben 1979], [Purkert & Ilgauds 1987].

[3] [Gauss 1906, vol. 8, 216]: "Was nun aber Ihren Beweis für 1) betrifft, so protestire ich zuvörderst gegen den Gebrauch einer Unendlichen Grösse als einer *Vollendeten*, welcher in der Mathematik niemals erlaubt ist. Das Unendliche ist nur eine façon de parler, indem man eigentlich von Grenzen spricht, denen gewisse Verhältnisse so nahe kommen als man will, während andern ohne Einschränkung zu wachsen verstattet ist."

assumptions about the behavior of geometrical figures in infinity, based upon mere analogy, and Gauss, led by his knowledge of non-Euclidean geometry, protested against such unjustified assumptions. However, Gauss's statements are sharp and general: he takes as a model the theory of limits, understood in the sense of a potential infinity. On the other hand, some authors have indicated that, at times, Gauss employs infinitesimal notions that would seem to be plainly rejected by the above quotation. For instance, his differential geometry, like Riemann's, can hardly be understood other than as infinitesimal mathematics (see Laugwitz in [König 1990, 26]).

Be it as it may, what is important for us is that Gauss was not the only German mathematician of that period. Others defended bolder positions. As early as 1788 Johann Schultz, a theologian and mathematician friend of Kant, developed a mathematical theory of the infinitely great (see [Schubring 1982; König 1990, 155–56]). A very important contribution, that unfortunately was scarcely known in its time, was that of the philosopher, theologian and mathematician Bolzano, not only in his *Paradoxien des Unendlichen* [Paradoxes of the infinite; 1851], but also in the earlier *Wissenschaftslehre* [Theory of science; 1837]. Bolzano introduced the notion of set in several different meanings: in general he talked about collections or concept-extensions [Inbegriffe], but he singled out those collections in which the ordering of elements is arbitrary [Mengen], and among these the ones whose elements are units, "multiplicities" [Vielheiten; Bolzano 1851, 2–4]. The notion of infinity is then carefully defined as follows: a multiplicity is infinite if it is greater than any finite multiplicity, i.e., if any finite *Menge* is only a part of it [*op.cit.*, 6]. Bolzano defended forcefully the actual infinite, showing that the "paradoxes" of infinity involved no contradiction at all, and attempted to elaborate a theory of infinite sets (see §II.6).

It should be emphasized that some 19th-century mathematical developments depended upon accepting the notions of point and line 'at infinity.' Cantor himself mentioned this kind of precedent when he introduced the transfinite ordinal numbers [Cantor 1883, 165–66]. That happened particularly in projective geometry, which played a central role in geometrical thinking all along the century. Sometimes the introduction of elements at infinity may have been just an instrumental move not implying an acceptance of actual infinity, but, as we shall see in the example of Steiner, at times it was accompanied by expressions that explicitly introduced the actual infinite. Another example is Riemann's function theory. In 1857 he took the step of 'completing' the complex plane with a point at infinity, thus turning it into a closed surface, which made it possible to reach general results in a simplified way.[1]

Jacob Steiner, the great representative of synthetic geometry, seems to have been a defender of actual infinity. Furthermore, he introduced notions that constitute quite clear precedents of the language of sets and mappings. Steiner was a professor at Berlin, and thus a colleague of Dirichlet and, later, Jacobi. We will

---

[1] Riemann's position *vis à vis* the infinite is analyzed in §II.4.2.

consider his main work, bearing the long title *Systematische Entwicklung der Abhängigkeit geometrischer Gestalten von einander*.[1] The first trait of this work that calls attention is the language employed, strongly reminiscent of neohumanism and idealism. In the preface, Steiner indicates that his aim is to go beyond proving some theorems, to discovering the "organically interconnected whole," the "organism" that gives a sense to the multiplicity of results. He aims at finding "the road followed by Nature" in forming the geometrical configurations and developing their properties.[2] Steiner's talk of "system" and "organism" may be compared, for instance, with Wolf's words quoted in §1. The similarity speaks for the free flux of ideas among mathematicians, philosophers and humanists, which in the case of Steiner is quite obvious, since in his younger years he was a teacher at the school of the famous Swiss pedagogue Pestalozzi.

More interesting for our present purposes is to observe the way in which Steiner emphasizes the conception of line, plane, bundle of lines, etc. as aggregates of infinitely many elements. Aristotle and other conscious partisans of the potential infinite had carefully avoided that move. It seems revealing to find that it was precisely authors like Steiner, close to neohumanistic and philosophical ideas, who broke with that restriction within a favorable intellectual context. Steiner says that in a straight line one may think "an innumerable amount" of points, and in the plane there are "innumerably many" lines and points.[3] Coming to specifically projective notions, he defines:

II. *The planar bundle of lines*. Through each point on a plane innumerably many straight lines are possible; the totality of all these lines will be called "planar bundle of lines" ...

The expression "the totality of all" will become prototypical of the works of Cantor and Dedekind that introduce sets. Later we read:

V. *The bundle of lines in space*. ... Such a bundle of lines does not only contain infinitely many lines, but also embraces numberless planar bundles of lines (II.) and bundles of planes (III.) as subordinated elements or configurations ...[4]

Steiner's viewpoint is thus close to the language of sets, although still far from a general viewpoint, not to mention a straightforward analysis of the notion of set.

---

[1] Systematic development of the interdependence among geometrical configurations [Steiner 1832].

[2] [Steiner 1832, v–vi]: "... organisch zusammenhängendes Ganze ... Gegenwartige Schrift hat es versucht, den Organismus aufzudecken ..." [*op.cit.*, vi]: "Wenn nun wirklich in diesem Werke gleichsam der Gang, den die Natur befolgt, aufgedeckt wird ..."

[3] Steiner [1832, xiii]: "In der Geraden sind eine unzählige Menge ... Punkte denkbar."

[4] Steiner [1832, xiii]: "Der ebene Strahlbüschel. Durch jeden Punkt in einer Ebene sind unzählige Gerade möglich; die Gesammtheit aller solcher Geraden soll „ebener Strahlbüschel" ... heissen." [*Op.cit.*, xiv]: "Der Strahlbüschel im Raum. ... Ein solcher Strahlbüschel enthält nicht nur unendlich viele Strahlen, sondern er umfasst auch zahllose ebene Strahlbüschel (II.) und Ebenenbüschel (III.) als untergeordnete Gebilde oder Elemente ..."

But it could be set in line with work of Riemann, Dedekind and other authors (see chapters II–V), insofar as it presents a new conceptualization of previous theories, based upon notions related to that of set. Moreover, the notion of a transformation, in the sense of a one-to-one correspondence, had become central for projective geometry. That notion it set out by Steiner in full clarity:

First a straight line and a planar bundle of lines will be related to each other, so that their elements get matched, that is, so that a certain line in the bundle corresponds to each point in the straight line.[1]

This interesting passage, and similar ones in Möbius and Plücker (see [Plücker 1828, vii]), obviously suggest the idea of one-to-one mapping. It is worth noting that Steiner regarded such correspondences as a key methodical element in order to show the "interdependence among the configurations," which in his work consti-tuted "the heart of the matter."[2]

The much less romantic Möbius, in *Der barycentrische Calcul*, set out very clearly the notion of a one-to-one correlation of points in two different figures or spaces [Möbius 1827, e.g. 169, 266].[3] He considered the simplest geometrical rela-tions that can be defined by means of such transformations, and ordered them sys-tematically: they were those of "similarity" (our congruence), "affinity," and the most general one of "collineation" (see [Möbius 1827, zweiter Abschnitt, 167ff; 1885, 519]). He argued that all such "relationships" belonged to elementary ge-ometry, because in all cases straight lines correspond to straight lines. A notable, but apparently little known fact, is that Felix Klein regarded Möbius's study of geometrical "relationships" a clear precedent of his own *Erlanger Programm*.[4] On the other hand, one does not find in Möbius any statement that might betray an admission of actual infinity, nothing comparable to the above-mentioned words of his more romantic contemporary Steiner.

Dedekind and Cantor were both aware of Steiner's ideas, and the first also read Möbius very carefully.[5] Dedekind had chosen geometry as the subject for his first

---

[1] [Steiner 1832, xiv–xv]: "Zuerst werden eine Gerade und ein ebener Strahlbüschel aufeinander bezogen, so dass ihre Elemente gepaart sind, d.h., dass jedem Punkt der Geraden ein bestimmter Strahl des Strahlbüschels entspricht."

[2] [*Op.cit.*, vi]: "den Kern der Sache ... der darin besteht, dass die Abhängigkeit der Gestalten von einander, und die Art und Weise aufgedeckt wird, wie ihre Eigenschaften von den einfachern Fig-uren zu den zusammengesetztern sich fortpflanzen."

[3] [*Op.cit.*, 266]: The "essence" of collineation consists in that "bei zwei ebenen oder körperli-chen Räumen, jedem Puncte des einen Raums ein Punct in dem anderen Raume dergestalt ent-spricht, dass, wenn man in dem einen Raume eine beliebige Gerade zieht, von allen Puncten, welche von dieser Geraden getroffen werden (collineantur), die entsprechenden Puncte in dem anderen Raume gleichfalls durch eine Gerade verbunden werden können."

[4] See [Klein 1926, vol. 1, 118], and also [Wussing 1969, 35–42], which includes a more detailed discussion of this aspect of Möbius's work.

[5] Although Riemann spent the years 1847–49 at Berlin, apparently he did not attend Steiner's lectures.

lecture course at Göttingen in 1854/55, when, as he said in a letter to Klein, he "made an effort to establish a parallelism between the modern analytic and synthetic methods" [Lorey 1916, 82]. To prepare that course, he borrowed from the Göttingen library Steiner's book, together with works by Chasles, Plücker [1828 and 1835], and the *barycentrische Calcul* of Möbius.[1] As regards Cantor, he himself [1932, 151] indicated that the term "Mächtigkeit" [power], which he used from 1877 to refer to the cardinality of a set, was taken from one of Steiner's works.[2] One may thus assume that Steiner played some role in the introduction of the notions of set and mapping, and the acceptance of the infinite, even if his statements are still vague and limited to a rather particular subject. We shall not, however, ascribe a particularly important role to his work in the events reported in the following chapters. More centrally, we may assume that, when Cantor and Dedekind expressed their confidence in the importance of sets and/or mappings for mathematics, they also had geometry in mind.

4. The Göttingen Group, 1855–1859

The University of Göttingen was the most advanced German one in the late eighteenth-century [McClelland 1980], and is famous in connection with mathematics, since it was here that Gauss worked in the early nineteenth-century, and by 1900 it had become a leading research center under Klein and Hilbert.[3] But the Göttingen of 1850 was quite different from such a center. The teaching of mathematics was far from advanced. Gauss was a professor of astronomy, and he was not attracted by the prospect of teaching poorly prepared and little interested students the basic elements of his preferred discipline. Thus, he only taught some lectures on a restricted field of applied mathematics, for instance on the method of least squares and on geodesy [Dedekind 1876, 512; Lorey 1916, 82].[4]

Gauss was a retiring man, and, strange as it may seem, he was particularly hard to approach for mathematicians, less so for astronomers and physicists. Therefore, although Riemann and Dedekind established some contact with him, and the second was his doctoral student, one should not expect much influence from the direct

---

[1] As can be seen in the volumes of Göttingen's *Ausleihregister* for the summer semester of 1854 and winter semester of 1854/55. He paid particular attention to Möbius, since he borrowed his [1827] in the semesters of 1850/51, 1853, 1854 and 1855. Notably, Möbius calls (finite) sets of points "Systeme von Puncten" [e.g., 1827, 170]; the word "System," also used by Riemann and by Dedekind himself in his algebraic work, will finally become his technical term for set in the 1870s (see chapter III and VII).

[2] Cantor refers to *Vorlesungen über synthetische Geometrie der Kegelschnitte*, § 2. Steiner used the term to indicate that two configurations were related by a one-to-one coordination.

[3] Klein came to Göttingen in 1886, strongly supported by the minister, with the purpose of building a center that could be compared with Berlin. He remained until 1913, while Hilbert came in 1895, until 1930. See [Rowe 1989].

[4] Lorey quotes in full a letter from Dedekind with reminiscences from his Göttingen time.

contact with him. Rather, it was through his writings that both were highly influenced and became his followers.

The teaching of mathematics proper was in the hands of a rather unimportant Ulrich and of Stern, who has already been named. The latter was a specialist in number theory, with a difficult career because of being a Jew, although he seems to have been a good professor [Lorey 1916, 81–82]. Ulrich and Stern founded in 1850 a mathematico-physical seminar for the training of *Gymnasium* teachers, also led by Wilhelm Weber and another physics professor. To sum up, despite the presence of the "princeps mathematicorum," Göttingen was still a rather traditional university as far as mathematics is concerned. University lectures were not adjusted to the high level of contemporary research, since the characteristically German combination of teaching and research had not yet arrived. Dedekind says that the teaching was perfectly sufficient for the purpose of preparing students for the *Gymnasium* entrance examination, but quite insufficient for a more thorough study [Lorey 1916, 82]. Projective geometry, advanced topics in number theory and algebra, the theory of elliptic functions, and mathematical physics were not available [*ibid.*].

The situation was different with physics. Wilhelm Weber had began to incorporate Göttingen to the modernizing trend:

Weber's extensive lecture course on experimental physics, in two semesters, made the most profound impression on me. The strict separation between the fundamental facts, discovered by means of the simplest experiments, and the hypotheses linked with them by the thinking human mind, afforded an unmatchable model of the truly scientific research, as I had never known until then. In particular, the development of electricity theory had an enormously stimulating effect ...[1]

Thus, Weber was teaching on his own research topics, which brought new standards of precision to electrical experimentation, and provided a first-rate theoretical contribution with his unification of electrostatics, electrodynamics and induction.[2] This came after his collaboration with Gauss on terrestrial magnetism in the 1830s, which, by the way, was the occasion for Gauss's research on potential theory. Interestingly, Weber's methodology, as described by Dedekind, coincides essentially with Riemann's in his famous work on geometry [Riemann 1854].

The mathematico-physical seminar soon went beyond its initial purpose of training secondary-school teachers, to assume the function of affording a better laboratory training. Both Dedekind and Riemann participated in it [Lorey 1916,

---

[1] Dedekind in [Lorey 1916, 82]: "hat mir die über zwei Semester vertheilte grosse Vorlesung von Weber über Experimentalphysik den tiefsten Eindruck gemacht; die strenge Scheidung zwischen den durch die einfachsten Versuche erkannten fundamentalen Tatsachen und den durch den menschlichen denkenden Geist daran geknüpften Hypothesen gab ein unübertreffliches Vorbild wahrhaft wissenschaftlicher Forschung, wie ich es bis dahin noch niemals kennen gelernt hatte, und namentlich war der Aufbau der Elektrizitätslehre von grossartiger begeisternder Wirkung ..."

[2] See [Jungnickel & McCormmach 1986, 138–48]. Weber's main work was his 'Elektrodynamische Maassbestimmungen' of 1846.

81], although it was the latter who got more involved in physical practices, eventually becoming Weber's teaching assistant around 1854 [Dedekind 1876, 512–13, 515]. Riemann's collaboration with Weber was the background for his many efforts to establish a unified treatment of the laws of nature. A manuscript note that must come from this or a later period, indicating Riemann's research topics, mentions his great research on Abelian and other transcendental functions, and on the integration of partial differential equations, and goes on:

My *main work* deals with a new conception of the known natural laws – expression of them by means of different basic notions – which would make it possible to employ the experimental data on the interactions between heats, light, magnetism and electricity, in order to investigate their interrelation.[1]

This was also consequential for his mathematical work. Among experts on Riemann it is commonplace that, in his mind, ideas of a physical origin found a natural development in pure mathematics, and conversely [Bottazzini 1977, 30].

The combination of teaching and research came to Göttingen mathematics after Gauss's death. In order to maintain the university's renown in mathematical research, his professorship was divided into one for astronomy and one for pure mathematics, and, through the good offices of Weber, Dirichlet received and accepted the call.[2] This "opened up a new era for mathematical studies at Göttingen," not because Dirichlet established new organizational arrangements (as Clebsch and Klein later), but because his brilliant lectures went all the way up to the research frontier. As Dedekind said, "through his teaching, as well as frequent conversations ..., he turned me into a new man. In this way he had an enlivening influence on his many students ...."[3]

The presence of Dirichlet, Riemann and Dedekind at Göttingen, and the courses they taught from 1855 onward, turned Göttingen into one of the most important mathematical centers, to be compared only with Berlin and Paris. The three mathematicians had common traits in that they all followed lines initiated by Gauss, and promoted an abstract, conceptual vision of mathematics. They valued each other very much. In 1852, one year after the death of Jacobi, Riemann wrote that Dirichlet was, with Gauss, the greatest mathematician alive [Butzer 1987, 58]. Dirichlet also regarded Riemann as the most promising young mathematician in

---

[1] [Riemann 1892, 507; emphasis added]: "Meine Hauptarbeit betrifft eine neue Auffassung der bekannten Naturgesetze – Ausdruck derselben mittelst anderer Grundbegriffe – wodurch die Benutzung der experimentellen Data über die Wechselwirkung zwischen Wärme, Licht, Magnetismus und Electricität zur Erforschung ihres Zusammenhangs möglich würde."

[2] Dedekind in [Lorey 1916, 82]. See [Jungnickel & McCormmach 1986, 170–72]. Dirichlet used the call to try getting freed from the heavy teaching at the military academy, but the Prussian ministry reacted too slowly.

[3] [Lorey 1916, 82–83]: "... womit für das mathematische Studium in Göttingen eine neue Zeit anbrach. ... er hat durch seine Lehre, wie durch häufige Gespräche in persönlichen Verkehr, der sich nach und nach immer vertrauter gestaltete, einen neuen Menschen aus mir gemacht. So wirkte er belebend auf seine zahlreiche Schüler ein ..." See also [Scharlau 1981, 35ff].

Germany. And after Gauss's death, in 1856, Dedekind wrote that Riemann was, after or even with Dirichlet, the most profound living mathematician [Scharlau 1981, 37]. Within the group, Dedekind had the role of the student who, even as a *Privatdozent*, enjoyed for the first time the opportunity of a first-level mathematical education. Both Dirichlet and Riemann left him important posthumous duties.

With regard to research fields, those of Riemann and Dedekind were quite far apart, since the former excelled in function theory and mathematical physics, while the latter devoted himself to algebra and number theory. In a sense, they divided among themselves the topics studied by Dirichlet, who was the most universal of them all. (This is understandable, since there was a growing trend towards specialization all along the century.) But, underlying these differences, it is possible to find meeting points at a deeper level, that of theoretical and methodological preferences. This, and the net of influences they exerted on each other, makes it justifiable to speak about a group with common traits.[1]

Although the point will become clearer in what follows (particularly chapters II, III, V), it is possible to give some examples at this point. The conceptual approach to mathematics is clear, for instance, in Cauchy, when he bases his treatment of analysis [Cauchy 1821] upon the notion of a continuous function, where continuity is defined independently of the analytical expressions which may represent the function.[2] Such a viewpoint is taken further by Dirichlet when, in a paper on Fourier series [1837, 135–36], he proposes to take a function to be any abstractly defined, perhaps arbitrary correlation between numerical values.[3] Already in 1829 he had given the famous example of the function $f(x)=0$ for rational x, $f(x)=1$ for irrational x [Dirichlet 1829, 169], which he (wrongly) took to be a function not representable by an analytical expression, and a non-integrable function. Riemann takes up Dirichlet's abstract notion of function in his function-theoretical thesis [Riemann 1851, 3], where he also makes reference to his teacher's work on the representability of piecewise continuous functions by means of Fourier series (see chap. V). This became the subject of his Habilitation thesis [Riemann 1854b], where we find the famous definition of the Riemann integral, a definition that finally consolidated the abstract notion of function, since it opened up the study of discontinuous real functions. Lastly, Dirichlet's notion of function was given its most general expression when Dedekind [1888, 348] defined, for the first time, the notion of mapping within a set-theoretical setting.

---

[1] In this case, I avoid the word "school" because there was not a relevant production of advanced students, that would later become research mathematicians (see below).

[2] Euler understood by "functiones continuae" those that corresponded to a single analytical expression throughout. See [Youschkevitch 1976].

[3] There has been some debate whether Dirichlet ever thought about applying this concept to functions more 'arbitrary' than piecewise continuous functions. The 1837 paper only considers continuous functions, but it is an expository paper published in a physics journal. It seems plausible to me that he entertained the abstract notion of function, but thought that in mathematics there is no need to consider highly arbitrary functions – except as counter-examples.

The emergence of the conceptual viewpoint was mentioned by Dirichlet in a frequently quoted passage in his obituary of Jacobi (see also Eisenstein's description in §1). According to him, there is an ever more prominent tendency in recent analysis, "to put thoughts in the place of calculations."[1] It is interesting, though, to consider the context of Dirichlet's quotation. He was calling attention to the fact that, in spite of that tendency, there were fields in which calculations preserved their legitimacy, and that Jacobi had obtained admirable results in this way. Also in the work of Dirichlet it is possible to find a clever combination of new thoughts, which sometimes are extremely simple (i.e., the box principle in number theory), with complex analytical calculations. One has the impression that it was above all Riemann who brought the conceptual trend to a new level, and his friend Dedekind followed his footsteps. Among other reasons to think so, we may consider the fact that the approach promoted by Riemann and Dedekind was quite specific in comparison with those of other Dirichlet students, such as Heine, Lipschitz or Eisenstein, not to mention Kronecker. Dedekind [1876, 512] tells us that, around 1848, Riemann discussed with his friend Eisenstein the issue of the introduction of complex numbers in the theory of functions, but they were of completely different opinions. Eisenstein favored the focus on "formal calculation," while Riemann saw the essential definition of an analytic function in the Cauchy–Riemann partial differential equations (see below).

In this connection, it is telling that Dedekind's most committed methodological statements consistently refer to Riemann's function theory as a model. For Dedekind always regarded himself as a disciple of Dirichlet. In 1859, when he had already left Göttingen, he wrote to his family that he owed to Dirichlet more than to any other man [Scharlau 1981, 47]. His debt to Dirichlet was particularly important in connection with his general formation as a mathematician, with number theory, and with the issue of mathematical rigor [Haubrich 1999, ch. 5]. Nevertheless, when it came to mathematical methodology and the conceptual approach, the main name he mentioned was Riemann. The following text, written in 1895, is interesting enough to warrant full quotation. Dedekind therein defended his approach to the foundations of algebraic number theory, in contrast to those of Kronecker and Hurwitz. This gives the occasion for a passage in which he opens his "mathematical heart" [Dugac 1976, 283]:

First, I will remember a beautiful passage in *Disquisitiones Arithmeticae*, which already in my youth made the most profound impression upon me. In art. 76 Gauss relates that Wilson's theorem was first made known by Waring, and goes on: "But neither could prove it, and Waring confesses the proof seems the more difficult, as one cannot imagine any notation that could express a prime number. – In our opinion, however, such truths should be extracted from concepts rather than notations." – In these last words, if they are taken in the most general sense, we find the expression of a great scientific thought, the decision for the inner in contrast to the outer. This contrast comes up again in almost all fields of mathemat-

---

[1] [Dirichlet 1889, vol. 2, 245]: "Wenn es die immer mehr hervortretende Tendenz der neueren Analysis ist, Gedanken an die Stelle der Rechnung zu setzen ..."

ics; it suffices to think about function theory, about the Riemannian definition of functions by means of characteristic inner properties, from which the outer forms of representation arise with necessity. But also in the much more limited and simple field of ideal theory both directions are effective ...[1]

At this point, Dedekind mentions that he has always set himself such requirements, and refers to a passage of an 1876 paper where he again takes Riemann's function theory as a model, and states that a theory "founded upon calculation would not offer the greatest degree of perfection."[2] A similar idea is found in a letter to Lipschitz written the same year:

My efforts in the theory of numbers are directed toward basing the investigation, not on accidental forms of representation (or expressions), but on simple basic notions, and thereby – though this comparison may perhaps seem pretentious – to attain in this field something similar to what Riemann did in the field of function theory.[3]

The relevant details of Riemann's function theory will be mentioned in chap. II, while those concerning Dedekind's ideal theory can be found in chap. III. At this point, however, we can give some simple examples which will be instructive, since they enable us to contrast the Göttingen abstract approach with the viewpoint adopted at Berlin (studied in the next section). Riemann sought, in his theory, to find global, abstract ways of determining complex functions by means of minimal sets of independent data [Riemann 1857, 97]. As he had written in his 1851 dissertation:

---

[1] [Dedekind 1930, vol. 2, 54–55]: "Ich erinnere zunächst an eine schöne Stelle der Disquisitiones Arithmeticae, die schon in meiner Jugend den tiefsten Eindruck auf mich gemacht hat. Im Art. 76 berichtet Gauss, dass der Wilsonsche Satz zuerst von Waring bekanntgemacht ist, und fährt fort: Sed neuter demonstrari potuit, et cel. Waring fatetur demonstrationem eo difficiliorem videri, quod nulla notatio fingi possit, quae numerum primum exprimat. – At nostro quidem judicio hujusmodi veritates ex notionibus quam ex notationibus hauriri debebant. – In diesen letzten Worten liegt, wenn sie im allgemeinsten Sinne genommen werden, der Ausspruch eines grossen wissenschaftlichen Gedankens, die Entscheidung für das Innerliche im Gegensatz zu dem Äusserlichen. Dieser Gegensatz wiederholt sich auch in der Mathematik auf fast allen Gebieten; man denke nur an die Funktionentheorie, an Riemanns Definition der Funktionen durch innerliche charakteristische Eigenschaften, aus welchen die äusserlichen Darstellungsformen mit Notwendigkeit entspringen. Aber auch auf dem bei weitem enger begrenzten und einfacheren Gebiete der Idealtheorie kommen beide Richtungen zur Geltung ..."

[2] [Dedekind 1930, vol. 3, 296]: "une telle théorie, fondée sur le calcul, n'offrirait pas encore, ce me semble, le plus haut degré de perfection; il est preférable, comme dans la théorie moderne des fonctions ..."

[3] Dedekind to Lipschitz, June 1876, in [Dedekind 1930/32, vol. 3, 468]: "Mein Streben in der Zahlentheorie geht dahin, die Forschung nicht auf zufällige Darstellungsformen oder Ausdrücke sondern auf einfache Grundbegriffe zu stützen und hierdurch – wenn diese Vergleichung auch vielleicht anmassend klingen mag – auf diesem Gebiete etwas Ähnliches zu erreichen, wie Riemann auf dem Gebiete der Functionentheorie."

A theory of those functions [algebraic, circular or exponential, elliptical and Abelian] on the basis of the foundations here established would determine the configuration of the function (i.e., its value for each value of the argument) independently of any definition by means of operations [analytical expressions]. Therefore one would add, to the general notion of a function of a complex variable, only those characteristics that are necessary for determining the function, and only then would one go over to the different expressions that the function can be given.[1]

Riemann starts with what is general and invariant, and from it the many possible analytical expressions for the function would be derived. He thus needed a general definition of analytic function which, as we have already mentioned, he found in the Cauchy–Riemann equations [Riemann 1851, 5–6]:

$$\frac{\partial u}{\partial x} = \frac{\partial v}{\partial y}; \quad \frac{\partial u}{\partial y} = -\frac{\partial v}{\partial x}$$

where u and v are the real and complex parts of $f(x+iy)$.[2] This is a perfect example of his preference for simple fundamental properties as a basis for the development of the theory, which contrasts nicely with Weierstrass's definition of analytic functions (next section). Further aspects of his theory, such as the introduction of Riemann surfaces, were consistent with that preference.

Similarly, Dedekind preferred to use basic notions of an abstract character, such as his fields and ideals. In this, more basic case, avoidance of forms of representation led to reliance on sets endowed with an algebraic structure. A field [Körper] was for him any subset of $\mathbb{C}$ that is closed with respect to the four basic algebraic operations [Dedekind 1871, 224]. An ideal was a set of algebraic integers which is characterized by two simple properties that can be stated in terms of sums, differences and products of algebraic integers [*op.cit.*, 251]. As in the case of Riemann, the decision to base the theory upon such basic notions deviated sharply from established tradition, since it was then customary to focus on the numbers themselves, not on sets of numbers. Besides, it was customary in number theory to use 'forms,' or algebraic equations, freely in the development of the theory, but Dedekind avoided that completely since it meant reliance on expressions. His main reason for preferring abstract notions was their generality and lack of arbitrariness, which meant that they immediately conveyed what was "invariant" (his term) in the object defined, be it a field or an ideal. Once again, this definition of the basic objects, and

---

[1] [Riemann 1892, 38–39]: "Eine Theorie dieser Functionen auf den hier gelieferten Grundlagen würde die Gestaltung der Function (d.h. ihren Werth für jeden Werth ihres Arguments) unabhängig von einer Bestimmungsweise derselben durch Grössenoperationen festlegen, indem zu den allgemeinen Begriffe einer Function einer veränderlichen complexen Grösse nur die zur Bestimmung der Function nothwendigen Merkmale hinzugefügt würden, und dann erst zu den verschiedenen Ausdrücken deren die Function fähig ist übergehen."

[2] The reason for the name is that Cauchy had already recognized that a complex function is analytic if and only if it is differentiable, although he did not use the differential equations as a definition.

the further development of the theory, is in sharp contrast to Kronecker's (next section).

At this point it should be clear that Riemann and Dedekind brought forward the conceptual tendencies that could be found, more or less clearly, in the work of Gauss, Cauchy and Dirichlet. In doing so, however, they effected a clear turn in the meaning of 'conceptual' which would be extremely consequential for mathematics, since their work had an enormous impact upon 20th-century function theory and algebra. They consistently attempted to frame mathematical theories within the most general appropriate setting, in such a way that "outer forms of representation" were avoided, new basic objects were chosen, and a definition of the characteristic "inner" properties of these objects (i.e., a fundamental concept) was placed at the very beginning of the theory. We may refer to this particular brand of the conceptual viewpoint as *abstract conceptual*. Thus, one of the characteristic traits of modern mathematics, which is frequently called its 'abstract' viewpoint, can be traced back to Riemann and Dedekind.

The emergence of a research school following the abstract-conceptual approach to mathematics, promoted by Riemann and Dedekind, was hindered by a confluence of events. Riemann, who succeeded Dirichlet at Göttingen in 1859, was incapacitated since 1862 due to a lung illness, spending most of the time in Italy and dying in 1866 at age forty. Dedekind was slow in publishing original research, so that he was offered almost no university position until 1870, and afterwards he consciously chose to remain at the Technical School in his birthplace Braunschweig. After Riemann's death, a great tradition of mathematicians consolidated at Göttingen with Clebsch, Fuchs and Schwarz, but these men came and went in rapid succession. It was only with the arrival of Klein in the 1880s, and later with Hilbert, that something that could be compared with a school was firmly established.[1] The diffusion of the methodological viewpoint favored by Riemann and Dedekind was difficult and slow since the only available means, given the circumstances, was their published work. Besides this, only Dedekind's rich correspondence with such mathematicians as Heinrich Weber, Frobenius and Cantor could have been instrumental in spreading that standpoint. It is worth mentioning that it might be quite interesting to study H. Weber as a key figure in the diffusion of the conceptual approach.[2]

---

[1] Klein tried to bring farther the tradition of Gauss and Riemann, but he conceived for himself a role that was much broader and more ambitious than that of a school leader. And the impressive number of mathematicians who studied with Hilbert may not constitute a school, strictly speaking. For details and nuances concerning this period, the reader should consult [Rowe 1989].

[2] He was strongly influenced by Riemann's work and by his collaboration with Dedekind (chap. III), and led an active academic career in Königsberg (where he counted Hilbert among his students), Göttingen and Strassburg, among other places. He was also extremely influential through several important textbooks.

5. The Berlin School, 1855–1870[1]

Berlin University, founded 1810, soon turned into the most important university in Prussia and all the German-speaking countries. Gauss might have ended up being a professor there, since Alexander von Humboldt made two attempts to bring him to Berlin, in the 1800s and 1820s.[2] The already mentioned Martin Ohm became an extraordinary professor in 1824 (and was named "ordinary" professor in 1839, at the same time as Dirichlet). But the situation in mathematics was not notable until 1828, when Humboldt and Crelle began to play an important role in their respective positions as court counselor and adviser on mathematics for the ministry. There was a tragically failed attempt to bring Abel to Berlin in 1829, the year of his death. But with Dirichlet and, from 1834, Jacob Steiner,[3] lectures of a high level began to be offered. Dirichlet is usually considered to have shaped modern-style mathematics lectures, and also established an informal seminar with selected groups of students. As we have seen, however, there was the grotesque situation that he could not take part in doctorates and Habilitationen until 1851. The situation for mathematics became even better in 1844, when Jacobi came from Königsberg to Berlin as a member of the Academy. The Academy of Sciences had been the most important scientific center in Berlin up to 1800, and it was a notable support for the new University, since its members enjoyed the right to impart lectures. This possibility was exploited by Jacobi during his Berlin period, as it was by other mathematicians before and after.[4]

Berlin had thus turned into an ever more important center for mathematics since about 1830. But the situation became even better and the actors changed completely in the 1850s, following the death of Jacobi in 1851 and Dirichlet's transfer to Göttingen in 1855. That same year, the specialist in number theory Ernst Eduard Kummer became Dirichlet's successor, on his proposal. Kummer regarded himself as a disciple of Dirichlet, although he never attended one of his lectures. He had studied mathematics and philosophy at Halle in the late 1820s, becoming a corresponding member of the Berlin Academy in 1839, through Dirichlet's proposal, on the basis of important work on the hypergeometric series. By then, he was still a *Gymnasium* teacher, but in 1842 he became professor at Breslau on the recommendation of Jacobi and Dirichlet. In this decade he began his path-breaking work on

---

[1] The name was coined in the 19th-century, and it is interesting to consider that, by then, the word "school" frequently had a negative ring, connoting a one-sided orientation. In this case, it may have also referred to the extremely powerful position of the school in academic matters.

[2] His purpose was to make him a professor at a new Polytechnical School that he was attempting to launch, as happened with Dirichlet and Abel later. Dirichlet was the only one who actually came to Berlin, becoming a professor at the University after Humboldt's plans failed. See [Biermann 1973, 21–27].

[3] Steiner (see §3) was also a protegé of Humboldt and Crelle. He became an extraordinary professor and member of the Berlin Academy of Sciences in 1834.

[4] University professors customarily were members of the Academy. Among the mathematicians, this was only false for M. Ohm [see Biermann 1973].

ideal numbers, that will be mentioned in chapter III; in the 1860s he made important contributions to geometry.[1] For our purposes, it is important to emphasize that Kummer stressed more the formal and calculational aspect of mathematics, than its conceptual side (see [Haubrich 1999]). One may conjecture that this would have been different, had he been a real student of Dirichlet.

A devoted teacher, Kummer became the driving force behind the new institutional arrangements at Berlin, which fully implemented the characteristically German combination of teaching and research. The institution of the seminar, created by the neohumanist philologists, had been adapted to the natural sciences at some places, as was the case with the famous Königsberg mathematico-physical seminar of Jacobi and Neumann, established in 1834. Kummer and his colleague Weierstrass were the first to create a seminar devoted to pure mathematics. This happened in 1861 and was, together with the high quality and novelty of Weierstrass's lectures, the reason for the immense appeal that Berlin exerted on young mathematicians throughout the world in the following decades. Karl Weierstrass, a completely unknown *Gymnasium* teacher, became a rising star after the publication of his first paper on Abelian functions in 1854. Until then, his mathematical education had been uncommon: he was basically self-taught, although he spent some time at an Academy in Munster, where the influence of his teacher Gudermann was notable. Interestingly, Gudermann was a follower of the combinatorial tradition (see [Manning 1975]), and some aspects of Weierstrass's work – particularly his eternal reliance on infinite series – are reminiscent of that tradition.[2] Weierstrass was also deeply influenced by the work of Jacobi and Abel on elliptic functions. In 1854 he received an *honoris causa* from Königsberg, and Kummer began to care for his becoming a professor at Berlin. In 1856 he was appointed to Berlin's Industrial Institute (later the Technical School), while rejecting offers from Austrian universities; months later Kummer had obtained him a position as extraordinary professor, and full membership in the Academy. But it was not until 1864 that he became full professor and abandoned the Industrial Institute.

Kummer and Weierstrass were in close personal and scientific contact with Leopold Kronecker, a wealthy man who lived privately at Berlin from 1855, and in 1861 also became member of the Academy with the right to teach. Kronecker knew Kummer from his *Gymnasium* years, when the latter stimulated his mathematical interests; he studied at the University of Berlin, where he took his doctorate in 1845, but also spent two semesters at Breslau, again with Kummer. In this way, Kronecker received a strong influence from both Dirichlet and Kummer, but some aspects of his work, particularly his interest in algorithms and effective calculation, bring him closer to the latter. In 1881, on the occasion of the 50th anniversary of

---

[1] See Biermann's biography in [Gillispie 1981, vol. 7], or [Bierman 1973].

[2] It would require a careful analysis to ascertain the depth of this influence, but it is clear from the outset that Weierstrass did not treat series in a purely formal way, as the combinatorialists, but rather viewed them in the 'conceptual' way of Abel and Cauchy (see [Jahnke 1987; 1991]). On the other hand, he was of the opinion that not all traits of the "combinatorial school" had been lost, as Hilbert recorded in his 1888 trip to Berlin (quoted in [Rowe 1995, 546]).

Kummer's doctorate, he said that Kummer had provided the "most essential portion" of "my mathematical, indeed ... my intellectual life."[1] In 1883 he became a professor, after his teacher retired, but already in the 1860s and 70s he had gained an ever more influential position on university affairs, in Berlin and elsewhere in Prussia – and, after 1871, the Empire.

The triumvirate turned Berlin into a world-renowned center. Weierstrass and Kummer established a coordinated, biennial structure of the university courses, which was in effect from 1864 to 1883. Kummer covered fundamental, well-established subjects such as elementary number theory, analytical geometry, the theory of surfaces, and mechanics, leaving his research topics for the seminar. His clear lectures were followed by as many as 250 students, an impressive number that Weierstrass eventually equaled. Weierstrass, like Kronecker, normally lectured on advanced research topics: analytic functions, elliptical and Abelian functions, and calculus of variations. With his sense for rigorous logical development and systematization, he arranged his lectures so that he could build on what he had already proven, and thus he hardly cited other sources. Kronecker, on the other hand, was a demanding and less careful teacher, who had few students and lectured on the theory of algebraic functions, number theory, determinants and integrals.

To some extent, there was a continuity between this and the previous generation, since Dirichlet's words on Jacobi can also be applied to the lectures of Weierstrass and Kronecker (and to Kummer's seminar activities):

It was not his style to transmit again the closed and the traditional; his courses moved totally outside the limits of textbooks, and dealt only with those parts of the discipline in which he worked creatively himself ...[2]

Such was the atmosphere in which Georg Cantor received his education as a mathematician, from 1863 to 1869. By this time, Berlin was reaching the height of its power, being led by a harmonious group of mathematicians who offered the most advanced course of studies in Germany. Cantor earned his doctorate in 1867 under Kummer, and his 1869 Habilitation was also on a topic in number theory, enjoying the guidance of Kummer and Kronecker. Soon, however, he would devote himself to the theory of trigonometric series, and the influence of Weierstrass played an essential role in all of his work.

As regards the characteristic methodological traits and mathematical style of the Berlin school, one may begin by saying that it was Kronecker and Weierstrass who expressed more openly and clearly their preferences. They shared a number of fundamental ideas, especially in the early period, up to about 1870. Both were stern partisans of the conception that mathematics must be rigorously developed starting

---

[1] As quoted by Bermann in [Gillispie 1981, vol. 7, 523].

[2] [Dirichlet 1889, vol. 2, 245]: "Es war nicht seine Sache, Fertiges und Ueberliefertes von neuem zu überliefern, seine Vorlesungen bewegten sich sämmtlich ausserhalb des Gebietes der Lehrbücher, und unfassten nur diejenigen Theile der Wissenschaft, in denen er selbst schaffend aufgetreten war ..."

from purely arithmetical notions. This was an idea that they shared with Dirichlet (see [Dedekind 1888, 338]), an idea that could also be found earlier in the work of Ohm (§2).[1] Nevertheless, the arithmetizing legacy of Dirichlet was taken up in diverging ways: in Dedekind it was tinged with abstract connotations, which express themselves in a determined acceptance of the set-theoretical standpoint; in Weierstrass it was adapted to the formal conceptual viewpoint – accepting the irrational numbers, but emphasizing the point of view of infinite series (see below and §IV.2); in the case of Kronecker, arithmetization came to be understood more radically, meaning reduction to the natural numbers, but without the use of any infinitary means, be they series or sets. This difference of opinion between Weierstrass and Kronecker began to show up around 1870, and led to a growing estrangement in the late 1870s and, above all, the 1880s.[2]

Weierstrass and Kronecker also shared a dissatisfaction with 'generalist' viewpoints in mathematics. They took pains to carefully consider the variety of particular cases that can show up in any mathematical topic. This can be seen in the attention that Weierstrass and his disciples paid to 'anomalous' functions – i.e., the famous examples of nowhere differentiable continuous functions. This feature has been considered in detail by Hawkins [1981] in connection with the work of Frobenius and Killing, both members of the Berlin school. Hawkins takes as a model Weiertrass's 1868 theory of elementary divisors. It had been common to deal with issues in algebraic analysis in a so-called 'general' way, as if there were no singular types of situations for particular values of the arguments. Weierstrass's theory of elementary divisors showed how to deal with those issues in a detailed way, paying attention to all possible special cases; as Kronecker said in an 1874 paper inspired by that theory,

It is common – especially in algebraic questions – to encounter essentially new difficulties when one breaks away from those cases which are customarily designated as general. As soon as one penetrates beneath the surface of the so-called singularities, the real difficulties of the investigation are usually first encountered but, at the same time, also the wealth of new viewpoints and phenomena contained in its depths.[3]

---

[1] It is likely that Ohm's ideas were influential on Weierstrass and Dedekind, in both cases before 1855, i.e., before they established closer contact with leading mathematicians. This would explain similarities in their treatment of the elements of arithmetic (chapter IV). As we saw, Weierstrass may have been influenced by the combinatorial tradition, to which Ohm was close; Dedekind's teacher Stern was also influenced by Ohm and the combinatorialists, and his Habilitation lecture of 1854 is strongly reminiscent of Ohm (see also VII.1).

[2] See [Biermann 1973] [Dugac 1973, 141–46, 161–63]. By 1884, Kronecker was promising to show the "incorrectness" [Unrichtigkeit] of all those reasonings with which "so-called analysis" [die sogenannte Analysis] works [Dauben 1979, 314]. Such comments affected Weierstrass strongly, and led him to fear that his mathematical style would disappear after his death, but he did not have the strength to defend his viewpoint publicly.

[3] As translated in [Hawkins 1981, 237].

Or, as he said in 1870, "all those general theorems have their hideout, where they are no longer valid."[1] This philosophy of paying a close attention to special cases may have reinforced constructive tendencies in the Berlin school.

Thus, the new group of Berlin professors pursued an approach to mathematics that is quite different from the abstract conceptual one, that we have seen in association with the names of Riemann and Dedekind. The Berlin standpoint can also be called 'conceptual,' particularly in the case of Weierstrass, who followed in the tradition of Cauchy and Dirichlet. But it did not share what we have called the 'abstract' turn, typical of Riemann. The differences between the Göttingen group and the Berlin school become evident when we consider the Berlin analogues of the basic notions, employed by Riemann and Dedekind, that we mentioned in the previous section. In his theory of analytic functions, Weierstrass defined them as those functions which are locally representable by power series. This allowed him to base the theory upon clear arithmetical notions, and to elaborate its first rigorous treatment. A necessary prerequisite was the principle of analytic continuation, which made it possible to 'reconstruct' the entire function from its local power-series representation. Weierstrass was able to establish that principle, thus creating a method for generating, from a given local representation or "element," a chain of new "analytic elements" defining the entire function.[2]

Of course, this is what Riemann would have called an approach which starts from "forms of representation" or "expressions," precisely what he was trying to avoid. Weierstrass, on the other hand, although admitting that Riemann's definition was essentially equivalent to his, criticized its reliance upon the notion of differentiable real function. This was not satisfactory because, at that time, the class of differentiable functions could not be precisely delimited (see [Pincherle 1880, 317–318]). Weierstrass's approach reduced the whole issue to representability by means of a perfectly specific class of series. Actually, it is clear in his work an interest in defining whole classes of functions by means of representability theorems using well-known simple functions. An interesting example is his 1885 theorem on the representation of continuous functions, in a closed interval, by an absolutely and uniformly convergent series of polynomials [Weierstrass 1894/1927, vol. 3, 1–37]. This is quite obviously a constructive trait in Weierstrass's approach to analysis.

Riemann, on the contrary, regarded differentiability and the Cauchy–Riemann equations as perfectly precise conditions, in an abstract sense. His approach was superior in that it yielded a direct, global overview of the multi-valuedness of complex functions. It was inferior insofar as it raised problematic issues and was not easily to amend or treat with complete rigor. To sum up, in contrast to Riemann's preference for global, abstract characterizations, Weierstrass was in favor of a local, relatively constructive approach. Borrowing the then-frequent terms 'form' and 'formal,' we may refer to Weierstrass's viewpoint as a *formal conceptual* one. In

---

[1] [Meschkowski 1969, 68]: "All solche allgemeinen Sätze haben ihre Schlupfwinkel, wo sie nicht mehr gelten." But here he was referring to the Bolzano–Weierstrass theorem!

[2] On Weierstrass's theory, see [Dugac 1973] and [Bottazzini 1986].

calling it 'conceptual,' I wish to emphasize that, in spite of his preference for representability theorems, he was no strict constructivist. The conceptual element was clearly present in the notions Weierstrass established as the foundation of analysis, for instance his definition of the real numbers (chap. IV) and the Bolzano–Weierstrass theorem (below and §VI.4.2).

The latter were elements that would come under severe criticism on the side of Kronecker. Although he and Weierstrass shared common points, they followed divergent lines of development, perhaps because of their dedication to such different fields. Kronecker came to advocate a radical arithmetization of the whole of mathematics, in the sense of acknowledging only the natural numbers and the algebra of polynomials, and requiring effective algorithms in connection with all the notions employed (see [Edwards 1989]). In a 1887 paper, Kronecker wrote:

And I also believe that some day we will succeed in "arithmetizing" the whole content of these mathematical disciplines [algebra, analysis], i.e., in basing them exclusively upon the notion of number, taken in the most restricted sense, and thus in eliminating again the modifications and extensions of this notion [note: I mean here especially the addition of irrational and continuous magnitudes], which have mostly been motivated by applications to geometry and mechanics.[1]

This is a sharp criticism of the theories of irrational numbers proposed by Weierstrass, Cantor and Dedekind (§IV.2). Kronecker regarded them as meaningless since they went beyond what is algorithmically definable from the natural numbers, depending upon the actual infinite instead. But the differences had already began to emerge around 1870. In his Berlin lectures, Weierstrass employed the Bolzano–Weierstrass theorem as a keystone; its proof was based on the principle that, given an infinite sequence of closed intervals of $\mathbb{R}$, embedded on each other, at least one real number belongs to all of them. Around 1870, Kronecker started to attack this principle and the conclusions drawn from it, as we know from Schwarz's letters to Cantor. He regarded the Bolzano–Weierstrass principle as an "obvious sophism," and was convinced that it would be possible to define functions "that are so unreasonable" that, in spite of satisfying all the conditions for the Bolzano–Weierstrass theorem, they would have "*no* upper limit."[2] Schwarz and Cantor, however, were on the side of Weierstrass and defended his principle as indispensable for analysis.

---

[1] [Kronecker 1887, 253]: "Und ich glaube auch, dass es dereinst gelingen wird, den gesammten Inhalt aller dieser mathematischen Disziplinen zu 'arithmetisieren,' d.h. einzig und allein auf den im engsten Sinne genommenen Zahlbegriff zu gründen, also die Modification und Erweiterungen dieses Begriffs [Ich meine hier namentlich die Hinzunahme der irrationalen sowie der continuirlichen Grössen] wieder abzustreifen, welche zumeist durch die Anwendungen auf die Geometrie und Mechanik veranlasst worden ist."

[2] [Meschkowski 1967, 68]: "Kronecker erklärte ... die Bolzanoschen Schlüsse als offenbare Trugschlüsse" "dass man Funktionen wird aufstellen können, die so unvernünftig sind, dass sie trotz des Zutreffens von Weierstrass' Voraussetzungen keine obere Grenze haben." See [*op.cit.*, 239–40].

Kronecker can thus be regarded as the first constructivist, and it is only natural that his approach to algebraic number theory would be extremely different from Dedekind's. The latter preferred a radically abstract, infinitistic approach employing the notion of set. The former, not less radically, favored a constructivist, finitistic viewpoint. At the same time, both were extremely concerned about questions of mathematical rigor. The first basic notion that both needed, in the context of algebraic number theory, was that of field. Whereas Dedekind defined it extensionally, as a certain set closed for algebraic operations, Kronecker defined it algorithmically: a "domain of rationality" [Rationalitätsbereich] is the totality of all "magnitudes" that are rationally representable over $\mathbb{N}$ by means of a finite set of generators R', R'', R''', ... Of course, that 'totality' is conceived as a potential infinity, not an actually existing set. Dedekind would have objected that Kronecker's definition is based on "form of representation ... that could, with the same right, be replaced by infinitely many other forms of representation," replacing R', R'', R''', ... by other sets of generators (see [Lipschitz 1986, 59–60; Dedekind 1895]).

The differences become even clearer when we consider the notions employed by both mathematicians in order to solve the main problem of algebraic number theory. In Kronecker's eyes, Dedekind's ideals were mere "symbols," and his approach was a "formal" one, since he did not grant the existence of infinite sets; meanwhile, Dedekind regarded them as "totally concrete objects" (see [Edwards, Neumann & Purkert 1982, 61]). Instead of ideals, i.e., sets of complex numbers with a certain structure, Kronecker relied on "divisors" that were not defined directly, but in association with certain "forms" or polynomials. Using modern language, we can say that, in order to study the integers of a certain field K, Kronecker relied on the ring of polynomials $K[x, x', x'', ...]$. The variables $x, x', ...$ are introduced only formally and attention is focused on the coefficients. What is important is that those "forms" or polynomials make it possible to elaborate a method for the effective construction of the required divisors (see [Edwards 1980]). As Dedekind said, the introduction of polynomials in the development of algebraic number theory "always seems to me an auxiliary means that is foreign to the issue" and "muddies the purity of the theory" [Dedekind 1895, 53, 55]. If such a method affords interesting results, there must be a deeper reason that can be formulated in 'pure' number-theoretical terms.

On the basis of Weierstrassian analysis, Cantor found the possibility of developing a theory of sets and infinity, but this led him to pursue a more abstract viewpoint, trespassing the limits set on mathematical research by the Berlin school. That provoked a strong negative reaction on the side of Kronecker, but even Weierstrass questioned Cantor's introduction of quantitative differences among infinite sets, and he never openly defended the work of his former student. As we shall see, Cantor's abstract turn can be related to the increasing influence of Riemann and Dedekind.[1]

---

[1] By the 1880s, as a result of theoretical differences mixed with personal difficulties, Cantor came to feel confronted with the Berlin school as a whole. In an 1895 letter to Hermite, he denied being a member of the school and stated that the mathematician he felt closer to was Dirichlet [Purkert & Ilgauds 1987, 195–96].

II   A New Fundamental Notion: Riemann's Manifolds

> Shyness, a natural consequence of his earlier sheltered life [as a child], ... never left him completely ... and frequently moved him to abandon himself to solitude and to his mental universe, in which he unfolded [his thoughts with] the greatest boldness and lack of prejudices.[1]

> In mathematics, the art of posing questions is more consequential than that of solving them.[2]

In this chapter we trace back the first influential appearance of a set-theoretical viewpoint to the work of Riemann. Of course, by speaking of "a set-theoretical viewpoint" I do not mean to suggest that Riemann reached technical results that we would classify today as belonging to set theory – only that he introduced set language substantially in his treatment of mathematical theories and regarded sets as a foundation for mathematics. This comes out in a public lecture given in 1854, on the occasion of his *Habilitation* as a professor at Göttingen, when he proposed a general notion of manifold – the famous 'On the Hypotheses upon which Geometry is Founded,' published posthumously by Dedekind in 1868. We shall refer to it as Riemann's *Habilitationsvortrag*.

Mentioning Riemann in connection with the history of sets is still quite uncommon, but there are indications that he played an important role in the early phases of development of the notion of set. From 1878 to 1890, his most creative period, Cantor referred to set theory as *Mannigfaltigkeitslehre*, the 'theory of manifolds,' employing the same word that Riemann had coined in his lecture of 1854. Notably, the 1878 paper in which Cantor first employs the word addresses a problem that is directly related to Riemann's *Habilitationsvortrag*, the characterization of dimension. And Dedekind understood Cantor's terminology to be related to the work of his close friend Riemann. In a letter of 1879, Dedekind proposed to

---

[1] [Dedekind 1876, 542]: "die Schüchternheit, ... eine natürliche Folge seines früheren abgeschlossenen Lebens, ... hat ihn auch später nie gänzlich verlassen und oft angetrieben, sich der Eisamkeit und seiner Gedankenwelt zu überlassen, in welcher er die grösste Kühnheit und Vorurtheilslosigkeit entfaltet hat."

[2] [Cantor 1932, 31]: "In re mathematica ars proponendi quaestionem pluris facienda est quam solvendi."

replace the clumsy word 'Mannigfaltigkeit' by the shorter 'Gebiet' [domain] which, he said, is "also Riemannian" [Cantor & Dedekind 1937, 47]. A decade later, Dedekind kept mentioning 'Mannigfaltigkeit' as a synonym for set [Dedekind 1888, 344]. So it seems clear that both Dedekind and Cantor interpreted Riemann's notion of manifold as the notion of set. This suggests that it may be important to analyze carefully the origins, scope and implications of Riemann's new concept, as will be done in the present chapter. It is actually a basic thesis of this work that Riemann's ideas, and above all his new vision of mathematics and its methods, influenced both Dedekind and Cantor (see chapters III, IV and VI).

In order to properly understand the origins of Riemann's new notion, we shall discuss the contributions of the two men that he regarded as his predecessors in this respect: Gauss and Herbart. We will also consider the traditional definition of mathematics as a theory of magnitudes [Grössenlehre], for Riemann explicitly presents his manifolds within this context. But the general definition of manifold, as given in the *Habilitationsvortrag* of 1854, is particularly difficult to interpret. In §2 we shall deal with a necessary prerequisite for satisfactory understanding, namely the ideas of traditional logic; in fact, that section constitutes an important background for much of the present book, since it explains how the notion of set was related to logic from 1850 to the early decades of the 20th-century.

Figure 2. *Bernhard Riemann (1826–1866) in 1863*

Having done that, it will become possible to analyze Riemann's contribution. §3 considers the mathematical context in which his new notion was forged, and the way in which it seems to have emerged. Then, we will consider Riemann's indications about how arithmetic and topology would be founded upon the notion of manifold, and, finally, the ways in which these contributions were influential upon the history of sets. The appendix discusses the impact of Riemann's ideas on his colleague and friend Dedekind, constituting a bridge to chapter III.

1. The Historical Context: Grössenlehre, Gauss, and Herbart

Part I of Riemann's *Habilitationsvortrag*, which sets out the notion of an *n*-dimensional manifold, begins by asking for indulgence, since the author lacks practice "in such tasks of a philosophical nature" [Riemann 1854, 273]. The immediate motivation for this incursion into the alleged realm of philosophy is the need of a deeper understanding of "multiply expanded magnitudes," the need to derive this concept from "general concepts of magnitude" [*op.cit.*, 272]. Riemann thus conceives of his manifolds as intimately related to the notion of magnitude, in fact he establishes the new notion of manifold as the basis for a general, abstract theory of magnitudes [*op.cit.*, 274]. On the other hand, the only brief indications on how to confront his task can be found in the work of Gauss and Herbart, two Göttingen professors. We proceed to analyze the elements of this historical context.

1.1. Mathematics as *Grössenlehre*. One should keep in mind that, by the mid-19th-century, it was still common to define mathematics as the science of magnitudes. That was the customary definition up to that century, following the ancient Greek conception. Aristotle, for instance, distinguished two kinds of magnitudes, discrete and continuous, including number among the former, line, surface and body among the latter (*Categories*, 4b 20). In his view, mathematical propositions deal with magnitudes and numbers (*Metaphysics*, 1077b, 18–20). This viewpoint offered a satisfactory overview of elementary mathematics, since it included the historical roots of this discipline, arithmetic and geometry, under a common conception. As is well known, beginning around 1600, with the work of Stevin and many other authors, magnitude and (real) number became coextensive [Gericke 1971].

This venerable definition of mathematics can be found again in Euler's *Algebra*, which begins with the following words:

First, everything will be said to be a magnitude, which is capable of increase or diminution, or to which something may be added or substracted. ... mathematics is nothing more than the *science of magnitudes*, which finds methods by which they can be measured.[1]

To give a couple more examples taken from German works of the 19th-century, we may refer to mathematical dictionaries. In the first decade of the century, Klügel, a professor at the University of Halle, defined "magnitude (quantitas, quantum)" as "that which is compound of homogeneous parts"; everything, in reality or in imagination, that possesses the property of being such a compound is an object of mathematics, which is thus properly called the "theory of magnitudes."[2] Even after 1850, Hoffmann's dictionary defines mathematics as the "theory of magnitudes, the science of the magnitudes, which may be numerical magnitudes or spatial magnitudes," corresponding to the distinction between the discrete and the continuous. With Hoffmann, a magnitude is again that which may be augmented or diminished.[3]

This traditional definition was not only common in dictionaries, but kept being employed by research mathematicians. We shall see that Gauss still spoke of a theory of magnitudes in connection with numbers and pure mathematics. We will see, however, that in emphasizing the possibility of an *abstract* theory of magnitudes, and the need for topological investigations, he was going beyond the traditional viewpoint, stretching it to include radical novelties. The same happens to a far greater extent with Riemann, but similar moves can also be found in authors such as Bolzano, Grassmann and Weierstrass. In fact, reconceiving the idea of a magnitude seems to have been one of the ways in which 19th-century mathematicians introduced novel abstract viewpoints and advanced toward the notion of set.

Riemann meant his manifolds to become a new, clearer and more abstract foundation for mathematics, which is consistent with his strong interest in philosophical issues and with his conception of mathematical methodology (§I.4). Unfortunately, this has normally not been taken into account by historians,[4] probably because they find difficulties in interpreting his – for us – obscure definition of a manifold (see §2), and because one can comfortably resort to modern concepts of differential

---

[1] [Euler 1796, 9]: "Erstlich wird alles dasjenige eine Grösse genennt, welches einer Vermehrung oder einer Verminderung fähig ist, oder wozu sich noch etwas hinzusetzen oder davon wegnehmen lässt. ... indem die Mathematic überhaupt nichts anders ist als eine Wissenschaft der Grössen, und welche Mittel ausfündig macht, wie man dieselben ausmessen soll."

[2] [Klügel 1803/08, vol. 2, 649]: "Grösse (Quantitas, Quantum) ist, was aus gleichartigen Theilen zusammengesetzt ist. ... Mathematik ... heisst daher ganz schicklich die Grössenlehre." Incidentally, we may mention that Klügel revealed some influence of the combinatorial school, for instance when he modified that definition under the entry "mathematics," saying that this was the science of the "forms of magnitudes" [*op.cit.*, vol. 3, 602], forms being equivalent to functions [*op.cit.*, vol. 1, 79].

[3] [Hoffmann 1858/67, vol. 4, 144]: "Mathematik ist Grössenlehre, die Wissenschaft von den Grössen deren es Zahlengrössen und Raumgrössen gibt." [*op.cit.*, vol. 3, 225]: "Grösse wird vielfach erklärt als Dasjenige, welches sich vermehren und vermindern lässt."

[4] With the exception of the recent work of Laugwitz [1996].

geometry while trying to interpret the *Habilitationsvortrag*. Nevertheless, Riemann's contemporaries had no option but to understand his own definitions, and, at a time when the foundations of mathematics were unclear, his comments on the issue should have caught the attention of at least some readers.

1.2. Gauss on complex numbers and "manifolds." In part I of his *Habilitationsvortrag* on geometry, Riemann [1854, 273] mentions that the only previous work of some relevance that he has had access to is a few short indications of Gauss, and some philosophical investigations of Herbart.

When Riemann quotes Gauss's works in his *Habilitationsvortrag*, he clearly differentiates those linked to differential geometry, and those related to the general notion of manifold [Riemann 1854, 273, 276]. According to him, some indications that are relevant to the issue of manifolds can be found in an 1832 paper on biquadratic residues, in the 1831 announcement of that paper [Gauss 1863/1929, vol. 2, 93–148 and 169–78], and in the 1849 proof of the fundamental theorem of algebra, read by Gauss on the occasion of his doctorate's golden jubilee [*op.cit.*, vol. 3, 71–102]. The common trait of these works is that all of them deal with the complex numbers. It seems likely that Riemann had carefully studied them already by the time of his dissertation, 1851. In §3.1 we will see Gauss indicating the need of a theory of topology, in the context of his 1849 proof; part I of Riemann's lecture was actually devoted to a discussion of fundamental concepts of topology on the basis of the notion of manifold (see [Riemann 1854, 274]).

It is well known that Gauss played an important role in the full acceptance of the complex numbers, with the above-mentioned 1831 and 1832 papers.[1] The immediate motivation for this contribution was a question in number theory, biquadratic residues, where Gauss found it necessary to expand the field of higher arithmetic and study the number theory of the Gaussian integers, $a+bi$ with $a,b \in \mathbb{Z}$ (§III.3). Anticipating criticism of this step, which might seem "shocking and unnatural" to some [1863/1929, vol. 2, 174], he decided to defend the full acceptability of complex numbers as mathematical objects. As we shall see in §3.1, Gauss seems to have relied on the idea of the complex plane since 1799.[2]

Gauss regarded the interpretation of complex numbers as points in a plane as a mere illustration of the much more abstract meaning of complex numbers. He argues that some physical situations afford an occasion for employing a particular kind of numbers, and some not. It suffices that there be situations where fractional parts or opposites occur, to make full sense of a theory of fractions or of negative numbers. The same happens with complex numbers, which, in his view, only find application when we are not dealing with substances, but with relations between

---

[1] The wide diffusion of the geometrical representation only took place around 1830, with the publication of some treatises in France and England, and then with the contribution of Gauss. See [Nagel 1935, 168–77; Pycior 1987, 153–56; Scholz 1990, 293–99]. Some interesting comments can be found in [Hamilton 1853, 135–37].

[2] The treatment of complex functions as conformal mappings, given by Gauss in 1825 (see §3.1), was also dependent on the idea of the complex plane [Gauss 1863/1929, vol. 2, 175].

substances [Gauss 1863/1929, vol. 2, 175–76]. The use of real and complex units for measurement is required

if the objects are such that they cannot be ordered into a single unlimited series, but only into a series of series, or, what comes to the same, if they form a manifold of two dimensions; and if there is a relation between the relations among the series, or between the transitions from one to the other, which is similar to the already mentioned transitions from one member of a series to another one belonging to the same series ... In this way, it will be possible to order the system doubly into series of series.

The mathematician abstracts entirely from the quality of the objects and the content of their relations; he only occupies himself with counting and comparing their relations to each other.[1]

We can here observe in some detail what Gauss meant by an abstract theory of magnitudes. In the 1832 treatise on biquadratic residues Gauss again uses the expression 'manifold of two dimensions.'[2]

In the previous quotation, Gauss understands by a manifold a system of objects linked by some relations, the dimensionality of the manifold depending on properties and interconnections of the relations. Though this is not the way in which Riemann conceived of manifolds in unpublished manuscripts of 1852/53 (§§1.3 and 3.2), Gauss was calling attention to the properties that a physical system must have in order to be regarded as a 2-dimensional manifold, and this is part of what Riemann tried to analyze. Gauss suggested the terminology, the topological point of view, and some embryonic ideas on dimensionality.

Toward the end of his 1831 paper, Gauss mentions the possibility of relations among things that give rise to a manifold of more than two dimensions [1863/1929, vol. 2, 178]. In lectures of the 1850s [Scholz 1980, 16–17] one can find indications of the possibility of n-dimensional manifolds, though without making explicit a satisfactory foundation. Actually, a move to notions of n-dimensional geometry was not infrequent in the early 19th-century, within the context of analytic or algebraic problems involving n variables [Scholz 1980, 15–19]. Several mathematicians

---

[1] This text is somewhat reminiscent of Cantor's work [1932, 420–39] on n-ply ordered sets. [Gauss 1863/1929, vol. 2, 176]: "Sind aber die Gegenstände von solcher Art, dass sie nicht in Eine, wenn gleich unbegrenzte, Reine geordnet werden können, sondern nur in Reihen von Reihen ordnen lassen, oder was dasselbe ist, bilden sie eine Mannigfaltigkeit von zwei Dimensionen; verhält es sich dann mit den Relationen einer Reihe zu einer andern oder den Uebergängen aus einer in die andere auf eine ähnliche Weise wie vorhin mit den Uebergängen von einem Gliede einer Reihe zu einem andern Gliede derselben Reihe, so bedarf es offenbar zur Abmessung ... ausser den vorigen Einheiten +1 und −1 noch zweier andern ... +i und −i. ... Auf diese Weise wird also das System auf eine doppelte Art in Reihen von Reihen geordnet werden können. / Der Mathematiker abstrahirt gänzlich von der Beschaffenheit der Gegenstände und dem Inhalt ihrer Relationen; er hat es bloss mit der Abzählung und Vergleichung der Relationen unter sich zu thun."

[2] [*Op.cit.*, vol. 2, 110]: "varietates duarum dimensionum." It is perhaps convenient to remind the reader that 'varietates' is the most adequate Latin translation of 'manifold;' accordingly, the name for this notion is 'variété' in French, 'variedad' in Spanish.

– Lagrange, Cauchy, Jacobi, Gauss – found it advantageous to employ geometrical language in order to clarify the analytical relations under study. By the mid-century, the British algebraists Cayley, Sylvester and Salmon made a similar move in their studies of homogeneous functions. In all of those cases *n*-dimensional language was introduced only by analogy and, so to say, metaphorically. A different case is that of Grassmann's attempt at a pure *n*-dimensional geometry [Grassmann 1844], since the move is here meant literally, just like in Riemann's lecture; but there is no reason to think that Riemann knew of Grassmann's work.

1.3. Herbart on objects as 'complexions' of properties. Originally, Riemann matriculated at Göttingen in theology and philology, although within a year he shifted to mathematics. Interest for philosophical topics, however, never abandoned him, and in the early 1850s he studied closely the work of Johann Friedrich Herbart, a professor at Göttingen until 1841. Reference to that work in Part I of the *Habilitationsvortrag* suggests how highly he valued Herbart's ideas; a manuscript note that he left is clear enough:

The author is a Herbartian in psychology and the theory of knowledge (methodology and eidology), but for the most part he cannot embrace Herbart's philosophy of nature and the metaphysical disciplines that are related to it (ontology and synechology).[1]

Riemann refers here to the peculiar names given by Herbart to the different parts of his doctrines. It should be noted, following Scholz [1982a, 415], that the part of Herbart's philosophy that he adhered to conforms a kind of epistemology. Among Riemann's philosophical texts, those on psychology and epistemology develop Herbartian viewpoints [Riemann 1892, 509–25].

Herbart was a disciple of the idealist Fichte, but by the end of his student time he had become very critical of Fichte's ideas. To mark his opposition to the idealist trend, so powerful in Germany, he always defined himself as a "realist", although some idealist remnants can be found in his psychological theories [see Scholz 1982a]. On a more positive note, he regarded himself as a follower of Kant, but not an orthodox Kantian, since he tried to avoid some *aprioristic* traits that were still present in the Königsberg philosopher. Many details in his doctrines were inspired by Leibniz, so that he became a link between the speculations of the great mathematician-philosopher and those of Riemann.

In a discussion of the possible influence of Herbart's ideas upon Riemann's geometrical thought, Erhard Scholz has denied that they may have affected the precise content of the notion of manifold.[2] More precisely, Scholz mentions some key elements of Riemann's notion that are absent from any related ideas of Herbart's: multidimensionality, separation between topological and metrical aspects,

---

[1] [Riemann 1892, 508]: "Der Verfasser ist Herbartianer in Psychologie und Erkenntnistheorie (Methodologie und Eidologie), Herbart's Naturhilosophie und den darauf bezüglichen metaphysischen Disciplinen (Ontologie und Synechologie) kann er meistens nicht sich anschliessen."

[2] In this he distances himself from Russell and Torretti [Scholz 1982a, 414].

and the opposition between a simple local behavior and a complex global one [Scholz 1982a, 423–24]. Herbart's influence would have been more on the level of general epistemology and, most importantly, of a conception of mathematical research. Riemann transformed some characteristic traits of Herbart's philosophy into guiding principles for his mathematical work [*op.cit.*, 428].

Herbart thought that mathematics is, among the scientific disciplines, the closest to philosophy. Treated philosophically, i.e., conceptually, mathematics can become a part of philosophy.[1] According to Scholz, Riemann's mathematics cannot be better characterized than as a "philosophical study of mathematics" in the Herbartian spirit, since he always searched for the elaboration of central concepts with which to reorganize and restructure the discipline and its different branches, as Herbart recommended [Scholz 1982a, 428; 1990a].

One can certainly grant the general correctness of Scholz's detailed analysis of the interrelations between the ideas of both authors, and still claim that there are a couple of more direct connections. Herbart's conception of space is developed in his theory of continuity [Synechology], which explains the emergence of the notions of space, time, number and matter, all of which involve continuity. Herbart proposes a more or less psychological explanation of continuity, which emerges from the "graded fusion" [abgestufte Verschmelzung] of some of our mental images [Vorstellungen; Herbart 1825, 192]. His preferred examples were those of the tones, which give rise to a line, and the colors, which produce a triangle with blue, red and yellow at the vertices, and mixed colors in between [*op.cit.*, 193]. As the quotation above makes clear, Riemann rejected the details of Herbart's theory of continuity. But he seems to have adopted some quite general aspects of Herbart's approach (compare [Scholz 1982a, 422–23]).

Like Leibniz, Herbart rejects the Newtonian (and Kantian) conception of space as an absolute receptacle for physical phenomena; rather it seems to be an "order of coexistence" of things (see [Leibniz & Clarke 1717; Herbart 1824, 429]). Space does not have an independent reality, but is a *form* which arises in our imagination as a result of specific traits of the mental images which we gain in experience. As a result, all kinds of mental images may give rise to continuous serial forms, and in all such cases the conception of space arises. This suggests that anything can be geometrized, and explains why Herbart offered a unified treatment of time, matter, number and space, since in all of these cases spatial forms arise. Herbart [1824, 428–29] made it explicit that spatial forms apply to all aspects of the physical world, and even to any domain of mental representation, including the unobservable. The conception of space as linked to the properties of physical objects is characteristic of Riemann's *Habilitationsvortrag*. From a historical point of view it is quite interesting to find that, apparently, there was a connection between Leibniz's and Riemann's proposals of such a conception, the link being Herbart's doctrines.

Two other interrelations between the ideas of Herbart and Riemann are linked with the latter's general definitions of the notion of a manifold. Here we shall ana-

---

[1] "Philosophisch behandelt, wird sie selbst ein Theil der Philosophie ..." [Scholz 1982a, 437].

lyze the influence of one aspect of Herbart's thoughts about objects and space, and in §2 we shall consider that of his treatment of logic.

In 1853, Riemann explained his idea of a manifold (see §3.2) by making reference to the totality of all possible outcomes of a measuring experiment in which the values of two, or perhaps *n*, physical magnitudes are determined for a given physical system. We may understand this as, essentially, the notion of the space of states for the given system. This is quite different from Gauss's explanation of manifolds in the 1831 paper, but it happens to be quite similar to a key idea of Herbart's. According to him, any object has to be considered as a bundle or "complexion" of properties, each of which can be regarded as located in a different qualitative continuum. The idea is natural given his approach to continuity outlined above, and it is found in two passages that seem to have attracted the attention of Riemann, since he excerpted them [Scholz 1982a, 416, 419]. Read by a person who was immersed in physical thinking, as Riemann was, that could only suggest the notion of a space of states, at least when we are talking about magnitudes that vary continuously, such as temperature and weight.

Apparently, then, Riemann's 1853 explanation of the notion of a manifold may have been suggested by Herbart, just like his 1854 definition must have been influenced by philosophical reading (see §2). Incidentally, it is worth mentioning that the word "manifold" [Mannigfaltigkeit] is extremely frequent in Herbart's writings, although he employs it in the common sense, not in any technical sense.

2. Logical Prerequisites

For the purpose of his *Habilitationsvortrag* of 1854, Riemann presented a definition of manifold that is clearly more general and abstract than that found in the earlier manuscripts. This definition seems to have puzzled modern commentators (see §4), the reason being that a proper understanding of it presupposes knowledge of the logical ideas current at the time. The reader may judge the difficulty by himself, in trying to understand Riemann's words without further help:

Notions of magnitude are only possible where there is an antecedent general concept which admits of different ways of determination. According as a continuous transition does or does not take place among these determinations, from one to another, they form a continuous or discrete manifold; the individual determinations are called points in the first case, in the last case elements, of the manifold.[1]

It is my contention that this somewhat obscure text will become quite clear by the end of this section. The difficulties encountered by commentators simply mirror the conceptual gap that separates traditional logic from contemporary logic, calling

---

[1] [Riemann 1854, 273]. The German text is given in §4. For the lecture of 1854 I employ Clifford's translation [1882, 55–71], with my own corrections.

attention to a dramatic shift in the meaning of "logic" during the century from 1850 to 1950.[1]

The situation in logic around 1850 constitutes a crucial background for much of the present work, particularly for our discussion of Riemann and Dedekind, and of developments in the early twentieth-century. Nineteenth-century Germany was surprisingly prolific in logical publications, and the meaning of "logic" varied enormously from author to author, ranging from an idealistic to a formal conception of logic [Ueberweg 1882, 47–79]. But, although there are substantial difficulties in identifying the exact sources of the logical knowledge possessed by most of the authors we will consider, it is possible to produce a schematic portrait of the basic logical doctrine of the time and this will prove sufficient for our purposes.

Internal evidence from Riemann's and Dedekind's writings strongly suggests that they only paid attention to logicians of the *formal* trend. During the modern era, logical doctrines had become confused with various epistemological, psychological, and metaphysical ideas, a tendency that persisted throughout the 19th-century. But Kant and his followers started a reaction which emphasized the formal character of logic, and the need to treat it as an autonomous discipline [Ueberweg 1882, 47–51].[2] Notably, Herbart was one of the most outstanding defenders of this viewpoint against the idealists. As a matter of fact, Ueberweg distinguishes a Kantian and a Herbartian school among logicians, the latter being even clearer in its preference for a formal standpoint [*op.cit.*, 52]. One of the most successful and widely read logic treatises of the century was written by a Herbartian, the Leipzig professor of mathematics Moritz Wilhelm Drobisch ([1836], with four eds. up to 1875). Ueberweg [1882, 53] says that this book was, "acknowledgedly," the best exposition of logic from the formal point of view.

Logicians of this formal trend championed a return to the Aristotelian conception and doctrines, and presented a core of logical knowledge that was also incorporated by all other authors (although embedded into a wider context). What is interesting for our purposes is that this core of traditional logic became the basis for the view that the theory of sets is properly a logical theory, promoting logicism, and the root for the famous principle of comprehension that played such a crucial role around 1900. It gave rise to a conception of logic that was widely influential among philosophers and mathematicians up to the early twentieth-century, when it had to be abandoned, or deeply modified, due to the impact of the so-called logical paradoxes. Riemann, too, rested upon this traditional doctrine in presenting his general definition of a manifold.

What we are interested in, then, is the import of the traditional conception of logic for the notion of set. The reason why it was natural to regard the concept of set as a purely logical notion can be explained simply on the basis of Aristotelian

---

[1] For the history of logic, the reader may consult [Bocheński 1956] and [Kneale & Kneale 1972]. Regarding 20th-century logic, it is convenient to consult more recent works, such as [van Heijenoort 1967; Goldfarb 1979; Moore 1980 and 1988].

[2] Interestingly, Kant's formal views influenced even British authors such as DeMorgan [1858].

syllogistics. As is well known, traditional logicians held that any complex reasoning is reducible to a chain of syllogisms, the arch-typical example of a syllogism being: "Every A is B, and every B is C; therefore, every A is C."[1] As we can see, a syllogism is a structured connection of judgments, i.e., affirmed or denied propositions like "every A is B," "some C is not D." A proposition is formed by a subject and a predicate, that Aristotle symbolized by letters, linked by the copula "is." Subject and predicate, in their turn, are simply concepts. This analysis of reasoning and propositions led to the usual structuring of traditional logic treatises, divided into three sections: "On Concepts," "On Judgements," and "On Conclusions."[2]

Reasoning appeared to be the result of different kinds of formal combination of concepts, by means of the logical particles "every," "some," "is," "not." It is not difficult to see how this conception of logic played some role as a source of the classical logic of our century, since it clearly identified the quantifiers as logical particles. According to Kant and his followers, concepts contained the whole matter or content of an argument, everything else being purely formal or logical.

One further element that belonged to the core of traditional logic was a crucial distinction concerning concepts: that of comprehension or intension [Inhalt] vs. extension [Umfang] of a concept. This was an inevitable component of the section "On Concepts" in 19th-century logic textbooks. Its roots can be traced back to Antiquity, to Porphyry's *Isagogé*, a third century commentary on Aristotle that was influential during the Middle Ages [Frisch 1969, 108–14; Walther-Klaus 1987, 9]. But the modern *locus classicus* for the distinction is the *Logique* of Port Royal, one of the most widely read treatises of the 17th-century:

In these universal ideas, it is very important to correctly distinguish *the comprehension and the extension*.
By the *comprehension* of the idea we understand the attributes which it involves and which cannot be withdrawn without destroying the idea, as the comprehension of the idea of triangle involves extension, figure, three lines, three angles, equality of those angles summed up to two right angles, etc.
By the *extension* of the idea we understand the subjects to which the idea applies, and which are also known as the inferiors of a general term which, in relation to them, is called superior; as the general idea of a triangle extends to all the different species of triangles. [Arnauld & Nicole 1662, 51]

The notion of extension was interesting as a tool for analyzing the theory of syllogisms: by considering the relations between the extensions of subject and predicate

---

[1] This is not the form in which Aristotle himself gave the argument, but is the usual modern formulation of syllogisms in Barbara (see [Łukasiewicz 1957]). As the reader probably knows, Aristotle analyzed 13 more modes of deduction besides Barbara, divided into three "figures." Traditional logicians added to Aristotle's "categorical syllogism" two other kinds of "conclusions" – "hypothetical" and "disjunctive" syllogism – which actually correspond to propositional logic.

[2] Frequently there was a fourth part, devoted to methodology. This was a consequence of the modern confusion of logic and epistemology, and the source of a frequent misunderstanding.

in a proposition it was possible to justify Aristotle's doctrines of the conversion of judgements, and of the syllogistic modes of deduction.

The expositions of Herbart and Drobisch followed that early practice. A summary of Herbart's way of dealing with logic can be found in [Herbart 1808], which includes the distinction between intension and extension [*op.cit.*, 218], and the employment of extensions for explaining the syllogistic conversion and deduction [*op.cit.*, 220–21, 223–26]. In an introductory text of 1837, he defined the intension of a concept as the sum of its attributes, the extension as the set [Menge] of the other concepts in which the first appears as an attribute [Herbart 1837, 71]. Herbart makes an application of these ideas to the notion of number that is interesting from the point of view of later developments, although it is typically vague. The natural numbers form the extension of the concept of number; moreover,

nobody would know what *number* is without first knowing what one, two, three, four are. The intension of this concept consists therefore in its extension.[1]

The traditional approach undoubtedly suggested that classes or sets are a logical matter, since they emerged from an extensional analysis of concepts, an analysis that might be regarded as formal. In this way, it prepared the way for the assumption that the theory of classes, or the theory of sets, was a part of logic. Consideration of concept-extensions was absolutely common in logic, and so, when classes and sets emerged in the practice of mathematicians, the scene was set for an understanding of these notions as belonging to logic. A notable confirmation of this is found in the introduction to Boole's epoch-making *The Mathematical Analysis of Logic* [1847]:

That which renders logic possible, is the existence in our minds of general notions, – our ability to conceive of a class, and to designate its individual members by a common name. [Boole 1847, 4]

The statement is unequivocal: according to Boole, concepts and their extensional counterpart, classes, are the very foundation of logic. The reader should recall that the main interpretation that Boole gave his logical calculus was in terms of classes.[2]

The argument for the logical character of sets is, then, based on the role of concepts and concept-extensions in logic. If we accept that logic is purely formal, this argument would depend upon the additional assumption that classes belong in the formal analysis of propositions and arguments. Naturally, there seems to be room for interpretation in this matter. For instance, it was perfectly possible to think that,

---

[1] [Herbart 1837, 73–74]: "Die Zahlen selbst bilden eine Reihe unter dem Begriff der Zahl ... niemand wissen würde, was *Zahl* sei, wenn er nicht zuvor wüsste, was Eins, Zwei, Drei, Vier ist. Dieses Begriffes Inhalt beruht demnach auf seinem Umfange."

[2] Similarly, post-Boolean logicians such as Schröder and even Peano designed their logical symbolism primarily for the purpose of applying it to classes, although of course they gave it the alternative propositional interpretation.

just like particular concepts do not belong to logic, neither does the extensional theory of concepts – since any concrete assumption concerning a class would be of a particular character, and logic ought to stay at a level of outmost generality. Something like this may have been Herbart's standpoint, since he stated that concept-extensions are useful but not essential in logic [Herbart 1808, 216, 222]. The main reason why Herbart took that position was, probably, that he regarded logic as a simple propaedeutics of knowledge [*op.cit.*, 267]. To accept that the theory of classes belongs to logic would mean to have a logical theory of too much content and power, no simple propaedeutics. A Herbartian might thus think that the auxiliary tool of classes belonged to a different sphere of knowledge, perhaps to mathematics. There are reasons to think that this was Riemann's own view, since he conceived of manifolds as the basic objects of the theory of magnitudes (§4). From this point of view, a class-theoretical analysis of logic would *ipso facto* be a mathematical analysis.

Nevertheless, a second argument seems to shortcut that interpretation. We have seen that traditional logicians analyzed the relation between the extensions of subject and predicate, in order to justify the Aristotelian doctrines. Now, "Every A is B" establishes that the class of As is included in that of Bs; "No A is B" says that both classes are disjoint; "Some A is B" means that both classes have an (non-void) intersection, while "Some A is not B" means that there is a (non-void) class of objects that are A but not B. The notions of inclusion, intersection, disjointness and others, including union ("Plato is A or B"), arise naturally in an analysis of the propositions of traditional logic.[1] Since, according to Kant and his followers, all that pertains to the effect of the logical particles upon concepts is purely formal, those set-theoretical notions are purely formal and belong to logic. In a word, one may say that assuming that the copula "is" is one of the logical particles (see above) makes set theory a part of logic.[2] This is essentially the argument that one can find, almost a century later, in the work of the famous philosopher and logician Quine [1940].

At this point, however, a clarification is in order. I am not arguing that the logical theory of classes incorporated no novelty or departure from tradition but only that it involved no essential reconception of logic. Herbart's definition of extension (above) shows a trait that is common to most logicians of the early 19th-century: there is no reference to classes of objects (individuals), only to classes of further concepts. This brings to mind the tendency of traditional logicians to concentrate upon hierarchies of concepts and to analyze the reciprocal relation genus–species. The hierarchy of concepts gives rise to a tree-like structure, the so-called *tree of*

---

[1] A more sophisticated analysis of extensional relations was given by the French mathematician Gergonne early in the 19th-century [Styazhkin 1969].

[2] Acknowledgedly the particle 'is' has several different meanings. A contemporary analysis of them would distinguish the meanings of identity, inclusion (or intensional subsumption) and membership (intensional predication). This analysis was first established by Peano and Frege late in the 19th-century, and as such is more advanced than what could be expected from traditional logicians.

Porphyry [Bocheński 1956, § 24.01–03]. This viewpoint suggested regarding extension and intension as reciprocal elements: at any point in the hierarchy, the intensional elements are the concepts found above, while the extension is constituted by the concepts found below. In this way, the traditional conception suggested problems that, in retrospect, seem to have distracted from the successful development of a logical calculus of classes and a theory of sets.[1] For instance, the first edition (and only the first) of Drobisch [1836] included an appendix with an attempt to establish a mathematical calculus of logic (see [Styazhkin 1969]), but this is quite different from Boole's later efforts since it was guided by the idea of the tree of Porphyry. Thus, the advances made by logicians and mathematicians in the second half of the 19th-century depended upon a simplification of the scene,[2] an indifference to the traditional problem of hierarchies coupled with a concentration on a purely extensional theory.

It is also important to mention that logical theory enjoyed a wide diffusion in 19th-century Germany.[3] Sometimes it was part of the secondary school syllabuses for "philosophical propaedeutics," as was the case in Austria and Bavaria. Sometimes it was an integral part of the teaching of language and grammar courses in *Gymnasia*, as happened in Prussia and other states. This was due to the influence of the linguistic movement of General Grammar, which took as an essential assumption the strict parallelism between linguistic sentences and the logical propositions of formal logic (see [Naumann 1986; Forsgren 1992]). The teaching of German was intended to be logical propaedeutics and ultimately to lead to philosophy. Emphasis on thinking and logic did not disappear even after General Grammar was superseded by the Historicist School [Naumann 1986, 110]. And, of course, logic was an important element of introductory philosophy courses given at the universities. It thus seems safe to assume that educated Germans, at the time of Riemann, were familiar with the basic doctrines of traditional logic, including the idea of concept-extensions.

To summarize, the conclusion that the theory of sets or classes belongs to formal logic was at least quite natural, judging from the traditional conception of the subject. Since the transition from a concept to its associated class was absolutely common, and taken for granted, it is also natural that the principle of comprehension would eventually be articulated explicitly. This principle says that, given a well-defined concept, there is the class of all objects that satisfy the concept. The first precise formulation of the principle is to be found in Frege [1893], which is understandable, since no such general principle was needed for the purposes of traditional logic – only for the purpose of founding a powerful theory of sets. But the principle of comprehension was assumed as self-evident by many authors, including Dedekind and the young Hilbert. It also underlies Bolzano's theory of sets

---

[1] I owe this insight to Gregory H. Moore.

[2] There were precedents for this step in tradition, for instance in Euler; see [Frisch 1969] and [Walther-Klaus 1987].

[3] This topic has not yet been sufficiently studied. A more detailed preliminary account, partly based on unpublished work done at Erlangen, can be found in [Ferreirós 1996, 13–15].

as presented in his *Paradoxien des Unendlichen* [1851]. Bolzano introduced several notions related to that of set, distinguishing between cases in which the order is or is not taken into account. In his theory, sets depend on concepts, since his sets [Mengen] are always defined by a concept [Bolzano 1851, §§3–4].

We are now in a position to understand Riemann's definition of a manifold, quoted above. That definition seems, quite unequivocally, to rely on the traditional relationship between a concept and its associated class, a manifold being simply a class, the extension of a general concept. In §4 we shall comment on the definition, analyze more closely the connection Riemann established between manifolds and magnitudes, and consider his position concerning the actual infinite. But first, we will consider the broader mathematical context in which Riemann developed his new notion.

3. The Mathematical Context of Riemann's Innovation

Although Riemann presented the notion of a manifold in his famous *Habilitations-vortrag*, the context from which that notion emerged was broader. Since he pro-posed the new notion as a foundation for the theory of magnitudes, he must have seen connections between manifolds and all branches of pure mathematics. But the available evidence suggests that the notion arose, more concretely, in relation to Riemann's function theory [Scholz 1980, 1982]. Shortly afterwards, it became the basic notion for his new approach to differential geometry and the question of space, which in its turn was related to Riemann's thoughts about a unified physical theory.

Riemann's function theory, known through an 1857 paper on Abelian functions, was the basis for the renown he enjoyed during his lifetime. That work was an outstanding feat, for it attempted to offer a general solution of the Jacobian inver-sion problem for integrals of arbitrary algebraic functions. This topic emerged from the fascinating competition that Abel and Jacobi sustained in the late 1820s on the subject of elliptic functions (the inverses of elliptic integrals). As for the importance that was attached to it, suffice it to say that Weierstrass became a rising star with his solution of the inversion problem for hyperelliptic integrals in 1854 and 1856. Riemann was tackling a much more general problem and his work, in spite of gaps in the proofs, aroused enormous excitement. His 1857 paper, together with the high opinion that Gauss, Weber and Dirichlet had of him, explains why Riemann was promoted to full professor at Göttingen, in 1859, without taking into account any other candidates. It also led to his election as a corresponding member of the Berlin Academy of Sciences that same year [Dedekind 1876, 522].

3.1. Function theory and topology.[1]   Riemann's approach to function theory has already been mentioned as a key example of his methodological preferences in mathematical research (§I.4). It is well known that complex analysis only consolidated during the nineteenth-century with the ground-breaking work of Cauchy and the general treatments proposed by Weierstrass and Riemann (see [Bottazzini 1986]). By 1850 known results afforded only a sketchy overview of the new branch of mathematics; Weierstrass and Riemann worked on trying to present a systematic development. As we have seen, both attempts at a synthesis were quite different, Weierstrass's being the first rigorous one, Riemann's being much more abstract and even visionary. It took a long time until his novel methods were given a sound foundation, which caused late-19th-century mathematicians to devote great efforts to reestablishing Riemann's results in different ways. His theory employed new "geometrical" considerations, namely topological notions, which only in the early decades of the twentieth-century received a satisfactory treatment [Weyl 1913]. Riemann also made essential use of what he called the "Dirichlet principle," which was severely criticized by Weierstrass in 1870, and reestablished by Hilbert in 1901 [Monna 1975]. No wonder that his methods seemed to be "a kind of arcanum" that other mathematicians looked at with distrust [Klein 1897, 79].

In Riemann's opinion, a satisfactory study of Abelian and other functions depended upon finding a system of conditions, independent from each other, that would be sufficient for determining the function [Riemann 1857, 97]. Since the global configuration of analytic functions depends on their local behavior, the traditional focus on formulas involved the use of redundant information. Riemann searched for a minimal set of determining conditions, which turned out to be partly analytical and partly geometrical [Scholz 1980, 62]. The analytical data consisted in conditions on the real and imaginary parts of the function, as well as its behavior at poles and singular points. The geometrical side of the information consists in his idea of the Riemann surface, which became a key element in the study of multi-valued functions.

Reliance on geometrical considerations was, to some extent, a legacy begetted by Gauss. In 1799 Gauss had taken advantage of the geometrical representation of complex numbers for his first proof of the so-called fundamental theorem of algebra (which from a modern viewpoint can be seen as part of function theory). That proof, however, obscured the fact by formulating the whole issue in terms of real numbers – one more example of Gauss's caution or, as he said, fear of the clamor of the Boeotians [Gauss 1863/1929, vol. 8, 200]. After having publicly defended the geometrical representation in 1831, Gauss came back to the fundamental theorem of algebra in 1849 and rewrote his first proof making free reference to the complex plane [Gauss 1863/1929, vol. 3, 74 and 114]. Given a complex polynomial of degree n, Gauss analyzed the behavior of its real and imaginary parts in a way that was essentially topological, to conclude the existence of n roots. He wrote:

---

[1] On this topic, see [Bottazzini 1986; Laugwitz 1996, chap. 1].

I will present the proof in a dressing taken from the geometry of position, since in this way it attains its maximum intuitiveness and simplicity. But in essence the true content of the whole argument belongs to a higher domain of the abstract theory of magnitudes, independent of the spatial, the object of which is the combinations among magnitudes linked by continuity, a domain which until now has been little cultivated, and where we cannot move without a language taken from geometrical images.[1]

Gauss's work had, therefore, begun to show the interest of a geometrico-topological approach to complex functions.

Riemann's doctoral dissertation also emphasized the usefulness of geometrical intuition for understanding complex functions. Departing from the geometrical representation of complex numbers, he regarded a complex function as a mapping [Abbildung] from one plane to another, and he showed that analytic functions determine conformal mappings [Riemann 1851, 6–7].[2] But soon Riemann began to move beyond what Gauss had suggested. His interest in multi-valued functions led him to introduce so-called Riemann surfaces and to develop topological methods for studying them. This was the reason why Riemann's theory was called "geometrical" at the time, although this characterization is far from satisfactory, since it overlooks many other important features of his approach.

The branching properties of multi-valued functions had just began to be studied by Cauchy and, especially, by Puiseux. The notion of a Riemann surface was quite a natural idea, although it posed some difficult problems, including foundational ones. Rather than taking the domain of the function to be the complex plane, he imagined a surface of several sheets which covers the plane; over this domain, the function becomes single-valued [Riemann 1851, 7–9; 1857, 89–91]. A very simple example is $f(z)^2 = z$, a function having two values at all points except $z = 0$ and $z = \infty$, which are the only branching points. The associated Riemann surface has two sheets that are continuously linked: one can pass from one to the other by describing a closed curve around the branching point. Riemann's geometrical 'invention' [Klein 1897, 75] amounted to a geometrization of the branching properties of the function. All the related information, including the location of branching points, was simply determined by the surface.[3]

---

[1] [Gauss 1863/1929, vol. 3, 79]: "Ich werde die Beweisführung in einer der Geometrie der Lage entnommenen Einkleidung darstellen, weil jene dadurch die grösste Anschaulichkeit und Einfachheit gewinnt. Im Grunde gehört aber der eigentliche Inhalt der ganzen Argumentation einem höhern von Räumlichen unabhängigen Gebiete der allgemeinen abstracten Grössenlehre an, dessen Gegenstand die nach der Stetigkeit zusammenhängenden Grössencombinationen sind, einem Gebiete, welches zur Zeit noch wenig angebauet ist, und in welchem man sich auch nicht bewegen kann ohne ein von räumlichen Bildern entlehnte Sprache."

[2] This linked again with work by Gauss, his 1825 treatise on conformal mappings, i.e., maps that involve "similarity in the least parts" of original and image [Gauss 1863/1929, vol. 4, 189–216; quoted in Riemann 1851, 6 note].

[3] Klein and Weyl [1913, vi–vii] emphasized that, far from being mere tools, Riemann surfaces are an indispensable component, and even the foundation, of the theory of analytic functions. See [Scholz 1980, 56] for details concerning how Riemann described the surfaces.

Riemann [1857, 91] found it almost indispensable, in order to study Abelian and related functions, to resort to topological considerations. He developed new methods that enabled him to define the "order of connectivity" of a surface – the Euler characteristic – and, later, what Clebsch would call the "genus" of the surface, which defines the second Betti number [Scholz 1980, 57–64]. In the 1851 dissertation he studied connected surfaces with a boundary, and analyzed their topological properties by means of dissection into simply connected components, making use of "transversal cuts" [Querschnitte] joining frontier points. In 1857 he analyzed closed surfaces, since he was now considering the complex plane completed by a 'point at infinity.' The method also had to change, and he studied them by considering the maximal number of closed curves that do not form a complete boundary for a part of the surface. We thus find here rudimentary forms of homological methods. The most astounding example of the intimate relations between topological notions and properties of functions was the Riemann–Roch theorem, which determines the number of linearly independent meromorphic functions on an

Zweifach zusammenhängende Fläche.

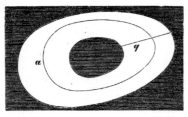

Sie wird durch jeden sie nicht zerstückelnden Querschnitt q in eine einfach zusammenhängende zerschnitten. Mit Zuziehung der Curve a kann in ihr jede geschlossene Curve die ganze Begrenzung eines Theils der Fläche bilden.

Dreifach zusammenhängende Fläche.

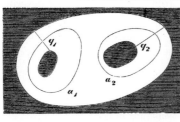

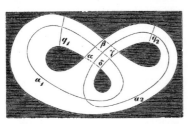

In dieser Fläche kann jede geschlossene Curve mit Zuziehung der Curven a_1 und a_2 die ganze Begrenzung eines Theils der Fläche bilden. Sie zerfällt durch jeden sie nicht zerstückelnden Querschnitt in eine zweifach zusammenhängende und durch zwei solche Querschnitte, q_1 und q_2, in eine einfach zusammenhängende.

In dem Theile $\alpha \beta \gamma \delta$ der Ebene ist die Fläche doppelt. Der a_1 enthaltende Arm der Fläche ist als unter dem andern fortgehend betrachtet und daher durch punktirte Linien angedeutet.

Figure 3. *Doubly and triply connected surfaces, from [Riemann 1857]. Riemann explains the behavior of transversal cuts and closed curves.*

algebraic surface, having a given finite number m of poles, as a function of the genus p of the surface.[1]

The connection between the topological methods developed in Riemann's function-theoretical work and the notion of manifold was obscured by the fact that, in order to make possible an easier understanding, his expositions in the former context were given in a geometrical dressing [Riemann 1857, 91]. Nevertheless, in the *Habilitationsvortrag* he made a clear reference to the existence of such a connection [1854, 274], and made it explicit that he was establishing the "preliminaries for contributions to analysis situs" [*op.cit.*, 286]. Likewise, in his work on function theory he made it clear that those methods could be developed in abstraction from metrical relations and belong to "analysis situs." Riemann wrote:

With this name, employed by Leibniz, though perhaps not exactly in the same sense, we may designate a part of the theory of continuous magnitudes, in which the magnitudes are not regarded as existing independently of position and as measurable by each other, but where, dispensing with metrical relations altogether, only the relations among places and among domains in them are submitted to investigation.[2]

Here we find a rather obscure definition of topology, which is however clear insofar as it emphasizes the absence of metrical considerations, and as it links with Gauss's earlier statements. Essentially the same definition was given in the famous *Habilitationsvortrag* [Riemann 1854, 274], to which we shall turn in §3.3.

3.2. From surfaces to manifolds. Until recently, all that was known about Riemann's notion of a manifold was the ideas presented in his 1854 *Habilitationsvortrag*. But some manuscript documents that have been published by Scholz [1982] enable us to trace the development of Riemann's ideas between his 1851 doctoral dissertation and 1854. These are four manuscripts dealing with continuous n-dimensional manifolds, n-dimensional topology, and the relation between manifolds and geometry. Scholz has been able to date them: all were written in the years 1851 to 1853. The manuscripts suggest that the idea of a manifold grew out of an attempt to find a satisfactory conceptualization of the Riemann surfaces that he had begun to employ in 1851.

The text that gives support for this conclusion was published by Scholz as appendix 4 to his paper [1982, 228–29]. In my opinion, there are reasons to believe that this is actually the earliest of the fragments published by Scholz. Riemann's

---

[1] Riemann estimated that number to be $\geq m - p + 1$, his student Gustav Roch established the precise result in *Journal für die reine und angewandte Mathematik* **64** (1864), 372–76.

[2] [Riemann 1857, 91]: "... sind einige der analysis situs angehörige Sätze fast unentbehrlich. Mit diesem von Leibnitz, wenn auch vielleicht nicht ganz in derselben Bedeutung, gebrauchten Namen darf wohl ein Theil der Lehre von den stetigen Grössen bezeichnet werden, welcher die Grössen nicht als unabhängig von der Lage existirend und durch einander messbar betrachtet, sondern von den Massverhältnissen ganz absehend, nur ihre Orts- und Gebietsverhältnisse der Untersuchung unterwirft."

idea of manifold is presented here in more concrete terms than in any other fragment. Other fragments define continuous manifolds in reference to a "variable object" that admits of different "forms of determination," i.e., that can be in different states. These states, or "forms of determination," constitute the "points" of the manifold, which is defined as the totality of all these points.[1] In appendix 4 this idea is illustrated with concrete examples: suppose we are making an experiment in which we measure one physical magnitude, a temperature; here all possible cases would be represented by the real numbers from $-\infty$ to $+\infty$, i.e., by a one-dimensional continuous manifold. But in case we were measuring two physical magnitudes, say a temperature and a weight, the possible results would be represented by two variables x and y, i.e., by a two-dimensional manifold [*op.cit.*, 229]. Since this approach entails no limitation of dimensions, we can similarly reach manifolds of an arbitrary number of dimensions by considering experiments in which a higher number of physical magnitudes have to be determined [*ibid.*].

Riemann emphasizes that the notion of a manifold, so defined, does not depend at all upon our geometrical intuitions [Scholz 1982, 228]. The spatial notions of space, plane and line are only the simplest, intuitive examples of three-, two- and one-dimensional manifolds. To this extent, Riemann is simply following in Gauss's footsteps, since Gauss had always emphasized the difference between an abstract theory of magnitudes and manifolds and its intuitive exemplification by means of spatial notions. But Riemann indicates that the notion in question is that of a multi-dimensional manifold and, moreover, that this notion affords a satisfactory basis for developing the whole of geometry without the least reliance on spatial intuition [*ibid.*]. In his opinion, on the basis of such manifolds it would be possible to give analytical definitions of the basic geometrical notions – he mentions that of a straight line – and to derive all axioms and propositions of Euclidean geometry as theorems.

By the time he wrote this manuscript, he regarded such an abstract approach to geometry only as a theoretically interesting possibility, but thought that it would be "extremely unfruitful." This new approach to the foundations of geometry would not yield a single new theorem, and it would make complex and obscure what appears as simple and clear when explained in intuitive, spatial language (in [Scholz 1982, 229]). It is plainly evident that Riemann was still a long way from realizing the novelties that he would be able to present in the 1854 lecture, after having understood that his continuous manifolds were a satisfactory foundation for a generalization of Gaussian differential geometry, and that a single underlying topology gave room for many different metrical geometries (§1.2).

---

[1] In appendix 1 [Scholz 1982, 222] we read: "Wenn unter einer Menge von verschiedenen Bestimmungsweisen eines veränderlichen Gegenstandes von jeder zu jeder anderen ein stetiger Übergang möglich ist, so bildet die Gesammtheit dieser eine stetig ausgedehnte Mannigfaltigkeit; jede einzelne heisst ein Punkt dieser Mannigfaltigkeit." Regarding the use of the word 'Menge,' see the Introduction.

While such an abstract approach at first appeared unfruitful to Riemann, the opposite held for the use of geometrical imagery as a tool in understanding multi-dimensional manifolds. Instead of developing geometry in an abstract way,

One has thus always followed the opposite path, and every time that one has stumbled upon manifolds of many dimensions in mathematics, as in the doctrine of definite integrals within the theory of imaginary magnitudes, one has had recourse to spatial intuition. It is well known, how one thus wins a true overview of the matter, and how only in that way the essential points become evident.[1]

Riemann refers here to the use of geometrical imagery and spatial intuition within the context of function theory as a paradigmatic example. The connection here established alongside the issues dealt with in the rest of the manuscript suggest a reconstruction of Riemann's intellectual journey in developing his conception of manifolds. It suffices to follow his reasoning backwards.

It seem natural to speculate that Riemann felt puzzled by his recourse to geometrical constructs, the Riemann surfaces, in function theory. There were actually several reasons to feel uneasy. First and foremost, the tendency in contemporary analysis was to avoid resorting to geometry and intuition; on the contrary, Cauchy and his followers reformulated previous geometrical ideas in abstract terms. Riemann had been educated in this tradition, where Dirichlet was one of the most prominent names, but the direction of his research seemed to contradict that tendency. Did function theory depend upon geometry in some way? Was the use of Riemann surfaces an *ad hoc* intuitive means for understanding multi-valued functions? And one might even ask, did the general theory outlined in his dissertation thus lack rigor?

In the second place, Riemann surfaces did not belong to traditional geometry, since they could only be conceived as objects in higher-dimensional space. Was there a satisfactory foundation for higher-dimensional geometry? Moreover, and third, Riemann had analyzed the behavior of Riemann surfaces from the viewpoint of *analysis situs*, but again this lacked a satisfactory foundation at the time. Could a new approach to geometry be sketched, that answered to all of the above problems? As we can see, many questions could have been asked which gave occasion for a "philosophical study" of this area of mathematics, in the Herbartian spirit (see Appendix).

The text we have commented, and the 1854 *Habilitationsvortrag*, give reasons to think that Riemann considered all of the above questions, and was able to answer them by locating a new fundamental concept on which to reformulate the whole

---

[1] [Scholz 1982, 229]: "Man hat daher auch überall den entgegengesetzen Weg eingeschlagen, und überall, wo man in der Geometrie [Heinrich Weber suggests that one should read 'Mathematik'] auf Mannigfaltigkeiten von mehreren Dimensionen stösst, wie in der Lehre von den bestimmten Integralen der Theorie der imaginären Grössen, nimmt man die räumliche Anschauung zu Hülfe. Es ist ja bekannt, wie man dadurch eine wahre Übersicht über den Gegenstand gewinnt und nur dadurch gerade die wesentlichen Punkte hervortreten."

issue. Scholz's appendix 4 indicates how Riemann was able to explain the relations between geometry and function theory. Riemann surfaces are obviously two-dimensional manifolds that can be embedded in some higher-dimensional space, i.e., in some higher-dimensional manifold. Far from making function theory, or, more generally, the theory of magnitudes, dependent upon geometry, the true situation was that the notion of manifold was independent from geometrical intuition and made possible an abstract derivation of geometry. It also explained naturally how higher-dimensional constructs emerged. Riemann's recourse to geometrical intuition in function theory simply became a ploy that greatly simplified the development of the theory. But he was convinced that his new approach could be given a completely rigorous abstract foundation.

The new notion, foreseen by Gauss, was actually indispensable for an abstract theory of magnitudes, and Riemann found in it the right concept on which to base topology. Around 1852/53, Riemann wrote his 'Fragment aus der analysis situs,' first published in the 1876 edition of his collected works [Riemann 1892, 479–82].[1] He presented here a fragment of the topological theory of n-dimensional manifolds, developing in an abstract way the homological method that he would publish in the 1857 paper on Abelian functions, while indicating its relation to the dissectional method already used for Riemann surfaces in 1851 (see §3.1). This adds to the plausibility of Scholz's reconstruction that manifolds emerged as a theoretical basis for Riemann surfaces, for, if so, it is only natural that the topological methods employed with the latter would be used for the former. Furthermore, it is remarkable that this text should have been written before the *Habilitationsvortrag*. One may say that Riemann had developed all of the basic aspects of his notion of a manifold, and of his topological ideas, by the time he delivered the famous lecture in 1854. But these ideas only became known gradually, with their publication from 1868 to 1876.[2] The abstract foundation for all this work was further refined, and explained in some more detail, in the first part of his *Habilitationsvortrag*.

3.3. Differential geometry and physical space.

The *Habilitationsvortrag* made two main contributions, a generalization of Gauss's differential geometry of surfaces, and a deep contribution to the question of physical space. Thus, it attests to the profound interrelation of mathematical and physical thought in Riemann.[3] The author was intent on making possible a deeper analysis of the concept of space and its presuppositions with the aim of freeing physical explanation from "conceptual limitations" and "traditional prejudices." Such conceptual problems were, in his view, the task of mathematics [Riemann 1854, 286]. Riemann proceeded by establishing a succession of conditions that gradually delimited the properties of space.

---

[1] The dating is Scholz's, see his [1982, 216 and 225–26].

[2] Riemann's ideas on n-dimensional topology also became known through Betti, who had discussed them with Riemann himself [Weil 1979].

[3] On Riemann's geometrical thought, the reader may consult [Gray 1989; Laugwitz 1996, chap. 3; Scholz 1980 and 1990a; Torretti 1984].

Rather than axioms, he called them "hypotheses," since he regarded it as an empirical task to determine their validity.[1]

Riemann found a satisfactory point of departure, possessing the necessary generality, in the notion of an *n*-dimensional continuous manifold. He hoped that an this notion would make it possible, for the first time, to offer the general analysis of the notion of space that he wished [1854, 272]. That notion he saw as basically topological in character, so the first part of the lecture is devoted to topological considerations (see §§5 and 6). The second part introduces the fundamental concepts of differential geometry, on the basis of a hypothesis that we may formulate as follows: one-dimensional measuring rods are freely movable without alteration of their lengths [*op.cit.*, 276]. This enabled him to introduce metrical notions – the fundamental quadratic form that defines the length of a line element in so-called Riemannian manifolds, and the Gaussian curvature of the space at each point – establishing a wide frame within which the properties of ordinary physical space can be adequately located and characterized.[2] Finally, in part three, he comes to the application of the previous ideas to physical space and the conceptual possibilities thus opened.

The basic insight that made possible Riemann's surprisingly new approach to the question of space was the following. Given an *n*-dimensional continuous manifold, one may endow it with many different metrics so that spaces with very diverse metrical properties may have the same topological substructure in common [1854, 283–84]. This includes not only the cases where the curvature is constant at all points, which yield the (now) well-known non-Euclidean geometries, but spaces of variable curvature too. With this insight, Riemann brought the discussion on geometry a long step farther from the work of Lobachevsky and Bolyai, which, by the way, he probably had not read [Scholz 1982, 217–221]. Although he conceded as certain that physical space is a 3-dimensional manifold, the experimental task of investigating its metrics was open [Riemann 1854, 255, 265–66].

The wideness of the new frame in which Riemann was conceiving geometry, and the freedom of thought that he promoted with regard to all of the possible "hypotheses," become clear when he comes from the abstract mathematical study of manifolds to the question of physical application [1854, 284–86]. Here we find the famous distinction between unlimited and infinite, the first time that the possibility of a finite space was seriously undertaken. But Riemann also points to the possibility that the expression for the line element be different, so that the manifold may not be a 'Riemannian' one, in present terminology. Likewise, he admits the possibility that space may suffer observationally unnoticeable changes in the local cur-

---

[1] Being no positivist, he thought that empirical determinations of the validity of such conditions never yield absolute truth, and so he preferred "hypotheses" to "facts" [*op.cit.*, 273], in contrast to Helmholtz [1868].

[2] Here Riemann built again on previous work of Gauss, the great 1828 paper on curved surfaces [Gauss 1863/1929, vol. 4, 217–58] showing that the curvature of a surface is an intrinsic property, invariant under isometric transformations ("theorema eggregium"). Riemann elaborated the intrinsic viewpoint directly.

vature – a possibility that he found promising for a unified theory of the physical forces and that quickly caught Clifford's attention.[1] He even regarded it as possible that physical space may not be continuous, but "a discrete manifold" [*op.cit.*, 286].

Riemann's geometrical ideas were novel and abstract, and their reception was slow, although they immediately caught the attention of some mathematicians.[2] The analytical side of his investigations was quickly taken up, leading to a development of the theory of differential invariants that eventually ended in the emergence of the tensor calculus. His topological ideas also found a continuation. Geometers accepted the differentiation of topological and metrical properties, and made use of Riemann's ideas concerning spaces of constant curvature in the context of the open discussion of non-Euclidean geometries around 1870. But they kept framing geometry within a more modest setting: the details of Riemann's introduction of the notion of curvature and, above all, the idea of manifolds of variable curvature were difficult to understand or accept. Helmholtz gave some arguments intending to show that the notion of a space of nonconstant curvature was necessarily wrong, and they were quite successful at the time [Freudenthal 1962; 1981, 455]. Only after Minkowski's interpretation of special relativity in terms of a 4-dimensional world, and after the advent of general relativity did the development of differential geometry receive a strong stimulus.

Nevertheless, a small group of mathematicians seems to have been able to understand Riemann's most basic ideas better. Prominent among them are Dedekind and Cantor, in whose work notions of set theory and point-set topology were developed. As we shall see, these men accepted the conceptual freedom with which Riemann had approached the "hypotheses upon which geometry is founded." They explored the possibility of a discontinuous geometry, respectively a discontinuous physical space, and they took up the notion of a manifold in its original sense (§§6 and V.4).

4. Riemann's General Definition

Riemann was addressing his lecture to members of the Philosophical Faculty, which is where mathematics and the sciences belonged. Thus, there was little need to make any explicit reference to the fact that his definition of a manifold depended upon basic ideas of contemporary logic, especially after having said that the task he was confronting was "philosophical" in character. He defined straightforwardly:

---

[1] See 'On the Space-theory of Matter,' in [Clifford 1882, 21–22]. For Riemann and physics, refer to §I.4 and [Laugwitz 1996, chap. 3].

[2] A review of the great number of 19th-century works that related to Riemann's mathematics can be found in Neuenschwander's appendix to the 1990 edition of Riemann's *Werke*.

Notions of magnitude are only possible where there is an antecedent general concept which admits of different ways of determination. According as a continuous transition does or does not take place among these determinations, from one to another, they form a continuous or discrete manifold; the individual determinations are called points in the first case, in the last case elements, of the manifold.[1]

I have chosen the literal translation 'ways of determination' for Riemann's 'Bestimmungsweisen;' Clifford wrote 'specialisations.' What Riemann meant can be gathered from the example he gives immediately below, a characteristically Herbartian example: if we take the concept of color, each particular color, each particular shade of blue or yellow, is a 'way of determination' of that general concept; the totality of these 'specialisations' forms a manifold. With §2 in mind, it should be easy to interpret the above explanation and the example. We find a clear reference to the traditional concept–class relation, the manifold being the class or set of all "determinations" that fall under the general concept.

Riemann's definition has puzzled some modern commentators, probably because they have had no contact with the ideas of traditional logic, and on this point the notion of logic has been crucially transformed by the impact of the paradoxes. Bourbaki [1969, 176] translates "element" where Riemann wrote 'general concept,' which may simplify his reader's task, but definitely alters the original text. Scholz [1980, 30] was misled by Riemann's definition into thinking that he called the general concept itself a manifold. Several details of Riemann's *Habilitationsvortrag* can be used to corroborate my interpretation. That his manifolds are not just topological objects is made clear by the fact that he accepts discrete and continuous manifolds (see below). Later in the lecture, we find a text which introduces Riemann's famous distinction between unlimitedness and infiniteness [1854, 284]. He says that when we try to determine a metrical relation by experiment, "the possible cases form a continuous manifold," giving rise to an unavoidable imprecision, while in an experimental determination of topological properties "the possible cases form a discrete manifold," which involves no inaccuracies. Obviously, he is talking here about the *set* of possible cases, or possible results of a measurement.

It is likely that the idea of connecting his manifolds of 1852/53 with concepts was suggested to Riemann by his reading of Herbart's works. We have seen that, in his theory of spatial concepts, Herbart used the examples of the continua that fall under the concepts of tone and color. He went so far as to say that each property of an object should be regarded as located in a "qualitative continuum," from which Riemann disagreed in his lecture [1854, 274]. In an early work we find Herbart's mention, within a discussion on the right way of teaching mathematics, of "*the*

---

[1] [Riemann 1854, 273]: "Grössenbegriffe sind nur da möglich, wo sich ein allgemeiner Begriff vorfindet, der verschiedene Bestimmungsweisen zulässt. Je nachdem unter diesen Bestimmungsweisen von einer zu einer andern ein stetiger Übergang stattfindet oder nicht, bilden sie eine stetige oder discrete Mannigfaltigkeit; die einzelnen Bestimmungsweisen heissen im erstern Falle Punkte, im letztern Elemente dieser Mannigfaltigkeit." For the lecture of 1854 I employ Clifford's translation [1882, 55–71], with my own corrections.

whole continuum that is contained under a *general concept*" [Herbart 1964, vol. 1, 174].[1] Such sentences may easily have suggested Riemann's new definition.

By identifying manifolds with the classes of logic, Riemann was stretching his new notion as far as possible, since any object of perception or thought can be an element of a class. There can be little doubt that the intention of generalizing as much as possible was one of the main reasons behind his new definition. In this way, he was also in a position to offer a reconception of the whole theory of magnitudes. Actually, the connection Riemann established between manifolds and magnitudes constitutes another difficulty for a satisfactory understanding of his general ideas. Part I of the *Habilitationsvortrag* is supposed to fulfill "the task of constructing the notion of a multiply extended magnitude out of general notions of magnitude."[2] We shall now turn to this second difficulty.

4.1. Manifolds and the theory of magnitudes. In searching for a solution of the problems mentioned in §3.2, Riemann stuck to the traditional definition of mathematics as the science or doctrine of magnitudes. As we saw at the beginning, this traditional definition can be traced back to Aristotle, who distinguished two kinds of magnitudes, the discrete and the continuous. The same conception is found in handbooks and even in the work of research mathematicians up to the mid-19th-century. Of course, there was no clear or univocal theory behind the definition of mathematics as the doctrine of magnitudes since the notion of magnitude was left rather vague. During the 19th-century, several authors tried to give it a precise sense, and this seems to have been one of the ways in which novel abstract viewpoints, and even the notion of set, began to be employed.

Let me give some examples. Bolzano [1851, 2] kept employing the traditional definition of mathematics, although he introduced the notions of ordered and unordered "sets" [Mengen] as a basis. Grassmann had formerly criticized the traditional definition, preferring to conceive of mathematics as the "doctrine of forms,"[3] but in his later textbook on arithmetic he defined mathematics as "the science of the connection of magnitudes" [Grassmann 1861, def. 1]. No basic change of viewpoint was involved here, simply a terminological change, as Grassmann was now advancing an altered, abstract notion of magnitude: magnitude is any thing that can be said to be equal or unequal to something else, where *a* equals *b* means that we can substitute *b* for *a* in any proposition [*op.cit.*, 1]. This extremely general definition employs the Leibnizian definition of equality, and seems to go far beyond the traditional. Weierstrass differentiated "numbers," meaning natural numbers, from magnitudes. His theory of rational and irrational numbers was formulated as a the-

---

[1] It is worth recalling that Riemann was particularly interested in Herbart's early writings [Riemann 1892, 507–08].

[2] [Riemann 1854, 272]: "Ich habe mir ... die Aufgabe gestellt, den Begriff einer mehrfach ausgedehnten Grösse aus allgemeinen Grössenbegriffen zu construiren." Part I bears the title: "Begriff einer *n*fach ausgedehnten Grösse."

[3] [Grassmann 1844, 65]: "Formenlehre."

ory of "numerical magnitudes," a terminology inherited by Cantor himself. Weierstrass defined his 'Zahlengrössen' as aggregates of certain units, but it is not quite clear to what extent he regarded those aggregates as sets, or simply as infinite series (see §IV.2.1).

Coming back to Riemann, the fact that he presented the notion of manifold as a new foundation for an abstract theory of magnitudes means nothing less than that he was proposing a new vision of the foundations of pure mathematics. His definition of manifold explicitly establishes links with discrete and continuous magnitudes, suggesting that arithmetic, geometry, and their higher developments can all be reestablished within the new framework. Like Gauss (see §3.1), he regarded quantitative or metrical relations as only a part of the general theory of magnitudes, the other being the topology of manifolds [Riemann 1854, 274]. The reader may be wondering what exactly is the relationship he establishes between magnitudes and manifolds. Actually, a careful reading of the *Habilitationsvortrag* shows that he simply employed both words as synonyms, although he tended to prefer "manifold" as a technical term. It seems plausible that he kept talking about magnitudes in order to let his audience grasp more easily what he intended to talk about, but at the same time he tended to substitute his new notion of manifold for that rather vague traditional term. A clear example is part I of the lecture, where he analyzes what he technically calls an "*n*-ply extended manifold" or manifold of *n* dimensions; this part is entitled "Notion of an *n*-ply extended magnitude" [1854, 273 and, e.g., 276].

By establishing the theory of magnitudes upon the foundation of manifolds, Riemann transgressed the limits of the traditional conception of mathematics, turning it into a discipline of unlimited extent and applicability, since it embraced all possible objects. Thus, his traditionalistic terminology hides strongly innovative viewpoints. In §5 we shall discuss the details of his embryonic theory.

4.2. Riemann on the infinite. Since we have set as a criterion for "serious" talk of sets the acceptance of the actual infinite (see Introduction), it is crucial to ask what Riemann's position on this issue was. The answer must be tentative, since there is no evidence that may be called direct, strictly speaking, and we can only reconstruct his viewpoint. But there is reason enough to think that he accepted the actual infinite.

First, we may recall that Leibniz was a partisan of the actual infinite at least in some of his work,[1] and that his ideas were quite influential in 19th-century Germany. One of the authors who fell under his influence was Herbart, who also spoke for the actual infinite in his early years. In the late 1830s the philosopher would declare that true infinity can only be regarded as undetermined and incomplete,[2] but

---

[1] Notably the *Monadologie* of 1714, see §I.3.

[2] There is a memorable passage in Cantor's works [1932, 392–93] where he presents, against Herbart, an argument that might be called the trip and road argument, intending to show that the potential infinite presupposes the actual infinite. There he speaks about the "Herbartian dogmatism" that only accepts the potential infinite, quoting extensively from [Herbart 1964, vol. 4, 88–89]; in particular, he gives the following quotation: "Hingegen ist bei einer *unendlichen* Menge

the reader should keep in mind that Riemann [1892, 507–08] preferred Herbart's early ideas to his later developments. In the 1800s, Herbart talked about the service that metaphysics had made mathematics by eliminating the aversion of the notion of infinity; such aversion had led mathematicians to teach

in strange ways, *without* that fundamental concept, that which *only* through it was accessible to the *discoverer* himself [Herbart 1964, vol. 1, 174].

By this time, Herbart devoted much time to mathematical studies, and he is most likely talking about the calculus and its discoverer Leibniz. Herbart also formulated the ontology of the 'Reale,' a version of Leibniz's monadology, which again is based on the actual infinite. All of this is evidence for an acceptance of the actual infinite in Herbart's early work, the one that Riemann found more convincing.

As regards Riemann, the notion of continuous manifolds as sets of "points," where these points are the counterparts of the "elements" of discrete manifolds (see his definition above), gives indirect evidence for a positive attitude towards actual infinity. Nevertheless, it might be argued that this does not go beyond the traditional, ambivalent position of geometers, even though Riemann's language appears to be more committed. More interesting information is given by a couple of passages, one from Riemann's philosophical manuscripts, the other from the *Habilitationsvortrag*.

A surprising passage in the 1854 lecture assumes the existence of infinite-dimensional manifolds (or spaces), at a time when the idea of going beyond the third dimension was already bold. The text also shows that Riemann did not identify the discrete with the finite, since he distinguishes the case of an infinite sequence (a discrete infinity) from that of a continuous set:

Nevertheless, there are manifolds in which the determination of position requires not a finite number, but either an infinite sequence or a continuous manifold of determinations of magnitude. Such manifolds are, for example, the possible determinations of a function for a given region, the possible shapes of a solid figure, etc.[1]

This text can hardly have been written by a person who has serious doubts about the acceptability of actual infinity. But, should the reader still be skeptical about Riemann's position, there is another piece of evidence that seems to be conclusive.

A philosophical text called 'Antinomien' is the most explicit evidence of Riemann's views regarding the infinite. Here he presents four pairs of contradictory propositions as theses and antitheses, no doubt following the famous example of the

---

die Möglichkeit des Zählens schlechthin ausgeschlossen, weil eben das *wahrhaft Unendliche nur als ein Unbestimmtes, Unfertiges gefasst werden kann.*"

[1] [Riemann 1854, 276]: "Es giebt indess auch Mannigfaltigkeiten, in welchen die Ortbestimmung nicht eine endliche Zahl, sondern entweder eine unendliche Reihe oder eine stetige Mannigfaltigkeit von Grössenbestimmungen erfordert. Solche Mannigfaltigkeiten bilden z. B. die möglichen Bestimmungen einer Function für ein gegebenes Gebiet, die möglichen Gestalten einer räumlichen Figur, u. s. w."

Kritik der reinen Vernunft [1787, 454–489]. Kant had tried to show that human reason falls naturally and inevitably into contradiction, when it considers some all-embracing notions – the world with its spatial and temporal limits, the simple or composite character of substances, the notions of causality and freedom, and the notion of God. (This, by the way, is the remote origin of the term "antinomy" as applied to the set-theoretical paradoxes, with the connotation that they are inevitable contradictions of our logic.) Riemann's antinomies are, thematically, quite close to Kant's: they deal with finite vs. continuous space and time, freedom vs. determinism, God as acting temporally vs. God as atemporal, and inmortality vs. a purely intelligible soul [Riemann 1892, 518–20].

What is interesting for our present purposes is that these antinomies are presented under the general headings "Thesis. Finite, representable" and "Antithesis. Infinite, conceptual systems which lie on the borders of the representable" [*op.cit.*, 518]. And a general comment on the relation between thesis and antithesis indicates that, under the latter, we find "concepts which are well determined by means of negative predicates but are not positively representable" [*op.cit.*, 519].[1] Thus, the notion of the infinite is well defined and seems to be completely acceptable – the same being the case for the notions of continuity, determinism, a providential God, and the soul.

At the same time, this text enables us to observe an important, while perfectly natural, difference between Riemann's position and the later results of Dedekind and Cantor. Certainly Riemann accepted the actual infinite, and regarded the notion of infinity as "well determined," but he did not think that it would be possible to define it directly. Hence, Dedekind's definition of infinite set (§§III.5 and VII.2) would most likely have been surprising to him, as it was to Cantor.[2] Furthermore, in saying that the infinite is not positively representable, Riemann implied that it is not possible to investigate it directly [1892, 520], to set forth a positive theory. Thus Cantor's achievements would have been even more surprising to him.

5. Manifolds, Arithmetic, and Topology

The manuscripts of 1851–53 only mentioned continuous manifolds, but in 1854 Riemann's interest in generalization was also shown by the fact that he included some spare comments on discrete manifolds. This is further evidence of his intention to place manifolds at the foundations of the theory of magnitudes, and thus of pure mathematics. His comments make it clear that the theory of discrete manifolds

---

[1] "Thesis. Endliches, Vorstellbares." "Antithesis. Unendliches, Begriffssysteme, die an der Grenze des Vorstellbaren liegen." "Die Begriffssysteme der Antithesis sind zwar durch negative Prädicate fest bestimmte Begriffe, aber nicht positiv vorstellbar."

[2] [Dedekind 1932, vol. 3, 488]: "doch bezweifelte er [Cantor] 1882 die Möglichkeit einer einfachen Definition [des Unendlichen] und war sehr überrascht, als ich ihm ... die meinige mittheilte."

includes that of natural number, in accordance with classical conceptions. Immediately after defining a manifold (see §5.1) he goes on:

Concepts whose determinations form a discrete manifold are so common that, at least in the cultivated languages, any things being given it is always possible to find a concept under which they are included (hence, in the theory of discrete magnitudes, mathematicians could unhesitatingly proceed from the postulate that certain given things are to be regarded as homogeneous), ...[1]

This text may again seem cryptic, but one should recall that Euclid defined number as a "collection of units," and that it was customary at the time to discuss the notion of unit in elementary arithmetic textbooks. Actually, mathematicians and philosophers were puzzled by the fact that number-units ought to be equal and unequal at the same time (see [Frege 1884, ch. 3]). Riemann is suggesting that a formulation in terms of sets eliminates the problem: diverse objects can be regarded as equal or homogeneous (though not identical), insofar as they fall under the same concept, that is, belong to the same manifold. The first sentence also makes it clear that *any* objects whatsoever may become elements of a manifold, i.e., mathematical objects.

The theory of numbers thus attains greater clarity thanks to the new notion. The next paragraph confirms that numbers express relations between manifolds:

Definite parts of a manifold, distinguished by a mark or by a boundary, are called quanta. Their comparison with regard to quantity is accomplished in the case of discrete magnitudes by counting, in the case of continuous magnitudes by measuring.[2]

The relation between manifolds and numbers is clearly suggested, though of course there is a great distance between this and a detailed, rigorous set-theoretical foundation of the number system. At any rate, the core idea of regarding sets as the basic referents for arithmetic was suggested by Riemann, and would be developed by his friend Dedekind, among other authors.

Everything points to the conclusion that Riemann mentioned discrete manifolds for the sake of completion; his main interest was in continuous manifolds, since these formed the basis for his work on function theory, topology, and geometry. A footnote at the end of the lecture indicates that the section we are discussing "also constitutes the preliminary work for contributions to analysis situs."[3] One can say that the topological viewpoint was Riemann's most important fundamental contri-

---

[1] [Riemann 1854, 273–74]: "Begriffe, deren Bestimmungsweisen eine discrete Mannigfaltigkeit bilden, sind so häufig, dass sich für beliebig gegebene Dinge wenigstens in den gebildeteren Sprachen immer ein Begriff auffinden lässt, unter welchem sie enthalten sind (und die Mathematiker konnten daher in der Lehre von den discreten Grössen unbedenklich von der Forderung ausgehen, gegebene Dinge als gleichartig zu betrachten), ..."

[2] [Riemann 1854, 274]: "Bestimmte, durch ein Merkmal oder eine Grenze unterschiedene Theile einer Mannigfaltigkeit heissen Quanta. Ihre Vergleichung der Quantität nach geschieht bei den discreten Grössen durch Zählung, bei den stetigen durch Messung."

[3] [Riemann 1854, 286]: "Art. I bildet zugleich die Vorarbeit für Beiträge zur analysis situs."

bution to mathematics. With the preliminary analysis of the notion of *n*-dimensional manifold presented in the 1854 lecture, and the beginnings of an abstract topology in the posthumously published 'Fragment aus der Analysis Situs,' Riemann established the program for an independent theory of topological spaces [Bourbaki 1976, 192–93].

From the manuscripts of 1851–53 on, it was perfectly clear that Riemann conceived of a non-metrical approach to the study of manifolds [Scholz 1982, 222–224]. This involves his most important rupture with the traditional conception of magnitudes, and of mathematics as the theory of magnitudes.[1] After the last sentence that we have quoted, he went on:

> Measuring consists in the superposition of the magnitudes to be compared; it therefore requires a means of transporting one magnitude as the standard for another. In the absence of this, two magnitudes can only be compared when one is a part of the other; in which case also we can only determine the more or less, and not the how much. The researches which can in this case be instituted about them form a general part of the theory of magnitudes, independent of metric determinations, in which magnitudes are regarded, not as existing independently of position nor as expressible in terms of a unit, but as domains in a manifold. Such researches have become a necessity for many branches of mathematics, e.g., for the treatment of many-valued analytic functions; and the want of them is no doubt a chief cause why the celebrated theorem of Abel, and the achievements of Lagrange, Pfaff, and Jacobi for the general theory of differential equations, have so long remained unfruitful.[2]

The relation between topology on the one hand, and function theory and differential equations on the other, is unequivocally stated, so that the connection with other parts of Riemann's work is clearly indicated. The above description of the topological viewpoint can be found again, almost word by word, in Riemann's celebrated paper on Abelian functions [1857, 91] (quoted in §1.1). Certainly it is only a suggestive description, not a precise definition, but the details of Riemann's topology of surfaces and manifolds served as further clarification. Quite clear was Riemann's intention of dispensing with metrical considerations altogether, but topological

---

[1] Such a rupture has some precedent in Gauss (§2.1), and other contemporary authors also emphasized that mathematics is not restricted to the study of the quantitative. This is the case of Grassmann [1844] and of the British tradition of symbolical algebra [e.g., Boole 1847, 42].

[2] [Riemann 1854, 274]: "Das Messen besteht in einem Aufeinanderlegen der zu vergleichenden Grössen; zum Messen wird also ein Mittel erfordert, die eine Grösse als Masstab für die andere fortzutragen. Fehlt dieses, so kann man zwei Grössen nur vergleichen, wenn die eine ein Theil der andern ist, und auch dann nur das Mehr oder Minder, nicht das Wieviel entscheiden. Die Untersuchungen, welche sich in diesem Falle über sie anstellen lassen, bilden einen allgemeinen von Massbestimmungen unabhängigen Theil der Grössenlehre, wo die Grössen nicht als unabhängig von der Lage existirend und nicht als durch eine Einheit ausdrückbar, sondern als Gebiete in einer Mannigfaltigkeit betrachtet werden. Solche Untersuchungen sind für mehrere Theile der Mathematik, namentlich für die Behandlung der mehrwertigen analytischen Functionen ein Bedürfnis geworden, und der Mangel derselben ist wohl eine Hauptursache, dass der berühmte Abel'sche Satz und die Leistungen von Lagrange, Pfaff, Jacobi für die allgemeine Theorie der Differentialgleichungen so lange unfruchtbar geblieben sind."

research up to the 20th-century was frequently framed within the setting of metric spaces, as was the case with Cantor's pioneering work on point-sets (subsets of $\mathbb{R}$ or $\mathbb{R}^n$; see §§VI.6–8).

Part I of Riemann's *Habilitationsvortrag* deals only with two points from the topological theory of manifolds: the concept of n-dimensionality, and the parametrization of an n-dimensional manifold [1854, 274]. Riemann clarified the notion of dimension by considering how the manifold might be "reconstructed" starting from a one-dimensional path, going up to a 2-dimensional manifold, ... to manifolds of $n–1$ and, finally, n dimensions. This is reminiscent of the traditional idea of a mechanical generation, which can be found in Aristotle, Proclus and Oresme, but mixed with the radically new acceptance of multi-dimensionality [Scholz 1980, 32–33]. More important was to establish the possibility of a local parametrization of points in the manifold, since this opened the way to the introduction of analytical concepts (fundamental metrics, Gaussian curvature) and therefore to differential geometry. The corresponding part of Riemann's lecture [1854, 275–76] leaves unclear whether the parametrization is intended to be local or global. But Riemann's awareness of the complex interrelation between local and global properties, elsewhere in the lecture, forces one to interpret these passages as referring to a local parametrization [Scholz 1980, 34–36].

In this way, Riemann discovered the "essential character" of n-dimensionality in the fact that the determination of position in an n-manifold requires n determinations of magnitude, i.e., n coordinates [Riemann 1854, 276]. This conclusion was accepted by most authors, particularly by Helmholtz in his influential papers on the foundations of geometry [1868, 612], but it was questioned by Cantor's work on one-to-one mappings from $\mathbb{R}$ to $\mathbb{R}^n$ (§VI.4).

6. Riemann's Influence on the Development of Set Theory

The development of Riemann's views affords a partial answer to the questions how the language of sets emerged from classical mathematics, and how sets came to be regarded as a foundation for mathematics (§3). Riemann understood the surfaces of his function theory, and the manifolds of his differential geometry, as based on the notion of concept-extension, i.e., of class or set. On this basis, he proposed a revision of the classical notion of magnitude; he regarded manifolds, i.e., classes, as a satisfactory foundation for arithmetic, topology and geometry – in a word, for pure mathematics (§§4–5).

In Riemann's work we do not find any development of an autonomous theory of sets, not even of the topology of point-sets, but only a rather general and still intuitive reconception of mathematics. It is natural, however, to think that his seminal ideas may have stimulated other authors to value the promise of sets, and to carry further the program of a reformulation of mathematics. The art of posing questions may not be "more consequential" than that of solving them, as Cantor believed, but no doubt it is equally important – and research programs are as important as technical results for the development of a new field. Were that correct, we should expect Riemann's ideas to have influenced further developments on a general programmatic level, not in the customary way of particular mathematical results or techniques. Their effectiveness lay in their potential to suggest interesting ways of inquiry leading to particular questions and technical developments. This is clear, for instance, in connection with the emergence of point-set topology. Although the issue of Riemann's influence will continue to be of our concern in other chapters, we shall explore it here in a preliminary way.

Riemann's new vision of mathematics remained unknown to the general public until 1868, when Dedekind published his works of 1854 in the *Abhandlungen* of the Göttingen Academy of Sciences. Before that, only close friends such as Dedekind himself, and perhaps the Italian mathematicians, could have learned about his speculations. The treatises published in 1868 are the *Habilitationsvortrag* and the famous paper on trigonometric series, including the definition of the integral. Both appeared simultaneously, and immediately caused a sensation in the German mathematical world (see [Klein 1926, vol. 1, 173]). One must take into account that both papers were published together and both were widely read. Thus, an author influenced by Riemann's work on real analysis may have also been influenced by his proposal of manifolds.

6.1. Reception of the notion of manifold. The impact of the paper on trigonometric series can be judged from work by Hankel, Heine, Cantor and du Bois Reymond that was published two or three years later (chapter V). Riemann's new definition of the integral opened the way for a systematic study of discontinuous functions, and thus constituted a most important background for the beginnings of the theory of point-sets. As for the paper on geometry, its impact can be gauged from the work of Helmholtz and Beltrami in 1868, and that of Klein in the 1870s, including his *Erlanger Programm* of 1872 [Scholz 1980, ch. 3].

As I have said, Riemann's sophisticated approach to differential geometry, particularly his conception of manifolds of variable curvature, was not taken up. Better fortune had the notion of manifold of constant curvature, the only viable one according to Helmholtz, and a very interesting proposal at a time when non-Euclidean geometries were in the midst of mathematical discussion. Helmholtz and Klein adopted the term "manifold," but they interpreted it in a restricted way: Helmholtz [1868] concentrated on the space problem, and referred to Riemann's paper for details regarding the idea of manifold, but he restricted his attention to continuous manifolds of constant curvature. Klein employed the word 'Mannigfaltigkeit' very frequently in his geometrical and function-theoretical work of the 1870s and 80s. The earliest mentions seem to occur in 1872, not only in the *Erlanger Programm*,[1] but also in work on line geometry and non-Euclidean geometry [Klein 1921/23, vol. 1, 106–26, 311–43, 460–97]. In the famous *Programm* [1893]

---

[1] The *Erlanger Programm* enjoyed some diffusion from the 1870s, but only in the 1890s, when it was properly published, did it exert a great influence [Hawkins 1984].

he understands by a manifold essentially an n-dimensional projective space endowed with a group of transformations. In work on algebraic surfaces, published in 1873, he employs the idea in the sense of a hyperspace [Klein 1921/23, vol. 2, 11–44].

In both cases, the influence of Riemann is clear in the abstract tendencies promoted by Helmholtz and particularly by Klein. The latter included in the *Erlanger Programm* a reference to the theory of invariants under one-to-one bicontinuous transformations, which is interesting in connection with nascent topology – that viewpoint constitutes an important complement to Riemann's seminal ideas (see [Klein 1893; Johnson 1979, 127]). Klein went on to promote Riemann's ideas, especially the notion of a Riemann surface, in practically all branches of mathematics. But he always understood the notion of a manifold in a specifically geometrical sense. The more general meaning of 'Mannigfaltigkeit' as set, and the connection between manifolds and the foundations of arithmetic and pure mathematics, were lost in these developments.

Nevertheless, one should not overlook the possibility that the most general aspects of Riemann's notion of a manifold may have had a powerful influence on authors related with the early development of the theory of point-sets. In the case of Cantor, it is notable that from 1878 to 1890 he termed his field of study "theory of manifolds" [Mannigfaltigkeitslehre]. And the first time it happened he established a direct relation to Riemann's 1854 *Habilitationsvortrag* – it was the paper devoted to prove that all continuous "manifolds" have the same cardinality (see §VI.4). By showing that $\mathbb{R}$ and $\mathbb{R}^n$ are equipollent, Cantor cast doubt on Riemann's idea that the "essential character" of n-dimensionality is the need of n coordinates for giving the position of a point. Cantor presented his results as directly related to Riemann's insufficient characterization of n-dimensionality.

It has been suggested that Cantor's use of the word 'manifold' could have come from Weierstrass's lectures [Johnson 1979, 128–129], but an analysis of the use of this term in extant transcriptions of those lectures suggests that Weierstrass never gave it the general meaning of set or class. Weierstrass seems to have called certain subspaces of $\mathbb{R}^n$ 'manifolds,' which is similar to Gauss's use of the word, and also to what a superficial reading of Riemann's *Habilitationsvortrag* might suggest.[1] Such a usage would not allow for calling 'manifold' a set of points scattered in a line or space, which is the sense given to the word by Cantor in his papers of the period 1879–84. On the other hand, Riemann's conception of manifolds as classes allows precisely this kind of use. Although his association of manifolds with concepts might seem to set narrow limits on acceptable classes, it should be noted that 'being a point at which a given function is discontinuous' would be considered as a concept by 19th-century logicians. It is in this sense that Riemann's definition

---

[1] Cf. [Weierstrass 1986; 1988], also [Weierstrass 1894/1927, vol.7, 55–60], and [Pincherle 1880, 234–237]. Actually, Gregory H. Moore has communicated to me that apparently Weierstrass never used that word before 1868, and this suggests strongly that he took it from Riemann [1854], which was published exactly in that year. By this time, Cantor was already working on his *Habilitation*.

is ampler than Weierstrass's or Gauss's use of the word.[1] Moreover, only Riemann talked about manifolds in a systematic way, in connection with arithmetic, analysis, and geometry. And it was only to Riemann that Cantor referred in his paper [1878].

As I have mentioned at the very beginning of this chapter, it is remarkable that Dedekind himself understood Cantor's terminology to be related to Riemann's work. In a letter of 1879, he proposed to replace the clumsy word 'Mannigfaltig-keit' by the shorter 'Gebiet' [domain], which, he said, is "also Riemannian" [Cantor & Dedekind 1937, 47]. Later on, he kept mentioning the word 'Mannigfaltigkeit' as a synonym for set [Dedekind 1888, 344], which he would probably not have done, had it meant a misunderstanding of Riemann's original notion. In some respects, the reception of Riemann's work [1854] by Cantor and Dedekind was better than that of any geometer; momentarily we shall see the case of Riemann's "hypothetical" understanding of the foundations of geometry.

6.2. Continuity and topology. Riemann based his discussion of manifolds on a distinction between discrete and continuous manifolds. This called for further elaboration, and there were open problems in connection with both sides of the distinction. The relation he established between discrete manifolds and numbers was still very rough and vague, but it may not be coincidental that Dedekind became the mathematician who elaborated more rigorously on the set-theoretical foundations of the number system. But the most pressing issues were the definition of continuity itself, and the development of the topological theory of continuous manifolds.

In his *Habilitationsvortrag*, Riemann explained the continuity of a manifold by reducing it to the possibility of continuous transitions from any point to any other (which seems similar to path-connectedness). Since he presupposed this notion of a continuous transition along a path, his explanation was almost purely verbal. Notably, Dedekind wrote some manuscripts on basic topological notions. In the 1860s he defined the notions of an open set, of its interior, exterior, and boundary, proving related theorems within the context of metric spaces (for a more detailed analysis, see §V.3). As he wrote in a letter to Cantor of 1879, this offers "a very good foundation" for a rigorous exposition of the elements of the "theory of manifolds," independently of geometrical intuition [Cantor & Dedekind 1937, 48]. Moreover, Dedekind's famous work on the real numbers [1872] includes the first abstract definition of continuity. The author himself emphasized that the definition was perfectly general, offering "a scientific foundation for the investigation of *all* continuous domains" (emphasis in the original, [Dedekind 1872, 322]). There is little doubt that he saw his definition as relevant to Riemann's manifolds and to his conception of topology (see §IV.3).

---

[1] A difficulty still exists, though, when we consider that no concept can be associated to an *arbitrary* set of points in the line. This might have been the reason why both Dedekind and Cantor later tended to dilute or even abandon the connection between concepts and sets (see chapters VII and VIII).

One more step along this line was given by Cantor in a famous paper where he offered, among other things, a new abstract definition of continuity [Cantor 1883, 190–94]. This definition, which is intimately related with Cantor's definition of the real numbers, showed the way to be followed by subsequent topological approaches (see §VI.7). Finally, it is worth mentioning that the influence of Riemann seems to offer a satisfactory explanation for a noteworthy coincidence in Cantor's and Dedekind's papers on the real numbers. Both, in contrast to contemporary authors, are explicit on the need to postulate the Cantor–Dedekind axiom of continuity of the line, and both think that the axiom does not represent a necessary property of geometrical or even physical space. But the idea of considering continuity as a *hypothesis* would seem natural to anyone influenced by Riemann's lecture on geometry (see §IV.3).

The important point here is that, in all of those contributions, both set theorists were working within Riemann's tradition. The same happens, of course, with Cantor's ground-breaking contributions to point-set topology, in a series of papers published from 1879 to 1884, entitled "On infinite, linear point-manifolds" (see §VI.6).

6.3. On the way to abstraction.   One last point that should be mentioned is that Riemann's abstract-conceptual approach to mathematics may have paved the way for the development of abstract set theory. This is particularly clear in connection with his conception of topology, which may be considered as a first step toward an abstract set theory. It is obvious that set theory emerged from the study of the concrete sets suggested by the usual topics of traditional mathematics. In my opinion, the early history of set theory, up to about 1890, should be regarded as a process of progressive differentiation of distinct kinds of abstract features (or structures) that appear intertwined in those concrete, traditional sets. The first such distinction, in connection with questions of geometry and analysis, was that of topological vs. metric aspects, and here the importance of Riemann's contribution is undeniable (§3). A second step was the beginnings of a study of algebraic structures, particularly clear in the context of Dedekind's work on Galois theory and algebraic number theory (chapter III). A third step was Cantor's 'discovery' of the transfinite realm and of the abstract properties of cardinality and order (chapters VI and VIII).

Naturally, these features were only gradually differentiated. In Cantor's work on sets, transfinite and topological aspects are not clearly distinguished until the mid-1880s (see chap. VIII). Likewise, it is doubtful that Dedekind may have differentiated algebraic and topological properties in the modern way.[1] As regards topology, certainly it was not the exclusive brain-child of Riemann. The appearance of non-metric geometries – projective geometry in particular – opened the way to topology, and important elements of the new theory appeared in the work of Gauss and Listing, in Weierstrass's lectures, and so on. But Riemann drew the most general

---

[1] For him, the number system was more basic than any abstract structure [Corry 1996, chap. 2], and the construction of the number system naturally led from sets endowed with algebraic properties to other sets with topological ones (but see §III.6.2).

consequences regarding the emerging topological viewpoint, and he did so in connection with the notion of a manifold. The effect of these new vistas on the development of set-theoretical ideas may be judged from a comparison of the work of Bernard Bolzano with that of Cantor.

Bolzano is usually named whenever the origins of set theory are discussed, even though his writings exerted almost no influence on further developments. As we have seen, he proposed to base mathematics on notions similar to that of set (§5.2). He made a clear defense of actual infinity [Bolzano 1851, 6–24], and he proposed precise notions for treating infinite sets. In this way he even came quite close to such a central notion of set theory as cardinality (power, in Cantor's terminology). But after having been close to the right point of view, he departed from it in quite a strange direction.[1] Bolzano recognized clearly the possibility of putting two infinite sets in a one-to-one correspondence while one of them is a subset of the other, and he argued that this involves no contradiction [Bolzano 1851, 27–28]. He gave two examples, the intervals [0,5] and [0,12], correlated by the function $5y = 12x$, and a second similar example expressed geometrically [*op.cit.*, 28–30]. In this way he came close to regarding equipollence as a criterion for measuring infinite "sizes," but he resisted the conclusion that those sets have equal cardinality or "size." He wrote:

from that circumstance alone we are not allowed to conclude that both sets, if they are infinite, are equal to each other with respect to the multiplicity of their parts (that is, if we abstract from all differences between them); ... Equality of those multiplicities can only be inferred when some other reason is added, for instance that both sets have absolutely equal grounds of determination, i.e., that their mode of formation is absolutely equal.[2]

It seems that the only right way to compare sets "with respect to the multiplicity of their parts," and abstracting from all other differences, is by means of equipollence and cardinality. Perhaps Bolzano was misled by his ambiguous usage of the word "part," which does not differentiate between element and subset, but it would be surprising that this alone might have been sufficient reason to lead astray his careful precision and logical rigor. Given the fact that he employed geometrical examples, it seems more plausible that the main source of his error was his familiarity with Euclidean geometry and classical analysis, which led him to give undue prominence to metric considerations.

---

[1] Here, I call 'right' the idea that cardinality is the only meaningful way to compare abstract sets "with respect to the multiplicity of their parts [elements] (that is, if we abstract from all differences between them)."

[2] [Bolzano 1851, 30–31]: "bloss aus diesem Umstande ist es – so sehen wir – noch keineswegs erlaubt zu schliessen, dass diese beiden Mengen, wenn sie unendlich sind, in Hinsicht auf die Vielheit ihrer Teile (d.h. wenn wir von allen Verschiedenheiten derselben absehen) einander gleich seien; ... Auf eine Gleichheit dieser Vielheiten wird erst geschlossen werden dürfen, wenn irgendein anderer Grund noch dazukommt, wie etwa, dass beide Mengen ganz gleiche Bestimmungsgründe, z.B. eine ganz gleiche Entstehungsweise haben."

Non-metric geometrical ideas seem to have been a prerequisite for the development of abstract set theory, particularly because the notion of cardinality would be applied to subsets of $\mathbb{R}$ and $\mathbb{R}^n$ before it could be abstractly formulated. The very fact that abstract and topological considerations were intertwined in the work of Cantor until about 1885 seems to reinforce this conclusion. Such non-metric considerations began to emerge within the work of projective geometers, and surfaced in the early evolution of topology. Judged from this viewpoint, Riemann's contribution, and his explicit differentiation of topological and metric aspects, would seem to have been crucial in the way to abstraction.

Appendix: Riemann and Dedekind

As we have seen (§I.4), Dedekind's crucial formative period was the years 1855 to 1858, when he already held his Habilitation. He emphasized above all the role played by Dirichlet in advancing and refining his knowledge of higher mathematics, but also the figure of Riemann, whom he regarded as one of the greatest mathematicians. As he wrote, intercourse with both of them was inestimable, and he could expect that it would bring fruits [Scharlau 1981, 37].

For one full year, 1855–56, Riemann lectured on Abelian and elliptic functions, having Schering, Bjerknes and Dedekind as his audience [Dedekind 1876, 519]. Both courses made a strong impression upon Dedekind. An 1856 letter to Riemann [Dugac 1976, 210], written after the second lecture had finished, mentions the "multiple teachings" that he owes him since a year ago, and asks Riemann to send the draft of his theory of Abelian functions. We have seen that Riemann's function theory became a methodological model of key importance for his younger colleague (§I.4). From that work, Dedekind learnt the principle that "accidental forms of representation" ought to be avoided in favor of "simple basic notions" [Dedekind 1930/32, vol. 3, 468]. One must look for fundamental concepts, in order to base any mathematical theory on characteristic, inner properties of the objects studied. External representations or notations, however useful for the purpose of calculation, should be relegated to a secondary role [*op.cit.*, vol. 2, 54–55; vol. 3, 296]. All fragments in Dedekind's work that deal with his basic methodological commitments happen to mention Riemann.

As we shall see (§III.1), the notion of set was absent from Dedekind's work up to 1855, even when he dealt with foundational issues. But it was conspicuously present both in his algebraic work of 1856–58, and his theory of irrational numbers, which dates from 1858. It seems that here, again, Riemann's example was decisive, although we do not have direct evidence for this claim. We do know that Dedekind was deeply involved with Riemann's approach to function theory for at least a year, and thus he faced the problem of understanding the notion of a Riemann surface. Since both mathematicians maintained frequent personal contact, it is only natural to conjecture that Riemann must have informed his friend of his thoughts on the notion of manifold and its foundational role.[1]

But Dedekind's admiration for Riemann's work, and his adherence to some of its basic tenets, does not mean that they both worked similarly in all respects. In particular, the sense of rigor was different, partly due to the fact that they worked in such different fields. In November 1874 Dedekind wrote to Weber, with whom he colaborated in editing Riemann's works:

---

[1] We know that both discussed freely Riemann's speculations, even those related to the "philosophy of nature," i.e., physical theory, as happened in the summer of 1857 [Dedekind 1876, 521].

I am not the profound expert on Riemann's works that you take me to be. I certainly know those works and I *believe* in them, but I do not master them, and I will not master them until having overcome in my way, with the rigor that is customary in number theory, a whole series of obscurities.[1]

It is notable how Dedekind emphasizes his belief in the correctness of Riemann's results, which calls to mind the lack of rigorous proofs of many of them – think of Weierstrass's then recent critique of the Dirichlet Principle, of the foundations for the topology of manifolds, etc. A passage in his biography of Riemann seems to offer Dedekind's explanation for his difficulties. He says that Riemann's brilliant power of thought and anticipatory imagination led him frequently to take very great steps that others could not follow so easily. And when one asked him to give a more detailed explanation of some intermediate steps of his conclusions, he might seem puzzled, and it caused him some effort to accommodate to the slower reasoning of others and to leave their doubts aside [Dedekind 1876, 518–19]. By contrast, Dedekind always characterized himself as a slow mind, a "step-wise understanding" [Treppenverstand], that needed to fully master the basics of a subject in order to be able to work on it [Dugac 1976, 179, 261].[2]

In 1863, having published Dirichlet's *Vorlesungen* on number theory, Dedekind started to prepare a publication of his lectures on potential theory. On this occasion he attempted to prove what Riemann called the Dirichlet Principle, concerning the existence of a minimal continuous function on a given domain. It was some years before the critique of Weierstrass. In order to accomplish that goal, he looked for basic notions upon which to base an abstract development of the topological theory of manifolds (a topic mentioned in §6.2, to be analyzed in §IV.3).

Three years later, in 1866, Dedekind was entrusted with Riemann's *Nachlass*, with the assignment of selecting those parts that could be published. He revised it completely, brought some order to the chaotic mass of papers, and transcribed those parts that he could understand [Dedekind 1930/32, vol. 3, 421–23]. Three pieces were ready for print, and he published them quickly – the already-mentioned Habilitation papers and a contribution to electrodynamics. There was also the so-called *Pariser Preisschrift*, related to the lecture on geometry, which became the motivation for a rather deep involvement with Riemann's differential geometry.[3] Dedekind wished to publish this paper accompanied by a commentary on its relations

---

[1] [Cod. Ms. Riemann, I, 2, 14a-r]: "namentlich bin ich nicht der gründliche Kenner der Riemann'schen Werke, für den Sie mich halten. Ich kenne zwar diese Werke und *glaube* an sie, aber ich beherrsche sie nicht, und ich werde sie nicht eher beherrschen, als bis ich eine ganze Reihe von Dunkelheiten mir auf meine Weise und mit der in der Zahlentheorie üblichen Strenge überwunden haben werde."

[2] Regarding questions of rigor, Dedekind's point of reference was no doubt Dirichlet, who taught him the meaning of number-theoretical rigor (see above and [Haubrich 1999]).

[3] 'Commentatio mathematica, ...,' in [Riemann 1892, 391–404]; see the 'Anmerkungen' on [405–23], and extracts from Dedekind's comments in the 1876 edition of the works. The other paper mentioned is 'Ein Beitrag zur Electrodynamik' [*op.cit.*, 288–93].

with the lecture, and a detailed development of its analytical aspects. An 1875 letter to Weber says as follows:

When I published [Riemann 1854] I expressed the intention of supplementing the analytical investigations, and in the following years (mostly 1867, I believe) I was quite busy with this topic, although later I renounced a publication completely, partly because others (Christoffel, Lipschitz, Beltrami) had taken up the subject, partly because in 1869 I was led, by the indispensable preparations for the second edition of Dirichlet's number theory, to devote myself to a completely different field – namely establishing a general theory of ideal numbers, free from exceptions. Yesterday I revised my manuscripts of the time, very extensive, and I found among them three drafts, partly developed in much detail, for that supplementary treatise ... I have with me an enormous amount of material containing some investigations on particular spaces, e.g., of constant curvature, and other more interesting ones, but they did not go into those drafts, which interrupt sooner.[1]

Thus, Dedekind devoted 2 or 3 years, immediately before elaborating the first version of his famous ideal theory, to Riemannian differential geometry. One may wonder whether this influenced in some way his radically new approach to algebraic number theory (see §§III.3-4).

Dedekind may have been the only mathematician in the last third of the 19th-century who worked on "more interesting" spaces than those of constant curvature, i.e., on spaces of variable Gaussian curvature.[2] As we saw, this part of Riemann's ideas was hardly understood by his contemporaries, most of whom followed Helmholtz in accepting only manifolds of a constant curvature. In Dedekind's *Nachlass* one can find several manuscripts with the title 'Ideale Geometrie,' which have not been carefully studied, and are not even well catalogued (see [Cod. Ms. Dedekind V, 8 and XII, 1].[3] The existence of these manuscripts is even a bit surprising, since Dedekind is normally taken to be the prototype of a pure algebraist and number

---

[1] [Cod. Ms. Riemann 1, 2, p. 23v]: "Bei der Herausgabe dieser letzteren Abhandlung habe ich die Absicht geäussert, die analytischen Untersuchungen nachzuliefern, und ich habe mich in den nächsten Jahren (hauptsächlich 1867, wie ich glaube) lange mit diesem Gegenstande beschäftigt, später aber die Publication ganz aufgegeben, theils weil Andere (Christoffel, Lipschitz, Beltrami) diesen Stoff ergriffen hatten, theils weil ich im Jahre 1869 durch die unerlässlichen Vorarbeiten zur zweiten Ausgabe der Dirichlet'schen Zahlentheorie gezwungen wurde, mich einem ganz anderen Felde, nämlich der Herstellung einer allgemeinen, ausnahmslosen Theorie der idealen Zahlen zu widmen. Ich habe nun gestern meine damaligen, sehr umfangreichen Papiere durchsucht, und zwischen denselben drei, zum Theil sehr genau ausgeführte Entwürfe zu einer solchen Nachtrags-Abhandlung vorgefunden ...; eine Menge von Untersuchungen von speciellen, z.B. constant gekrümmten und anderen interessanteren Räumen liegen bergehoch bei mir, sind aber in diese Entwürfe, die vorher abbrechen, nicht mehr eingegangen."

[2] Clifford's famous contribution is just two pages presenting an interesting conjecture; see [Clifford 1882; Farwell & Knee 1990].

[3] The paper that probably contains what he intended to print was published a few years ago: 'Analytische Untersuchungen über Bernhard's Riemann Abhandlung über die Hypothesen welche der Geometrie zu Grunde liegen,' in [Sinaceur 1990]. In [Cod. Ms. Dedekind XII, 16] there is also a paper on congruence under constant curvature, and a short commentary on [Helmholtz 1868].

theorist. That impression is partly due to his extreme thoroughness when preparing a publication. As we shall see, by 1860 he could have published important research on algebra and algebraic number theory, but he refrained from going into print until he developed a completely general theory. Likewise, around 1870 he might have made important contributions to topology, differential geometry and the study of differential invariants. But he seems to have shared Gauss's motto: 'pauca sed matura.'

Figure 4. *Richard Dedekind (1831–1916) in 1868.*

III   Dedekind and the Set-theoretical Approach to Algebra

As almost no other in the history of mathematics, Dedekind made an effort to develop his discipline systematically, and in particular he prepared the ground for present-day 'abstract' mathematics – above all 'modern algebra' in the sense of van der Waerden's book. He contributed in an essential way to clarifying the most important basic notions of algebra – fields, rings, modules, ideals, groups – and he dealt with the foundations of mathematics – real numbers, Cantorian set theory, set-theoretical topology. In this sense, we can regard Dedekind as an antecessor and important precursor of Bourbaki.[1]

The connections between the work of Dedekind and that of such figures as Noether or Bourbaki explain the close attention that historians have given him in the last 25 years. It is certainly true that Dedekind's work is the outcome of a serious and deep attempt to reconceive and systematize classical mathematics, and that it prepared the ground in an essential way for modern abstract mathematics. But the idea that Dedekind (and Galois) gave modern algebra its structure, which can be found here and there,[2] is too simplifying – the emergence of the structural viewpoint in algebra was a lengthy process, and there is reason to doubt that Dedekind ever viewed mathematics from a strictly structural perspective. For him, pure mathematics was the science of numbers in all its extension and derivations, number systems being more basic than any possible abstract structure. Nonetheless, the abstract-conceptual viewpoint, that he took from Riemann and pursued in a new direction (§I.4), led him to prefer a kind of approach and methods that would prove to be

---

[1] [Scharlau 1981, 2–3]: "Erstens hat sich Dedekind wie kaum ein zweiter in der Geschichte der Mathematik um einen systematischen Aufbau seiner Wissenschaft bemüht und insbesondere die heutige 'abstrakte' Mathematik – vor allem die ‚moderne Algebra' im Sinne des Buches von van der Waerden – vorbereitet. Er hat wesentlich zur Klärung der wichtigsten algebraischen Grundbegriffe – Körper, Ringe, Moduln, Ideale, Gruppen – beigetragen und sich mit Grundlagenfragen der Mathematik – reelle Zahlen, Cantors Mengenlehre, mengentheoretische Topologie – beschäftigt. In diesem Sinne können wir Dedekind als Vorfahren und wichtigen Wegbereiter Bourbakis ansehen. ... Zweitens war Dedekind ... geprägt von bedeutenden Vorgängern."

[2] E.g., van der Waerden in his introduction to Dedekind, *Über die Theorie der ganzen algebraischen Zahlen* (Braunschweig, Vieweg, 1963). On Dedekind and structural algebra, see [Corry 1996, particularly 70–71, 79].

extremely fruitful in the context of 20th-century structural mathematics. This accounts for his influence on Noether and others, and makes it particularly interesting to explore the methodological and conceptual traits of Dedekind's work that underlay his preferred mathematical style.

An essential part of Dedekind's approach was the set-theoretical viewpoint, that he employed in all kinds of settings, but particularly for the development of the number system and in his algebraic and number-theoretical research. These two issues correspond, respectively, to his foundational interests and his main research; the former will be analyzed in chapters IV and VII, while the latter constitutes the topic of the present chapter. This division is observed here for expository reasons, and also in order to follow the chronological sequence of publications. However, the reader should not overlook that, with Dedekind, work on a particular mathematical theory was never independent from reflections on its place within an overall view and systematization of the discipline.

Dedekind regarded mathematics as an edifice built on set-theoretical foundations. This applies to arithmetic, algebra and analysis, but also to geometry, at least when treated in Riemann's way (§II.7). Therefore, in a summary of his work such as Scharlau's (quoted above) one misses the mention of his 1888 book on the natural numbers, which was an attempt to lay down the foundation for pure mathematics as a whole (see §VII.3). The origins of Dedekind's set-theoretical approach are quite old: he was clearly moving along that path by the late 1850s, and he deepened his understanding of, and confidence in, that approach with the formulation of ideal theory in 1871. Our present purpose is to analyze the evolution of his conceptions throughout this period, particularly – but not exclusively – in connection with his work on algebra and algebraic number theory.[1] Obtaining a clear vision of the state of his conceptions by 1871 is particularly important here, since Dedekind met Cantor in 1872, and they had occasion to comment on these issues (see §VI.1).

We begin considering the algebraic origins of Dedekind's set-theoretical approach in the late 1850s. Then we analyze the emergence of his new notion of field and of algebraic number theory. § 4 deals with his notion of ideal, together with the ideals that informed its formulation. After considering the roots of Dedekind's infinitism in §5, we analyze briefly the diffusion of his conceptions and approach.

1. The Algebraic Origins of Dedekind's Set Theory, 1856–58

Dedekind took his Habilitation in 1854, but by then it was still unclear what his research field would be. The dissertations presented for his Ph.D. and Habilitation had been occasional work and did not reveal his talents.[2] Dedekind's crucial for-

---

[1] Such a task would be almost impossible, were it not for a good number of contributions that have rescued and analyzed important manuscripts shedding new light on Dedekind's development between 1857 and 1871 [Purkert 1977; Scharlau 1981, 1981a, 1982; Haubrich 1999].

[2] 'Über die Elemente der Theorie der Eulerschen Integrale' (1852) and 'Über die Transformationsformeln für rechtwinklige Coordinatensysteme' (1854).

mative period was the years 1855 to 1858 (§I.4), and he emphasized, above all, the role played by Dirichlet in advancing and refining his knowledge of higher mathematics. Besides, there was also the figure of Riemann (§II.7).

Dedekind maintained very close relations with Dirichlet, attending all of his lectures, particularly those on number theory. Dirichlet indicated the gaps in his knowledge and gave him the means to fill them·[Scharlau 1981, 35; also 37, 40, 47]. After the lectures, they entered into extremely detailed discussion of the topics and proofs. In relation to this, Dedekind mentions that "the variety of methods that can be employed for the proof of one and the same theorem constitutes one of the main attractions of number theory."[1] Dirichlet showed him how number theory could be approached in a way that avoided formal steps, by focusing directly on the arithmetical properties of algebraic numbers. This is to say that Dedekind learnt from his cherished teacher the meaning of number-theoretical rigor, which also included the need to analyze carefully every proof in order to discover its true kernel and the most convenient proof method in order to approach it directly. So Dirichlet became his reference point regarding questions of rigor.[2]

After some time devoted to projective geometry and probability theory, in 1855 Dedekind started a study of recent algebraic work that would mark his future career. Beginning with Gauss's work on cyclotomic equations in *Disquisitiones arithmeticae*, he quickly went on to the research of Abel and Galois on the theory of equations. By the end of the year he began serious study of so-called "higher arithmetic," that is, works by Kummer, Eisenstein and others that are regarded today as contributions to algebraic number theory. As a result of this, he lectured on the Gaussian theory of cyclotomic equations and on "higher algebra" – which he essentially identified with Galois' theory – in the winter semesters of 1856/57 and 1857/58. It was the first time that a university course included a substantial discussion of the work of Galois [Purkert 1977].

In this new field, Dedekind was able to obtain his first significant results. Most of his work of the period can be seen as a reformulation, systematization and completion of the great contributions of his predecessors. It was a highly significant reformulation: Dedekind worked out independently and systematically the group-theoretical prerequisites for Galois theory, recognized that the theory was essentially concerned with field-extensions, and presented for the first time what is now regarded as the core of the theory – the relations between (in modern language) subfields of the splitting field and subgroups of the Galois group of a polynomial.[3]

---

[1] [Scharlau 1981, 48]: They have "... in den täglichen Zussamenkünften die kleinsten Einzelheiten von Neuem besprochen." [Dedekind 1930/32, vol. 3, 394]: "Da die Mannigfaltigkeit der Methoden, welche zum Beweise eines und desselben Satzes dienen, einen Hauptreiz der Zahlentheorie bildet, ... so lag es nicht im Sinne Dirichlets, sich ... auf den Inhalt dieser Vorlesung zu beschräncken."

[2] On Dedekind and Dirichlet, see [Haubrich 1999, chap. 5].

[3] See the manuscript published in [Scharlau 1981] as 'Eine Vorlesung über Algebra,' and the comments by Scharlau himself [1981a, 107], and by Scholz [1990, 386–94] and Haubrich [1999, chap. 6]. Dedekind was much less interested in, or perhaps unable to make progress on, specific problems such as criteria of solvability.

Dedekind's treatment of 1858, it has been said, could have become the first text-book on "modern algebra," 40 years before Weber, and 75 years before van der Waerden [Scharlau 1982, 341; but see Corry 1996]. Dedekind was also led to study some aspects of abstract group theory, including a very precise proof of the homo-morphism theorem (see below). His involvement with the theory of equations made him aware that it was extremely fruitful and clarifying to base it on field-theoretical notions, and this prepared his 1871 definition of fields. It also led to original re-search on the reciprocal decomposition of two irreducible polynomials.[1]

Dedekind did not publish any of the above-mentioned reformulations and re-sults. There is little doubt that his work would have obtained recognition even if his abstract orientation probably would not have been followed. But he saw little value in mere reformulations of Gauss or even Galois, and, as regards original results, he preferred to wait until he had brought them to completion (see [Haubrich 1999, ch. 6]). His extreme thoroughness when preparing a publication went, in a way, against his own interests.[2]

To sum up, in the late 1850s Dedekind began to move substantially in the di-rection of modern structural algebra, or – speaking more cautiously – advanced toward an abstract-conceptual reformulation of previous algebraic and number-theoretic work. To that end, he employed the language of sets. His attainments must be seen as triumphs of his methodological principles and preferences in the domain of algebra. By trying to understand "higher algebra" in a way that satisfied him, he was led to theoretical constructions that we recognize as essentially identical with the modern orientation. Our purpose in the present section is to analyze the impli-cations of his innovations for the emergence of set language. No attempt will be made to satisfactorily depict his algebraic and number-theoretical work, for which the reader is referred to the above mentioned contributions (particularly the general overview in [Haubrich 1999]).

1.1. The *Habilitationsvortrag* of 1854. The lecture Dedekind gave for his Ha-bilitation, 'On the Introduction of New Functions in Mathematics' [Dedekind 1854], is interesting here for two reasons. First, it already shows some traits of Dedekind's thought that were to last for his whole life. He was deeply interested in the problem of rigor, and he also aimed to understand the historical evolution of mathematics. Some sentences could be used to characterize Dedekind's work as a whole, for instance when he says that "turning definitions over and over again, for love of the laws or truths in which they play a role, constitutes the greatest art of the

---

[1] See [Scharlau 1982]. Since this can be reformulated as the reciprocal reduction of two fields, the editor regards it as the first significant result in field theory [*op.cit.*, 343]; notably, the result dates back to December 1855.

[2] On the occassion of a vacant position at the University of Giessen, in 1868, Clebsch had to exclude Dedekind – "unfortunately," as he wrote – from those who have contributed something from the scientific point of view [Dugac 1976, 155].

systematician."[1] The systematician's art of Dedekind led him to transform many of the fundamental concepts in which the mathematics of his time were based.

In his lecture, Dedekind claimed that the introduction of new functions, or new operations, is the key to the development of mathematics, and he analyzed the peculiarities of this process in mathematics – where, in contrast to other sciences, there is no room for arbitrariness [Dedekind 1854, 428, 430]. He discussed carefully the foundations of arithmetic, offering an excellent summary of his ideas at the time (§VII.1). Dedekind presented the idea of gradually developing arithmetic, from the sequence of natural numbers to $\mathbb{C}$, through successive steps in which new numbers and operations are defined. This was the program the he would carry out in his later foundational works [Dedekind 1872, 1888], but there is one important difference. From 1872 Dedekind would emphasize the definition of the new numbers themselves, while in 1854 he emphasized the problem of extending the operations to the expanded realm. The change is especially important in light of the fact that, in 1872 and later, sets will be the means that allow him to define or 'create' new numbers.

Thus, the second important point is that in Dedekind's *Habilitationsvortrag* we do not find the slightest indication of the notion of set. This is the more noteworthy when we consider that his definition of the reals by means of cuts dates back to 1858 [Dedekind 1872, 315], and that the notion of set is used time and again in his algebraic work of 1856–58. All of this supports the view that Riemann's ideas were instrumental in convincing Dedekind of the usefulness of the notion of set. Recall that he attended and followed carefully the former's courses on function theory in 1855/56, which he always regard as a model for his own research.[2]

1.2. Dedekind's set-theoretical approach. The abstract orientation of Dedekind's mathematics and his preference for a set-structural approach are particularly clear in his exposition of group theory, both in 'Eine Vorlesung über Algebra' [Scharlau 1981, 60–70] and 'Aus den Gruppen-Studien, 1855–58' [Dedekind 1930/32, vol. 3, 439–45]. Dedekind himself stressed that, in his 1856–58 lectures, he had presented group theory "in such a way that it could be applied to groups Π of arbitrary elements π" [Dedekind 1894, 484 note]. In fact his original explanations, as registered in the manuscript edited by Scharlau, are noteworthy because they essentially constitute an axiomatization of group theory. After proving two theorems about the product of "substitutions," which establish the associative law and the law of simplification (from any two of the equations: $\phi=\theta$, $\phi'=\theta'$, $\phi\phi'=\theta\theta'$, the third follows), he wrote:

---

[1] [Dedekind 1854, 430]: "Dieses Drehen und Wenden der Definitionen, den aufgefundenen Gesetzen oder Wahrheiten zuliebe, in denen sie eine Rolle spielen, bildet die grösste Kunst des Systematikers."

[2] It was by trying to understand Dedekind's path to the notion of set, in the period 1854–58, that I started to ponder Riemann's important role in the early history of sets.

The following investigations are exclusively based on the two fundamental theorems which we have proved, and on the fact that the number of substitutions is finite: therefore, their results will be equally valid for *any domain* of a finite number of *elements, things, concepts* θ, θ', θ"..., which from θ, θ' admit a composition θθ', defined arbitrarily but in such a way that θθ' is again a member of that domain, and that this kind of composition obeys the laws expressed in both fundamental theorems. In many parts of mathematics, but especially in number theory and algebra, we continuously find examples of this theory; the same methods of proof are valid here as there.[1]

In fact, those two laws suffice to insure the existence of neutral and inverse elements when we require, with Dedekind, the group to be finite.[2] Some pages later, Dedekind applied the above remark to the particular case given by the law of composition induced in the partition of a group by a normal subgroup, since this satisfies the two laws in question [Scharlau 1981, 68].

The standpoint taken here by Dedekind is noteworthy for its abstractness, being only comparable in the 1850s to that of Cayley; the mathematical community would only adopt a similar viewpoint in the 1890s [Wussing 1969]. Nevertheless, one has to emphasize that Dedekind was not advancing the axiomatic or the abstract structural standpoint. Historians tend to think that his standpoint regarding groups was peculiar precisely because he was dealing with groups, not with more familiar mathematical objects; as we shall see, he never dealt with fields in an analogous way. Groups were a tool for the investigation of traditional objects, and the limited familiarity with them seems to have motivated his abstract axiomatic approach [Corry 1996, 77–80; Haubrich 1999, ch. 6].

At any rate, what is particularly interesting for us is that Dedekind consciously employs the language of sets: the word "domain" in the passage above refers to sets, while in some other paragraphs of the manuscript he used "complex" with apparently the same meaning (see also [Dedekind 1930/32, vol. 3, 440–45]). Moreover, in working with composition laws on the classes which form the partition of a group he was employing sets as concrete objects, submitting them to operations that are analogous to the traditional ones. One should note that Dedekind's treatment of groups was formulated after having come under the

---

[1] [Scharlau 1981, 63; emphasis added]: "Die nun folgenden Untersuchungen beruhen lediglich auf den beiden so eben bewiesenen Fundamentalsätzen und darauf, dass die Anzahl der Substitutionen eine endliche ist: Die Resultate derselben werden deshalb genau ebenso für ein Gebiet von einer endlichen Anzahl von Elementen, Dingen, Begriffen θ, θ', θ"... gelten, die eine irgendwie definirte Composition θθ' aus θ, θ' zulassen, in der Weise, dass θθ' wieder ein Glied dieses Gebietes ist, und dass diese Art der Composition den Gesetzen gehorcht, welche in den beiden Fundamentalsätzen ausgesprochen sind. In vielen Theilen der Mathematik, namentlich aber in der Zahlentheorie und Algebra finden sich fortwährend Beispiele zu dieser Theorie; dieselben Methoden der Beweise gelten hier wie dort."

[2] Modern axiomatization of groups (and fields) began with [Weber 1893], a paper strongly influenced by Dedekind, and written after Lie's work made new axioms for infinite groups necessary [Wussing 1969, 223–251].

influence of Riemann's function theory, and, most likely, of his abstract conception of manifolds.[1] Significantly, many years later Dedekind would recommend to Cantor the word "domain" [Gebiet] as a substitute for the clumsy term 'Mannig-faltigkeit' [manifold], a substitute that is "also Riemannian" [Cantor & Dedekind 1937, 47]. Riemann actually uses the word in the lecture on geometry (see, e.g., [Riemann 1854, 274, 275]). It is likely that a direct influence is acting here.

The reader may suspect that Dedekind's use of set language might have been incidental, or perhaps restricted to finite sets as above. Quite the contrary. The manuscript on Galois theory reveals a clear awareness of the role played by num-ber-fields, conceived as infinite sets (see §3). And other work of the period, par-ticularly a paper on higher congruences, written in 1856 and published the year after, makes the point indisputable. The reader need not understand the precise content of that paper in order to grasp the importance of statements like the fol-lowing for the emergence of the language of sets:

The preceding theorems correspond exactly to those of number divisibility, in the sense that the whole system of infinitely many functions of a variable, congruent to each other modulo p, behaves here as a single concrete number in number theory, for each function of that system substitutes completely for any other in any respect; such a function is the representa-tive of the whole class; each class possesses its definite degree, its divisors, etc., and all those traits correspond in the same manner to each particular member of the class. The system of infinitely many incongruent classes – infinitely many, since the degree may grow indefi-nitely – corresponds to the series of whole numbers in number theory. To number congru-ence corresponds here the congruence of classes of functions with respect to a double modulus.[2]

Dedekind employs here the words 'System' and 'Klasse'; the first had also been used by Riemann in his lecture [e.g., Riemann 1854, 275, 279]. According to Dedekind himself [Dirichlet 1894, 36], the second had been used for the first time in the sense of equivalence class by Gauss in his theory of the composition of quad-ratic forms. Riemann [1857, 101, 119] had also used the notion of "class of alge-

---

[1] Riemann's definition ensured that manifolds could be formed by any elements whatsoever, see §II.5.

[2] [Dedekind 1930/32, vol.1, 46–47]: "Die vorhergehenden Sätze entsprechen vollständig denen über die Teilbarkeit der Zahlen in der Weise, dass das ganze System der unendlich vielen einander nach dem Modulus p kongruenten Funktionen einer Variabeln sich hier verhält, wie eine einzige bestimmte Zahl in der Zahlentheorie, indem jede einzelne Funktion eines solchen Systems jede beliebige andere desselben Systems in jeder Beziehung vollständig ersetzt; eine solche Funktion ist der Repräsentant der ganzen Klasse; jede Klasse hat ihren bestimmten Grad, ihre bestimmten Divi-soren usw., und alle diese Merkmale kommen jedem einzelnen Gliede einer Klasse in derselben Weise zu. Das System der unendlich vielen inkongruenten Klassen – unendlich vielen, da der Grad unbegrenzt wachsen kann – entspricht der Reihe der ganzen Zahlen in der Zahlentheorie. Der Kon-gruenz der Zahlen entspricht hier Kongruenz von Funktionenklassen nach einem doppelten Modulus."

braic functions" in his function theory.[1] Dedekind presents very clearly the notion of equivalence class and the role played by representatives; equivalence classes are submitted to operations that he regards as perfect analogues of ordinary arithmetical operations. His emphasis on the fact that all functions in the class have all their characteristic traits in common is noteworthy in the light of traditional logic: such traits define the "intension" [Inhalt] of a concept, while the class constitutes its "extension" [Umfang].

While speaking about classes of quadratic forms, Gauss had been careful to express himself in such a way that the implication of the existence of actual infinities was avoided. But Dedekind (and Riemann) had no such philosophical prejudices. Dedekind chose rather to emphasize the fact that we are dealing with infinitely many classes, *each of which contains infinitely many elements* (functions). And he went so far as to trace an analogy between these infinite classes and the natural numbers, which were most concrete objects for a traditional mathematician. That makes plainly clear that he had no philosophical objection to make against actual infinity; quite the contrary: he regarded infinite classes as natural objects for a mathematician. With the exception of Bolzano [1851], no other mathematician would have gone so far in the 1850s. It thus seems that Riemann and Dedekind are the most significant early representatives of the introduction of the language of sets in mathematical research.

1.3. The notion of mapping. Dedekind's set theory, as developed in the 1870s and 80s, is not just a theory of sets. This may sound paradoxical, but the reason is simply that Dedekind employed the notion of mapping as a primitive idea and developed carefully a theory of mappings (see §VII.2). The surprising fact is that this notion, too, emerged in his work of the 1850s. It seems quite likely that his reading of geometrical work by Möbius and Steiner prepared him to conceive of mappings in a general way (§I.3), but we find the notion clearly stated in the manuscript on Galois theory edited by Scharlau. This is obscured by the fact that Dedekind used a rather strange name for maps, namely "substitutions," but there is substantial evidence that avails a set-theoretical interpretation of this notion. The manuscript begins as follows:

Article 1
Definition. By a *substitution* one understands, in general, any process by which certain elements *a, b, c,* ... are transformed into others *a', b', c',* ..., or are replaced by these; in what follows we shall consider only those substitutions in which the complex of replacing elements *a', b', c'* is identical with that of the replaced *a, b, c.*[2]

---

[1] This corresponds to the "fields of algebraic functions" of Dedekind and Weber in their paper of 1882; see [Dedekind 1930/32, vol. 1, 239].

[2] [Scharlau 1981, 60]: "*Erklärung.* Unter *Substitution* versteht man im Allgemeinen jeden Process, durch welchen gewisse Elemente *a, b, c,* ... in andere *a', b', c',* ... übergehen oder durch diese ersetzt werden; wir betrachten im Folgenden nur die Substitutionen, bei welchen der Complex der ersetzenden Elemente *a', b', c'* mit dem der ersetzten *a, b, c* identisch ist."

The general meaning of "substitution," in the above definition, is that of mapping. Of course, one might doubt whether Dedekind had in mind only one-to-one maps, but there is also evidence that he dealt with non-injective mappings at the time.

'Aus den Gruppen-Studien' includes a section on "Equivalence of groups," which contains the homomorphism theorem.[1] Dedekind assumes a given correspondence between the objects of a group M and those of a "complex" M_1, such that the image of a product of elements in M is the product of their images. He shows that M_1 is then a group, considers what we call the kernel N of the homomorphism, and shows that the partition M/N gives rise to an isomorphism [Äquivalenz] between the quotient group M/N and the image M_1. He also indicates that the "equivalence" so defined is transitive [Dedekind 1930/32, vol. 3, 440–41]. Dedekind shows a very clear awareness that several elements in M may correspond to the same object in M_1, and therefore that a homomorphism is not an injective mapping. All of this is further confirmation that by 1858 at most he was employing the notion of mapping.

A text written in 1879 is further evidence that his "substitutions" of the 1850s are simply what we now call maps. The text was not written with the purpose of establishing any kind of priority, it simply attempted to clarify notions introduced in the second version (1879) of his ideal theory, in the light of his (unpublished) work on set theory. It is also interesting because it includes the first public announcement of the 1888 booklet, that will occupy us in chapter VII:

It happens very frequently, in mathematics and other sciences, that when we find a system Ω of things or elements ω, each definite element ω is replaced by a definite element ω' which is made to correspond to it according to a certain law; we use to call such an act a substitution, and we say that by means of this substitution the element ω is transformed into the element ω', and also the system Ω is transformed into the system Ω' of the elements ω'. Terminology becomes somewhat more convenient if, as we shall do, one conceives of that substitution as a mapping of the system Ω, and accordingly one calls ω' the image of ω, and also Ω' the image of Ω. [Note:] Upon this mental faculty of comparing a thing ω with a thing ω', or relating ω with ω', or making ω' correspond to ω, without which it is not at all possible to think, rests also the entire science of numbers, as I shall try to show elsewhere.[2]

---

[1] This fragment was, most likely, written before the manuscript on Galois theory, since only at the end it introduces the word "substitution." It begins talking simply about "objects" that "correspond to" [entspricht] the "objects" of a given group.

[2] [Dedekind 1879, 470]: "Es geschieht in der Mathematik und in anderen Wissenschaften sehr häufig, dass, wenn ein System Ω von Dingen oder Elementen ω vorliegt, jedes bestimmte Element ω nach einem gewissen Gesetze durch ein bestimmtes, ihm entsprechendes Element ω' ersetzt wird; einen solchen Act pflegt man eine Substitution zu nennen, und man sagt, dass durch diese Substitution das Element ω in das Element ω', und ebenso das System Ω in das System Ω' der Elemente ω' übergeht. Die Ausdrucksweise gestaltet sich noch etwas bequemer, wenn man, was wir thun wollen, diese Substitution wie eine Abbildung des Systems Ω auffasst und demgemäss ω' das Bild von ω, ebenso Ω' das Bild von Ω nennt." "Auf dieser Fähigkeit des Geistes, ein Ding ω mit einem Ding ω' zu vergleichen, oder ω auf ω' zu beziehen, oder dem ω ein ω' entsprechen zu lassen, ohne welche ein Denken überhaupt nicht möglich ist, beruht, wie ich an einem anderen Orte nachzuweisen versuchen werde, auch die gesammte Wissenschaft der Zahlen."

With his remarks on the importance of this notion for "other sciences" as well as its relation to basic mental faculties, Dedekind is already here suggesting his logicistic standpoint. The word 'Abbildung' can be found in Riemann's function theory [1851, 5–6], and in a paper of 1825 by Gauss, but in the specifically geometrical meaning of conformal mapping. The word was later employed in general for functional relationships; according to Cantor, this usage seems to have started with Clebsch (see [Grattan-Guinness 1970, 87]). Dedekind followed these precedents, and from 1872 he systematically employed that word in the meaning of mapping ('Abbildung' has this meaning in German still today). It would be tempting to translate that word it by "representation," which captures some connotations of the German word and makes it understandable that Dedekind could consider it as a basic logical operation.

2. A New Fundamental Notion for Algebra: Fields

Set and mapping would become the central notions for Dedekind's understanding of arithmetic, algebraic number theory, algebra, and also, one may safely conjecture, analysis. But in the context of algebra and algebraic number theory, the focus would be on sets (mostly number-sets) with a given structure, and on structure-preserving mappings.[1]

We have already mentioned how Dedekind, while working on Galois theory, arrived at a novel understanding of the role played by fields in that theory. In 1871 he reminisced that, in the course of his lectures of 1856/58, he had become convinced that "the study of the algebraical relationship among numbers" was most conveniently based upon "a concept that is immediately related to the simplest arithmetical principles."[2] Correspondingly, in the manuscript Dedekind employs the name "rational domain,"[3] and the notation S for the ground field of the polynomial under study. The 'rational domain' comprises all numbers that are "rational functions" of the roots of the polynomial, or that are "rationally representable" by means of those roots [Scharlau 1981, 84, 89]. As Edwards [1980, 343] has remarked, Dedekind's idea of employing a single capital letter S, or K later, instead of a notation like $\mathbb{Q}(\alpha)$ that suggests a particular basis, – or form of representation, in his terminology – characterizes his philosophy of mathematics.

---

[1] The history of the notion of field has been analyzed by Purkert [1973], but the reader should compare his exposition with [Haubrich 1999, chap. 6], on which I rely to some extent.

[2] [Dedekind 1930/32, vol. 3, 400]: "das Studium der algebraischen Verwandtschaft der Zahlen am zweckmässigsten auf einen Begriff gegründet wird, welcher unmittelbar an die einfachsten arithmetischen Prinzipien anknüpt."

[3] See, e.g., [Scharlau 1981, 83]: "rationales Gebiet."

In the late 1860s, Dedekind came to employ the name 'Körper' [field, literally 'body'] for his former "rational domains," and he made explicit their connection with the "simplest arithmetical principles": fields "reproduce" through the four basic operations, the traditional four "species" of elementary arithmetic [Dedekind 1930/32, vol. 1, 239, 242; vol. 3, 409]. He chose the name 'Körper' because a number field constitutes a system possessing a certain completeness and closeness, an "organic totality" or a "natural unity," analogous to those entities we call bodies in natural science, in geometry, and in the life of human society [Dedekind 1894, 452 note]. As is well known, continental European languages follow Dedekind's choice, while 'Körper' translates into the English term 'field.'

It must have been by the late 1850s when Dedekind became convinced that the notion of field, together with operations on fields and "substitutions" or field homomorphisms, lead on the one hand to Galois theory, and on the other to ideal theory [Dedekind 1930/32, vol. 3, 401]. Actually he went on to identify algebra with field theory [Haubrich 1999, ch. 6]. In 1873, he tentatively proposed to define "algebra proper" as "the science of the relationships among fields," explaining that relations among equations can be translated into relations among fields.[1] The same conception can be found again in his last version of ideal theory [Dedekind 1894, 466, 482], where it becomes clear that he is equating algebra with the field- and group-theoretical content of Galois theory. We find here an interesting parallelism with Riemann: the fundamental role that fields play in algebra, according to Dedekind, is similar to that of abstractly defined analytic functions in Riemann's function theory, or of his manifolds in the theory of magnitudes (including topology) and in geometry. The author himself points to this connection, without naming Riemann, in the preface of 1871 [Dedekind 1930/32, vol. 3, 396–97]. Guided by his conviction, shared with Riemann, that any branch of mathematics should be based upon one fundamental concept, Dedekind came to propose a definition of algebra that is too restrictive, both judged from contemporary and from 20th-century standards.

Dedekind's first public presentation of the notion of field can be found right at the beginning of the 1871 exposition of ideal theory. This was contained in the tenth supplement to Dirichlet's *Vorlesungen*, 'On the composition of binary quadratic forms.'[2] This supplement begins with an exposition of work by Gauss and Dirichlet, to which Dedekind made some original contributions, and then goes on to an extreme generalization of it. He writes that it will be convenient to adopt a higher standpoint and introduce a notion which seems very appropriate to serve as a foundation for higher algebra and those parts of number theory connected with it:

---

[1] [Dedekind 1930/32, vol. 3, 409]: "... gegenwärtigen Stande der Algebra ... um in einem Nichtkenner wenigstens eine dunkle Vorstellung von ihrem Charakter zu erwecken, vielleicht als ... die Wissenschaft von der Verwandtschaft der Körper bezeichnen könnte." He is referring to the "Entwicklung der eigentlichen Algebra" in recent times, through the ideas of Abel and Galois, as he indicates below.

[2] The reasons for this title, and a brief description of the subject matter, can be found in §3.1.

By a *field* we shall understand every set of infinitely many real or complex numbers, which is so closed and complete in itself, that addition, substraction, multiplication, and division of any two of those numbers yields always a number of the same set. The simplest field is constituted by all rational numbers, the greatest field by all [complex] numbers. We call a field A *divisor* of field M, and this a *multiple* of that, if all the numbers contained in A are also found in M; it is easily seen that the field of rational numbers is a divisor of all other fields. The collection of all numbers simultaneously contained in two fields A, B constitutes again a field D, which may be called the *greatest* common divisor of both fields A, B, for it is evident that any divisor common to A and B is necessarily a divisor of D; similarly, there always exists a field M which may be called the *least* common multiple of A and B, for it is a divisor of all other common multiples of both fields. Moreover, if to any number a in the field A, there corresponds a number $b = \phi(a)$, in such a way that $\phi(a+a') = \phi(a) + \phi(a')$, and $\phi(aa') = \phi(a)\phi(a')$, the numbers b constitute also (if not all of them are zero) a field $B = \phi(A)$, which is *conjugate* to A and results from A through the *substitution* ϕ; inversely, in this case $A = \psi(B)$ is also a conjugate of B. Two fields conjugate to a third are also conjugates of each other, and every field is a conjugate of itself.[1]

As Dugac [1976, 29] emphasized, one can hardly overestimate the significance of this rich passage for the history of sets and set-language. It incorporates all crucial ideas related to the notions of set and map as used in algebra, and Dedekind gives abundant proof of the mastery he had attained of the set-theoretical viewpoint by 1871. There is little doubt that contemporary readers must have found it difficult to follow, accustomed as they were to an algebra and "higher" number theory formulated in terms of numbers and equations or forms (compare §§I.4 and I.5). They were completely foreign to such a strong reliance on the notion of set in this domain, and for this very reason Dedekind's exposition had to attract strongly the attention of readers to his new standpoint – whether to accept it or refuse it.

The terminology employed by Dedekind may seem strange, and so it will be worthwhile to devote a couple of paragraphs to commenting on the text. It is clear

---

[1] [Dedekind 1871, 223–224]: "Unter einem *Körper* wollen wir jedes System von unendlich vielen reellen oder komplexen Zahlen verstehen, welches in sich so abgeschlossen und vollständig ist, dass die Addition, Substraktion, Multiplikation und Division von je zwei dieser Zahlen immer wieder eine Zahl desselben Systems hervorbringt. Der einfachste Körper wird durch alle rationalen, der grösste Körper durch alle Zahlen gebildet. Wir nennen einen Körper A einen *Divisor* des Körpers M, diesen ein *Multiplum* von jenem, wenn alle in A enthaltenen Zahlen sich auch in M vorfinden; man findet leicht, dass der Körper der rationalen Zahlen ein Divisor von jedem anderen Körper ist. Der Inbegriff aller Zahlen, welche gleichzeitig in zwei Körpern A, B enthalten sind, bildet wieder einen Körper D, welcher der *grösste* gemeinschaftliche Divisor der beiden Körper A, B genannt werden kann, weil offenbar jeder gemeinschaftliche Divisor von A und B notwendig ein Divisor von D ist; ebenso existiert immer ein Körper M, welcher das *kleinste* gemeinschaftliche Multiplum von A und B heissen soll, weil er ein Divisor von jedem andern gemeinschaftlichen Multiplum der beiden Körper ist. Entspricht ferner einer jeden Zahl a des Körpers A eine Zahl $b = \phi(a)$ in der Weise, dass $\phi(a+a') = \phi(a) + \phi(a')$, und $\phi(aa') = \phi(a)\phi(a')$ ist, so bilden die Zahlen b (falls sie nicht sämtlich verschwinden) ebenfalls ein Körper $B = \phi(A)$, welcher mit A *konjugiert* ist und durch die *Substitution* ϕ aus A hervorgeht; dann ist rückwärts auch $A = \psi(B)$ mit B konjugiert. Zwei mit einem dritten konjugierte Körper sind auch miteinander konjugiert, und jeder Körper ist mit sich selbst konjugiert."

from his definitions that "divisor" and "multiple" refer to the two sides of an inclusion, "greatest common divisor" denotes the intersection of two fields, and "least common multiple" the union, in the sense of the smallest field containing two given fields. Terminology for fields was analogous to that employed for modules and ideals,[1] and here the reason for selecting those expressions was the intention to preserve an exact parallelism between the wording of elementary number-theoretical theorems, and their analogues in algebraic number theory. In the case of $\mathbb{Z}$, inclusion of principal ideals corresponds to the divisibility of their generators, and this led Dedekind to the analogy between inclusion and division that is employed throughout his work on ideal theory and algebra. (Dedekind was very much aware of the arbitrariness of mathematical terminology and of the need to state explicitly all relevant properties of the notions involved.[2]) His terminology kept being employed until the 1930s: in 1935 Krull proposed to modify it, arguing that its motivation was historically clear, but it had become no longer tenable since it conflicted with the more basic set-theoretical language (see [Corry 1996, 230]).

Dedekind's treatment of "substitutions" is equally a model. The fact that he is considering non-injective morphisms is indicated by his comment that not all of the images should be zero, i.e., that the trivial case of a unitary field is discarded. We have seen that homomorphisms were already present in his group-theoretical work of the 1850s. The precision and conciseness with which he sets forth the reflexive, symmetric and transitive properties of field "conjugation" are likewise noteworthy. In later versions of his ideal theory, Dedekind emphasized the differences between the general notion of mapping, or "substitution," and the algebraic notion of structure-preserving maps (see [Dedekind 1879, 470–71; 1894, 456–57]). In the 1894 edition, he used the word "permutation" in the sense of a field homomorphism [Dedekind 1894, 457]. The novelty and difficulty of his emphasis on morphisms can be judged from Frobenius's comment to Weber, immediately after this publication, that "his permutations are too incorporeal, and it is certainly unnecessary to take the abstraction so far."[3]

A little known fact is that Dedekind's terminology for algebraic operations was taken up by Cantor in his decisive series of papers on point-sets of the years 1879 to 1884. Cantor used the terminology of divisor and multiple to denote the set-theoretical operations of inclusion, intersection and union, thus revealing that he, for one, understood very well the set-theoretical underpinnings of Dedekind's ideal theory. The terminology is introduced in the second paper of the series [Cantor 1879/84, 145–46] and the latest instances are found in the sixth and last part [*op.cit.*, e.g. 214, 226, 228]. This is striking because that terminology seems rather inappropriate in a general set-theoretical setting – Dedekind himself replaced it

---

[1] Although for fields A is a divisor of B means $A \subset B$, while for ideals it means $B \subset A$.

[2] See his letter to Lipschitz of 1876, quoted in §IV.1, also [Dedekind 1888, 360, 377–78] and the 1890 letter to Keferstein [Sinaceur 1974, 274].

[3] [Dugac 1976, 269]: "Seine neuste Auflage enthält so viel Schönheiten, ... aber seine Permutationen sind zu körperlos, und es ist doch auch unnöthig, die Abstraktion so weit zu treiben."

when doing general set theory, starting in 1872 (see §6) – , and above all because it constitutes evidence that Cantor paid careful attention to the set-theoretical aspects of Dedekind's work in the 1870s.

The set-theoretical basis of Dedekind's fields parallels the set-theoretical underpinnings of Riemann's manifolds, and Riemann's avoidance of external "forms of representation" (see above and §I.4) has its parallels here too. In a paper of 1877, Dedekind defined a "finite field" [finite extension of $\mathbb{Q}$] to be the set of all numbers of the form:

$$\phi(\theta) = x_0 + x_1\theta + x_2\theta^2 + \ldots + x_{n-1}\theta^{n-1}$$

where $x_i \in \mathbb{Q}$; this is similar to the modern notation $\mathbb{Q}(\alpha)$ for finite extensions of $\mathbb{Q}$. Concerning this definition, Dedekind wrote to Lipschitz that it is "spoiled" [verunziert], since it relies on a form of representation that is somewhat arbitrary – θ could be replaced by many other numbers, and the definition presupposes that such changes leave the field invariant [Dedekind 1930/32, vol. 3, 468–69]. Hence, he thought, one should prefer for reasons of principle the definition employed in 1871: "a finite field is one that only possesses a finite number of divisors [subfields]."[1]

Although we have only given the example of fields as evidence, the whole exposition of Dedekind's ideal theory in 1871 (and even more so in later editions) called attention to the importance of sets and set-structures in algebraic number theory. The very standpoint adopted required a constant exercise in the translation of problems formulated for numbers to the new and more abstract set formulation. One should finally mention that inclusion, union, intersection and mapping are all the basic notions presented by Dedekind in his set theory [1888], which makes even clearer its algebraic origins. When he began writing the first draft for his 1888 book, in the year 1872, Dedekind presented the same notions he had employed the year before in his ideal theory, with the only difference that now they were freed from the natural algebraic restrictions (see §6).

3. The Emergence of Algebraic Number Theory

The theory of algebraic integers, which emerged in the period 1830–1871, gives a perfect example of the typical 19th-century orientation toward pure mathematics in Germany. The problems studied here had no practical application, they were pursued purely for knowledge's sake. It has frequently been said that, with Dedekind's publication of 1871, the theory at once reached its maturity, leaving far behind all previous attempts and sketches (see, e.g., [Bourbaki 1994, 101]). Dedekind introduced the notions of field, module, and ideal in the 1871 version of ideal theory,

---

[1] [Dedekind 1930/32, vol. 3, 468–69]: "ein endlicher Körper ist ein solcher, der nur eine endliche Anzahl von Divisoren besitzt".

and the notion of a ring (under the name 'Ordnung') in the 1879 version; these, together with Galois's groups, would constitute the core of modern algebra.[1] But Dedekind seems to have done more than simply elaborating the adequate solution for a noteworthy problem, since there are reasons to say that he created a new discipline. On the other hand, the story of the names that he introduced in mathematical research – "field," "module," "ideal" – is more complex than might appear.[2]

3.1. Ideal factors. The theory of divisibility of natural numbers, or of the integers, builds upon a well-known fundamental theorem: any natural number (integer) has a unique decomposition into a product of prime numbers. Building stones for the proof of this result, though not the theorem itself, can already be found in book seven of Euclid's *Elements*, but the first detailed proof was given by Gauss in his *Disquisitiones arithmeticae* [1801]. Up to this time, number theory was conceived as concerned only with the integers, although occasionally, in the work of Euler and others, complex numbers appeared as tools for the calculation of number-theoretical results. In 1832, Gauss was led to a ground-breaking innovation: he proposed an "extension" of the "field of arithmetic" to the imaginary numbers,[3] and studied the divisibility properties of the Gaussian integers $a+bi$, with $a, b \in \mathbb{Z}$. In this new domain, he defined the notions of a unit, prime number, etc., and was able to prove that an analogue of the fundamental theorem of elementary number theory was in effect.

The motivation behind Gauss's bold step was the proof of a certain result concerning biquadratic residues, a topic related to the study of so-called "higher reciprocity laws." These are laws for the relationship between the solvability of two congruences $x^\lambda \equiv p$ (mod q) and $x^\lambda \equiv q$ (mod p); in case $\lambda=2$ we have quadratic reciprocity, if $\lambda=3$ cubic reciprocity, if $\lambda=4$ biquadratic reciprocity, and so on. Gauss himself established the law of quadratic reciprocity in the *Disquisitiones*, publishing up to six different proofs during his life. In trying to prove higher reciprocity laws, Gauss recognized that their formulation and proof required the use of certain complex numbers. Actually, in 1832 he stated that the Gaussian integers were indispensable for biquadratic reciprocity, and he suggested that other laws would require the introduction of other kinds of complex numbers. This is what required an expansion of the field of arithmetic.

It should be mentioned that reciprocity laws were one of the main topics in so-called "higher" number theory during the 19th-century, and, according to Kummer in 1850, the most interesting open problem [Haubrich 1999, ch. 1]. Also important,

---

[1] The notion of *Ordnung* [order; Dedekind 1930/32, vol. 3, 305] denotes rings of algebraic integers, while the ring of all integers, characterized by its integral closure, was called *Hauptordnung* [main order]. Actually it was first used by Dedekind in 1877, while "ring" was first used by Hilbert [1897].

[2] As regards the emergence of algebraic number theory, I follow the reconstruction given by Haubrich [1999, chap. 4 and 7]. On the second topic, see [Corry 1996] and §6.

[3] [Gauss 1863/1929, vol. 2, 102]: "ita theoremata ... resplendent, quando campus arithmeticae ad quantitates *imaginarias* extenditur."

and actually more central as seen by most mathematicians, was the study of so-called "forms," for instance the binary quadratic forms

$$ax^2 + bxy + cy^2, \qquad a, b, c \in \mathbb{Z},$$

that were carefully studied by Gauss too. Lagrange had introduced the notion of "equivalent" forms; Gauss studied the classes of equivalent forms and defined a "composition" of forms, showing that (in modern language) form-classes with that composition constitute an Abelian group. An exposition of these results was the context in which Dedekind presented his first version of ideal theory in 1871. The theory of congruences, including reciprocity laws, and the theory of forms were the two main branches of "higher" number theory around the mid-century (actually, from 1801 until almost the end of the century; see [Haubrich 1999]).

After Gauss's paper of 1832, some of the most important German mathematicians, including Jacobi, Kummer and Eisenstein, tried to prove higher reciprocity laws. The next decisive step would be given by Kummer, who was eventually able (in 1861) to prove the λth reciprocity law for all regular prime λ. He began working on the cubic case in the 1840s, but in 1844 he found a difficult obstacle: cubic reciprocity required employing the algebraic integers of $\mathbb{Z}[\alpha]$, where α is a root of unity ($\alpha^\mu = 1$), but Kummer found that, in general, decomposition of these complex numbers into prime factors ceased to be unique. To give a simple example, the numbers 2, 3, $1+\sqrt{-5}$ and $1-\sqrt{-5}$ belong to $\mathbb{Z}[\sqrt{-5}]$, and they are not further decomposable into factors in this set. But they do not behave as primes: e.g., we have two different decompositions $6=2\cdot3=(1+\sqrt{-5})\cdot(1-\sqrt{-5})$. Kummer's brilliant idea was to postulate the existence of "ideal" prime numbers such that unique decomposition was reestablished. In the previous case, we need three ideal primes p, q_1, q_2, with $2=p^2$, $3=q_1\cdot q_2$, etc.

Kummer expressed himself in a rather disconcerting way regarding those ideal numbers. He symbolized them by means of the same symbols employed for the given or "real" numbers of $\mathbb{Z}[\alpha]$, and assumed that they satisfied exactly the same formal laws; hence, they were taken to be numbers complying with certain well-determined conditions, "except for [their] existence" [quoted in Edwards 1980, 342]. His way of confronting this issue can be charged with having originated a certain lack of rigor in his presentation, and even some important gaps in his proofs [Haubrich 1999, ch. 2]. But with him we find the seminal idea that would eventually lead to a new theory of ideal numbers, attempting to explain the factorization properties of algebraic numbers in general, and to establish an analogue of the fundamental theorem of elementary number theory in that general case.

The process that led to this new theory of algebraic numbers was in no way straightforward or continuous (see [Haubrich 1999, ch. 4 and 7]). For Kummer and his successors, ideal numbers were simply a tool that was necessary for the proof of higher reciprocity laws. Number theory was conceived in the traditional way, complex numbers were still regarded with some distrust, and there was a very limited understanding of the arithmetic of algebraic numbers. The focus was on reciprocity laws, binary forms, etc., not on the auxiliary means of ideal numbers, nor on algebraic numbers. It would be Dedekind and Kronecker, some years later, who introduced two crucial novelties: they expanded number theory, as Gauss suggested, to all kinds of algebraic numbers; and they also felt concerned with the foundational problems raised by ideal numbers, trying to give that general theory a completely satisfactory foundation.

By 1847, Kummer had solved the problem of the factorization of the cyclotomic algebraic integers, which in modern terms are the integers of cyclotomic fields $\mathbb{Q}(\alpha)$, with $\alpha^\mu = 1$. At this point, only Kronecker and Dedekind undertook the project of developing a general theory of algebraic integers. There is evidence that Kronecker had obtained a general theory by 1858, but he refrained from publishing even an indication of his methods until 1882 (see [Edwards 1980, 329–30]). As regards Dedekind, he strove with that project throughout a period of 14 years – though with some interruptions – until he finally found a satisfactory generalization in 1870. The theory of Kronecker embraces also the more general case of fields of algebraic functions, studied by Dedekind and Weber in their [1882] (see §6.2).

3.2. Algebraic integers. With Dedekind, we encounter a new discipline that has algebraic numbers, in particular algebraic integers, as its subject matter; algebraic numbers and sets of algebraic numbers (fields, rings) are at the focus of attention. This involved quite a radical departure from tradition, which emerged with the publication of Dedekind's ideal theory [1871]. Nineteenth-century "higher" number theory was put upside down: the theory of algebraic integers, a generalization of Kummer's contribution, became the foundation and core, while the topic of binary and other forms became secondary, and reciprocity laws were not even mentioned by Dedekind.[1] The reason why the new step was left to Kronecker and Dedekind seems to be that they had in common a number of motivations, including methodological orientations and a certain vision of the relations between algebra and number theory.[2] One may conjecture that Dirichlet fostered that orientation in both of them, since he partially followed Gauss's calling for an expansion of the field of arithmetic, and he always placed special emphasis on the number-theoretical underpinnings of the theory of forms [Haubrich 1999, ch. 1].

One serious difficulty in generalizing Kummer's theory was the definition of algebraic integer, that is, of the very objects one should investigate.[3] Kummer and his predecessors defined algebraic integers in a *formal* way, as numbers built in a certain way. To begin with, they did not talk about algebraic numbers, but used

---

[1] This is partly due to the fact that Dedekind deals mainly with the foundations of algebraic number theory; in other work, for instance on cubic fields (1900), Dedekind studies reciprocity laws (I thank Ralf Haubrich for this remark).

[2] See [Haubrich 1993; 1999]. Both worked on Galois theory in the 1850s, which led to a field-theoretical orientation in their number-theoretical work; both conceived of the new discipline as intimately related to algebra; both were deeply interested in foundational issues.

[3] Edwards [1980, 331, 337] went as far as to say that this was the only real difficulty, but it is clear that Dedekind was not at all of the same opinion; see [Haubrich 1999].

expressions like "complex numbers composed from third roots of unity,"[1] and similar ones, to denote their subject matter. In the case of cyclotomic integers, it was possible to show that the numbers in question were of the form

$$f(\alpha) = r_0 + r_1 \cdot \alpha + \dots + r_{\mu-2} \cdot \alpha^{\mu-2},$$

with $\alpha^\mu = 1$, and $r_i \in \mathbb{Z}$. Thus, Kummer described his theory, in his crucial papers of 1847 and 1851, as dealing with complex numbers formed from roots of unity and integer numbers [Kummer 1975, vol. 1, 165–92, 363–484].

In the case of cyclotomic fields, the integers happen to coincide with the numbers of the above form, and, according to Dedekind, this fortunate circumstance was one of the reasons for Kummer's success (see his letter in [Lipschitz 1986, 61]). But in an attempt to extend the theory to more general cases, the formal approach became extremely confusing. The obstacle could only be eliminated when number theorists began to think within the framework of algebraic number fields. Rather than going from the study of divisibility properties in $\mathbb{Z}$ to its formal expansion $\mathbb{Z}[\alpha]$, one had to think of extending $\mathbb{Q}$ to the field $\mathbb{Q}(\alpha)$, then determine the (ring of) integers in $\mathbb{Q}(\alpha)$, and finally study their factorization. To put it differently, the theory would only work if one called integers some *quotients* of numbers of the form $f(\alpha)$ above. The problem, then, was to define algebraic integers: what conditions should an algebraic number satisfy, in order to be called an integer?

Kronecker and Dedekind would adopt radically opposite views on almost all issues in algebraic number theory, but here both proposed the same solution. A number in $\mathbb{Q}(\alpha)$ is an integer just in case it is the root of a monic polynomial with coefficients in $\mathbb{Z}$ [Dedekind 1871, 236]. Neither Dedekind nor Kronecker gave reasons for their definition, and neither indicated the historical path that led to it. Nevertheless, it is not difficult to conjecture how it came about [Scharlau 1981; Haubrich 1999, ch. 7]. In the 1850s, both mathematicians were busy with Galois theory, and both came to consider the role played by fields in it, implicitly in the case of Kronecker, explicitly in Dedekind. Dedekind studied the work of Kummer after a rather detailed study of Galois, and it seems that he read his results from the standpoint of number fields right from the start. This must have suggested the idea of determining the set of integers starting from the field, and it seems that both mathematicians were thus easily led to the right definition of the integers [Haubrich 1999, ch. 7].

Around 1856 or 1857, Dedekind recognized that in general the subset of algebraic integers of a given number field is not of the form $\mathbb{Z}[\alpha]$. And, to believe his later statements, he found the correct definition of algebraic integer almost immediately, guided by the field-theoretical orientation of his thoughts. In the period 1858–62, Dedekind was at work on his first attempt to formulate a completely

---

[1] The title of an 1844 paper by Eisenstein is: "Beweis des Reciprocitätssatzes für die cubischen Reste in der Theorie der aus dritten Wurzeln der Einheit zusammengesetzten complexen Zahlen" (in *Journal für die reine und angewandte Mathematik* **27**, 289–310).

general theory of the factorization of algebraic integers. To this end, it was necessary to study the algebraic properties of the set of integers $\mathfrak{o}$ in a field. Some time around 1860, he was in the possession of a number of important results concerning algebraic integers: the sums, differences, and products of algebraic integers are integers (which implies that $\mathfrak{o}$ is a ring); the roots of a polynomial with coefficients in $\mathfrak{o}$ are likewise algebraic integers ($\mathfrak{o}$ is integrally closed); and $\mathfrak{o}$ has an integral basis $(\beta_1, \ldots \beta_n)$, such that $\omega = \Sigma b_i \cdot \beta_i$, with $b_i \in \mathbb{Z}$, for all $\omega \in \mathfrak{o}$ [Haubrich 1999, ch. 7]. In order to prove this last result, he probably had to introduce the notion of module and develop some basic results concerning it.[1] This completes our picture of Dedekind's state of development by 1860: he must have already mastered set-theoretical language as applied in algebra, since he developed basic results in the theory of fields, rings of integers [Hauptordnungen], and modules.

Having solved the problem of the definition of integers, it was possible to attempt the generalization of Kummer's theory in several ways. Kronecker looked for an approach that made possible the effective determination of ideal factors, in accordance with his constructivist tendencies [Edwards 1980 and 1990]. Dedekind, in accordance with his abstract-conceptual orientation, finally came to avoid the postulation of ideal factors and established a theory that deals with certain sets of algebraic numbers, which he called *ideals* to honor Kummer's seminal work. But in 1860 he was still far from this viewpoint.

4. Ideals and Methodology

In the autumn of 1858 Dedekind left Göttingen and accepted a position at the Zürich Polytechnic (the later ETH). During the period at Zürich he seems to have kept working – though perhaps somewhat sporadically – on algebraic number theory, but around 1862 he abandoned original research and concentrated on preparing Dirichlet's lectures on number theory for the print.[2] This opened a period in which he mainly devoted his time to the publication of the work of his mathematical masters – Gauss, Dirichlet, and Riemann. Some seven years went by before he made a second attempt to formulate a general theory of ideal factorization, while preparing a second edition of the *Vorlesungen* in 1869/70.

---

[1] A number set is a module when it is closed under addition and substraction [Dedekind 1871, 242]. An abstract conception of modules, including their relation to groups, can be found in [Dedekind 1877, 274].

[2] In a letter to the Göttingen physiologist Henle and his wife, he speaks of having renounced his earlier aspirations to fame: "eifrigem Arbeiten, nicht um berühmt zu werden – das habe ich aufgegeben – sondern um Dirichlet's Vorlesungen zum Druck fertig zu machen" [Haubrich 1999].

His first attempt, in the period 1856–62, was based on higher congruences and led him quite far away, although never to the desired goal. It followed Kummer rather closely, insofar as it still depended upon the postulation of ideal prime numbers and approached the determination of ideal factors by considering the irreducible polynomial $F(x)$ associated to every cyclotomic integer α. The ideal prime factors of any prime $p \in \mathbb{Z}$ are essentially determined by means of certain higher congruences having to do with $F(x)$ (see [Edwards 1980, 324–28; Haubrich 1999, ch. 2 and 9]). But around 1862 Dedekind still lacked a way of proving the factorization theorem that could satisfy him, by being appropriately general. It had become possible to approach the decomposition into ideal factors of algebraic integers of any kind, but Dedekind found that there are fields which contain prime numbers whose ideal prime factors cannot be studied by means of higher congruences.

It is unclear whether he might have been able to sidestep this problem, but the fact is that by 1862 Dedekind had abandoned the whole issue and, due to his involvement with the publication of Dirichlet's and Riemann's work (§II.7), he abandoned the problem for several years. In 1869 he began preparing a second edition of Dirichlet's *Vorlesungen*, and this motivated a new attempt to establish a general theory of ideal numbers. This time, after about a year of work, he became convinced that a new formulation of the kernel of the theory itself in set-theoretical language was the only way of finding a satisfactory way out. (Interestingly, it has been possible to establish that most technical details of the 1871 theory must have been already in his possession by 1860–62, see §6.1.) The reformulation made possible a completely general proof of the fundamental theorem of factorization.

4.1. Methodological demands.   A detailed study of Dedekind's motives for preferring his ideal theory yields two main conclusions.[1] First, it reinforces the analysis of common methodological traits of the "Göttingen group" that was offered in §I.4; and second, it reveals that Dedekind regarded the ideals of his algebraic number theory on a par with the elements of the number system, real numbers, etc. Here, as with the irrational numbers, the issue was "the introduction or creation of new arithmetical elements."[2] His theory of algebraic integers was thus guided by the abstract conceptual approach that he had in common with Riemann, and also by the preference for arithmetization that he shared with Dirichlet and led farther.

It will be convenient to start with Dedekind's reasons for abandoning his first approach employing higher congruences. He gives two main arguments, mentioning two imperfections that marred that approach. The first is simply the fundamental principle that particular forms of representation ought to be avoided in favor of abstract concepts, the second consists in the exceptions mentioned above:

---

[1] Such a detailed study was initiated by Mehrtens [1979a] and Edwards [1980, 1983]. See also [Haubrich 1999, chap. 5].

[2] [Dedekind 1877, 269]: "l'introduction ou la création de nouveaux éléments arithmétiques".

but although this investigation brought me quite close to the desired goal, I could not resolve to its publication, because the ... theory suffers mainly from two imperfections. One consists in that the investigation of a domain of algebraic integer numbers was based upon consideration of a certain number and of the corresponding equation, which is conceived as a congruence, and that the definitions of the ideal numbers (or better of divisibility by the ideal numbers) that are thus obtained do not immediately show the character of *invariance* that actually corresponds to those concepts, as a consequence of the particular form of representation selected; the second imperfection of this kind of foundation consists in that sometimes peculiar exceptions appear, which call for special treatment.[1]

By contrast, the notions of field, algebraic integer, and ideal are defined in an abstract way and do not require "forms of representation," immediately suggesting the required "invariance" of the objects defined; and these "extremely simple notions" make a completely general proof of the factorization theorem possible [Dedekind 1930/32, vol. 1, 203].

The second objection points to an important difficulty in the way to a completely general theory,[2] but it also has to do with a difference between Dedekind and Kummer. The latter was interested in calculating the prime factors of cyclotomic integers, and was not worried by the need to employ several different methods that may depend upon our knowledge of particular properties of the concrete kind of numbers under investigation. This difference was partly due to the fact that Kummer regarded his theory of ideal numbers as a mere tool for research on reciprocity laws, and so he was not interested in the tool itself, but in the applications. But, in fact, Kummer even thought that the differentiation of several cases was inevitable and positive, and criticized Dedekind's theory precisely for its abstract generality (see [Dedekind 1930/32, vol. 3, 481]; this is reminiscent of the differences between Dedekind and Kummer's disciple Kronecker, we find here the earliest tensions between constructive and abstract approaches in mathematics). From Dedekind's viewpoint, a theory was only completely satisfying when it attained the utmost generality, and this may have been the only reason why he did not publish the results of his first approach.

---

[1] [Dedekind 1930/32, vol. 1, 202]: "allein obgleich diese Untersuchungen mich dem erstrebten Ziele sehr nahe brachten, so konnte ich mich zu ihrer Veröffentlichung doch nicht entschliessen, weil die so entstandene Theorie hauptsächlich an zwei Unvollkommenheiten leidet. Die eine besteht darin, dass die Untersuchung eines Gebietes von ganzen algebraischen Zahlen sich zunächst auf die Betrachtung einer bestimmten Zahl und der ihr entsprechenden Gleichung gründet, welche als Kongruenz aufgefasst wird, und dass die so erhaltenen Definitionen der idealen Zahlen (oder vielmehr der Teilbarkeit durch die idealen Zahlen) zufolge dieser bestimmt gewählten Darstellungsform nicht von vornherein den Charakter der *Invarianz* erkennen lassen, welcher in Wahrheit diesen Begriffen zukommt; die zweite Unvollkommenheit dieser Begründungsart besteht darin, dass bisweilen eigentümliche Ausnahmefälle auftreten, welche eine besondere Behandlung verlangen." The exceptions mentioned in the second place are discussed in [*op.cit.*, 218–30].

[2] The exceptions arise in connection with divisors of $k = (\mathfrak{o}:\mathbb{Z}[\alpha])$, the index of $\mathbb{Z}[\alpha]$ in the ring of integers $\mathfrak{o}$; see [Haubrich 1999, chap. 9]. They also affect the prime number μ in the case of Kummer's cyclotomic numbers, but Kummer knew how to give *ad hoc* an explicit decomposition.

Also important were some issues that affect the adequacy of the tools employed to the theoretical goals. Dedekind always separated clearly the problems that had to do with polynomials and algebraic equations, from those connected with number theory, even when he acknowledged that both kinds of questions were intimately related. The main criticism he had of Kronecker's standpoint was that it mixed questions in number theory with other, totally different questions having to do with polynomials. Kronecker employed the so-called "methodical means of indeterminate coefficients," which in modern terms can be seen as a way of determining ideal factors by embedding the ring of integers of the field under study in a ring of polynomials [Edwards 1980, 353–54]. Dedekind regarded the employment of polynomials as an auxiliary means that is completely foreign to the question [Dedekind 1930/32, vol. 2, 53]. In the 1890s he employed similar criticisms against Hurwitz's way of dealing with the theory of ideals, an approach influenced by Kronecker to some extent [*op.cit.*, 50–58].

Dedekind's demand was that each theory ought to be developed with its own characteristic means, without employing foreign elements taken from other theories; in a word, each theory should be stated and developed in a "pure" form. This requirement appears frequently in his writings, particularly those that have to do with the foundations of arithmetic: "I require that arithmetic ought to be developed out of itself" [Dedekind 1872, 321; also see 1888, 338]. The same can be found in the joint paper with Weber on algebraic functions [Dedekind & Weber 1882, 240].

In order to satisfy this requirement of "purity," Dedekind strove to analyze any new result that might employ any kind of auxiliary means: the result had to be understood at a deeper level; the proofs had to be transformed until the kernel was found that expressed the pure content behind the first result. He described in these terms the way in which he transformed an approach to ideal theory somewhat similar to Hurwitz's into the fourth version of his ideal theory (see [Dedekind 1930/32, vol. 2, 50–58]). In the context of number theory, "pure" meant expressed in terms of numbers, sets of numbers, and homomorphisms; here, he was always guided by the point of view of arithmetization: all complex objects, or concepts, ought to be genetically reducible to the natural numbers. Moreover, the theory ought to be strictly deductive, and of course auxiliary forms of representation should be replaced by abstract concepts. A reformulation of any rough first result in these terms would end in definitions of basic concepts characterizing the main properties of the objects under study, and would thus offer a sound basis for a deductive development of the whole theory. Once the main definitions had been found, a final effort had to be made in order to adapt, as closely and directly as possible, all means of proof to those basic properties and language. This is essentially the method that he applied in the cases of groups, ideals, Dedekind cuts, simply infinite systems, etc. Of course, it required a disposition to accept new definitions, theoretical approaches, and means of proof – which, as Dedekind learnt, most mathematicians would reject, preferring the familiar language in which they had been trained.[1]

The methodological demands that affected Dedekind's approach to ideal theory were in effect also applied to the definition of the number system. Dedekind develops this relationship in a French paper of 1877, where he compares the case of ideals with that of real numbers as defined by means of cuts. In both cases, the issue is the introduction of new "arithmetical elements" in the progressive "construction" of the number system [Dedekind 1877, 268–69]. He summarized the demands in four points:

1. "Arithmetic ought to be developed out of itself" [Dedekind 1872, 321], avoiding foreign elements and auxiliary means – the notion of magnitude in the case of the reals, polynomials in that of ideals.

2. When new elements are introduced, these must be defined in terms of operations and phenomena that can be found in the previously given "arithmetical elements": the arithmetic of $\mathbb{Q}$ in the case of the reals, the arithmetic of $\mathbb{C}$ in the case of the ideals.[2]

3. The new definitions must be completely general, applying "invariantly" to all relevant cases: one should not define some real numbers as roots, some as logarithms, etc., and one should not employ different theoretical means in order to determine ideal factors in different cases, as Kummer (as a result of his different focus and orientation) had done.

4. The new definitions must offer a solid foundation for the deductive structure of the whole theory: they ought to enable a sound definition of operations on the new "elements," and the proof of all relevant theorems (see also [Lipschitz 1986, 65]).

The same requirements are operative in the context of Dedekind's theory of natural numbers, which, as will be seen in chapter VII, was conceived as the general frame for a satisfactory development of arithmetic, algebra, and analysis.

As regards Dedekind's insistent talk of 'creation', he was always convinced that mathematical objects and concepts are creations of the human mind. This was perhaps his most persistent philosophical conviction, one that he entertained from youth to death.[3] In his eyes, the prototype mathematical object is number, and numbers are free creations [freie Schöpfungen] of the mind [1888, 335, 360]. In a letter to Weber of 1888, Dedekind writes that we have the right to claim for ourselves such a creative power:

---

[1] See an 1895 letter to Frobenius in [Dugac 1976, 283], and also [Dedekind 1888, 337].

[2] This requirement is particularly critical. Today we assume that it is better to treat the number system axiomatically, but this is done (explicitly or implicitly) within the frame of set theory. It may be argued that Dedekind's viewpoint was almost unavoidable, as an intermediate step in the historical development (see §IV.1).

[3] From the 1854 *Habilitationsvortrag* [Dedekind 1930/32, vol. 3, 431] to the third preface to *Zahlen* [1988, 343] written in 1911, five years before his death. See also [1872, 317, 323, 325].

We are of divine lineage and there is no doubt that we possess creative power, not only in material things (railways, telegraphs), but quite specially in mental things.[1]

It is interesting to see how he reflects in a simple way, but seriously and coherently, about the impressive material changes in the world surrounding him, and how he links them to his own activities and his conception of mathematics.

4.2. The heuristic way towards ideals. The strict methodological demands established by Dedekind would be satisfied by the new set-theoretical notion of ideal. Actually, the language of sets was always a most useful tool for him, entering in his attempts to apply his principles within the most diverse contexts. But, as we saw in §II.4, in the last half of the nineteenth-century sets were primarily conceived as logical classes, i.e., the extensional counterparts of concepts. This was Dedekind's conception for a long time, too,[2] and it turns out that the new notion of ideal emerged from considerations in which the logical transition from concept to class – the comprehension principle – was central.

This, at least, is what the introduction to a paper of 1876/77, 'Sur la théorie des nombres entiers algébriques,' indicates. Here he described the path which led him to formulate the theory of the factorization of algebraic integers in terms of ideals. The introduction was meant to be a historical description of the process of his own reflections in which he "wrote each word only after the most careful reflection" [Lipschitz 1986, 59]. Of course, any such account involves a good measure of rationalization, although Dedekind was writing only six years after the innovation in question. His description is of special interest as it shows his implicit assumptions, including underlying conceptions in relation to the notion of set.

Dedekind begins with a reference to his first approach of around 1860:

I have not arrived at a general theory without exceptions ... until having abandoned completely the old, more formal approach, and having replaced it by another which departs from the simplest basic conception, and fixes the eyes directly on the end. Within this approach, I do not have any more need of new creations, as that of Kummer's *ideal number*, and it is entirely sufficient to consider the *system of really existing numbers* which I call an *ideal*. The power of this notion resting on its extreme simplicity, and being my wish to inspire confidence in this concept, I shall try to develop the series of ideas which led me to this notion.[3]

---

[1] [Dedekind 1930/32, vol. 3, 489]: "Wir sind göttlichen Geschlechtes und besitzen ohne jeden Zweifel schöpferische Kraft nicht blos in materiellen Dingen (Eisenbahnen, Telegraphen), sondern ganz besonders in geistigen Dingen."

[2] Until he accepted the implications of the set-theoretical paradoxes, apparently in 1897 or 1899. Already in 1888 he supressed mention of concepts in favor of a more abstract approach to sets, perhaps due to his methodological preferences (see §VII.2).

[3] [Dedekind 1877, 268]: "Je ne suis parvenu à la théorie générale et sans exceptions ... qu'après avoir entièrement abandonné l'ancienne marche plus formelle, et l'avoir remplacée par une autre partant de la conception fondamentale la plus simple, et fixant le regard immédiatement sur le but. Dans cette marche, je n'ai plus besoin d'aucune création nouvelle, comme celle du *nombre idéal* de Kummer, et il suffit complétement de la considération de ce *système de nombres réellement existants*, que j'appelle un *idéal*. La puissance de ce concept reposant sur son extrême

It is noteworthy how Dedekind states that the set-theoretical approach to ideal theory "departs from the simplest basic conception, and fixes the eyes directly on the end." In my opinion, this is not just a metaphor, but can be interpreted literally: the "simplest basic conception" involves the notions of number, number-operation, and set; the definition of the number-sets called ideals is determined by "fixing the eyes directly" on the goal of founding a theory of the divisibility of the integers in a field, as will be seen below.

Dedekind wished to have a general, precise definition of all ideal factors to be considered, and also a general definition of the multiplication of ideal factors [1877, 268]. This is consistent with his methodological demands (§5.1), but was also an outcome of his previous experiences with Kummer's and his own first theory. None of them offered a general definition of all ideal prime factors, and Kummer had not defined satisfactorily the product of ideal factors, which Dedekind regarded as the reason for some deficiencies in his proofs [Haubrich 1999, ch. 2]. Dedekind hoped that a joint definition of all the ideals would make possible a completely general proof of the fundamental theorem, which is what he could not reach in his first theory.

In his first theory, Dedekind introduced ideal factors – or better, defined divisibility by the ideal factors – in connection with one or more congruence relations that an algebraic integer might satisfy or not. To reach his desired new definitions, it was necessary and sufficient to determine what is common to all congruence properties A, B, C ... which are thus used, and to establish how two properties A, B associated with two ideal factors determine the property C which defines their product [Dedekind 1877, 268–69]. At this point, the comprehension principle offered a way of simplifying the task:

This problem is essentially simplified by the following reflections. As such a characteristic property A serves to define, not the ideal number in itself, but only the divisibility of numbers contained in o [the 'Hauptordnung' or ring of integers] by an ideal number, one is naturally led to consider the set a of *all* those numbers α in the domain o which are divisible by a certain ideal number; from now on I shall call such a system a, for short, an *ideal*, and so to each particular ideal number corresponds a certain *ideal* a. But as, reciprocally, the property A ... consists only in that α belongs to the corresponding ideal a, one may, instead of properties A, B, C... ..., consider the corresponding ideals a, b, c... in order to establish their common and exclusive character.[1]

---

simplicité, et mon dessein étant avant tout d'inspirer la confiance en cette notion, je vais essayer de développer la suite des idées qui m'ont conduit à ce concept."

[1] [Dedekind 1877, 270]: "Ce problème est essentiellement simplifié par les réflections suivantes. Comme une telle propriété caractéristique A sert à définir, non un nombre idéal lui-même, mais seulement la divisibilité des nombres contenus dans o par un nombre idéal, on est conduit naturellement à considérer l'ensemble a de *tous* ces nombres α du domaine o qui sont divisibles par un nombre idéal déterminé; j'ai appellerai dès maintenant, pour abréger, un tel système a un *idéal*, de sorte que, à tout nombre idéal déterminé, correspond un *idéal* déterminé a. Maintenant comme, réciproquement, la propriété A ... consiste uniquement en ce que α appartient à l'idéal correspondant a, on pourra, au lieu des propriétés A, B, C... ..., considérer les idéaux correspondants a, b, c... pour établir leur caractére commun et exclusif."

The step towards set language, which Dedekind regarded as suggested "naturally" by the problem in question, was quite strange and difficult for his contemporaries; it was one of the reasons why ideal theory was not generally accepted until the 1890s. What made it natural for him was his familiarity with the conceptions of traditional logic, where the transition from a property to a class was absolutely natural; and above all his confidence in the set-theoretical approach (the abstract-logical conception) of mathematics. Without the latter, the connection between sets, or any logical conceptions, and mathematics would seem completely unclear.

At this point, Dedekind needed to find the properties that characterize the sets $\mathfrak{a}$, $\mathfrak{b}$, ... mentioned above, i.e., a general definition of ideals consisting in necessary and sufficient conditions for a set of integers to be an ideal. Since the goal of the theory was to establish the laws of divisibility for algebraic integers, the "really existing" integers had also to be viewed from the set-theoretical viewpoint. This was easily done by considering principal ideals, sets of integers that are divisible by a given algebraic integer α. Through a study of the "phenomena" that one can find in this simple, pre-existing case, Dedekind was able to find two conditions that he judged apt for a general definition of ideals. The set of multiples of α satisfies the following two conditions:

I. sums and differences of any two numbers in the set are again numbers belonging to it, and

II. products of any number in the set by any integer of the field are again numbers belonging to the set.

One simply had to turn this two properties of the simplest case into a definition of the general case: an *ideal* was defined to be *any* system or set of algebraic integers in a field Ω that satisfies conditions I. and II. [Dedekind 1871, 251; 1877, 271].

Thus, he obtained a general definition that was based exclusively on the arithmetic of $\mathfrak{C}$ and the notion of set, and hence satisfied the requirements that we analyzed in the previous section. But Dedekind did not stop here. For many years he had been working with ideal *numbers*, and he felt the need to check whether the new definition made sense against that background. This was also necessary, because he was still employing methods of proof taken from his first theory, and so had to prove that there is a perfect overlap between both approaches [Haubrich 1999, ch. 10]. Dedekind [1877, 271] took pains to prove, after many vain attempts and with great difficulties, that any ideal, in the sense defined above, was either the set of multiples of a given integer (a principal ideal) or of an ideal number in the old sense. Actually, this is the content of the "fundamental theorem" that can be found in his first version of the theory [Dedekind 1871, 258]. The basic idea of going from an ideal factor to its corresponding ideal suggested to Dedekind, for the first time, a way of proving the fundamental theorem in its full generality. The set-theoretical viewpoint was crucial not only for the new basic definition, but especially for the new proof strategy that finally satisfied Dedekind's requirement of generality.[1]

---

[1] The key idea was to construct a finite succession of ideals $\mathfrak{a}_i$ embedded on one another, such that: $\mathfrak{o}\mu \subset \mathfrak{a}_0 \subset \mathfrak{a}_1 \subset ... \subset \mathfrak{o}$ (where $\mu \in \mathfrak{o}$), and ending with a maximal ideal [Haubrich 1999, *ibid.*].

Ideals satisfied all of Dedekind's demands concerning definitions, proofs, and the abstract conceptual approach to mathematics generally. This seems to have been the reason why he adhered to them, and to the set-theoretical viewpoint in algebraic number theory, ever after 1870, in spite of the novelty of his approach and the difficulties it posed for his contemporaries.

5. Dedekind's Infinitism

In 1872, one year after the publication of his ideal theory, Dedekind began to write a draft containing an "attempt to analyze the number-concept from a naive standpoint." It already bears the original, and quite characteristic, motto: "Nothing capable of proof ought to be accepted in science without proof."[1] The draft was continued sporadically until 1878, rewritten in 1887, and then published as the famous booklet *Was sind und was sollen die Zahlen?* [1888]. The very first part of the draft, which has to be dated 1872, contains most of the basic notions in the later book: a definition of set or "system," definitions of "part" (the abstract analogue to the "divisor" of algebraic number theory) and of the union operation, definitions of mapping and of infinite set, and a basic analysis of the series of natural numbers, including the crucial notion of "chain" [Dedekind 1872/78, 293–97]. Most of these ideas will be studied in chapter VII, but here we shall consider the notions of set and mapping, and the definition of infinite set, as Dedekind presented them in 1872.

The very idea that Dedekind's analysis was "naive" already calls for attention. In the draft he elaborated an abstract definition of the natural numbers, based on the notions of set and mapping. Apparently, the reason why this was regarded as "naive" is that both notions belong to logic, hence to the innermost constitution of sound common sense. Dedekind writes that both notions, on which he bases the number concept, would be indispensable for arithmetic even if the concept of cardinal number were taken as given in so-called "inner intuition" [1872/78, 293]. Most likely, the word "arithmetic" is understood here, as 16 years later, in a broad sense that embraces algebra and analysis [Dedekind 1888, 335]. The author's experience in algebra, algebraic number theory, and the foundations of analysis is reflected in that statement.

As regards the notion of set, Dedekind explains it as follows:

A *thing* is any object of our thought ...
A *system* or *collection* [or manifold] *S* of things is determined, when of any thing it is possible to judge whether it belongs to the system or not.

---

[1] The draft was published as appendix LVI to [Dugac 1976, 293–309]. "Versuch einer Analyse des Zahl-Begriffs vom naiven Standpuncte aus." "Motto (eigenes): 'Was beweisbar ist, soll in der Wissenschaft nicht ohne Beweis geglaubt werden.'" [*op.cit.*, 293].

The words "or manifold" were written on the right hand column, which Dedekind left blank initially, in order to employ it for later corrections or additions. On the same column we find a definition of set:

We use to treat all those things that possess a common property, insofar as the differences between them are not important, as a new thing in front of all other things. It is called a system, or a collection of all these things.[1]

The correlation between sets and properties comes up again in the immediately following pages [1872/78, 295–96]. Hence, in 1872 Dedekind employed the comprehension principle to explain the notion of set; the word "Inbegriff" [collection] directly expresses the relationship between "Begriffe" [concepts] and sets. Likewise, he seems to have been the first to express clearly the principle of extensionality, that can be found in the first quotation. At any rate, it seems quite indisputable that one should not hesitate to interpret his technical word "system" as denoting the notion of set.

Next Dedekind explains what he means by "part" and "proper part," introduces the union of two sets under the names "least common multiple" (taken from algebraic number theory) or "compound system," and finally presents the crucial notions of mapping and injective mapping:

A system S is called *distinctly mappable in a system T*, when to every thing contained in S (original) one can determine a (corresponding) thing contained in T (image), so that different images correspond to different originals.

On the right hand side we read a more straightforward definition of mapping ϕ of A in B, by saying that to each element a of A "*corresponds*" an element $a|\phi$ of B, and that the mapping is *distinct* when $a'|\phi \neq a''|\phi$ whenever a' and a'' are different (today we write $\phi(a)$ instead of $a|\phi$).[2] The notion of mapping that he has in mind here is the general one, since on [1872/78, 296] he starts an "investigation of a (distinct or undistinct) mapping of a system S in itself." This can be seen as a consequence of his long experience with homomorphisms in the context of algebra.

---

[1] [Dedekind 1872/78, 293]: "Ein *Ding* ist jeder Gegenstand unseres Denkens ... Ein *System* oder *Inbegriff* [oder Mannigfaltigkeit] S von Dingen ist bestimmt, wenn von jedem Ding sich beurtheilen lässt, ob es dem System angehört oder nicht." "Alle diejenigen Dinge, welche eine gemeinschaftliche Eigenschaft besitzen, pflegt man solange die Unterscheidung derselben nicht wichtig ist, den anderen Dingen gegenüber wie ein neues Ding zu behandeln. Dasselbe heisst System, oder Inbegriff alles dieser Dinge."

[2] [Dedekind 1872/78, 294]: "Ein System S heisst *deutlich abbildbar in einem System T*, wenn man für jedes in S enthaltene Ding (Original) ein (zugehöriges, correspondirendes) in T enthaltenes Ding (Bild) so angeben kann, dass verschiedenen Originalen auch verschiedene Bilder entsprechen." "*Abbildung* ϕ des Systems A in dem System B. Jedem Ding a des A entspricht (durch a ist *bestimmt*) ein Ding $a | \phi = b$ des Systems B. *Deutlichkeit* einer Abbildung ϕ: $a' | \phi$ und $a'' | \phi$ verschieden abgebildet, wenn a', a'' verschieden."

Dedekind can now state his famous explanation of the infinite:

The infinite and the finite.

A system S is called *infinite* (or: the *number* of the things contained in S is called *infinitely great*), when there is a part U of S, which is different from S, and in which S can be mapped distinctly; the system S is called *finite* (or: S consists of a *finite number* of things), when there is no part U of S different from S, in which S is distinctly mappable.[1]

It is notable that this text was written before any of Cantor's papers on set theory, for it can be seen as the first noteworthy and influential attempt to elaborate an abstract theory of finite and infinite sets.[2] One should perhaps add that, of course, there is nothing in Dedekind's draft that prefigures Cantor's path-breaking non-denumerability results of 1873 and later.

We have no indication of the origins of Dedekind's definition of infinity: on its first appearance it emerges in a fully developed form, so it might well have originated some years earlier. What we know is that he had been relying on infinite sets from the late 1850s. Evidence includes the paper on higher congruences, written in 1856 (§2), the 1858 theory of irrational numbers that was published in [1872], his work on fields and rings of integers which must date back to about 1860, and the ideal theory of 1871. Nevertheless, the question might be raised, why was Dedekind so convinced of the acceptability of the infinite? What were the roots of his infinitism? This question was actually asked some years ago by Harold M. Edwards [1983], in a paper called 'Dedekind's Invention of Ideals.'

Edwards poses the question as follows. The version of ideal theory that Dedekind published in 1871 can easily be reformulated (today, at least) in a way that avoids all reliance on infinite sets, by means of the concepts of divisor theory. Edwards regards that first version as Dedekind's best, even though the great mathematician was dissatisfied with it and kept reformulating it in accordance with his "set-theoretic prejudices" [Edwards 1980, 321]. But, with the goal of establishing a theory of the factorization of ideal numbers in mind (and forgetting about their influence on the further development of modern algebra), Dedekind's later versions are unnecessarily complex. More generally, Edwards considers that, in all of elementary algebra, the infinite sets involved are so "tame" that they can always be described in a finitistic way. In his opinion, Dedekind must have been prompted to "fly in the face of the doctrine against completed infinities" by "something from analysis" [1983, 12]. He knew that his theory of real numbers could not be devel-

---

[1] [Dedekind 1872/78, 294]: "Das Unendliche und Endliche. / Ein System S heisst ein *unendliches* (oder: die *Anzahl* der in S enthaltenen Dinge heisst *unendlich gross*), wenn es einen Theil U von S giebt, welcher von S verschieden ist, und in welchem S sich deutlich abbilden lässt; das System S heisst *endliches* (oder: S besteht aus einer *endlichen Anzahl* von Dingen), wenn es keinen von S verschiedenen Theil U von S giebt, in welchem S deutlich abbildbar ist."

[2] The words "and influential" are written with Bolzano in mind.

oped constructively, hence he avoided taking steps in this direction within the context of ideal theory [*op.cit.*, 14].

Edwards' explanation does not look plausible. First, we know that Dedekind employed infinite sets in his algebraic work of 1856–58: we have the examples of the 'rational domains' (fields) in his early version of Galois theory, and of the paper on higher congruences. Here he stressed the fact that he was considering infinitely many classes, each of which contain infinitely many elements, but in spite of this he established an analogy between those infinite sets and the natural numbers, concrete objects for a traditional mathematician (§2). And all of this happened before the formulation of his theory of irrational numbers, late in 1858. (It would be more plausible to try finding the roots of Dedekind's infinitism in Riemann's 'prejudices.')

Second, Dedekind placed no special value on his theory of real numbers: he regarded it as a rather straightforward contribution, that simply came to fill a gap in the elements of the number system; other mathematicians would have formulated something similar had they devoted some effort to it [Dedekind 1930/32, vol. 3, 470, 475]. Therefore, it is highly implausible that he might have conditioned his cherished ideal theory on a trait of the much less valuable theory of real numbers.

Third, the only text that I have been able to find where Dedekind tried to justify his use of infinite sets in ideal theory, establishes no relation between this and the real numbers. Instead, he traces an interesting parallel between his work and Gauss's theory of the composition of quadratic forms in *Disquisitiones Arithmeticae*. In a letter to Lipschitz of June 1876, he wrote:

Just as we can conceive of a collection of *infinitely many* functions, which are still dependent on variables, as *one* whole, e.g., when we collect all equivalent forms in a form-class, denote this by a single letter, and submit it to composition, with the same right I can conceive of a system A of infinitely many, completely determined numbers in [the ring of integers] o, which satisfies two extremely simple conditions I. and II., as *one* whole, and name it an ideal.[1]

Gauss had been very careful to express himself in a way that did *not* imply the existence of actual infinities, but it is clear that his reader Dedekind was not worried by philosophical subtleties.

Dedekind's acceptance of the infinite does not have the appearance of a more or less *ad hoc* position, adopted to safeguard his theory of the real numbers. It rather looks like a deep-rooted conviction: infinite sets seemed to him perfectly acceptable objects of thought, that involve no contradiction, and that play a crucial role in

---

[1] [Lipschitz 1986, 62]: "So gut, wie man einen Inbegriff von *unendlich vielen* Functionen, die sogar noch von Variablen abhängen, als *ein* Ganzes auffasst, wie man z.B. alle äquivalenten Formen zu einer Formen-Classe vereinigt, diese wieder mit einem einfachen Buchstaben bezeichnet und einer Composition unterwirft, mit demselben Rechte darf ich ein System A von unendlich vielen, aber vollständig bestimmten Zahlen in o, welches zwei höchst einfachen Bedingungen I. und II. genügt, als *ein* Ganzes auffassen und ein Ideal nennen."

mathematics. This seems to be the reason why he took the side of the actual infinite. From the 1850s on, he accepted actual infinities as natural mathematical objects, and in the 1880s he even tried to *prove* the existence of infinite sets, thus establishing the soundness and consistency of the notion of infinity (§§VII.3 and 5).

6. The Diffusion of Dedekind's Views

In his tenth supplement to Dirichlet's *Vorlesungen* of 1871, Dedekind proposed a new conception of "higher" number theory, understood as the theory of algebraic numbers, and approached it in a revolutionary way, namely, set-theoretically. He also suggested that a similar approach was appropriate for algebra, on the basis of the field concept. Given these rather strong changes in conceptualization and theoretical orientation, it is not strange that the reception of his ideas was slow. Actually, only in the 1890s did the theory of algebraic numbers begin to be generally treated more or less along his lines. In 1876, five years after publishing the new theory, Dedekind expressed surprise and happiness to know that Lipschitz had an interest in his work, for he was, with Weber, the only such mathematician (see [Lipschitz 1986, 48]).[1] Dedekind had become convinced that his tenth supplement was just not read by anyone [Edwards 1980, 349]. Nevertheless, in 1880 he was named a member of the Berlin Academy of Sciences for his number-theoretical work, on the proposal of Kronecker, who on this occasion expressed his difficulties with Dedekind's terminology and methods (see [Biermann 1966]).

6.1. Subsequent versions of Dedekind's ideal theory. Not surprisingly, the main instrument for the diffusion of Dedekind's views were his own refined versions of ideal theory. Three different versions were published, the first in 1871, the second in 1876–77 (French) and 1879 (German), and the third in 1894; each one diverges widely from its predecessor (plans for a new fourth version never reached completion).

Dedekind was not satisfied with his first presentation of the theory, because it was a kind of compromise. Due to the fact that the second edition of Dirichlet's *Vorlesungen* was scheduled for late 1870, and the notion of ideal was first formulated in August of that year, he had to work under great time pressure in developing his new ideas. It was impossible to adapt the development of the theory to the set-theoretical characteristics of its central notion, as his methodological preference for a "pure" development of the theory demanded. Dedekind had recourse to the tools developed in his first approach to the factorization of algebraic integers. This is reflected in the fact that the proof of the fundamental theorem follows a detour

---

[1] As we have seen (§2), Cantor may be added to the list of those who had carefully read Dedekind's ideal theory in the 1870s. One may add that he lectured on the theory of algebraic numbers at Halle [Purkert & Ilgauds 1987, 103]; see also [Cantor 1877], where he cites Dedekind's fields of algebraic numbers as examples of denumerably infinite sets.

through so-called "simple ideals," which are defined by means of certain number-congruences, just like the ideal factors were in his first approach.[1] Finally, all prime ideals are shown to be simple ideals, which eliminates this somewhat *ad hoc* tool. But the fact is that his proofs depended on notions defined by means of a form of representation, that always involves a degree of arbitrariness – precisely what the abstract conceptual methodology that he shared with Riemann recommended to avoid.

The same shortcomings, as Dedekind saw them, were reflected in the fact that the product of ideals is only defined *after* the fundamental theorem has been proven. This notion ought to become the true basis for the whole theory, as he was finally able to do in the third version. Given the set-theoretical formulation of the theory, and particularly of the notion of ideal, Dedekind saw the main issue in the compatibility of two notions that emerge quite naturally. If two algebraic integers are such that α is divisible by β, it is clear that α will belong to the principal ideal of β, denoted $o(\beta)$; likewise, if $o(\alpha)$ is included in $o(\beta)$, it is clear that α is divisible by β. Hence, there is a natural analogy between the inclusion of ideals and number-divisibility, which leads Dedekind to the following definition. An ideal $\mathfrak{A}$ is divisible by the ideal $\mathfrak{B}$ if and only if $\mathfrak{A} \subset \mathfrak{B}$. Accordingly, the natural definition of a prime ideal is the following: $\mathfrak{P}$ is a prime ideal if it is only divisible by o and $\mathfrak{P}$ itself [Dedekind 1871, 252–53]. But, on the other hand, it is natural to define ideal multiplication in the obvious algebraic way: for all $\alpha \in \mathfrak{A}$ and $\beta \in \mathfrak{B}$, the additive closure of the set of all products $\alpha\beta$, which is again an ideal, is the product $\mathfrak{A}\mathfrak{B}$ [*op.cit.*, 259]. Given these two notions, the main difficulty in the theory of ideals is to prove the following theorem: If the ideal $\mathfrak{C}$ is divisible by the ideal $\mathfrak{A}$ ($\mathfrak{C} \subset \mathfrak{A}$), there is one and only one ideal $\mathfrak{B}$ such that $\mathfrak{A}\mathfrak{B} = \mathfrak{C}$ [Dedekind 1877, 272–73; 1895, 50]. Once this has been proven, one can reach the required generalization of the fundamental theorem of factorization: Every ideal, that is different from o, either is a prime ideal or can be uniquely represented as a product of prime ideals.

Dedekind's aim was to establish these two theorems as directly as possible, with the simplest possible means, and taking as little as possible from other mathematical theories. His reasonings should avoid reliance on particular numbers, instead trying to exploit directly the possibility of considering set-theoretical or algebraic relationships between number-sets, taken as wholes.[2] This is consistent with Dedekind's understanding of the deductive structure of mathematical theories: a set-theoretical definition of ideals implied a strict reliance on sets and set-theoretical relations throughout the theory. This was attained in the second version, published

---

[1] See [Dedekind 1871, 255–58]. That made it possible for Edwards [1980, 337–42] to reformulate Dedekind's theory of 1871 in terms of the divisor theory of Borewicz and Safarevic, avoiding sets completely; and it enabled Haubrich [1999, chap. 9] to attempt a reconstruction of Dedekind's first approach.

[2] See [Corry 1996, 103–20], where a more detailed analysis of Dedekind's later versions can also be found.

in [Dedekind 1877] and in a new supplement XI to the 1879 edition of the *Vorlesungen*, now bearing the straightforward title 'On the theory of algebraic integral numbers' [1879]. Here, forms of representation were avoided in favor of a "purer" derivation of the main results. Nevertheless, the proofs were quite complex, for Dedekind was only able to reach the fundamental theorem by a step-by-step elimination of limiting assumptions. As he wrote, it was "a *long* chain of theorems!"[1]

By contrast, the third version of 1894 offered a much more direct and powerful method for proving the fundamental theorem, that was based on a development of the theory of modules. Dedekind elaborated an algebra of modules, based on definitions for the addition (algebraic union), substraction (intersection), product and division of modules. He established some characteristic results, including the following modular identity [Dedekind 1894, 503]:

$$(A+B+C) \ (BC+CA+AB) = (B+C)(C+A)(A+B)$$

which became an essential tool in his new proof of the factorization theorem.[2] On this occasion he also employed module-theoretical definitions of ideals and of *Ordnungen*, which, however, stemmed in the 1870s; these definitions made possible the direct application of the algebra of modules. In the preface to this edition of the *Vorlesungen*, he wrote:

Only the last supplement, that deals with the general theory of algebraic integer numbers, has gone through a complete reworking; both the algebraic foundations [Galois theory] and those that are properly number-theoretical receive a more detailed exposition, from the point of view that I regard, with that conviction afforded by many years, as the simplest, basically because it presupposes only a clear comprehension of the number domain and the knowledge of the basic rational operations.[3]

This, again, was an expression of his ideal of purity and of the autonomy of number theory, as well as a recognition of the intimate links between Dedekind's foundational work on the "number domain" and his ideal theory. Dedekind's ideal was to base ideal theory directly on a theory of the number system and the number operations, for which the general basis was given in [Dedekind 1888] (see chap. VII). This is also made clear by the frequent footnote references to that booklet in [Dedekind 1894].

---

[1] Letter to Lipschitz, [Dedekind 1930/32, vol.3, 468]: "eine *lange* Kette von Sätzen!"

[2] His new research on modules also led to path-breaking work on lattice theory, which, however, had little direct influence (see [Mehrtens 1979; Corry 1996, chap. 2]).

[3] [Dedekind 1930, vol. 3, 426; or Dirichlet 1894, v–vi]: "Nur das letzte Supplement, welches die allgemeine Theorie der ganzen algebraischen Zahlen behandelt, hat eine vollständige Umarbeitung erfahren; sowohl die algebraischen als auch die eigentlich zahlentheoretischen Grundlagen sind in grösster Ausführlichkeit und in derjenigen Auffassung dargestellt, welche ich nach langjähriger Überzeugung für die einfachste halte, weil sie hauptsächlich nur einen deutlichen Überblick über das Reich der Zahlen und die Kenntnis der rationalen Grundoperationen voraussetzt."

6.2. Later contributions: Weber and Hilbert.   As we shall not come back to the history of algebra and algebraic number theory again, it seems convenient to end this chapter with a few indications of the reception and further elaboration of Dedekind's ideas. No attempt is made to offer a general overview, I will content myself with some remarks on the relation between Dedekind and the influential work of Weber and Hilbert.

In §II.7 we mentioned that Dedekind met Weber in 1873 and collaborated with him closely in the first edition of Riemann's works, published 1876. This was the beginning of a deep friendship and scientific collaboration, that would be decisive for Weber's later contributions to algebraic topics. Between 1878 and 1880, both corresponded on the theory of algebraic functions, which resulted in the publication of [Dedekind & Weber 1882]. This paper offered a purely algebraic treatment of an important part of Riemann's function theory, including a novel definition of the Riemann surfaces, and a proof of the Riemann–Roch theorem. To this end, Dedekind and Weber established a strict analogue of ideal theory for fields of algebraic functions, and the corresponding ring of "integral functions." For this reason, the 1882 paper has been considered as the first important example of a structural unification of widely different mathematical domains.[1] It presented new notions and viewpoints that would be of crucial importance for the development of algebraic geometry in the 20th-century. Also noteworthy is the authors' remark that considerations of continuity concerning the algebraic functions are not required at all in their work, and therefore the theory is "treated exclusively through means belonging to its own sphere."[2] This is, again, Dedekind's demand for purity, but also one of the very first appearances of the idea that in algebraic work one should avoid topological considerations.

The many-sided Weber continued to be interested in algebra and number theory. In [1893] he presented the general foundations of Galois theory, stimulated by the manuscript of Dedekind's lectures in the 1850s (see [Weber 1895/96, vol. 1]).[3] In the paper of 1893 one can find, for the first time, an explicit recognition of the relationship between groups and fields, regarded as abstract structures, together with the first axiomatization of the notion of field. While Dedekind had only considered finite groups and infinite fields, Weber went beyond in analyzing both the finite and infinite cases in some detail. Like his friend, he emphasized the role played by extensions of fields, and the interplay between group-theoretical and field-theoretical notions in Galois theory, giving more attention to the nature of the

---

[1] Jean Dieudonné in a 1969 article on Dedekind for the *Encyclopaedia Universalis*, quoted in [Dugac 1976, 77].

[2] [Dedekind & Weber 1882, 240]: "Bis zu[m Beweis des Riemann–Rochschen Satzes] kommt die Stetigkeit und Entwickelbarkeit der untersuchten Funktionen in keiner Weise in Betracht. ... Dadurch wird ... [die] Theorie ... lediglich durch die seiner eigenen Sphäre angehörigen Mittel behandelt."

[3] In 1881, Paul Bachmann had already developed Dedekind's suggestions in some detail, in a paper published in vol. 18 of *Mathematische Annalen*. Bachmann was one of the few students that attended Dedekind's lectures in the 1850s.

solution of an equation than to actual calculations [Kiernan 1971, 141]. Dedekind would also contribute to the subject the year after, when he proposed, for the first time, to view the group of the equation as group of the automorphisms of the splitting field [Dedekind 1894]. The modern treatment of Galois theory, in the hands of Emil Artin, would follow this lead [Kiernan 1971, 144–51].

But perhaps Weber's most important role, as long as algebra is concerned, was as a textbook-writer and teacher.[1] Particularly important is the fact that David Hilbert attended his lectures at Königsberg until Weber's departure in 1883. Weber had occasion to lecture on algebraic number theory, which he probably presented in Dedekind's way, and this early exposure had a strong influence in Hilbert's formation. In the mid-1890s Weber published his textbook on algebra, which became a reference work for the new generation of algebraists. The introduction to his *Algebra* [1895/96, vol. 1] was a lengthy presentation of Dedekind's ideas on the foundations of the number system. Vol. 1 dealt with algebra as the theory of equations, beginning with the elements and working up to Galois theory, presented in a rather concrete way, not abstractly as in [1893]. Vol. 2 dealt with more advanced material: the abstract theory of groups, and algebraic number theory. Interestingly, though, Weber's approach to algebraic number theory is not close to Dedekind's, but to Kronecker's: he determines ideal factors by means of so-called "functionals," which are essentially the same as Kronecker's 'forms' or polynomials; only later he proceeds to show that his reformulation of Kronecker's approach is equivalent to Dedekind's. It seems likely that he chose this approach in order to offer a synthetic overview of previous work. After all, Dedekind had done more than enough to explain and promote his viewpoint, while Kronecker's ideas had remained hardly accessible to other mathematicians. Meanwhile, Weber's textbook became an important instrument for the diffusion of some basic ideas of Dedekind's, such as his notion of field, and even of his conception of algebra.[2]

Although in the 1890s Dedekind's conception of algebraic number theory began to enjoy wide acceptance, his strict adherence to the "structural" viewpoint, i.e., his preference for simple number- and set-relations, was not adopted by his immediate followers, with the exception of Weber. To give an example, Hurwitz defined an ideal as the course-of-values of a homogeneous linear form, the coefficients and values of which are integers in the field under study [Hurwitz 1894]. Similar, although closer to Dedekind in that it avoids 'forms' in favor of number-relations, is the case of Hilbert: ideals are defined as "systems" of algebraic integers such that any linear combination of them, with coefficients in the ring, also belongs to the ideal [Hilbert 1897].

---

[1] He enjoyed a very successful career, teaching at the universities of Königsberg, Marburg, Göttingen and Strassbourg, and the technical schools of Zürich and Charlottenburg (Berlin).

[2] A detailed analysis of Weber's *Algebra* can be found in [Corry 1996, 34–45]. However, one must take into account that Weber's presentation of algebraic number theory did *not* follow Dedekind.

Hilbert's famous *Zahlbericht* [Number-report; Hilbert 1897] became the standard reference work for the next generation of number theorists, strongly promoting the new set-theoretical concepts.[1] Nevertheless, Hilbert did not follow Dedekind in the way the theory was structured either. Already in his work on invariant theory, he had taken suggestions from [Kronecker 1882] and combined them with the abstract approach of [Dedekind & Weber 1882], a noteworthy combination of widely different viewpoints that is characteristic of the author. In his eyes, that contribution effected a subsumption of invariant theory under the theory of fields of algebraic functions (see [Corry 1996, ch. 3]). In developing ideal theory, Hilbert abandoned the 'purism' of his predecessor, employing theoretical means that Dedekind would have regarded as formal and alien to the subject. In particular, he employed a theorem that Dedekind had found by re-working on his own Kronecker's approach to the theory of algebraic integers, a theorem independently proven by Hurwitz.[2] Hilbert's choice of arguments was guided by considerations of generalizability and usefulness in further research, which he judged on the basis of extensive research of his own [Weyl 1944]. His most important contribution was the formulation of new and prolific problems within algebraic number theory, leading to class field theory.

Dedekind's original definition of ideal, together with his style of reasoning and proof, reappeared in the textbook literature with van der Waerden's *Moderne Algebra* of 1930. One of van der Waerden's main influences was Emmy Noether, a central figure in the movement of abstract algebra of the 1920s and 1930s, who explored ideal theory in the context of commutative and non-commutative abstract rings. Noether expressed an open preference for Dedekind's mathematical style:

For Emmy Noether, the eleventh supplement was an inexhaustible source of stimuli and methods. On all occasions she used to say: 'It is already in Dedekind.'[3]

In her edition of Dedekind's works, she wrote that the conceptual constructions of supplement XI cross through the whole of abstract algebra [Dedekind 1930/32, vol. 3, 314].

---

[1] There is no detailed historical analysis of Hilbert's *Zahlbericht* yet. See W. & F. Ellison in [Dieudonné 1978, 191–92], who refer to Weyl's obituary, that can be found as an appendix to [Reid 1970].

[2] See [Hurwitz 1894, 1895; Dedekind 1895; Edwards 1980, 364–68].

[3] Van der Waerden, introduction to Dedekind's ideal theory (Braunschweig, Vieweg, 1963): "Für Emmy Noether was das elfte Supplement eine unerschöpfliche Quelle von Anregungen und Methoden. Bei jeder Gelegenheit pflegte sie zu sagen: 'Es steht schon bei Dedekind.'" On this topic, see [Corry 1996, chap. 5].

IV   The Real Number System

The precise distinction of ideas of Extension does not consist in *magnitude*: for in order to recognize distinctly the magnitude, one must resort to the integers, or to the other numbers that are known by means of the integers, so that from *continuous quantity* one has resort to *discrete quantity*, in order to have a distinct knowledge of *magnitude*.[1]

The more beautiful it seems to me that man can rise to the creation of the pure, continuous number domain, without any idea of the measurable magnitudes, and in fact by means of a finite system of simple thought steps; and it is first by this auxiliary means that it is possible to him, in my opinion, to turn the idea of continuous space into a distinct one.[2]

From what we have seen in the preceding chapters, around 1870 there were several indications that the notion of set might prove of fundamental importance for algebra, function theory, and even geometry. In the next chapter we shall consider another line of development, consolidated around that time, which led mathematicians working on real functions to pay attention to point-sets. In the present chapter we consider more elementary questions in analysis that also stimulated the emergence of a theory of sets, and which firmly established the conception of pure mathematics as the science of number. This conception was crucial for Weierstrass, Dedekind and Cantor, the central (German) figures in the rigorization of the real number system. Each one of them presented a rigorous definition of the real numbers, together with basic notions and results on the topology of $\mathbb{R}$.

It is well known that the need for a sound treatment of the real numbers was first felt in connection with the rigorization of analysis. The approach of Cauchy, based on the notions of limit and continuous function, can be taken to have been

---

[1] [Leibniz 1704, book II, chap. 16]: "La distinction precise des idées dans l'Etendue ne consiste pas dans la *grandeur*: car pour reconnoistre distinctement la grandeur, il faut recourir aux nombres entiers, ou aux autres connus par le moyen des entiers, ainsi de la *quantité continue* il faut recourir à la *quantité discrete*, pour avoir une connaissance distincte de la *grandeur*."

[2] [Dedekind 1888, 340]: "Um so schöner scheint es mir, dass der Mensch ohne jede Vorstellung von messbaren Grössen, und zwar durch ein endliches System einfacher Denkschritte sich zur Schöpfung des reinen, stetigen Zahlenreiches aufschwingen kann; und erst mit diesem Hilfsmittel wird es ihm nach meiner Ansicht möglich, die Vorstellung vom stetigen Raume zu einer deutlichen auszubilden."

firmly implanted from the mid-century.[1] Its characteristic tendency was to build upon notions of a geometric origin, e.g., the continuity of a function, but giving them a rigorous and abstract formulation in terms of numerical relations. The constant use of numerical conditions, and the notion of limit, made possible an important step forward in the elimination of the obscurities and inconsistencies of the calculus of the 18th-century. But, of course, absolute rigor had not been attained. Some basic theorems of an *existential character* still lacked an adequate foundation. Examples are the intermediate value theorem – if a continuous function takes positive and negative values at both ends of an interval, *there is* a real number in the interval which is a zero of the function; the theorem that, given a monotonically increasing and bounded sequence of real numbers, *there is* a unique real number that is the limit of the sequence; or the principle that, given an infinite sequence of embedded intervals of $\mathbb{R}$, *there is* at least one real number that belongs to all those intervals.

The gap is clearly visible, for instance, in the pioneering paper by Bolzano [1817] devoted to proving the intermediate value theorem. Bolzano anticipates the approach of Cauchy by attributing a decisive importance to limits and by giving the modern definition of a continuous function – with a proof of the continuity of polynomial functions – and the Cauchy condition for the convergence of sequences. This was the basis for his correct proof of the intermediate value theorem; but his attempt to justify the Cauchy condition turned out to be circular, because it lacked an arithmetical definition of real numbers. Bolzano's memoir remained almost unknown at the time, but it is convenient to mention it here, because apparently Weierstrass himself relied on that work.[2]

What was needed was a satisfactory theory of the real numbers, establishing the continuity (completeness) of $\mathbb{R}$. This was accomplished by Weierstrass, Dedekind, Mèray and Cantor, who presented definitions of the real number system on the basis of the rational numbers. In doing so, they seem to have sharpened and given a clear sense to the obscure declarations of Leibniz, quoted above. From a modern viewpoint, those definitions have to be interpreted as set-theoretical "constructions," but it is a different issue how the authors themselves understood them, and how they fit into the historical context. The four mathematicians seem to differ as to their consciousness of the role played by the notion of set as a foundation for their theories. As a matter of fact, it was only gradually and after the publication of those theories, with the early development of set theory, that the mathematical community became aware of the foundational role played by the notion of set.

---

[1] For the approach of Cauchy, see [Kline 1972; Grattan-Guinness 1980; Bottazzini 1986]. A more specialized discussion can be found in [Grabiner 1981]. On its implantation in Germany, [Jahnke 1987].

[2] It seems that, in the 1860s and 70s, he used to refer to it in his lectures, and admitted that he had taken up (and perfected) Bolzano's methods; see the 1870 letter from Schwarz to Cantor in [Meschkowski 1967, 239], and the Hettner redaction [Weierstrass *unp.*, 304]; Schwarz is also mentioned in connection with Bolzano in [Dedekind 1930/32, vol. 2, 356–57].

Here we shall center on the German contributions, in accordance with the limits set for the present work in the Introduction. At any rate, the work of Mèray, first published in 1869 and similar to Cantor's, did not exert an influence on the three German authors. The theories of Dedekind and Weierstrass seem to be the oldest, dating back to the late 1850s; Weierstrass presented his ideas regularly in his courses at Berlin. But all three were first made known to the public in 1872, when Cantor and Dedekind published their famous articles, and Kossak dealt with the ideas of Weierstrass in a book which, however, was disowned by the great Berlin mathematician [Kossak 1872].

The rigorous definitions of the real numbers were, quite naturally, accompanied by work on the topology of the real line. The last section of this chapter will be devoted to related basic work by the three German mathematicians, including a short analysis of work by Dedekind that was only published in 1931.

1. 'Construction' vs. Axiomatization

The traditional definition of the real numbers relied on the notion of magnitude as previously given. Number was the ratio or proportion between two homogeneous magnitudes. This definition, based on the Greek theory of ratios, was first proposed by Stevin in the late 16th-century, and defended by such men as Newton and even Cauchy [Gericke 1971]. Such an approach had its shortcomings: it did not account for complex numbers, nor even for negative numbers, and, above all, the continuity of $\mathbb{R}$ was neither justified, nor explicitly required. This last point was particularly emphasized by Dedekind in his correspondence with Lipschitz: the continuity of the previously given domain of magnitudes was an *implicit* assumption, and the notion of magnitude was never precisely defined [Dedekind 1872, 316; 1930/32, vol. 3, 477; or Lipschitz 1986, 77–78]. In fact, for the traditional conception of mathematics there was no question of postulating that the real numbers, or the magnitudes, have the desired properties. Reference to magnitudes involved the idea that mathematical objects exist in physical reality, and so there is no problem of existence.

When it became clear that a rigorous development of analysis required a precise theory of the real numbers, mathematicians might have taken recourse to the axiomatic method, which in the form given to it in Euclid's *Elements* had always been taken to be prototypical of mathematics. As a matter of fact, none of the authors we are considering followed that path. What they did was in essence to assume the rational numbers as given, and build the system of real numbers on top of the rationals by means of certain infinitary "constructions." Hilbert referred to this as the "genetic" approach, when he proposed the axiomatic approach, instead, as more convenient and precise [Hilbert 1900]. One obvious reason for the option of the authors we shall consider is that the axiomatic method was not a paradigm for 19th-century mathematics, in the sense of a model followed in the actual conduct of research and writing. However, a number of mathematicians made contributions that can be regarded as forerunners of modern axiomatics; this includes not only authors working on geometry, but also a good number of mathematicians that investigated arithmetic and algebra. Among them we find the British school of symbolical algebra, and German authors such as Ohm, Grassmann, Hankel and Dedekind.

Actually, the fact that the foundations of analysis required an axiom of continuity (or completeness, in modern terminology) only became clear after the publications of Cantor and Dedekind in 1872 that we shall review. Both were convinced that continuity had to be axiomatically postulated in the realm of geometry, but not so in arithmetic. Taking this into account, a historiographical question emerges: Why was the "construction" approach preferred by all authors around 1870? Why none of them opted for an axiomatic viewpoint?

To some extent, we can point to reasons that were explicit in the mind of some of the historical actors. Dedekind regarded arithmetic as a development from the laws of pure thought, as a part of logic, and there are reasons to think that Weierstrass (and also Cantor early in his career) essentially agreed with that viewpoint. This, joined with the old conception of axioms as true propositions that cannot be proven, led them to the idea that arithmetic needs no axioms – everything can be rigorously proved starting from purely logical notions (see §VII.5). But, instead of following this issue farther, I would like to consider in this section some further reasons that are also interesting from a historiographical and philosophical viewpoint. The factors we shall review seem to have underlain the decisions taken by the historical actors, but they may appear to be meta-historical more than simply historical. Readers who are not interested in this kind of deeper reflection on the reasons for the limitations of thought in a period may safely skip the reminder of the present section and go directly to our exposition of the definitions of the real numbers (§2).

Answering a question such as the above – why the genetic approach and not the axiomatic one – is not easy, because most likely a host of factors have played a role in determining the attitude of 19th-century mathematicians. We shall consider some plausible reasons. To begin with, there probably was the psychological need to somehow justify the fact that we talk about real *numbers*. Had they simply postulated the basic properties that were needed in order to obtain the desired theorems in arithmetic and analysis, the question might have arisen, why not talk about points or anything else, instead of numbers? There had to be something in common that all of the different number systems shared, and 19th-century mathematicians looked for a common genealogy, a 'genetic' process by which the more complex systems emerge from the simpler. The means by which the genetic process took place was some infinitistic 'construction' that in retrospect can be characterized as set-theoretical (although in all cases one had to work with sets of some more or less complex structure).

The origins of this genetic program can be found in the work of Martin Ohm, already mentioned in §§I.3 and III.1. Ohm gave it a very general formulation, for he attempted to give a consistent (i.e., unitary, systematic) presentation of arithmetic, algebra and analysis on the only basis of natural numbers. Natural numbers were regarded as given, and the basic operations on them were established by taking into account their intuitive meaning. This enabled Ohm to formulate a set of fundamental equations as a starting point, from which the systematic development is carried out as follows. Ohm introduced the inverse operations – substraction, division, radication – and observed that they cannot always be carried out in the limited domain of the natural numbers. The requirement that it ought to be possible to realize them motivates an extension of the number system. In order to be rigorous, this extension must be accompanied by new definitions of equality and of the basic operations that can be applied to the new numbers; the process is guided by the requirement that the fundamental equations that were valid for natural numbers should be preserved.[1] Interestingly, this basic program can be found again in Dedekind (see §III.1) and also in Weierstrass;[2] the same may be true of Heine [1872, 174–75]. It thus seems that there was a German tradition, a rudimentary program for the foundations of pure mathematics, that led to important results in the hands of Weierstrass and Dedekind.

Ohm treated the new kinds of numbers as intellectual objects or mental symbols: given natural numbers a, b, the negatives were defined as symbols $a-b$; given integers a, b, the fractions were defined as symbols a/b. Ohm's definitions of equality and the operations [Bekemeier 1987, chap. II] were correct in these cases, and so his treatment turns out to be equivalent to later ones relying on ordered pairs (especially when this is presented in a formal way, not purely set-theoretically). Ohm was more interested in finding satisfactory definitions of the operations on the new numbers, than in giving an explicit justification for our right to "create," if only mentally, the new objects. On this point, his successors were more careful – actually, they tended to focus on the definition of the new numbers, rather than on defining the new operations.

A more radical departure happened in connection with the reals, for Ohm introduced them successively as roots, logarithms, etc., while the 1872 definitions are completely general. We have seen (§III.5.1) that Dedekind explicitly criticized this point, a shortcoming related to the fact that with $\mathbb{R}$ we are not facing an algebraic completion, but a topological one. Ohm's more advanced proposals in analysis were based on calculations with purely symbolical, and equivocal, expressions. While Ohm took symbolic forms as the objects of mathematics, Cauchy was treating symbols (numerical conditions, etc.) as mere tools for analyzing the properties of mathematical objects [Bekemeier 1987; Jahnke 1987]. Ohm's successors opted for a compromise with the conception of analysis favored by Cauchy, what we have called the conceptual approach (§I.3), and so the German tradition was gradually adjusted to the new rigorous analysis.

---

[1] This is the well-known principle of the permanence of equivalent forms, that was formulated around the same time by Peacock in Britain, and which can still be found in Hankel [1867].

[2] See, e.g., his critical comments on [Kossak 1872] in a letter to Schwarz quoted in [Dugac 1973, 144].

The factors that we have just reviewed can also be read in terms of a more theoretical issue, the problem of rigor. The genetic approach made possible a rigorous treatment of the number systems, because the central requirement adopted by Weierstrass, Dedekind, and Cantor (and Ohm) was that the new numbers, together with relations and operations on them, had to be defined in exclusive reference to previous numbers, their relations and operations. The "construction" of the real numbers presupposed only, or so it was said, the rational numbers. As these authors presented the matter, it seemed that the infinitary means employed by them were obvious or even transparent, not worthy of a more careful and explicit consideration. Perhaps the reason for this attitude was that they regarded them as logically admissible 'constructions' or procedures. This is certainly the case with Dedekind, and plausibly also with Weierstrass;[1] Cantor would have inherited the attitude as a member of the Berlin school. In the long run, it became clear that the infinitary means employed called for a foundation in set theory and that this entailed peculiar difficulties.

But, beyond questions of rigor, tradition, or psychological needs, I wish to point out a deeper and more urgent conceptual reason that seems to give a satisfactory explanation to the question why "construction" was preferred to axiomatization. During the last three quarters of the 19th-century, starting with Ohm in Germany and Peacock in England, we can observe a small trend of developments that gradually prepared the emergence of the axiomatic mentality. These authors ceased to rely on empirical assumptions regarding mathematical objects, and tried to offer purely deductive developments of their theories, based on careful analysis of the assumptions involved.[2] The British school became completely aware of the free interplay between formal conditions (abstract systems of algebraic laws) and models or interpretations, a key insight that prepared the emergence of modern axiomatics and logic (see, e.g., the introduction to [Boole 1847]). But, throughout this period, mathematicians had a real problem with simply establishing a system of laws as a basis for subsequent work; they could not just decree that mathematics considers this or that axiom system.

Mathematics had traditionally been regarded as a science that deals with some quite abstract, but no less real, objects – the magnitudes (see §II.1). The new trend advanced in the direction of superseding that outdated conception of mathematics, but the process was gradual, and they always paid careful attention to the question of *interpretations*. To grasp the limits of their viewpoints, we must recall that at that time the notion of consistency had not yet been formulated, the first proofs of relative consistency by means of models were still to come, and, of course, nobody had

---

[1] Klein, who knew him personally, used to call him a "logician" [Klein 1926, vol. 1, 152, 246].

[2] It is in this sense that they prepare the axiomatic orientation, though of course their approach is still far from 20th-century axiomatics with its emphasis on absolute freedom to set up consistent axiom systems. Even the notion of axiom evolved in the process (§3).

even dreamed of a formal proof of consistency in Hilbertian style.[1] As late as 1910, Russell was still writing:

freedom from contradiction can never be proved except by first proving existence: it is impossible to perform *all* deductions from a given hypothesis, and show that none of them involve a contradiction![2]

Only by exhibiting a model of the system can we show its consistency; for, as Frege wrote Hilbert in the 1900s, truth implies consistency [Frege 1980].

In the early phases of the development, a purely axiomatic approach seemed to be a matter of merely playing with symbols. In 1835, De Morgan confessed that, at first sight, Peacock's approach "appeared to us something like symbols bewitched, and running about the world in search of a meaning."[3] Although their explicit methodology placed more emphasis on other points, authors belonging to the British tradition of symbolical algebra always paid careful attention to the possibility of giving interesting mathematical interpretations for their abstract systems of laws. The books of Peacock and his followers are full of such examples of interpretations, and De Morgan even came to distinguish two parts of algebra, one devoted to the formal manipulation of laws, the other devoted to interpretation [Pycior 1983]. Analogous considerations apply to similar developments in Germany, to the writings of Ohm, Grassmann and Hankel. Weierstrass and Dedekind, being exponents of the conceptual approach, were even more interested in avoiding purely formalistic games with symbols.

Lastly, during most of the 19th-century, available means for the construction of models or interpretations were extremely limited; they were basically restricted to arithmetical or geometrical interpretations. The rise of set theory and of an abstract notion of structure were crucial in palliating those limitations. But, as a matter of fact, the 'constructions' of the real number system were pioneering examples of the use of (quasi) set-theoretic means for the construction of models, in a particularly complicated case. We thus come to the conclusion that an axiom system for the real numbers would not have been seen as a solution to the problem that mathematicians faced around 1870. To simply postulate an axiom system would have meant begging the question by means of an arbitrary formalistic trick. The models of $\mathbb{R}$ that our authors elaborated on the basis of $\mathbb{Q}$ played a key historical role in preparing the modern axiomatic mentality. Only after those models had been formulated, and with the modern set-theoretic viewpoint as a background, could Hilbert advance toward the characteristically 20th-century axiomatic standpoint. Besides, it was

---

[1] Euclidian or projective models for non-Euclidean geometry were given by Beltrami in 1868 and Klein in 1871; see [Kline 1972, chap. 38], [Gray 1989]. Hilbert's approach to the reals grew out of his work on the foundations of geometry; it seems that in 1899 he was still far from focusing on consistency, not to mention formal consistency proofs.

[2] Quoted in [Grattan-Guinness 1980, 438].

[3] Quoted in [Pycior 1983, 216]. I have to say that I do not agree with Pycior's interpretation of De Morgan's attitudes toward symbolical algebra; see [Ferreirós 1990; 1991].

Cantor and Dedekind who clarified the need for an axiom of continuity (completeness) in the axiomatization of the system of the real numbers.[1]

2. The Definitions of the Real Numbers

We proceed to a succinct analysis and comparison of the theories that were developed by Weierstrass, Dedekind and Cantor. In recent times, this topic has been analyzed above all by Dugac, but early in the century it was carefully studied by several authors, due to its undeniable historical importance, and because the definitions of $\mathbb{R}$ show quite clearly the conceptions of rigor promoted by each mathematician.[2] But perhaps the best joint presentation was that of Cantor himself [1883, §9, 183–190], though of course it does not offer a modern perspective. We have already mentioned that the infinitary means employed are of different kinds, and the authors do not always explicitly presuppose a foundation in the notion of set. Weiertrass employs infinite series, Cantor uses Cauchy sequences, and Dedekind resorts to his cuts; in all cases we have sets with some more or less complex structure, that of cuts probably being the simpler.

2.1. Weierstrass: series.   Weiertrass's theory was never published by himself, it has to be analyzed through redactions of his courses made by his students; from the 1860s, such redactions enjoyed a wide diffusion within the German mathematical community. Every two years, Weierstrass gave a course 'Introduction to the theory of analytic functions' at Berlin. Since he was convinced that the lack of rigor in analysis was due to a hasty and imprecise handling of the basic notions [Dugac 1973, 77], he devoted about one quarter of the course to a careful exposition of the notion of number and the arithmetical operations [Ullrich 1989, 150]. On the basis of several manuscripts originating in different years, both Dugac and Ullrich come to the conclusion that Weierstrass's theory of numbers remained essentially unchanged from the early 1860s.[3]

---

[1] Thus, in my opinion, the rise of axiomatics cannot be seen as a direct development of formalistic approaches; moreover, it is incorrect and anachronistic to regard the British school of symbolical algebra as a purely formalistic movement (see [Ferreirós 1990], and compare [Pycior 1987]).

[2] See [Dugac 1970, 1973, 1976] and his summarizing presentation in [Dieudonné 1978]. For early works, see [Pringsheim 1898], [Jourdain 1910], [Cavaillès 1962]. Perhaps I should add that the discussion in [Kline 1972, chap. 41] is not reliable.

[3] Dugac employs redactions by Schwarz (1861), Hettner (1874), Hurwitz (1878) and Thiéme (1886), and also the expositions of Kossak [1872] (based on the course of 1865/66) and Pincherle [1880] (based on the 1878 course). Ullrich has studied redactions by Killing (1868) and Kneser (1880/81), beyond the already mentioned ones of Hettner and Hurwitz, which seem to be particularly reliable [Ullrich 1989, 149]. Hurwitz's redaction has been published as [Weierstrass 1988].

Like Ohm, Weierstrass assumed the natural numbers as given, and presented the operations on them in an intuitive manner; he also motivated the introduction of new numbers in a similar way, and like his predecessor he required the preservation of formal laws (see above and [Dugac 1973, 144]). But, in spite of assuming $\mathbb{N}$ to be given, his explanations of the notion of natural number seem to prefigure the notion of set. According to him, the notion of number is formed by gathering in thought several things in which a common trait has been discovered;[1] particularly, it is formed when the things are identical for thought, in which case they are called the units of number (Hurwitz redaction, 1878, in [Dugac 1973, 96]).

Rational and irrational numbers are presented as 'Zahlengrössen,' numerical magnitudes, but under the traditional denomination lies an abstract conception of magnitudes that also seems a prefiguration of the set-theoretical viewpoint. Taking into account the need to introduce "new elements" in order to be able to carry out the operation of division without limits, Weierstrass introduces the "exact parts of the unity," numbers of the form $1/a$ with $a \in \mathbb{N}$ (Hurwitz redaction, 1878, in [Dugac 1973, 97–98]). This way of dealing with the rationals opens the way for a satisfactory treatment of the irrational numbers, since both are defined as aggregates of the unit and its exact parts. Weierstrass defines a numerical magnitude, 'Zahlgrösse,' to be a "number whose elements are the unity and its exact parts, of which there are infinitely many" [*op.cit.*, 98]. One may have numbers containing infinitely many elements of that kind, as long as they are given through a well determined law [*op.cit.*, 101]; like all other authors at the time, Weierstrass emphasizes the need of a defining property or concept. By introducing not just one unit, but four (which are called positive, negative, imaginary positive, and imaginary negative), Weierstrass is also able to solve the classical problems with the negative and complex numbers [*op.cit.*, 96].

The intuitive idea will be to treat real and complex numbers as finite or infinite series built from units and their exact parts. But the comparison of 'Zahlengrössen' takes place through *finite* sums of their elements, which is a logical requirement, since we only presuppose the finite arithmetic of the naturals and rationals. A key point is that the series ought to converge, and Weierstrass gives as a defining property that there should exist an integer n greater than *any* finite sum of elements of the 'Zahlgrösse.' Rigor is thus attained by employing only finite sums of rational numbers; as Cantor explained:

One sees that the productive moment connecting the set with the number which it defines, can be found here in *summation*; but one must emphasize as *essential* that one only employs the summation of a quantity of rational elements that is always *finite*, and that one does *not* set beforehand the number b to be defined as the sum Σa_n of the infinite series (a_n). This would be a *logical error*, because one rather obtains the definition of the sum Σa_n only when it is equated to the *given* number b, which by necessity has to be defined previously. I believe that this logical error, which Mr. Weierstrass avoided for the first time, had been

---

[1] Compare Dedekind's explanation of the notion of set as the mental gathering of different things, which, for any reason, are regarded from a common viewpoint [Dedekind 1888, 344].

committed by almost all in previous times, and was not noticed because it is one of the rare cases in which a true error cannot give rise to any important problem in calculations.[1]

Each 'Zahlengrösse' may be represented in (infinitely many) different ways, since there are different collections of units and exact parts of the unit which turn out to be equivalent. In order to define equality, Weierstrass allows some transformations that essentially make it possible to go from one representation to any other [Dugac 1973, 97–103]. He also gives the following definition: a 'Zahlengrösse' a' having a finite number of elements is called a *part* of another (arbitrary) 'Zahlengrösse' a if a' can be transformed into some a'', so that all elements of a'' are found as many times in a as well. Two numerical magnitudes a, b are equal if every part of a is a part of b, and inversely [*op.cit.*, 103]. Essentially, this means that the equality or inequality of real numbers are defined in terms of rational numbers, taken as their parts; finite sums are again the basis. Finally, the sum of a and b is the 'Zahlengrösse' that has as its elements all elements of a and b, taken together; and the product of a and b is the 'Zahlengrösse' whose elements are all possible products of elements of a times elements of b.

Thus, Weierstrass was able to give a logically rigorous definition of the real numbers, although a rather prolix and complex one. Dedekind and Cantor will simplify the matter by taking the arithmetic of the rational numbers as given. In Weierstrass's presentation, one can only object to the fact that the infinitely many "exact parts" of the unity are, apparently, introduced without any justification. Perhaps this is due to an oversight of Hurwitz in the redaction of the course; otherwise, we would have to think that, by sticking to the terminology of magnitudes, Weierstrass was not able to achieve total control of the implicit assumptions of his theory. His position tries to find a subtle compromise between the finite and the infinite, in line with his semi-constructivistic, 'formal conceptual' approach to mathematics (see §I.5). The infinite plays an essential role in the theory, since one departs from the infinitely many exact parts of the unity and since the irrational 'Zahlengrössen' have infinitely many elements. But one never operates with infinitely many elements, since the arithmetic of $\mathbb{R}$ is defined by means of finite sums of rational numbers.[2]

---

[1] [Cantor 1883, 184–85]: "Man sieht, dass hier das Erzeugungsmoment, welches die Menge mit der durch sie zu definirenden Zahl verknüpft, in der Summenbildung liegt; doch muss als *wesentlich* hervorgehoben werden, dass nur die Summation einer stets endlichen Anzahl von rationalen Elementen zur Anwendung kommt und *nicht* etwa von vornherein die zu definirende Zahl b als die Summe $\sum a_n$ der unendlichen Reihe (a_n) gesetzt wird; es würde hierin ein *logischer Fehler* liegen, weil vielmehr die Definition der Summe $\sum a_n$ erst durch Gleichsetzung mit der nothwendig vorher schon definirten, fertigen Zahl b gewonnen wird. Ich glaube, dass dieser erst von Herrn Weierstrass vermiedene logische Fehler in früheren Zeiten fast allgemein begangen und aus dem Grunde nicht bemerkt worden ist, weil er zu den seltenen Fällen gehört, in welchen wirkliche Fehler keinen bedeutenderen Schaden im Calcül anrichten können."

[2] On a subtler level, one may notice that Weierstrass employs freely the universal quantifier (see, e.g., the convergence criterion that he employed). Most likely, no author of this early period was able to understand the implications of such a move (see chapter X).

A different matter is whether and to what extent Cantor is faithful to the Berlin master when he calls the 'Zahlengrössen' "sets" in the text that we have quoted. Weierstrass's definition avoids using any word like 'set' or 'aggregate,' and it is not clear whether he conceives of the 'Zahlengrössen' as infinite sets or as infinite series. The fact that a 'Zahlengrösse' may contain the same number several times (see definition of 'part' above) suggests that the latter is the case. (Incidentally, this would be consistent with the central role that series play in his presentation of real and complex analysis.) But perhaps more important is the fact that Weierstrass does not seem to have employed any common notion, be it denoted by 'set' or 'manifold' or any other word, for the two cases of points and numbers (see §3.2). He neither seems to have completed the step from concrete notions to an general concept of set, nor to have advanced from sets with some structure to abstract sets; Dedekind and Cantor will make both steps, though not in their published work of 1872.

2.2. Cantor: fundamental sequences.   The theory of Cantor is closely related to that of Weierstrass, even in the terminology employed: his purpose is to define what is to be understood by a 'Zahlgrösse.' Since he was a student of Weierstrass, the influence of the latter is beyond question;[1] but Cantor was able to simplify the ideas of his teacher and give a much more elegant, concise, and practical presentation, that enjoyed great success. Where Weierstrass employed convergent series, Cantor uses Cauchy sequences of rational numbers; the relation is quite obvious: any infinite series can be associated to the sequence of its partial sums, so that a Cauchy sequence corresponds to a convergent series.

There is evidence that Cantor presented his theory in lectures of 1870 [Purkert & Ilgauds 1987, 37], but the occasion for publication came with a paper of 1872 on trigonometric series (to be analyzed in §V.3). Here it is enough to know that, in that paper, Cantor studied in detail some complex distributions of points on the real line, and felt the need to clarify the notion of real number in order to be able to present that work. It is doubtful whether he really needed to present his theory of irrational numbers for the purposes of the paper (see below). Perhaps he wished to take the opportunity of this article for an exposition of his ideas, knowing that Heine was going to give an exposition of them in a paper of his own [Heine 1872].[2]

Cantor departs from the domain of rational numbers, which he denotes by A, in order to define the domain B of the reals. The basic means is what he would call some years later "fundamental sequences" ('Fundamentalreihen,' [Cantor 1883, 186]); here he still talked of numerical magnitudes:

---

[1] Heine [1872, 173] and the disciples of Weierstrass usually regarded Cantor's theory as a particularly fortunate version of Weierstrass's theory. For a detailed study, see [Dauben 1979].

[2] Heine published his paper in the *Journal für die reine und angewandte Mathematik*, giving due credit to his younger colleague; interestingly, Cantor published in the *Mathematische Annalen*, then the semi-official journal of the Göttingen school of Clebsch.

When I speak of a numerical magnitude in general, it happens above all in the case that there is present an infinite series, given by means of a law, of rational numbers

(1) $a_1, a_2, a_3, ...$

which has the property that the difference $a_{n+m}-a_n$ becomes infinitely small with increasing n, whatever the positive integer m may be; or in other words, that given an arbitrary (positive, rational) ε one can find an integer n_1 such that $|a_{n+m}-a_n| < \varepsilon$, if $n \geq n_1$ and m is an arbitrary positive integer.

This property of the series (1) I will express by means of the words: "*The series (1) has a certain limit b.*"[1]

The symbols b will be the 'Zahlengrössen' that constitute the domain B of the real numbers. Mathematical terminology had not yet consolidated, so that the same word, 'Reihe,' was employed for series and sequences, but Cantor's definition is unequivocal as to the intended meaning. The fundamental sequences have also been called 'Cauchy sequences,' for their defining property agrees with the criterion of convergence given by Cauchy (and Bolzano).

Cantor goes on to make clear that the words "has a certain limit b" have in principle no other meaning than that of expressing the above mentioned property of (1). To every sequence one should in principle associate a different symbol $b, b', b'', ...$ and only then proceed to define the equality of 'Zahlengrössen' b and b'; this can be done by considering the behavior of the corresponding sequences (a_n), (a'_n), as follows: $b=b'$ if $lim\ (a_n-a'_n) \to 0$ when $n \to \infty$. (Cantor never employed equivalence classes in connection with his theory of $\mathbb{R}$, not even in [Cantor 1883, 185–86].[2]) Similarly, Cantor defines the order relations among the "symbols" [Zeichen] of the domain B, the relations between numbers of B and rational numbers, and the basic arithmetical operations [Cantor 1872, 93–94]. For example, one says that $b+b' = b''$ if among the corresponding sequences there is the relation

$$lim\ (a_n+a'_n-a''_n) \to 0 \text{ when } n \to \infty.$$

The real numbers take here the form of purely mental symbols associated to the fundamental sequences. Cantor tries to emphasize the abstractness of his approach when he writes that, in his theory, the numerical magnitude, "which in principle and in general lacks an object in itself, only appears as an element in propositions

---

[1] [Cantor 1872, 92–93]: "Wenn ich von einer Zahlengrösse im weiteren Sinne rede, so geschieht es zunächst in dem Falle, dass eine durch ein Gesetz gegebene unendliche Reihe von rationalen Zahlen: ... vorliegt, welche die Beschaffenheit hat, dass die Differenz $a_{n+m}-a_n$ mit wachsendem n unendlich klein wird, was auch die positive ganze Zahl m sei, oder mit anderen Worten, dass bei beliebig angenommenem (positiven, rationalen) ε eine ganze Zahl n_1 vorhanden ist, so dass $|a_{n+m}-a_n| < \varepsilon$, wenn $n \geq n_1$ und wenn m eine beliebige positive ganze Zahl ist. / Diese Beschaffenheit der Reihe (1) drücke ich in den Worten aus: "*Die Reihe (1) hat eine bestimmte Grenze b.*""

[2] As we saw, Dedekind employed them in his algebraic work of 1857, and later (1870s and 80s) in connection with his theory of the integers; see §VII.3.

that are endowed with objectivity."[1] The sentence makes apparent the difficulties one faced at the time in trying to express such a viewpoint.

Let us now discuss the use to which Cantor put his theory of irrational numbers in this particular paper. What he really needed for his theorem on trigonometric series was the notions of limit point and derived set of a point-set (see §V.3), that he went on to present. The theory of irrational numbers was used, however, in order to give an example of a point set whose nth derived set consists of a single point. What is most notable is that, to this end, Cantor felt the need to introduce real numbers of "higher kinds," as Dedekind said [1872, 317]. This is a peculiar trait of his theory, which he preserved in the second version of 1883, and which distinguishes it from the modern version. Having defined B as the totality of all 'Zahlengrössen' b [Cantor 1872, 93], he went on to define a new domain [Gebiet] C in analogous way, namely by means of fundamental sequences of elements of A and B [op.cit., 95]. Continuing the process analogously, we arrive at domains $D, E, F, \ldots$

Cantor was careful to point out that, while there are elements in B which have no corresponding member in A, the same is not true for the higher kinds: to each c we can find a b, to each b a c, that we can regard as equal to each other. But, in his theory,

it is ... essential to maintain the conceptual distinction between both domains B and C, just like the equality of two numerical magnitudes b, b' from B does not include their identity, but only expresses a certain relation that takes place between the sequences to which they are related.[2]

The distinction was essential for him, because Cantor used it in order to give an example of a point-set of the nth kind; to this end, he takes a single point in the real line, and considers its abscissa as determined by a numerical magnitude of the nth domain, say L, that complies with some conditions. By resolving this 'Zahlengrösse' into its elements of the $(n-1)$th, $(n-2)$th, ... domains, say $K, J, \ldots$, he finally gets to infinitely many rationals; the corresponding point-set, of points with rational abscissa, is a point-set of the nth kind [1872, 98–99].

We may regard the need that Cantor felt to preserve that distinction as a purely psychological one, because one can easily get the same results by directly employing sequences of real numbers. But this seems to be a fine example of the difficulties encountered in the early phase of development of set theory. Lacking a general theory of sets, there was no possibility of working with sets of sets, building unions

---

[1] [Cantor 1872, 95]: "in [der hier dargelegten Theorie] die Zahlengrösse, zunächst an sich im Allgemeinen gegenstandlos, nur als Bestandtheil von Sätzen erscheint, welchen Gegenständlichkeit zukommt."

[2] [Cantor 1872, 95]: "bei der hier dargelegten Theorie ... [ist es] wesentlich, an dem begrifflichen Unterschiede der beiden Gebiete B und C festzuhalten, indem ja schon die Gleichsetzung zweier Zahlengrössen b, b' aus B ihre Identität nicht einschliesst, sondern nur eine bestimmte Relation ausdrückt, welche zwischen den Reihen stattfindent, auf welche sie sich beziehen."

of families of sets, and the like.[1] Cantor circumvented this limitation by means of the formal trick of distinguishing several different kinds of real numbers, according to the way in which they are determined. Besides, he kept finding that distinction useful in order to apply his conception of the real numbers directly to the analysis of the "conceptual content" [gedankliche Inhalt] of the formulas of analysis [Cantor 1883, 189; 1872, 95]. A 'Zahlengrösse' l can always be set equal to a 'Zahlengrösse' $k, i, \dots c, b$, and inversely; and

> The results of analysis (disregarding a few known cases) can be brought into the form of such equalities, although the notion of number, as it has been developed here, carries in itself the germ for a necessary and absolutely infinite expansion (which is here touched upon only in reference to those exceptions).[2]

This is the most notable point in this whole matter: Cantor's thoughts on real numbers of higher kinds, and their relation with the theorems of analysis, led him to consider an extension of the higher kinds to infinity. In a way, this is an amazing anticipation of his later introduction of the transfinite numbers, although the context makes it clear that there is no direct relation between both.[3] From his student years, Cantor was fond of philosophy and theology, and he was particularly interested in the philosophy of Spinoza, which ascribes a central role to the idea of absolute infinity (see §VIII.1–2). This may have been one of the reasons why he showed an interest in expanding the domain of mathematics beyond the infinite, from such an early period.

Cantor's second exposition [1883] of the theory of real numbers includes a new theorem, that was not formulated in the first. After having defined the real numbers, one can rigorously prove that, given a fundamental sequence of real numbers (b_n), there is a real number b (determined through a fundamental sequence of rational numbers) such that lim $(b_n) \to b$ when $n \to \infty$ [Cantor 1883, 187]. Of course, it is not that Cantor was unaware of this fundamental property of $\mathbb{R}$ in 1872, simply that his exposition had not been completely rounded off from a logical viewpoint. Dedekind's 1872 paper, on the other hand, took the corresponding theorem as its culminating point. It thus seems that Cantor followed the model of his older colleague in the 1883 exposition. This difference is indicative of the different turn of mind of both mathematicians: Dedekind is the great systematician, who loved to

---

[1] As a matter of fact, Cantor did not normally employ sets whose elements are in turn sets in his work, not even in his mature work of the 1890s. His tendency to work with sets whose elements are (intuitively considered) simple, may explain why he never rounded off his theory of $\mathbb{R}$ by using equivalence classes.

[2] [Cantor 1872, 95]: "Auf die Form solcher Gleichsetzungen lassen sich die Resultate der Analysis (abgesehen von wenigen bekannten Fällen) zurückführen, obgleich (was hier nur mit Rücksicht auf jene Ausnahmen berührt sein mag) der Zahlenbegriff, soweit er hier entwickelt ist, den Keim zu einer in sich nothwendigen und absolut unendlichen Erweiterung in sich trägt."

[3] The issue of the real numbers of higher kinds and their relation with the theorems and formulas of analysis would be worthy of a detailed study, which certainly would prove difficult.

pay careful attention to logico-mathematical details; in his attention to rigor, he can only be compared with Weierstrass. By contrast, Cantor is a more creative mathematician, who is interested above all in difficult problems and results, is less inclined to engross himself in careful system-building, and tailors his methods to the needs of problem-solving.

2.3. Dedekind: continuity and cuts. Dedekind was led to search for a theory of the irrational numbers by the unsatisfactory experience of having to resort to geometrical intuition in teaching the elements of the calculus. When he taught analysis for the first time, in 1858, he felt "more than ever before the lack of a truly scientific foundation of arithmetic" [Dedekind 1872, 315], and resolved to work steadily on the problem; he arrived at his theory in that same year.[1] The fact that Dedekind's theory is completely independent from the other two is further confirmed by the great technical differences. Weierstrass and Cantor employ infinitistic "constructs" that were usual in analysis, series and sequences respectively; Dedekind chooses to rely on a new means for "construction." The resulting theory is simpler in that every real number corresponds to only one, or at most two, cuts, while it corresponds to infinitely many fundamental sequences or Weierstrass series. On the other hand, many analysts have preferred Cantor's presentation, precisely because it only employs notions that are of frequent use in analysis and thus can be applied more directly.

After careful research, Dedekind was convinced that the following theorem was a sufficient basis for analysis: if a variable magnitude x grows steadily, but not beyond all limits, it approximates a limit value [Dedekind 1872, 316, 332]; in modern terminology: a monotonically increasing and bounded sequence of real numbers has a unique limit.[2] In his first lectures on analysis he could only give a geometrical justification for that theorem, although he was convinced that one could prove it abstractly in terms of the basic property of continuity. It ought to be possible to "discover its true origin in the elements of arithmetic and to obtain thereby a real definition of the essence of continuity."[3] The notion of continuity had a fundamental role in analysis, as it had in Riemann's work on manifolds (chap. II), but it was never defined, not even actually used for the proof of basic theorems [*op.cit.*, 316]. Dedekind's search will be guided by the ideals of purity and autonomy for arithmetic, and the methodological principles, that we have reviewed in §III.5.

In fact, Dedekind was convinced that continuity was not a requirement of Euclidean geometry, that the need for a continuous domain of real numbers (or of

---

[1] Actually, on Nov. 24, 1858. The pedantic precision in Dedekind's datings becomes less surprising when one knows that this meticulous man kept a journal in which he even noted down daily temperatures and correspondence received.

[2] The equivocal use of "magnitude" both for numbers and functions was also typical of the period; Dedekind followed this usage also in lectures.

[3] [Dedekind 1872, 316]: "Es kam nur noch daran, seinen eigentlichen Ursprung in den Elementen der Arithmetik zu entdecken, und hiermit zugleich eine wirkliche Definition von dem Wesen der Stetigkeit zu gewinnen."

magnitudes) arises first in analysis. In his view, this explained why a theory of real numbers based on the Greek theory of ratios was insufficient and in lack of basic postulates – Euclid was not responsible, for he never aimed at that ad he did not need the axiom in question for his real purposes. The paragraph in which Dedekind communicates his opinion to Lipschitz, in 1876, is worthy of full quotation, for it brings the whole issue close to the modern axiomatic viewpoint:[1]

Let us analyze all assumptions, both explicitly and tacitly made, on which the whole edifice of Euclidean geometry rests; let us grant the truth of all its theorems, the possibility of carrying through all its constructions. (An infallible method for such analysis consists, in my opinion, in replacing all technical expressions by words that have just been invented (until then senseless); the building should not thereby collapse, if it is well constructed, and I assert that, e.g., my theory of the real numbers bears this test.) *Never*, so far as I have investigated, do we reach in that way the *continuity* of space as a condition that is indissolubly linked with Euclidean geometry. The whole system stands up even without continuity – a result that will certainly be astonishing to many, and which for that reason seemed to me well worthy of being mentioned.[2]

The same idea comes up in the preface to [Dedekind 1888], where he affirms that even if we eliminate all spatial points that have a transcendent coordinate for a given coordinate system, leaving only those whose coordinates are algebraic numbers, we still obtain a model where all Euclidean constructions can be carried out, and all Euclidean theorems preserve their validity [*op.cit.*, 339–40]. The correspondence gives evidence for the difficulties that the emergence of an axiomatic viewpoint had to face in this early period, for Lipschitz, a very apt mathematician, was unable to understand that axiomatic reading of the *Elements*.

Dedekind's article is carefully structured. §1 includes a brief analysis of the properties of $\mathbb{Q}$, in which the structure of field and the linear dense ordering are emphasized. In §2, the fact that points in a line satisfy the same order properties is employed as a justification for the correspondence between points and rational numbers. §3 includes a critique of the traditional definition of the reals, and attacks the key difficulty by giving a definition of continuity. In §4, an original proof for the existence of irrational numbers is given, and cuts on $\mathbb{Q}$ are employed in order to

---

[1] And prefigures a famous sentence of Hilbert, the one that talks about tables, chairs and mugs [Weyl 1944, 153].

[2] [Lipschitz 1986, 79; Dedekind 1930/32, vol. 3, 479]: "Man analysire alle Annahmen, sowohl die ausdrücklich als die stillschweigend gemachten, auf welchen das gesammte Gebäude der Geometrie Euklid's beruht, man gebe die Wahrheit aller seiner Sätze, die Ausführbarkeit aller seiner Constructionen zu (eine untrügliche Methode einer solchen Analyse besteht für mich darin, alle Kunstausdrücke durch beliebige neu erfundene (bisher sinnlose) Worte zu ersetzen, das Gebäude darf, wenn es richtig construirt ist, dadurch nicht einstürzen, und ich behaupte z.B., dass meine Theorie der reellen Zahlen diese Probe aushält): *niemals*, so weit ich geforscht habe, gelangt man auf diese Weise zu der *Stetigkeit* des Raums als einer mit Euklid's Geometrie untrennbar verbundenen Bedingung; sein ganzes System bleibt bestehen auch ohne die Stetigkeit – ein Resultat, was gewiss für Viele überraschend ist und mir deshalb wohl erwähnenswerth schien."

define the real numbers, and the natural order of ℝ. Then, §5 presents the order properties of ℝ: a linear dense ordering that is also continuous; this last theorem is proved. Operations on the real numbers are defined in §6 in terms of cuts, and, finally, §7 includes proofs of the basic theorem on monotonically increasing sequences, and of the Cauchy condition of convergence.

The title and the whole exposition emphasize the fact that the definition of continuity, and the proof of continuity of ℝ, constitute the core of the matter. This is a way of presenting the whole issue that differs from the expositions of Weierstrass, Heine and Cantor. Dedekind writes that only a precise definition of continuity will offer a sound foundation for "the investigation of *all* continuous domains." This has to be read not only in the context of analysis, but also in that of Riemann's work. Dedekind's radically deductive methodology shows up in the remark that vague statements about "the uninterrupted connection between the least parts" would be useless, and that one needs a precise definition that may be actually used in real proofs.[1] The basis for such a definition is found in the "phenomenon of the cut in its logical purity."[2] Already in ℚ we find that any rational number q determines a partition of the set [System] into two disjoint classes A_1, A_2, such that all numbers in the first are less than any number in the second class; q itself can be ascribed to any class, at will (thus, we have two different but equivalent cases).

In the case of a line, the reciprocal is also true, and here Dedekind sees the kernel of his theory:

I find the essence of continuity in the inverse, therefore in the following principle:

"If all points in the line are decomposed into two classes, such that each point in the first class is to the left of any point in the second class, then there exists one and only one point, which produces this division of all points in two classes, this cutting of the line in two parts."[3]

He takes this to be an axiom, an unprovable proposition (see below). Dedekind goes on to present a proof to the effect that, if D is a natural number but not a square, $\sqrt{D}$ is not a rational number; this establishes the result that there are cuts on ℚ which are not produced by a rational number, and opens the way to the introduction of new numbers:

---

[1] [Dedekind 1872, 322]: "nur durch [die Beantwortung dieser Frage] wird man eine wissenschaftliche Grundlage für die Untersuchung *aller* stetigen Gebieten gewinnen. Mit vagen Reden über den ununterbrochenen Zusammenhang in den kleinsten Teilen ist natürlich nichts erreicht; es kommt darauf an, ein präzises Merkmal der Stetigkeit anzugeben, welches als Basis für wirkliche Deduktionen gebraucht werden kann."

[2] [Dedekind 1888, 341]: "die Erscheinung des Schnittes in ihrer logischen Reinheit."

[3] [Dedekind 1872, 322]: "Ich finde nun das Wesen der Stetigkeit in der Umkehrung, also in dem folgenden Prinzip: 'Zerfallen alle Punkte der Geraden in zwei Klassen von der Art, dass jeder Punkt der ersten Klasse links von jedem Punkte der zweiten Klasse liegt, so existiert ein und nur ein Punkt, welcher diese Einteilung aller Punkte in zwei Klassen, diese Zerschneidung der Geraden in zwei Stücke hervorbringt.'"

Whenever one finds a cut (A_1,A_2), which is not produced by any rational number, we *create* a new, *irrational* number α, which we regard as completely defined by this cut (A_1,A_2); we shall say that the number α corresponds to this cut, or that it produces this cut.[1]

Much has been said about the emphasis that Dedekind places on the "creation" of new numbers. It is essentially equivalent to Cantor's introduction of the real numbers as "symbols" associated with the fundamental sequences, and, one should add, it is made rigorous by the requirement that the new numbers are to be regarded as "completely defined" by the cuts. Dedekind was always convinced that in mathematics we create notions and objects (see §III.1); this belonged in his basic convictions, his peculiar philosophy of mathematics.[2] He thus takes a position that seems somewhat surprising from a 20th-century viewpoint: mathematical objects are human creations, but no constructivist limitations apply to that process; it may be called a non-constructivistic intellectualism. As late as 1911, he regarded the step from some elements to the corresponding set as the quintessence of our creative mathematical powers [Dedekind 1888, 343].

One further reason to introduce the irrationals as new objects was the goal of preserving the homogeneity of numbers: he attempted to preserve some of our intuitive ideas concerning numbers, and keep the notion of number free from foreign traits (e.g., the fact that a cut contains infinitely many elements is completely foreign to our intuitive ideas regarding the reals; see the letter to Weber, [Dedekind 1930/32, vol. 3, 488–90]). This explains why Dedekind behaves here in the opposite way as with ideals, in spite of his claim that his methodological principles are the same. Perhaps one might say that here we deal with elementary mathematics, while there we treat an advanced problem. At any rate, it was just a matter of preference, as he wrote to Lipschitz in 1876:

if one does not wish to introduce new numbers, I have nothing to object; the theorem which I have proved (§5, IV) says then: the system of all cuts in the domain of rational numbers – discontinuous in itself – constitutes a continuous manifold.[3]

If we adopt this formulation, his treatment of the theory of cuts is essentially identical with the theory of ideals. Incidentally, it is noteworthy that Dedekind employs here the Riemannian term 'manifold,' and not his 'system,' particularly in view of

---

[1] [Dedekind 1872, 325]: "Jedesmal nun, wenn ein Schnitt (A_1, A_2) vorliegt, welcher durch keine rationale Zahl hervorgebracht wird, so *erschaffen* wir eine neue, eine *irrationale* Zahl α, welche wir als durch diesen Schnitt (A_1, A_2) vollständig definiert ansehen; wir werden sagen, dass die Zahl α diesem Schnitt entspricht, oder dass sie diesen Schnitt hervorbringt."

[2] See, e.g., [Dedekind 1854] and, a third century later, the 1888 letter to Weber in [1930/32, vol. 3, 488–90].

[3] [Lipschitz 1986, 64–65; Dedekind 1930/32, vol. 3, 471]: "will man keine neue Zahlen einführen, so habe ich nichts dagegen; der von mir bewiesene Satz (§5, IV) lautet dann so: das System aller Schnitte in dem für sich unstetigen Gebiete der rationalen Zahlen bildet eine stetige Mannigfaltigkeit."

the fact that Lipschitz had made important contributions to the theory of differential invariants, in connection with Riemann's Habilitationsvortrag.[1]

Once the irrationals have been "created," Dedekind can proceed to define the order relations and arithmetical operations, exclusively on the basis of cuts [1872, 326–27, 329–30]. To give just an example, let us see how the sum of α and β is defined. We consider the corresponding cuts (A_1, A_2), (B_1, B_2) and define a new cut (C_1, C_2) as follows: any rational number c will belong to the class C_1 if there are numbers a in A_1 and b in B_1 such that $a + b \geq c$; the real number γ that produces (C_1, C_2) is defined to be the sum $\alpha + \beta$.

The work culminates in the proof that $\mathbb{R}$ is continuous, according to the definition given above. This means that, given any cut on $\mathbb{R}$, there is one and only one number, rational or irrational, that produces it. The proof is extremely simple, for it suffices to consider the real number defined by the cut on $\mathbb{Q}$ which is contained in the cut on $\mathbb{R}$ [Dedekind 1872, 329]. Finally, Dedekind goes on to prove the theorem on increasing bounded sequences that motivated the whole investigation, and also the validity of the Cauchy condition, proved directly on the basis of the continuity of $\mathbb{R}$ [*op.cit.*, 332–33].

3. The Influence of Riemann: Continuity in Arithmetic and Geometry

Dedekind employs the comparison between numbers and points in the line as an organizing thread for the presentation of his paper. This makes his exposition different from those of Heine [1872] or Weierstrass, though notably not from that of Cantor. For Cantor, too, discusses the relation between his "numerical magnitudes" and the geometry of the straight line [Cantor 1872, 96–97]. More surprisingly, both mathematicians coincide in their break with the traditional conception of the matter. It had been customary to assume that the continuity of space or of the basic domain of magnitudes induces, through the definition of real numbers as ratios, the continuity of the number system. But now we find two mathematicians who emphasize the point that it is possible to define abstractly a continuous number system, while geometrical space is not necessarily continuous. One needs an axiom, sometimes called the *axiom of Cantor–Dedekind*, to postulate that space is continuous.

Both Cantor and Dedekind make it explicit that here we are talking about an axiom in the old sense of the expression – an unprovable proposition that is needed as a basis for the theory of space. They do not yet use the word "axiom" in the sense of modern axiomatics. Cantor puts the point more succinctly; one has to postulate that to each real number there is a corresponding point, and this is an axiom:

---

[1] After discussing the introduction of the real numbers, their correspondence goes on to deal with Riemannian topics [Lipschitz 1986, 82–85].

I call this proposition an *axiom*, because not being generally provable belongs in its nature.[1]

Dedekind and Cantor think that a discontinuous space is perfectly conceivable, and both of them returned to this issue later. Cantor comes back to it in a paper of 1882, where he states more explicitly the possibility of a discontinuous space, and uses his non-denumerability results for an interesting application: he elaborates on the possibility of a modified mechanics in which the underlying space would be discontinuous, but continuous motion would be possible [Cantor 1879/84, 156–57] (see §VI.4.3). Dedekind discussed it again in his correspondence with Lipschitz (§1.3) and explains the matter in more detail in the preface to [Dedekind 1888]. He asserts that the set of all points in $\mathbb{R}^3$ which have only algebraic numbers as coordinates is a model of Euclidean geometry; the discontinuity of this space would not be noticed or experienced by a Euclidean geometer [*op.cit.*, 339–40; also Dedekind 1930/32, vol. 3, 478].

In order to explain this surprising coincidence, I believe we must take into account Riemann's lecture on the hypotheses of geometry. This helps round the picture that we started to draw in chapter II, for, if I am right, we find confirmation that [Riemann 1854] was not simply a pioneering contribution, but a really acting factor in the early development of set theory. In his lecture, Riemann regarded the properties of space as consequences of some hypotheses which are to be experimentally tested; they can be very probable, but never absolutely certain. The continuity of physical space is taken as the first such hypothesis, so that the possibility that real space may be discontinuous, i.e., may be a discrete manifold, is never discarded. For instance, in the last section Riemann discusses the "internal cause" [innerer Grund] for the actual metrics of space. He remarks that the issue looks quite different if space is a discrete manifold, for then the principle of the metrics would be contained in the very notion of that manifold; while, if space is continuous, the cause of the metrics must be found somewhere else, in the linking forces that act on the manifold [Riemann 1854, 286].

Therefore, one should expect that a mathematician deeply influenced by Riemann would assume the hypothetical character of spatial continuity, and this is what happens with Cantor and Dedekind. As we have remarked time and again, Riemann's influence on Dedekind is beyond question, both because of their intimate friendship and because of Dedekind's involvement in the publication and further development of [Riemann 1854]. The coincidence and the explanation I have just presented is more interesting in case it is accepted as evidence that Cantor too was under the influence of Riemann's ideas. As regards this, one must take into account that Riemann's lecture was published together with his Habilitation thesis on trigonometric series, and that this last work was crucial for Cantor's work in the period 1870–72, which culminated in the paper that we have discussed. Since the

---

[1] [Cantor 1872, 97]: "Ich nenne diesen Satz ein *Axiom*, weil es in seiner Natur liegt, nicht allgemein beweisbar zu sein." Compare [Dedekind 1872, 322–23]: "Die Annahme dieser Eigenschaft der Linie ist nichts als ein Axiom ... Hat überhaupt der Raum eine reale Existenz, so braucht er doch *nicht* nothwendig stetig zu sein."

details will be discussed in chapter V, it is enough here to know that the techniques applied by Cantor in his work on trigonometric series were based on Riemann's work on the same topic. Thus, Cantor was very familiar with at least one of the two papers; most likely, he was very familiar with both, since in 1877 he acknowledged that, for some years, he had been following the discussion on the foundations of geometry motivated by the work of Gauss, Riemann, Helmholtz and others [Cantor & Dedekind 1937, 33]. In the next chapters we shall go on presenting further data that seem to justify my emphasis on the important role played by Riemann in the evolution of Cantor's thought.

One last comment regarding the axiom of continuity. As Dedekind and Cantor saw it, geometry depends upon an axiom that can be avoided in arithmetic, for an abstract *definition* of a continuous number domain is possible. In a way, this can be interpreted as proof that arithmetic is 'superior' to geometry, since it is a 'purer' discipline (reliance on axioms, in the old meaning of the word, means reliance on intuition or experience). Above all, the definition of the real numbers, coupled with the traditional conception of logic (§II.4), could lead rather directly to a logicist viewpoint. If number is purely a product of our minds, as Gauss thought (§I.2.2), one may easily conclude that arithmetic is *a priori* and a good candidate to being purely logical,[1] especially in case one has a sufficiently wide conception of logic – for, then, all that arithmetic requires is the natural numbers plus logical conceptual means. Everything points to the interesting speculation that the abstract definition of the real numbers was the crucial novelty that triggered the emergence of logicism, not only in the case of Dedekind, but also of Frege. Of course, it remained to explain the natural numbers as purely logical, and this is what both authors attempted; both started to carry out this program in the 1870s, which is in good agreement with that hypothesis.

4. Elements of the Topology of ℝ

Detailed study of the real numbers was accompanied by research on the topology of the real line and spaces. As we shall see in the next chapter, the theory of integration and of trigonometric series showed the need for such research, and led Cantor to the notion of derived set. But his teacher Weierstrass had already prepared the stage with investigations that he presented in his lectures on function theory; the theory of point-sets emerged from the work of both mathematicians. Dedekind, too, did some basic research on topology in connection with the work of Dirichlet and Riemann, but he never came to publish, so that his work never had an effect on other mathematicians. Since the work of Weierstrass and Cantor connects directly with the next chapter, we shall begin commenting on that unpublished work.

---

[1] Schröder wrote that Weierstrass, Cantor and Dedekind had established the purely analytical character of the truths of arithmetic, and therefore Kant's famous question: how are synthetic *a priori* judgements possible?, simply lacked an object [Schröder 1890/95, vol. 1, 441].

4.1. Dedekind and topology.   In order to complement our discussion of Dedekind's conceptions as of 1872, it is interesting to consider a manuscript that was first published by Emmy Noether in the collected works of 1930/32. The manuscript bears the title 'Allgemeine Sätze über Räume' [General theorems on spaces] and stems from the 1860s (my dating).

In the 1860s, Dedekind was intensely engaged in the publication of the works of Dirichlet and Riemann. After the first edition of Dirichlet's *Vorlesungen über Zahlentheorie* in 1863, he prepared for publication his lecture on forces that are inversely proportional to the square root of the distance [Dedekind 1930/32, vol. 3, 393], where Dirichlet made important contributions to potential theory. Dirichlet employed here what Riemann would call the "Dirichlet principle." Already in Göttingen, Dedekind had noticed that there were "difficulties" in Dirichlet's proof of the principle, and "occasionally" he discussed them with Dirichlet and Riemann.[1] While preparing the lecture for publication, Dedekind tried to find a satisfactory foundation for the principle, and actually he thought he was in a position to prove it [Dugac 1976, 177–78]. The problem is to establish the existence of a minimal continuous function on a given domain and under given boundary conditions; in order to prove it, Dedekind needed some basic notions on continuous domains, or, what comes to the same, he had to develop some elements of the topological theory of manifolds.[2]

After 1866 Dedekind was absorbed by his work on Riemann's *Nachlass*, and then, in 1869, he resumed work on ideal theory on the occasion of the second edition of Dirichlet's *Zahlentheorie*. He then abandoned the project of publishing Dirichlet's lectures on potential theory, and thus it is clear that his manuscript was written some time between 1863 and 1869. This is further confirmed by a letter to Cantor of January 1879, that Noether quotes in full in her editorial comment [Dedekind 1930/32, vol. 2, 355; Cantor & Dedekind 1937, 47–48]. Dedekind speaks about the need for precise definitions, independent of geometrical intuition, of the names or technical expressions of the theory of manifolds, and says that

many years ago, while I still intended to publish Dirichlet's lecture on potential, and to give a more rigorous foundation for the so-called Dirichlet principle, I occupied myself very much with such questions. I have some such definitions which, it seems to me, offer a very good foundation; but later I left the whole issue lie, and at present I could only offer something incomplete, because I was extremely busy with the reworking of Dirichlet's Zahlentheorie.[3]

---

[1] See the letter to Heine, probably not sent, in [Dugac 1976, 177–78]. Most likely, and taking into account one of Heine's publications, the letter was written before 1871, see [*op.cit.*, 107].

[2] See [Sinaceur 1990, 239], where he remarks on the connection between the manuscript under discussion and his work on Riemann spaces ("ideal geometry").

[3] [Cantor & Dedekind 1937, 47–48]: "Ich würde mir ein solches Urteil nicht erlauben, wenn ich nicht vor vielen Jahren, als ich noch die Dirichletsche Potentialvorlesung herausgeben und dabei das sogenannte Dirichletsche Prinzip strenger begründen wollte, mich schon recht viel mit solchen Fragen beschäftigt hätte. Ich habe einige solche Definitionen, die mir eine recht gute Grundlage zu geben scheinen; aber ich habe später die ganze Sache liegen lassen, und könnte für

Dedekind works within the context of metric spaces, without mentioning it explicitly, and presents the notions of open set, interior, exterior and boundary, together with elementary theorems. Once more, Dedekind's technical term for sets is 'System;' the word 'Körper,' which from the second edition of the *Zahlentheorie* he employed exclusively for number fields, denotes open sets here:[1]

A system of points p, p' ... forms a body [Körper], if for every point p one can determine a length δ such that all points, whose distance from p is less than δ, also belong to the system P. The points p, p' ... lie *inside P*.[2]

Next, Dedekind proves that the set of all points whose distance from a given point p is less than a given length δ form an open set, which is called a "sphere" [Kugel]. Dedekind goes on to define that p "lies *outside*" [liegt *ausserhalb*] of the body (open set) P if there is a sphere with center p such that all of its points are not inside P; and he proves that if there is one point outside of the open set P, there are infinitely many such points and they form a 'Körper.' If P is an open set and π a point which does not lie inside, nor outside of P, one says that π is a "limit point" [Grenzpunkt] of P; the set [System] of all limit points of an open set P is called the "*boundary*" [Begrenzung] of P. The last theorem says that the boundary of an open set is not an open set [Dedekind 1930/32, vol. 2, 353–55].

One may safely say that, had Dedekind published these ideas, together with his work on the Dirichlet principle, around 1870, the history of the theory of point-sets and of function theory would have been somewhat different. As always, his presentation is concise, clear, systematic and strictly deductive. Interestingly, Cantor, whose work inaugurated the study of the topology of point-sets, never formulated the notion of an open set – although he presented that of a closed set in 1884 [Cantor 1879/84, 226]. Dedekind's approach would be rediscovered by Peano, who presented it in his *Applicazioni geometriche del calcolo infinitesimale* of 1887, and taken up by Jordan in his influential *Cours d'analyse* of 1893/96 (see [Hawkins 1970, 86–90]). But the most important point, here, is that Dedekind applied the set-theoretical approach and terminology not only in his work on the number system, on algebra and on algebraic number theory, but also in the realms of topology, analysis, and Riemannian geometry.

4.2. The principle of Bolzano–Weierstrass. In his already mentioned, biannual course on the theory of analytic functions, Weierstrass presented basic notions of the topology of ℝ and ℝⁿ. This was done after his careful discussion of the number

---

den Augenblick nur Unvollständiges geben, da ich durch die Umarbeitung der Dirichletschen Zahlentheorie ganz in Anspruch genommen bin."

[1] This is further, internal evidence for my dating.

[2] [Dedekind 1930/32, vol. 2, 353]: "Ein System von Punkten p, p' ... bildet einen Körper, wenn für jeden Punkt p desselben sich eine Länge δ von der Beschaffenheit angeben lässt, dass alle Punkte, deren Abstand von p kleiner als δ ist, ebenfalls dem System P angehören. Die Punkte p, p' ... liegen innerhalb P."

system, and before the notion of analytic function was introduced. As prerequisites, Weierstrass showed that a bounded set of real numbers has greatest lower and least upper bounds, and proved a number of theorems by means of variations of the well-known method of bisection that Bolzano had employed in 1817 [Ullrich 1989, 153, 156]. The theorems in question are: the Bolzano–Weierstrass theorem, the intermediate value theorem (mentioned above), and the theorem that a continuous function, that is bounded inside a closed interval, reaches its maximum and minimum. We shall discuss the basic notions related with the Bolzano–Weierstrass theorem.

I have mentioned that it is quite unclear whether Weierstrass's approach can be characterized as set-theoretical. The Bolzano–Weierstrass theorem is presently formulated as the theorem that an infinite, bounded set of real numbers has a limit point. But even in 1874, when he knew of Cantor's early work, Weierstrass formulated it as follows:

Suppose that, within the *domain* of a real magnitude x, one has defined in a certain way another magnitude x', but in such a way that it can take infinitely many values, all of which fall within two definite limits; then, one can prove that within the domain of x there is at least one place a, such that in any arbitrarily small neighborhood of a there are infinitely many values of x'.[1]

Certainly, Weierstrass employs the word "domain" [Gebiet], which denotes the set of possible values of x; he has the real line in mind, for immediately below he makes clear that x can take as values all real numbers from $-\infty$ to $+\infty$. Later on, he even speaks of n-dimensional manifolds ["*eine nfache Mannigfaltigkeit*"], meaning n-tuples of real numbers, that is, the 'domain' $\mathbb{R}^n$ (Hettner redaction, [Weierstrass *unp.*, 313]).[2] In a characteristic way, Weierstrass only uses those words for embedding spaces, but when it comes to arbitrary point sets, he prefers the old terminology of variable magnitudes, and so he seems to refrain from adopting a set-theoretical viewpoint. It is worth noting that the variable magnitudes have to be "defined in a certain way" – we must have a defining property in order to be able to talk about a variable magnitude, a function.

The notion of neighborhood [Umgebung] is used in the modern sense, and, according to Dugac [1973, 120], it appears already in the lectures of 1861. On the other hand, the notion of limit point is clearly implicit in the formulation of the

---

[1] Hettner redaction, [Weierstrass *unp.*, p. 305] (I thank Gregory H. Moore for making a copy available to me): "Es sei im *Gebiet* einer reellen Grösse x eine andere Grösse x' auf bestimmte Weise definiert, jedoch so, dass sie unendlich viele Werte annehmen kann, die sämtlich zwischen zwei bestimmten Grenzen liegen, dann kann bewiesen werden, dass im Gebiete von x mindestens ein Stelle a gibt, die so beschaffen ist, dass in jeder noch so kleinen Umgebung von ihr es unendlich viele Werte von x' giebt."

[2] According to Gregory Moore, who is studying the origins of Weierstrass's notion of limit point, the Berlin mathematician seems to have started using the words 'Gebiet' and 'Mannigfaltigkeit' after 1868, which makes it likely that he was influenced by the publication of [Riemann 1854].

theorem, but Weierstrass – in contrast with Cantor – does not seem to have introduced special terminology for it. This may be related to his methodological preferences: instead of talking about limits points in an abstract way, Weierstrass prefers to explain carefully how the point a can be explicitly given by means of an arithmetical expression.[1] Due to his extremely careful and explicitly arithmetical way of proceeding, the proof of the Bolzano–Weierstrass theorem for ℝ occupies slightly more than 5 pages in Hettner's redaction, with 5 more pages devoted to the case of $\mathbb{R}^n$.

The proof method is that of bisecting intervals, or, better, n-secting an interval. Given that the magnitude x' takes infinitely many values in an interval, if we divide it into n subintervals (of equal length), at least one of them must present infinitely many values of x', etc. (Hettner redaction, [Weierstrass *unp.*, 305–10]). The method rests on the principle that, given a sequence of closed intervals, embedded on each other, there must be at least one point that belongs to every interval. This can be taken to be a continuity or completeness principle; given its importance for Weierstrass's students, and in particular for Cantor, we shall give it a name: the '*Bolzano–Weierstrass principle*.' (As we have already said, it seems that Weierstrass himself referred to Bolzano is his lectures.) However, Weierstrass did not formulate it as a particular principle: he just showed how to construct an expression for the point a from a given partition of the real line into intervals of equal length, so that a is characterized by means of a series [*ibid.*]. This, of course, agrees perfectly well with his definition of the real numbers.

Already in 1870 Kronecker objected to that proof method, calling the proofs based on it "obvious sophisms" [offenbare Trugschlüsse], and saying that Kummer, Borchardt and Heine agreed with him.[2] Schwarz and Cantor believed that the Bolzano–Weierstrass principle was unobjectionable, and admitted that they used it time and again. In work of 1884, Cantor comes back to this matter and points out – without mentioning his name – that the criticisms of Kronecker are purely skeptical arguments, comparable with the "paralogisms" of Zeno of Elea [Cantor 1879/84, 212].[3] Cantor also suggests that the "essence" of the principle can be traced back to work in number theory by Lagrange, Legendre and Dirichlet, and that it can also be found in Cauchy, so that there is no reason to link it exclusively to Bolzano and Weierstrass.

4.3. Cantor's derived sets.   The notion of limit point and the Bolzano–Weierstrass principle became two basic pillars for Cantor's work on point-sets. Most likely he took both from Weierstrass's lectures; he attended the lecture on analytic functions in the winter of 1863–64. Cantor used those tools in connection with

---

[1] Hettner redaction, [Weierstrass *unp.*, 308, 317].

[2] See the letter from Schwarz to Cantor in [Meschkowski 1967, 68; also 239]. It is difficult to believe that Heine agreed, in the light of his [1872].

[3] Compare a letter to Mittag-Leffler, written in that same year, where he compares Kronecker's arguments with those of the skeptics [Schoenflies 1927, 12–13].

radically new problems: his 1874 proof that the continuum is uncountable rests on the above mentioned principle (see §VI.2). But, interestingly enough, he gave the ideas of Weierstrass a characteristically abstract turn, and there are reasons to think that the influence of Riemann was here in action.

[Cantor 1872] deals with a generalization of a theorem on trigonometric series, in fact the theorem that the function represented by a trigonometric series is unique. As we have said, the basic techniques employed come from Riemann [1854a]. Heine and Cantor had seen the way of generalizing that theorem to the case where there are finitely many exceptional points, at which the trigonometric series is not convergent or does not coincide with the represented function. But it was Cantor's merit to have noticed that the result admitted of a further generalization to some infinite sets of exceptional points. The notion of derived set was introduced in order to characterize in a precise way these infinite point-sets.

To begin with, Cantor introduces some convenient terminology. When there are finitely or infinitely many 'Zahlengrössen' (resp., points), he will speak, "for brevity's sake," of a "*set of values*" [Wertmenge] (resp., a "*set of points*" [Punktmenge]) [Cantor 1872, 97]. Curiously, the terminology employed here is different from that used in the introduction of the real numbers, where he speaks of the "domain" *B* [*op.cit.*, 94–95], in accordance with Weierstrass's terminology. It seems reasonable to conclude that Cantor still lacked an overall set-theoretical viewpoint; this would be the reason why he distinguished the case of the embedding space ℝ, called a 'Gebiet,' from that of the point-sets within it, called 'Mengen.' Nevertheless, it is clear that Cantor was quite close to such a set-theoretical viewpoint. Disregarding Dedekind, no other German mathematician employed more consciously and explicitly the notion of set as of 1872. The reader should recall, from the Introduction, that the word 'Menge' had initially a rough meaning: to speak of a 'Punktmenge' would sound like speaking of a mass of points. Perhaps for this reason, later on Cantor preferred to speak of 'manifolds' and 'theory of manifolds,' although he kept employing the terminology of 1872 (probably for brevity's sake). It was from 1895 that he finally decided to refer to set theory as 'Mengenlehre' throughout.

In order to prepare the reader for the notion of derived set, Cantor next presents the idea of limit point and mentions in passing the Bolzano–Weierstrass theorem. His presentation of these basic notions differs from that of his teacher, who did not introduce a particular name for limit points, nor define them *in abstracto* (at least not in the 1874 lectures). Immediately thereafter, Cantor goes on to introduce the notion of derived set, which widens the gap that separates him from Weierstrass. In my opinion, the divergence is aptly described as the difference between the formal conceptual approach pursued at Berlin, and another viewpoint that is quite close to the abstract conceptual one promoted by Riemann. This is the relevant text:

By a limit point of a point-set *P* I understand a point in the line whose position is such that in every one of its neighborhoods there are infinitely many points of *P*, where it may also happen that that same point belongs to the set too. By a neighborhood of a point one should understand here any interval which has the point *in its interior*. According to that, it is easy to prove that a ["bounded," Zermelo] point-set formed by an infinite number of points always has at least *one* limit point.

It is a well determined relation between any point in the line and a given set P, to be either a limit point of it or no such point, and therefore with the point-set P the set of its limit points is conceptually co-determined; this I will denote P' and call the *first derived point-set* of P.

If the point-set P' does not consist of only a finite number of points, it also has a derived point-set P'', which I call *the second derived* [set] *of P*. Through ν such transitions one finds the concept of the νth derived point-set $P^{(\nu)}$ of P.[1]

Cantor's mastery of set-theoretical language is evident in this passage, which clearly marks the difference in approach with Weierstrass. What is really original in this contribution is that Cantor does not consider limit points in isolation, so to say, as Weierstrass had done, but makes the step toward a set-theoretical perspective. As a result, 'set derivation' is conceived as an operation on sets. This, more than his theory of the real numbers, is what makes the Cantor of 1872 a mathematician that is clearly placed on the road to set theory. On the other hand, the reader should notice that Cantor motivates his introduction of the derived set P' by observing that the relation 'being a limit point of P' is well determined, and so the set P' is *conceptually* determined, once P has been given. This confirms that his standpoint is analogous to that of Riemann in his [1854]: a set is understood as the extension of a concept.[2] Cantor adds, of a 'well defined' concept, and perhaps this emphasis reveals the influence of Weierstrass.

For his theorem on trigonometric series, Cantor considered those point-sets whose nth derived set, for a natural number n, consists of finitely many points [Cantor 1872, 99]. Seven years later, in order to express that condition more conveniently, he introduced the following terminology: if $P^n = \varnothing$ for some $n \in \mathbb{N}$, P is a point-set "of the *first species* and of the nth type" [von der *ersten Gattung* und von der n^{ten} Art]; if the derivation process goes on for all finite n, P is a point-set "of the *second species*" [von der *zweiten Gattung*; Cantor 1879/84, 140]. An obvious example of a set of the second species is any dense set, for instance the set of

---

[1] [Cantor 1872, 98]: "Um diese abgeleiteten Punktmengen zu definiren, haben wir den Begriff *Grenzpunkt einer Punktmenge* vorauszuschicken. / Unter einem Grenzpunkt einer Puntmenge P verstehe ich einen Punkt der Geraden von solcher Lage, dass in jeder Umgebung desselben unendlich viele Punkte aus P sich befinden, wobei es vorkommen kann, dass er ausserdem selbst zu der Menge gehört. Unter Umgebung eines Punktes sei aber hier ein jedes Intervall verstanden, welches den Punkt *in seinem Innern* hat. Darnach ist es leicht zu beweisen, dass eine aus einer unendlichen Anzahl von Punkten bestehende ["beschränkte"] Punktmenge stets zum Wenigsten *einen* Grenzpunkt hat. / Es ist nun ein bestimmtes Verhalten eines jeden Punktes der Geraden zu einer gegebenen Menge P, entweder ein Grenzpunkt derselben oder kein solcher zu sein, und es ist daher mit der Puktmenge P die Menge ihrer Grenzpunkte begrifflich mit gegeben, welche ich mit P' bezeichnen und die *erste abgeleitete Punktmenge* von P nennen will. / Besteht die Punktmenge P' nicht aus einer blos endlichen Anzahl von Punkten, so hat sie gleichfalls eine abgeleitete Punktmenge P'', ich nenne sie *die zweite abgeleitete von P*. Man findet durch ν solcher Uebergänge den Begriff der ν^{ten} abgeleiteten Punktmenge $P^{(\nu)}$ von P."

[2] Recall that, in defining the real numbers, Cantor says that the fundamental sequence is "given by a law" [*op.cit.*, 92].

rationals in the interval $(0,1)$, whose first derived set is all of $[0,1]$; here we have derived sets P'', P''', ..., P^∞ which are all equal to each other. It was by considering this and more interesting examples of point-sets of the second species, that Cantor was led to the idea of employing "symbols of infinity" in order to determine the "type" of such sets (in the above sense). These symbols of infinity, which were introduced in print in 1880, are the forerunners of the transfinite numbers (see chap. VIII).

Cantor's derived sets, building upon ideas developed by Weierstrass in the context of the Bolzano–Weierstrass theorem, constituted a decisive step toward the theory of point-sets, which will be the topic of the next chapter. The arithmetizing approach incorporated in the definitions of the real numbers, that came to public knowledge in 1872, became the basis for a rigorous study of the topology of $\mathbb{R}$. For some decades, the context of real functions and integration theory would give impulse to this new line of developments, which in the long run contributed essentially to the emergence of topology as a core, autonomous branch of 20th-century mathematics.[1] We thus turn to real functions, integration, and point-sets.

---

[1] A development that seems to have been foreseen, however vaguely, by Riemann (§II.5).

V   Origins of the Theory of Point-Sets

> For the stimulus to pursue these studies I am indebted essentially to the writings of Riemann, in particular his brilliant work on trigonometric series, after the publication of which one needs no apology for devoting oneself to these questions, which, as this author remarks in agreement with Dirichlet, 'stand in the most intimate connection with the principles of the infinitesimal calculus and may contribute to bring them to greater clarity and precision.'[1]

In the 1870s the theory of functions of a real variable consolidated into an autonomous branch of mathematics. Its initial development was intimately related to the theory of trigonometric series, a subject in which Dirichlet's work was a milestone. Point-set theory was initially developed as a tool for the study of trigonometric series and real functions. Early steps in this direction were taken by Dirichlet, Lipschitz and Hankel, but Cantor's work on derived sets was considerably more sophisticated than the previous rather rough ideas regarding possibilities for point-sets (i.e., subsets of $\mathbb{R}$).

A key factor in the emergence of real variable theory was the publication of papers that acquainted the public with the ideas of Riemann and Weierstrass. The publication in 1868 of Riemann's *Habilitationsschrift*, 'On the Representability of a Function by a Trigonometric Series' [Riemann 1854a], had an immediate impact, as can be seen from the quotation above. It seemed to most mathematicians that Riemann had found the most general conception of the integral, and his work immediately stimulated deeper research on real functions.[2] Immediately afterwards, authors such as Heine, Schwarz, Cantor and du Bois-Reymond revealed several aspects of Weierstrass's work on the foundations of analysis – e.g., the notion of uniform convergence and some examples of continuous nondifferentiable functions.

---

[1] [Hankel 1870, 70]: "Ich verdanke die Anregung zu diesen Studien wesentlich Riemann's Schriften, in's Besondere seiner glänzenden Arbeit über die trigonometrischen Reihen, nach deren Erscheinen es keiner Entschuldigung mehr bedarf, sich mit diesen Fragen zu beschäftigen, welche, wie ihr Verfasser in Uebereinstimmung mit Dirichlet bemerkt, 'mit den Principien der Infinitesimalrechnung in der engsten Verbindung stehen und dazu dienen können, diese zu grösserer Klarheit und Bestimmtheit zu bringen.'" The quote, not completely literal, comes from [Riemann 1854a, 238]; compare [Dirichlet 1829, 169].

[2] One of the very few critics of the Riemann integral would be Weierstrass in the 1880s, see [Dugac 1973, 141].

Figure 5. *Georg Cantor (1845–1918) around 1870.*

These ideas were made known outside of Germany too, thanks to Darboux in France, or Dini and Pincherle in Italy.[1]

Initially, these authors seem to have worked under the assumption that the topology of point-sets would be sufficient for the theory of real functions. As time went by, however, it became clear that integration theory required its own peculiar notions, a realization reached around 1880 by du Bois-Reymond and Harnack. Then, in 1884, Stolz and Cantor presented the notion of 'content' (Cantor's terminology, referring to outer content), which served as an immediate antecedent of the Jordan content and measure theory.[2] Thus, in the 1880s a differentiation between

---

[1] Ulisse Dini published the first advanced textbook on the topic, *Fondamenti per la teorica delle funzioni di variabili reali* [1878]. The work includes a modern definition of the real numbers, following Dedekind, a detailed discussion of Cantor's derived sets, and original contributions to the theory of derivation.

[2] Still, there was a long intellectual journey before a satisfactory notion of measure was reached that would make it possible to supersede the Riemann integral. On this topic, see [Hawkins 1970] and his brief summary [1980].

the topology of point-sets and notions of measure theory began to be made. Here, however, we study the common origins of both theories, for in this case the history of sets is again the history of a series of distinctions that were introduced only gradually. Besides, even after 1900 point-set topology and measure theory were regarded as integral parts of general set theory (see §IX.6).

The history of real functions and point-sets has been studied by numerous authors; it has been the most exploited lode in the history of set theory. Relevant works range from those of Schoenflies [1900] and Jourdain [1906], to the recent paper [Cooke 1993]. The monographs of Hawkins [1970] and Dauben [1979] have been an invaluable guide in covering this ground, particularly the more detailed account of Hawkins. Taking into account the availability of these works, I have striven for conciseness in my discussion of these issues.

1. Dirichlet and Riemann: Transformations in the Theory of Real Functions

Starting with Fourier's *Théorie analytique de la chaleur* [1822], trigonometric series became the most general mathematical tool for the representation of real functions.[1] This is one of the most famous cases of an influence of physical problems (e.g., the diffusion of heat in solids) on the history of mathematical ideas. Fourier convinced himself that all functions that one might find in connection with physical problems could be represented by a series of the form

$$f(x) = a_0 + \sum_{r=1}^{\infty} (a_r \cos rx + b_r \sin rx), \qquad x \in [-\pi, \pi]$$

where the coefficients a_i and b_i were obtained by integration of $f(x)$, $f(x) \cdot \cos rx$ and $f(x) \cdot \sin rx$ within the interval of representation. Fourier was able to give plausible arguments to justify his viewpoint, but his work lacked a purely mathematical foundation. In the age of Cauchy, it could only be a matter of time before other authors tackled this problem.

1.1. Arbitrary functions and the convergence problem. Throughout its early history, the study of trigonometric functions was linked with the very question: what is a function?[2] In the 18th-century, d'Alembert suggested that a curve can only be called a function of a variable when it is governed by a single analytical expression throughout. Euler replied that one should accept more general functions, so-called 'discontinuous' functions, which may be represented by different laws in

---

[1] As is well known, Fourier series have an extremely interesting prehistory in connection with the 18th-century problem of the vibrating string, see the above-mentioned books by Hawkins and Dauben, [Grattan-Guinness 1980], [Bottazzini 1986], and even [Riemann 1854a].

[2] See the already mentioned works, especially [Hawkins 1970] and [Bottazzini 1986], and also [Youschkevitch 1976] and [Dugac 1981a].

different intervals or even drawn freely by the hand. 'Continuous' meant that a function obeyed a single analytical law, but everything points to the conclusion that both Euler and his interlocutors presupposed continuity, in the sense of Cauchy, for all of the functions they had in mind. The real issue was, then, whether one should admit *arbitrary* (continuous) functions. Fourier, with his more sophisticated series in hand, was unequivocally for arbitrary functions, i.e., for admitting the idea that a function is any correspondence by which ordinates are assigned to abscissas; there was no need to assume that the correspondence ought to follow a common law [Fourier 1822, 430]. But he, again, seems to have assumed that one is talking about functions that are, in general, continuous in the modern sense [Hawkins 1970, 6].

Dirichlet too was radically in favor of the conception of functions as arbitrary laws; this is clear in the very title of his papers. As he wrote in the second,

it is not at all necessary that *y* depend on *x* throughout the interval according to the same law, one does not even need to think of a dependence that can be expressed by mathematical operations. ... This definition does not prescribe for the different parts of the curve any common law; it can be thought of as composed of the most different parts or totally without a law. ... Insofar as a function has only been determined in one part of the interval, the manner in which it is continued in the rest of the interval is left completely arbitrary.[1]

It must be noted that, here, Dirichlet is defining the notion of a continuous function, and it has been discussed whether he ever seriously entertained the concept of a completely arbitrary function. One may safely assume that he did not see the need to develop a research program on discontinuous functions; this step would only be done after the publication of Riemann's work (§1.2).

The real issue in the case of Dirichlet is the unequivocal promotion of a purely conceptual approach to the notion of function. In the light of the above text, and using the terminology of chapter I, we might even say that he promoted an 'abstract conceptual' approach, in diametric opposition to the formal standpoint that was common at his time. This is clearly important in the context of our story, for it seems, in retrospect at least, that such an approach could only be adequately understood under a set-theoretical perspective: a function is any (arbitrary) one-one correspondence between the numbers of a domain and the numbers of a codomain.[2] But one has to acknowledge that Dirichlet never expressed himself in this way, nor did he make the slightest step, in print at least, in the direction of introducing the language of sets. The question whether he may have come closer to the set-

---

[1] [Dirichlet 1837, 135–136]: "Es ist dabei gar nicht nöthig, dass *y* in diesem ganzen Intervalle nach demselben Gesetze von *x* abhängig sei, ja man braucht nicht einmal an eine durch mathematische Operationen ausdrückbare Abhängigkeit zu denken. ... Diese Definition schreibt den einzelnen Theilen der Curve gar kein gemeinsames Gesetz vor; man kann sich dieselbe aus den verschiedenartigsten Theilen zusammengesetzt oder ganz gesetzlos gezeichnet denken. ... So lange man über eine Function nur für einen Theil des Intervalls bestimmt hat, bleibt die Art ihrer Fortsetzung für das übrige Intervall ganz der Willkür überlassen."

[2] This formulation, couched in slightly different language, was given for the first time by Dedekind [1888]; see §§III.6 and VII.2.

theoretical conception of mathematics and thereby influenced Riemann and Dedekind through personal discussions, must be left unsettled. At any rate, it is clear that his mathematical style and methodology helped prepare the way for the new conception.

Dirichlet's proposal was not readily accepted by mathematicians who tended to prefer a formal approach. Weierstrass criticized his notion of function for being too general, so that it did not allow any conclusions to be drawn concerning the properties of functions (see [Dugac 1973, 70–71]). His purpose was to determine a sufficiently wide class of functions that are analytically representable [*op.cit.*, 70–71, 77]. Hermann Hankel, who attended both Riemann's courses at Göttingen and Weierstrass's at Berlin, echoed this criticism in a paper of 1870, when he said that Dirichlet's definition of function was "purely nominal" and insufficient for the needs of analysis, since his functions failed to possess "general properties."[1] Still, he began with Dirichlet's definition, and then went on to define continuous functions and investigate possibilities of analytical representation. This way of proceeding is also found, in a more sophisticated way, in du Bois-Reymond (see end of §1.2).

Let us now come to Fourier series. In the 1820s, criticism of divergent series was a lively issue, to which Gauss, Abel and Cauchy contributed. Fourier had not shown whether his series were always convergent, a problem that Poisson and Cauchy tried to solve, the latter in 1826.[2] Their attempts were insufficient, and it was only Dirichlet [1829] who was able to offer a correct proof of the convergence of Fourier series. Dirichlet criticized Cauchy's attempt on the basis of the distinction between absolute and conditional convergence of series. His own approach was to determine a set of sufficient conditions on $f(x)$ for the resulting Fourier series to converge. His proof transformed the nth partial sum of the series into an integral expression, and studied its behavior in the limit; historically it was extremely important because it introduced new levels of rigor in analysis. What is of interest here is that Dirichlet showed that the Fourier series for $f(x)$ converges to the mean value of its left and right limits, for all x in $[-\pi,\pi]$, in case:

(1) $f(x)$ is defined and bounded for all x in the interval,
(2) the number of maxima and minima of $f(x)$ is finite, and
(3) $f(x)$ is continuous except perhaps at a finite number of points.

The next step in the theoretical development would be marked by attempts to weaken Dirichlet's conditions. This problem inspired a good number of contributions, in which mathematicians faced the study of functions with infinitely many maxima and minima, or infinitely many discontinuities, in a finite interval (see §2). Dirichlet himself suggested, in 1829, that all other cases would prove reducible to

---

[1] [Hankel 1870, 67]: "Diese reine Nominaldefinition, der ich im Folgenden den Namen Dirichlet's beilegen werde, ... reicht nun aber für die Bedürfnisse der Analysis nicht aus, da Functionen dieser Art allgemeine Eigenschaften nicht besitzen."

[2] [Grattan-Guinness 1980, chap. 3], [Bottazzini 1986, chap. 5].

the one he had solved, and proposed a conjecture for the case in which the function has infinitely many discontinuities (§2). In a new expository paper, Dirichlet [1837] offered some reflections on how to treat functions that become infinite for certain values of *x*. Subsequently, it became crucial to study the distribution of such singular points of a function, and so the topic led to the study of point-sets.

Toward the end of his 1829 paper, Dirichlet commented that in his opinion the representability of a function depends exclusively on whether it makes sense to speak of its definite integral within the interval of representation. For this ensures that the Fourier coefficients will have a meaning. Thinking about the notion of the integral, it seemed to him that one single condition is enough to guarantee the existence of the integral:

> it is necessary that the function $\phi(x)$ be such that, if *a* and *b* designate any two arbitrary quantities within $-\pi$ and π, one may always locate between *a* and *b* two other quantities *r* and *s* that are close enough, so that the function remains continuous within the interval from *r* to *s*. ... It will now be apparent that the integral of a function does not mean anything unless the function satisfies the condition we have enunciated.[1]

Not all arbitrary functions satisfy that requirement: Dirichlet gave the well-known example of the function *f(x)* that is 0 for rational *x* and 1 for irrational *x*. His words suggest that he regarded that defining condition not just as necessary but as sufficient, and so it was interpreted by Lipschitz and others. The condition is purely topological in character; to put it in modern terminology, it is necessary and sufficient that the set of discontinuities of the function be nowhere dense in $\mathbb{R}$. The conjecture that integration theory only required the study of topological properties of point-sets was a pervasive and influential idea that remained very much alive up to the 1880s.

1.2. The Riemann integral.   For arbitrary functions to win a secure place in mathematics, they would have to be subjected to the operations of analysis, which implied the need to refine Cauchy's ideas. As Dirichlet said [1829, 169], an attempt to generalize his theorem on the representability of functions by trigonometric series would require some details relative to the fundamental principles of infinitesimal analysis. Although it was thought that discontinuous functions would never be used in the study of nature, work on them became interesting because of that connection with foundational issues, as well as their possible applications in analytical number theory [Riemann 1854a, 237–38].

---

[1] [Dirichlet 1829, 169]: "Il est nécessaire qu'alors la fonction $\phi(x)$ soit telle que, si l'on désigne par *a* et *b* deux quantités quelconques comprises entre $-\pi$ et π, on puisse toujours placer entre *a* et *b* d'autres quantités *r* et *s* assez rapprochées pour que la fonction reste continue dans l'intervalle de *r* à *s*. ... On verra alors que l'intégrale d'une fonction ne signifie quelque chose qu'autant que la fonction satisfait à la condition précédemment énoncée."

While preparing his dissertation, Riemann had the opportunity to discuss the matter with Dirichlet [Dedekind 1876, 546]. In fact, his investigation of trigono-metric series is preceded by a discussion of the concept of the integral. Dirichlet's condition (3), that the function be continuous except at a finite number of points, does not really enter into his 1829 proof; it merely ensures that the Fourier integral coefficients make sense [Hawkins 1970, 13]. Riemann [1854a, 238–39] realized this, or perhaps Dirichlet himself pointed it out to him. He reformulated Dirichlet's representability conditions as follows: a function is required to be integrable throughout the domain of representation, and it should not have infinitely many maxima and minima [1854a, 235, 237]. From this standpoint, it would seem possi-ble to study the class of representable functions in full generality by means of a more careful analysis of the concept of the integral (or, as we would say, by ex-panding the definition of integrability). This is why Riemann's work begins with an analysis of the concept of definite integral and the extension of its validity.[1]

Cauchy defined the integral as the limit of a sum: given a partition of interval $[a,b]$ into $n-1$ subintervals, $a = x_0 < x_1 < ... < x_{n-1} < x_n = b$, the integral of a continuous function $f(x)$ is the limit to which the sum

$$S = (x_1 - x_0)f(x_0) + ... + (x_n x_{n-1})f(x_{n-1})$$

tends when the norm of the partition (the greatest of the $|x_i - x_{i-1}|$) tends to zero. Riemann took the abstract notion of function seriously, accepting the challenge of expanding the notion of integral accordingly. To this end, he simply abandoned the continuity condition that Cauchy imposed, and required that the 'Cauchy sum' S tend to a unique limit value when the norm of the partition decreases. That limit value is then the integral of $f(x)$ between a and b [Riemann 1854a, 239].

Next, Riemann stated without proof a different criterion for integrability, based on considering the maximum oscillation of $f(x)$ within each subinterval, i.e., the difference between the maximal and minimal value of $f(x)$ in $[x_{i-1}, x_i]$. Finally he showed that this criterion is equivalent to the famous Riemann integrability condi-tion. If each subinterval $[x_{i-1}, x_i]$ has length $<d$, this condition can be stated as fol-lows:

For the sum S to converge, when all the [subintervals] become infinitely small, one has to require, beyond the boundedness of the function $f(x)$, that, by adequately choosing d, the total magnitude of the intervals in which the oscillations are $>\sigma$ may be made arbitrarily small, whatever σ may be.[2]

---

[1] [Riemann 1854a, 239]: "Ueber den Begriff eines bestimmten Integral und den Umfang seiner Gültigkeit." Here we find, again, the traditional logical terminology; see §II.2.

[2] [Riemann 1854a, 241]: "Damit die Summe S, wenn sämtliche δ unendlich klein werden, convergirt, ist ausser der Endlichkeit der Function $f(x)$ noch erforderlich, dass die Ge-sammtgrösse der Intervalle, in welchen die Schwankungen $>\sigma$ sind, was auch σ sei, durch geeignete Wahl von d beliebig klein gemacht werden kann."

Riemann remarked that this is a necessary and sufficient condition, and showed that such a definition of the integral made possible the integration of highly discontinuous functions.

He thought it convenient to present an example of an integrable function with a dense distribution of discontinuities, "since these functions have never been considered before" [Riemann 1854a, 242]. The function was somewhat similar to Dirichlet's example, but it did not comply with the integrability condition Dirichlet had conjectured. Riemann defines the function $(x) = x-n$ as the difference between the real number x and the closest integer n, except when $x = n/2$ for odd n, in which case $(x) = 0$. Then he considers the function

$$f(x) = (x) + (2x)/2^2 + \ldots + (nx)/n^2 + \ldots$$

which is discontinuous for all $x = m/2n$ with m and $2n$ relatively prime, and therefore has a dense set of discontinuities. Nevertheless, as an effect of the denominators n^2, $f(x)$ satisfies the above integrability condition: for all σ there are only finitely many points $x = m/2n$ where the jump is greater than σ [Riemann 1854a, 242].

Riemann's memoir was rich in examples of such 'pathological' functions. Thus, he justified the notion of arbitrary function and stimulated study of pathological functions by showing how the notion of definite integral could be applied to them. It has been said that only after the publication of his work (1868) did the notion of arbitrary function emerge in mathematical practice in its full generality [Bottazzini 1986, 217].

With the new analysis of the concept of integral, Riemann went on to his investigation of trigonometric functions. It is convenient to comment on some points of this investigation, since they are directly related to the issues we shall discuss in what follows. Dirichlet looked for sufficient conditions for a function to be represented by a convergent Fourier series. Abandoning that approach, Riemann tried to analyze in the most general way the *necessary* conditions a representable function must satisfy;[1] his ultimate aim was, given a detailed analysis of such necessary conditions, to select sufficient conditions among them. This approach enabled him to forget about the way in which the coefficients of the series may be given, that is, it enabled him to introduce a distinction between trigonometric series and Fourier series. Riemann reduced the treatment of convergent trigonometric series to the case in which the coefficients a_n, b_n tend to zero when $n \to \infty$. He approached this case by considering an auxiliary function $F(x)$ obtained by double formal integration of the trigonometric series:

---

[1] [Riemann 1854a, 244]: "Wenn eine Function durch eine trigonometrische Reihe darstellbar ist, was folgt daraus über ihren Gang, über die Aenderung ihres Werthes bei stetiger Aenderung des Arguments?"

$$F(x) = C + C'x + A_o\frac{xx}{2} - A_1 - \frac{A_2}{4} - \frac{A_3}{9} - \dots$$

with $A_o = \frac{1}{2}a_o$; $A_n = a_n \sin nx + b_n \cos nx$. $F(x)$ is a continuous function that converges absolutely and uniformly, though Riemann did not remark on the latter point.[1] Riemann's main theorems referred to that function, and among them we find the following:

Theorem 2.
$$\frac{F(x+2\alpha) + F(x-2\alpha) - 2F(x)}{2\alpha}$$

always becomes infinitely small with α.[2]

Riemann's approach and methods had an immediate impact upon other mathematicians, particularly on Heine and Cantor, who employed the function $F(x)$ and theorem 2 in an essential way in their work on the uniqueness of the representation by means of trigonometric series.

The reformulation of the notion of integral opened up a wide new field for analysis, the realm of discontinuous functions, which had never been seriously considered before. The impact of Riemann's ideas and the way in which they were received is reflected in a paper published by Paul du Bois-Reymond in 1875. Du Bois-Reymond [1875] made an attempt to classify arbitrary real functions according to their behavior "in the least intervals." He accepted the notion of an arbitrary function and was a fervent supporter of the Riemann integral. His classification is based on a series of ever stronger requirements: the weakest condition is Riemann integrability; then comes continuity, and then the requirement of differentiability. Thus, within the realm of arbitrary functions he obtained three classes of functions, each embedded in the former: integrable functions, continuous functions, differentiable and "customary" functions. In order to show that differentiability is a stronger condition than continuity, du Bois-Reymond gave Weierstrass's famous example of a continuous nowhere differentiable function. By the end of the paper, he referred to the issue of the distribution of singularities of a real function; he took up Cantor's idea of derived sets, and also an erroneous conjecture due to Dirichlet that we shall consider in detail in the next section.

---

[1] The notion of uniform convergence would be introduced by Weierstrass.

[2] [Riemann 1854a, 248]: "Lehrsatz 2. ... wird stets mit α unendlich klein."

2. Lipschitz and Hankel on Nowhere Dense Sets and Integration

With the notion of arbitrary function and the Riemann integral, the need to develop a theory of point-sets emerged, as can be seen from the work of Lipschitz and Hankel. Their contributions were still primitive and naive, but they started a trend that would lead to sophisticated results in integration theory and the topology of point-sets.

Rudolf Lipschitz studied at Königsberg and Berlin, regarding himself as a disciple of Dirichlet. He himself suggested that his article [1864] is a continuation of Dirichlet's memoirs, reconstructed from suggestions that can be found scattered in his work. Lipschitz discussed how to weaken Dirichlet's representability conditions, considering three cases: points where the function is unbounded, infinitely many discontinuities, and infinitely many maxima and minima [Lipschitz 1864, 283]. He thought that the first two cases were essentially solved on the basis of Dirichlet's ideas, and so he focused on the third, which had created difficulties for Riemann. The long Latin title of his work states that he aims to investigate the development in trigonometric series of arbitrary functions of a variable, and mainly those which have an infinite number of maxima and minima within an interval. In this connection he contributed the so-called Lipschitz condition (see [Dauben 1979, 10–11]), but here we are interested above all in his reconstruction of what may be called the *Dirichlet conjecture*.

Dirichlet had indicated one important, albeit rather obvious, distinction to be made when talking about the distribution of singularities of a function: the set could be dense or 'scattered,' i.e., nowhere dense. The modern terminology suggests that the dichotomy does not cover all cases, but the mathematicians we shall review argued as if it was. Dirichlet conjectured that $\phi(x)$ would be integrable in $[a,b]$ whenever the discontinuities of $\phi(x)$ are nowhere dense in the interval. This is the conjecture that Lipschitz tried to justify. It was accepted by many mathematicians until about 1880, when it came to be regarded as false. Lipschitz considered the case in which a function has infinitely many points of discontinuity 'scattered' in an interval. After quoting Dirichlet (see end of §1.1), he wrote that "by an appropriate reasoning" it is possible to divide the interval $(-\pi,\pi)$ into finitely many intervals, in such a way that some subintervals, whose total length is arbitrarily small, contain infinitely many discontinuities; and the remaining subintervals contain only finitely many discontinuities of the function, thus satisfying Dirichlet's representability conditions [Lipschitz 1864, 284]. The reasoning involves an implicit assumption that Hawkins locates in the idea that the derived set P' of a nowhere dense set of discontinuities is at most finite.[1] If so, one can cover the points of P' with intervals of arbitrarily small total length, in accordance with Lipschitz's reasoning.

---

[1] In this he follows Montel, who translated Lipschitz's paper into French (*Acta Mathematica*, vol. 36, 1912); see [Hawkins 1970, 14–15].

That implicit assumption is interesting because of its connections with Cantor's work on derived sets (§IV.4.3). If Cantor read Lipschitz, which is likely,[1] he would probably have interpreted his words in the above mentioned sense. Lipschitz does not mention the idea of a limit point, but there seems to be no other way to interpret his statement, and Cantor was acquainted with this notion from his student years in Berlin. Be that as it may, in 1872 Cantor introduced essential corrections in realizing that, in general, P' will be an infinite point-set; Lipschitz's naive assumption was thus superseded.

The realization that nowhere dense sets have no bearing on integrability was the key of Hankel's [1870], dealing with functions with infinitely many oscillations (maxima and minima) or discontinuities. Hankel was interested in clarifying the notion of a function of a real variable. He wrote that the writings of Riemann were the stimulus for his studies (see quotation at the beginning of this chapter) and expressed his admiration for Riemann's way of dealing with complex function theory [Hankel 1870, 68, 70]. Showing a certain degree of dissatisfaction with Dirichlet's definition of function, Hankel emphasized the need to impose some conventional restrictions on the class of 'legitimate' functions [*op.cit.*, 101–02] and looked for explicit analytical representations of the legitimate functions.[2] Hankel presented a method of 'condensation of singularities' that he developed from Riemann's example of integrable discontinuous function. His method enabled him to offer an analytical representation for the characteristic function of the irrationals, that Dirichlet had presented in an abstract way.[3]

The novelty in Hankel's paper was the attempt to systematically study "linearly discontinuous functions," that is, functions with infinitely many discontinuities in a finite interval [Hankel 1870, §6]. In treating this question, the author felt the need to formulate in an explicit and general way some basic notions having to do with point-sets:

Whenever within a segment there lies a multitude of points possessing a certain property, I say that these points

fill the segment if no interval, however small, can be given within the segment in which one does not find at least one point of that multitude;

that on the contrary this multitude of points does *not* fill the segment, but that the points lie *scattered* on it, if between any two arbitrarily close points of the segment one can always give an interval, in which no point of the multitude lies.[4]

---

[1] Cantor came to work on this topic through Heine, who quotes Lipschitz in his [1870].

[2] In this he reminds one of Weierstrass, whose classes he attended.

[3] Cantor would propose a refinement of this principle of condensation of singularities in 1882; see [Cantor 1932, 106–13].

[4] [Hankel 1870, 87]: "Wenn auf einer Strecke eine Schaar von Punkten liegt, denen eine gewisse Eigenschaft zukommt, so sage ich, dass diese Punkte / *die Strecke erfüllen*, wenn in der Strecke kein noch so kleines Intervall angegeben werden kann, in dem nicht wenigstens Ein Punkt jener Schaar läge; / dass dagegen diese Schaar von Punkten die Strecke *nicht* erfüllen, sondern die Punkte *zerstreut* auf ihr liegen, wenn zwischen je zwei beliebig nahen Punkten der Strecke immer ein Intervall angegeben werden kann, in dem kein Punkt jener Schaar liegt."

Thus, Hankel seems to have been the first author who explicitly introduced notions of point-set theory – dense and nowhere dense – , although these ideas had been suggested by Dirichlet. The examples of point-sets he considered were very simple, e.g., the points of [0,1] with abscissa $1/n$ for $n \in \mathbb{N}$ [Hankel 1870, 86], and he was led to hasty conclusions by them. In general, the errors and confusion of these early years were simply due to the scarcity of known examples of point-sets.

Hankel went so far as to present an alleged proof of the theorem that, whenever the set of points in which the value of a function has a jump $>\sigma$ is nowhere dense, the total length of the intervals in which the oscillations are $>2\sigma$ can be made arbitrarily small [Hankel 1870, 87]. This faulty theorem offered necessary and sufficient conditions for integrability, leading to a simple and harmonious overall scheme. Hankel had two kinds of 'linearly discontinuous functions,' "*pointwise discontinuous*" ones, which satisfy the theorem and therefore are integrable, and "*totally discontinuous*" functions which are not [*op.cit.*, 89, 91–92].

Hankel's work exerted a good measure of influence and was widely read, in spite of its somewhat unusual form of publication.[1] Cantor published a review early in 1871, and in 1882 he wrote that it contained the first attempts, worthy of attention, to make distinctions upon which a natural classification of the general concept of function could be based [Cantor 1932, 107–08]. Neither Cantor nor other mathematicians criticized Hankel's incorrect theorem early in the 1870s, which suggests that these confusions reflected the state of knowledge at the time [Hawkins 1970, 34–37].

Up to this point, we have found three different conditions on point-sets: 1) being nowhere dense, 2) having a cover of arbitrarily small total length, and 3) having a finite derived set. Only the second condition is crucial for integration theory; it gave rise to the notion of outer content. Cantor substituted for Lipschitz's condition 3) the more general requirement of being a point-set of the *first species*, i.e., a set P such that its nth derived set P^n is finite, for $n \in \mathbb{N}$. Even so, none of those three conditions is equivalent: there are nowhere dense sets of the second species (i.e., having a derived set P^∞) and also nowhere dense sets with positive outer content. But mathematicians still tended to regard them as equivalent.

---

[1] It was published as a *Gratulationsprogramm* of Tübingen University; as late as 1882 it was reprinted in *Mathematische Annalen*.

3. Cantor on Sets of the First Species

Cantor's early work, including his dissertation and *Habilitation* under Kummer, dealt with number theory. In 1869 he went to the University of Halle as a *Privatdozent* and came into contact with his senior colleague Eduard Heine, who made important contributions to the theory of real functions and who was interested in foundational issues. At the time, Heine was exploring the implications for trigonometric series of Weierstrass's work on uniform convergence and term-by-term integration. His paper of 1870 was important in calling attention to these issues. Heine oriented his young colleague toward the theory of trigonometric series, the topic of his first really important work, with five papers published between 1870 and 1872.

Although the problem had been suggested by Dirichlet many years before, with the exception of Riemann's work, the contributions of Heine and Cantor were the first substantial ones. The new ideas on point-sets that the subject suggested to Cantor were definitely new and went far beyond Dirichlet, Lipschitz and Hankel. It was the first time that a wide class of infinite point-sets having no influence on the integral, and therefore on Fourier series, was precisely delimited. The impact of Cantor's work on others was almost immediate, and some people were even led to according an undue prominence to Cantor's sets of the first species [Hawkins 1970, 36].

3.1. The uniqueness theorem. Fourier based the existence and uniqueness of representations of arbitrary functions by means of a trigonometric series, on the validity of term-by-term integration of the series. But Weierstrass realized that uniform convergence was a necessary condition for term-by-term integrability of a series. Heine [1870] was the first to consider the implications this had for the theory of trigonometric series. Dirichlet, Lipschitz and Riemann had only determined that, under rather general conditions, a function can be developed into a trigonometric series with known coefficients, "but *not, in how many different ways the development could take place.*"[1] Heine was able to prove the uniqueness of representation under the condition of uniform convergence of the series, within the interval of representation, except at a finite number of points.

It was Cantor who suggested to Heine the possibility of generalizing his result by admitting finitely many exceptional points [Heine 1870, 355]. This went along the line of Dirichlet's work and prefigured Cantor's later efforts, that led to his work on point sets.

Acknowledging his debt to the Halle professor, Cantor went on in 1870 to generalize Heine's result by requiring only simple convergence of the series. First, he

---

[1] [Heine 1870, 353]: "durch die Arbeiten von Dirichlet, Lipschitz und Riemann ist daher nur festgestellt, *dass* eine Function von *x* in einer Anzahl von Fällen sehr allgemeiner Natur sich in eine Reihe von der Form (∀.), deren Coefficienten bekannt sind, entwickeln lasse, aber *nicht, auf wie viele Arten die Entwickelung geschehen könne.*"

proved that if the limit of $A_n = a_n \sin nx + b_n \cos nx$ is zero when $n{\to}\infty$, then the coefficients a_n and b_n tend to zero; this is a particular version of the Cantor–Lebesgue theorem [Cantor 1932, 71–79, 87–91]. The result complemented Riemann's work in an important way, since it justified the introduction of Riemann's function $F(x)$ for any convergent trigonometric series (see §1.2). Next, Cantor conjectured that $F(x)$ would be a linear function, and this would make it possible to transform the initial trigonometric series into a uniformly convergent one, which would lead to the desired result. In February 1870 he communicated his conjecture to Schwarz, who was able to prove the desired lemma on the basis of properties of $F(x)$ that Riemann had established. With Schwarz's lemma, Cantor established the uniqueness of representation for any convergent trigonometric series.[1]

The whole episode can be regarded as a typical example of the results obtained by members of the Berlin school (see §I.5). Heine was not a member, but he enjoyed good relations with the Berliners and was acquainted with their viewpoints. In another letter of 1870, Schwarz indicated that the main merit in the proof of his lemma was should be credited to the principles of "Bolzano and Weierstrass" [Meschkowski 1967, 239]. After Cantor told him that he had discussed the matter with Weierstrass, who found the proof irreproachable, Schwarz wrote:

You can believe me, I am proud that the Berlin mathematical school, to which we both belong, can celebrate a triumph, once again a palpable result by which an important scientific question can be answered *completely*. You will recognize ever more, if you have not noticed it already, the great significance that one has to attribute to the distinguished school that we have enjoyed and the superiority we have gained by familiarity with the need for painstaking care in proofs, in contrast to the mathematical "romantics" and "poets." At present I do not know of any mathematical school that offers its students such a solid foundation as the Berlin school.[2]

This is one of the clearest statements about the atmosphere surrounding the Berlin school, as about the competition among schools that reigned in Germany at the time. It would be interesting to know whom Schwarz had in mind when he talked about "romantics" – perhaps Riemann himself.

---

[1] See [Cantor 1932, 82–83]. The correspondence between Schwarz and Cantor can be found in [Purkert & Ilgauds 1987, 21].

[2] [Meschkowski 1967, 240]: "Du kannst mir glauben, ich bin stolz darauf, dass die Berliner mathematische Schule, der wir beide angehören, einen Triumph feiern kann, wieder ein greifbares Resultat, durch welches eine wichtige wissenschaftliche Frage *vollständig* beantwortet wird. Du wirst, wenn Du es nicht schon bemerkt hast, immer mehr wahrnehmen, welche grosse Bedeutung der ausgezeichneten Schule beizulegen ist, die wir genossen haben und welches Übergewicht uns diese Angewöhnung an peinliche Sorgsamkeit bei Beweisen gewährt, den mathematischen "Romantikern" und "Poeten" gegenüber. Gegenwärtig ist mir keine mathematische Schule bekannt, welche ihren Schülern ein so solides Fundament zu geben vermag, wie die Berliner."

3.2. Point-sets of the first species. The most important aspect, in the present context, is that Heine and Cantor set out, at the suggestion of the latter, to work on generalizing the representability result to discontinuous functions. Heine considered the possibility that the function be discontinuous, or the series not convergent, at a finite number of points within the interval of representation. He dealt with this case by means of a method based on Riemann's theorem 2, a method that could also be applied to Cantor's more general theorem [Heine 1870, 359]. Cantor's argument above remained valid for every subinterval determined by adjacent exceptional points, so that $F(x)$ is linear within each subinterval. By continuity of $F(x)$ and Riemann's theorem 2, if $F(x)$ has a right or left derivative at a point, then it has both and they are equal. Therefore, $F(x)$ is a single linear function throughout the interval (see Cantor's 1871 paper in [Cantor 1932, 85]).

Already in 1871 Cantor announced new extensions of this theorem. He had found a way to apply the same method to infinite collections of exceptional points, by considering the limit points and by reasoning inductively. The crucial idea was related to what Lipschitz had suggested, but Cantor gave it a perfectly clear sense and, more important, he was able to avoid the errors of Lipschitz. The new ideas were published in his fifth paper on the topic, which we have already analyzed in part; this was the article containing the theory of real numbers, the notion of limit point and the new idea of derived set (§IV.2.2 and §IV.4.3). Cantor went beyond the customary approach to analysis within the Berlin school, with its close attention to explicit analytical representations – what we have called Weierstrass's 'formal conceptual' approach. He presented the notion of limit point as an abstract one, and took the crucial steps of considering sets of limit points and forming the derived set of a point-set. In so doing, Cantor was turning toward an abstract approach to mathematics that employed the language of sets. Perhaps he wanted to mark this change outwardly as well, for he published his new paper in *Mathematische Annalen*, the journal associated with the school of Clebsch, and not in *Crelle's Journal*.

Like Hankel, Cantor turned to consider point-sets as such, but he went far beyond his predecessor. Given a bounded infinite point-set P, the Bolzano–Weierstrass theorem ensures that it will have limit points, and Cantor takes the derived set P' of all such points as "*conceptually*" given [Cantor 1872, 98]. P' will also, in general, be infinite, and therefore Cantor considers P'', and so on. The examples he presented in 1872 were still quite simple: if P is the set of points of abscissa 1, 1/2, ... 1/n, ... then P' consists of a single point, 0; if P is the set of points of rational abscissa, then P' is the interval [0,1] and the next derived sets P^n, ... coincide with P' [*ibid.*]. Cantor also indicated how his *Zahlengrössen* of higher kinds could be used to give examples of point-sets with a (nonempty) nth derived set, but he did not go into further details [*op.cit.*, 98–99].

Cantor was interested in what he would call, years later, point-sets of the *first species*, which are sets that have an empty derived set for some n; the remaining point-sets would be called of the *second species* [Cantor 1879/84, 140]. Or as he put it in 1872:

It can happen, and this is the only case which interests us here, that after v steps the set $P^{(v)}$ consists of a finite number of points, so that it has no derived set itself. In this case we shall say that the original point-set P is *of the vth kind*, from which it follows that P', P'', ... are of the $v{-}1^{th}$, $v{-}2^{nd}$, ... kinds.

From this standpoint, the domain of all point-sets of a definite kind will be regarded as a particular genus inside the domain of all conceivable point-sets, of which genus the so-called point-sets of the vth kind form a particular kind.[1]

Cantor's opinion, at this time, regarding those point-sets which are not "of a definite kind" is unclear. Later he wrote that as early as 1870 he had considered new symbols of infinity which he used in order to designate the 'kinds' of sets of the second species.[2]

Cantor was now in a position to give a succinct formulation of his generalization of the theorem on trigonometric series: the series is unique if it converges to the represented function for all values of x except those belonging to a point-set of the first species [Cantor 1872, 99; see 92]. The basic idea of the proof is simple, once one has the above prerequisites in hand. Cantor proceeds inductively on the order $<$ of the derived set, applying Heine's method each time [*op.cit.*, 99–101]. In this way he is able to show that, under the specified conditions, $F(x)$ is a single linear function over the whole interval of representation. The essential condition is that the inductive process must terminate, i.e., that $P^{(v)}$ be finite for some $v \in \mathbb{N}$ or, alternatively, that P constitute a point-set of the first species. Actually, the theorem can be extended to a set of exceptional points such that P^{α} is empty for transfinite α, in which case one has to use transfinite induction. But Cantor never came to this extension of the theorem, not even in the 1880s or 90s, after he had introduced transfinite numbers.

3.3. Refined confusions.   Cantor's contribution represented the emergence of a sophisticated theory of point-sets; the ideas of Lipschitz and Hankel, by contrast, were either rather obvious or erroneous. But some previous confusions remained in a subtler form. If Lipschitz thought that a nowhere dense set must have at most a finite number of limit points, his successors came to think that first species sets

---

[1] [Cantor 1872, 98]: "Es kann eintreffen, und dieser Fall ist es, welcher uns hier ausschliesslich interessirt, dass nach v Übergängen die Menge $P^{(v)}$ aus einer endlichen Anzahl von Punkten besteht, mithin selbst keine abgeleitete Menge hat; in diesem Falle wollen wir die ursprüngliche Punktmenge P *von der v^{ten} Art* nennen, woraus folgt, dass alsdann P', P'', ... von der $v{-}1^{ten}$, $v{-}2^{ten}$, ... Art sind. / Es wird also bei dieser Auffassungsweise das Gebiet aller Punktmengen bestimmter Art als ein besonderes Genus innerhalb des Gebietes aller denkbaren Punktmengen betrachtet, von welchem Genus die sogennanten Punktmengen v^{ter} Art eine besondere Art ausmachen."

[2] [Cantor 1984, 55] (this is a footnote to the last page of [1879/84], part 2, which did not find its place in the *Abhandlungen*). Since Cantor showed at times an eagerness to anticipate the dates of his findings and publications, his word on this issue cannot be trusted completely. An example of such eagerness is his crucial *Grundlagen* [1883], which Cantor signed October 1882 even though we know from letters to Mittag-Leffler and Klein that he kept sending new material during the early months of 1883 [Cantor 1991, 95, 107, 109, 111].

exhaust nowhere dense sets. Around 1880, several mathematicians from different countries came to realize that there are nowhere dense sets of positive outer content, which therefore have to be of the second species (§4.1). But up to then, most mathematicians thought that point-sets of the first species are the most general kind of sets having no influence on the integral. This was a refined version of the Dirichlet conjecture [Hawkins 1970, 34–37].

Examples can be found in the contributions of du Bois-Reymond and Dini. After formulating the distinction between dense and 'scattered' sets, first given by Hankel, du Bois went on to identify nowhere dense sets with Cantor's sets of the first species [du Bois-Reymond 1875, 35–36]. He maintained the Dirichlet conjecture in the following refined version: whenever the set of discontinuities of a function is of the first species, the function is integrable within the corresponding interval. Dini proved that sets of the first species can be covered by means of finitely many intervals with arbitrarily small total length, i.e., using later terminology, that they have outer content zero [Dini 1878, §14].

Cantor showed a great interest in Dini's work, which he mentioned in the first of his series of papers [1879/84, 139]; this may even have motivated him to publish this work. He also wished to translate Dini's book into German, and tried unsuccessfully to engage the collaboration of Dedekind.[1]

Dini proved also a number of theorems that emphasized the importance of first species sets for integration theory,[2] and thus gave "undue prominence" to them [Hawkins 1970, 36]. In 1881/82, Axel Harnack would emphasize that one can accept a wider class of point-sets, which includes first species sets, in Dini's theorems. But in 1880 Harnack himself contributed to the confusion, in the worst possible way, by *defining* sets of the first species as those sets that have outer content zero [Harnack 1880, 128].

4. Nowhere Dense Sets of the Second Species

The main thrust behind the developments that we have reviewed thus far came from the idea that integration theory could be developed on the basis of purely topological notions. This assumption found clear expression in the erroneous conjecture of Dirichlet, which Hankel turned into a theorem (§2). With Cantor's [1872], knowledge of the subsets of $\mathbb{R}$ became much more refined. In this section we shall see how even more sophisticated examples were given by several mathematicians, which made it clear that the Dirichlet conjecture was false and that integration theory required special new notions.

---

[1] See letters of 1878 and 1880 in [Cantor & Dedekind 1937, 42–43; Cantor & Dedekind 1976, 233–34, 238].

[2] An example is the following. If f is an integrable function and g is a bounded function such that $f(x) = g(x)$ except at a point-set of the first species, then g too is integrable and has the same integral as f [Dini 1878, §18ss].

Mathematicians realized that several different, non-equivalent conditions were being considered, and in particular that sets of the first species do not exhaust nowhere dense sets. As it turned out, the crucial property for integration theory was – in the terminology of 1885 – that of having (outer) content zero. Several mathematicians in different countries were able to make this step: Smith in England, du Bois-Reymond and Cantor in Germany, Volterra in Italy. It seems that the German mathematicians were unaware of the contributions of Smith and Volterra. All of them employed distributions of intervals as a method for defining sophisticated point-sets.

The context in which the new ideas were elaborated is interesting, because from 1879 to 1884 a great number of articles touching on the question of point-sets appeared in *Mathematische Annalen*. Among them we find Cantor's papers 'On infinite linear point-manifolds' [1879/84], the first that studied point-set theory as an independent topic. This marks a clear difference between Cantor's work and those of du Bois-Reymond and Harnack. But it is quite unclear whether du Bois-Reymond or Cantor was the first to give an example of nowhere dense set of positive outer content, or even an example of nowhere-dense sets of the second species. As we shall see, these mathematicians had a clear feeling of competition at the time.

Before we discuss German contributions, however, it will be convenient to consider those of H.J.S. Smith, who was the first to offer examples, and very clear ones. Unfortunately, his paper was not properly appreciated on the Continent, otherwise it would have accelerated even more the development of the theory of point-sets and integration theory.[1]

4.1. Smith on the integration of discontinuous functions. A professor at Oxford University, Smith was an expert in number theory, and as such he was known to Cantor and Dedekind.[2] In 1875 he published a paper 'On the Integration of Discontinuous Functions,' motivated by Riemann's work [1854a] and also by the intention to criticize some points in [Hankel 1870]. He was the first to criticize his alleged proof having to do with 'pointwise discontinuous' functions and offer a counterexample, the first counterexample to the Dirichlet conjecture. To this end, he employed two methods for the construction of nowhere dense sets that are extremely clear and rigorous. It will be convenient to explain them here, in order to clarify the whole issue.

The first method served to define examples of point-sets of the first species. Smith began with the set P_1 of all points of the form $1/n$ with $n \in \mathbb{N}$, and went on to the set P_2 of all points of the form $1/n_1 + 1/n_2$, ... In general we have $P_s = \{1/n_1 + 1/n_2 + ... + 1/n_s\}$, which has as limit points all points of P_{s-1} and therefore is a set of the first species and sth kind [Smith 1875, 145–47]. Smith formulated a correct

---

[1] As indicated by Hawkins [1870, 40], the reviewer for *Fortschritte der Mathematik* did not realize the theoretical importance of Smith's examples; he did not even mention them.

[2] See Cantor's letter of Jan. 1879 in [Cantor & Dedekind 1976, 232].

inductive proof that P_s is nowhere dense. Here we have a very elegant way of giving examples of the sets considered by Cantor.

The second method employed a distribution of intervals.[1] In his own words:

Let m be any given integral number greater than 2. Divide the interval from 0 to 1 into m equal parts; and exempt the last segment from any subsequent division. Divide each of the remaining $m–1$ segments into m equal parts; and exempt the last segment of each from any subsequent division. If this operation be continued *ad infinitum*, we shall obtain an infinite number of points of division P upon the line from 0 to 1. These points are in loose order. [Smith 1875, 147]

By 'being in loose order,' Smith means that the point-set is nowhere dense in [0,1]. After k steps, the total length of the unexempted segments is $(1 – 1/m)^k$, and so Smith remarked that when $k{\to}\infty$ the points of P "occupy only an infinitesimal portion" of [0,1]. The union of P and P' is an example of what we call today a 'Cantor set.' This and the coming examples show that Cantor was not the first to give examples of Cantor sets.

Both of the foregoing examples are sets of zero outer content, but a variation of the last method enabled Smith to define a point–set of positive outer content. One proceeds as before, but at each step one diminishes much quicker the length of the exempted segments. First we divide [0,1] into m segments and exempt the last one from subsequent division; each of the remaining segments is divided into m^2 parts, exempting the last one from further division; each of the remaining $(m–1)(m^2–1)$ segments is divided into m^3 parts, and so on [Smith 1875, 148]. After k steps the total length of the unexempted segments is $(1–1/m)\cdot(1–1/m^2)\cdot\ ...\ \cdot(1–1/m^k)$, which tends to a positive limit when $k{\to}\infty$. This limit is, in fact, the outer content of the set in question. On the basis of this example, Smith showed that the Dirichlet conjecture, and Hankel's theorem, are wrong; he also criticized some other points in [Hankel 1870]. Smith's paper serves as a model for what number-theoretical rigor meant at the time; on the other hand, it only clarified previous ideas and did not open any essentially new perspectives.

4.2. A German competition. Cantor's first paper of the series on linear "point-manifolds," as he now preferred to call them, was published in vol. 15 of *Mathematische Annalen*. It did not present new results, but rather gave a systematic exposition of ideas developed in previous papers. Among other things, he analyzed the relation between the property of being an "everywhere-dense" set, as he said, and the behavior of derived sets. In this connection, he promised to come back, in a future paper, to the question whether "*every* point-set of the *second species* is so constituted, that there exists an interval (α ... β) in which it is *everywhere-dense*" [Cantor 1879/84, 141]. The contents of Cantor's paper, and especially this com-

---

[1] The idea may have been suggested by an example of non-integrable function given by Hankel, see [Hawkins 1970, 30–31].

ment, seem to have alarmed du Bois-Reymond. Apparently, he had corresponded on the subject with Cantor a few years earlier and had communicated to him a method for the construction of sets of the second species that solved this question in the negative.

To establish his priority, du Bois-Reymond hastened to include a lengthy footnote toward the end of a paper on the fundamental theorem of the integral calculus which he published in the next volume (no. 16) of the *Annalen*. Although the paper dealt with a different issue, du Bois found a natural way to introduce the topic by beginning with a critique of the condition of integrability that Dirichlet had established [du Bois-Reymond 1880, 128]. He wrote that one can distribute "*intervals* pantachically" ('pantachisch' was his peculiar term for 'dense'), so that one may cover $[-\pi,\pi]$ densely with parts of a segment $D < 2\pi$. This made it possible to give an example of a function $f(x)$ such that, for any interval within $[-\pi,\pi]$, there is a subinterval where $f(x)$ is continuous, but nevertheless the function is not integrable. Du Bois took $f(x) = 0$ over the densely distributed subsegments of D, and $f(x) = 1$ at all other points of $[-\pi,\pi]$. The set of points at which the function takes the value 1 has a positive outer content $2\pi-D$, and so the function does not satisfy Riemann's integrability condition.

Du Bois-Reymond went on:

One is led to this kind of distribution of intervals, of which I have several examples at hand, if one looks for points of condensation of order ∞ [derived sets $P^{(\infty)}$, J.F.], the existence of which I announced to Mr. Cantor at Halle by letter some years ago. I plan to deal on another occasion with this distribution, with condensation points of finite and infinite order of *segments* that always become smaller, and finally with my choice of the expression 'pantachically' in comparison with that adopted by Mr. Cantor's later, everywhere-dense.[1]

The rivalry between du Bois-Reymond and Cantor is plainly evident. This text is sufficient to establish that du Bois-Reymond discovered examples of sets of the second species independently of Cantor. On the other hand, it may well be that Cantor arrived at essentially similar examples before du Bois corresponded with him. As regards the notion of dense set, it originated in Dirichlet, and Hankel formulated it clearly (under the words 'filling a segment') well before Cantor or du Bois-Reymond.

---

[1] [du Bois 1880, 128]: "Auf diese Art der Intervallvertheilung, zu der ich verschiedene Beispiele bei der Hand habe, wird man geführt, wenn man die Verdichtungspunkte der Ordnung ∞ aufsucht, deren Vorhandensein ich vor jahren Herrn Cantor in Halle brieflich anzeige. Auf diese Vertheilung, ferner auf die Verdichtungspunkte endlicher und unendlicher Ordnung von immer kleiner werdenden *Strecken*, endlich auf meine Wahl des Ausdrucks 'pantachisch' verglichen mit dem später von Herrn Cantor angenommenen überalldicht gedenke ich bei einer anderen Gelegenheit einzugehen."

If du Bois-Reymond feared that Cantor would publish what he regarded as his idea without mentioning him, his fear was realized in the next volume of the *Annalen* (vol. 17, 1880). This was the second installment of Cantor's series of papers, dealing exclusively with derived sets 'of an infinite order.' Cantor defined derived sets such as $P^{(\infty)}$, $P^{(2\infty)}$, $P^{(n\infty+m)}$, etc., and toward the end of the paper indicated how to build a set with a single limit point of order ∞. His method was to take a sequence of disjoint intervals, bordering on each other and converging toward a point p, and to take within each interval a point-set of the first species, so that the "orders" of the point-sets "grow beyond all bounds" as the intervals approach p. Thus, for intervals (a_n, b_n) such that the sequence of extremes converges to p, take a point-set P_n of the nth kind or 'order' within each interval. The union of these point-sets constitutes an example of the kind sought [Cantor 1879/84, 148].

This episode seems to have caused the rivalry that developed between Cantor and du Bois-Reymond. But the dispute died out rather quickly, probably because of the importance and originality of Cantor's new contributions, starting in 1882. One should not overemphasize the importance of the above episode as a priority dispute. It seems more interesting to consider it as an indication that the development of point-set theory was already a community enterprise by this time. In 1880 a third German mathematician of lower rate, Axel Harnack, was joining the competition for new results in point-set theory and integration theory.[1]

Cantor's example of 1880, like those of Smith and others,[2] was enough to correct the previous misunderstanding that nowhere dense point-sets must be of the first species; for it is easy to show that the former point-set of the second species is nowhere dense. On the other hand, it has outer content zero, as did the second (but not the third) of Smith's examples. By 1880, mathematicians had reached a sufficiently refined understanding of point-sets, so that the scene was ready for the advancement of sound notions for integration theory.

5. Crystallization of the Notion of Content

With the publications mentioned in the last section, it became clear that the crucial notion for integration theory was that of point-sets that can be covered by means of finitely many intervals, whose total length is arbitrarily small. This was explicitly emphasized by Axel Harnack in a textbook on the calculus [Harnack 1881] and in a paper published in the *Annalen* [1882]. This insight would lead to the notion of

---

[1] The frictions between du Bois, Harnack and Cantor left traces in the latter's correpondence; see particularly [Cantor 1991].

[2] Volterra published in *Giornale di Matematiche*, in 1881, a paper on 'pointwise discontinuous' functions; the title shows the influence of Hankel. He criticized the Dirichlet conjecture essentially as Smith, and on the same basis he gave an example of a function g whose derivative g' is not Riemann integrable. This confirmed a conjecture of Dini, see [Hawkins 1970, 52–53].

(outer) content, formulated by Stolz and Cantor in 1884, and to the Jordan content in the early 1890s [Hawkins 1970; 1980].

Harnack tried to introduce new terminology for point-sets, useful for integration theory. He termed 'discrete sets' those that we have been calling sets of outer content zero; other point-sets, with positive content, were called 'linear' [Harnack 1882, 238–39]. This terminology was certainly inadequate, for 'discrete' had always referred to the opposite of continuous. Harnack showed that Dini's results could be readily generalized to 'discrete' sets of exceptional points [1881], proved that 'discrete' sets are nowhere dense, and gave an example of a nowhere dense 'linear' set [1882, 239]. Some of Harnack's 'theorems' turned out to be false, but his contributions were nevertheless important for the development of the notion of content. He also showed that Riemann's integrability condition could be formulated in terms of outer content [Hawkins 1970, 59].

In order to apply the notion of content to integration theory, it was necessary to assign a number to every point-set in an interval, in accordance with the total length of the (finitely many) intervals needed to cover it. This was done simultaneously by Otto Stolz and by Cantor. Stolz called this number 'interval-limit' [Intervallgrenze] and defined it as follows. Take a point set P in (a,b) and consider a finite partition T_n of the interval; calling S_n the sum of the subintervals that contain points of P, and making the norm of the partition decrease with increasing n, it turns out that $L = \lim_{n \to \infty} S_n$ is a well-defined number measuring P [Stolz 1884, 152]. Stolz, in fact, showed that the content of a point-set does not depend on the way in which the partition is refined. He also showed that if $L = 0$, then P is a 'discrete' set in Harnack's sense, whereas if $L > 0$, then P in 'linear' [*op.cit.*, 154]. Finally, he generalized those ideas to define the content of point-sets in the plane (subsets of $\mathbb{R}^2$).

Cantor mentioned the work of Harnack and du Bois-Reymond in the fourth installment of his series 'On infinite, linear point-manifolds.' As in the work of those authors, the notion of content is only implicit here. Cantor mentioned that a necessary, but not sufficient, condition for (outer) content zero is that the point-set be nowhere dense. Looking for sufficient conditions, and in order to show the importance of his set-theoretical notions (see chapter VI), he proved that a point-set P with denumerable derived set P' always has content zero [Cantor 1879/84, 160–64]. The sixth installment of the series contains a subsection devoted to giving an explicit definition of "*content*" [Inhalt], so called for the first time. Cantor's definition is more general than Stolz's, for it deals with subsets of $\mathbb{R}^n$. He also presented the first notable theorems having to do with outer content.[1]

Without entering into details, it suffices to say that Cantor defined the content of a bounded set $P \subset \mathbb{R}^n$ by considering the closed covering of P, that is, the closed set $P \cup P'$, and taking the n-tuple integral.[2] This is a remarkable peculiarity of his

---

[1] This time he does not mention other mathematicians; see his letter to Klein, having to do with Harnack, where he says that it was due to 'Nachlässigkeit' and not to 'bösem Willen' [Purkert & Ilgauds 1987, 191–93].

[2] See [Cantor 1879/84, 229–36] and [Hawkins 1970, 61–63].

approach, which seems quite interesting from our viewpoint. Stolz and Harnack wanted to use the notion of content as a basis for integration theory, and wished to define the integral in its terms. This is the natural way of proceeding if one is interested in foundational problems, and especially in case one regards sets as a foundation for mathematics. On the contrary, Cantor is not interested in the foundations of integration theory, just the opposite: he uses the integral to study a property of sets, their content. This happened in 1884, and is consistent with the approach taken by Cantor in all his papers up to then. Thus, everything points to the conclusion that Cantor was far from taking set theory to be the foundation of mathematics or, at least, that he was far from being interested in developing the implications of that idea.[1]

Cantor went on to prove some properties of the content, so defined, in connection with other notions he had developed until then. For instance, what he called the *"fundamental theorem,"* that the content of a point-set P equals that of its derived set P' [Cantor 1879/84, 231]; this property was taken for granted at the time, which forced mathematicians to consider only finite coverings. Another interesting result of Cantor's is that the content of a non-denumerable set P equals the content of the perfect set included in P' ([Cantor 1879/84, 235]; for the notion of perfect set, see §VI.6).

One year later, Harnack published another paper that connected with those of Stolz and especially of Cantor. He presented his notion of discrete set again, emphasizing that it was defined on the basis of *finite* coverings, and proving some general theorems in the style of Cantor's. He emphasized that his definition of content did not presuppose the integral [Harnack 1885, 246–47]. A notable aspect of the paper is that Harnack mentions a 'paradox' that one encounters as soon as one considers the idea of covering a point-set by means of *infinitely many* intervals: one then finds that every denumerable point-set can be enclosed by intervals with arbitrarily small total length [*op.cit.*, 242]. As this remark shows, at this early phase in the development of measure theory, a definition of measure such as Borel's, which employs denumerably infinite coverings, seemed quite paradoxical for it entailed that even *dense* denumerable sets have measure zero. Such a definition of measure or content would have contradicted Cantor's fundamental theorem (see [Hawkins 1970; 1980]).

In the origins of the theory of point-sets we find a clear example of the phenomenon that I mentioned in the introduction: the gradual differentiation of notions or structures that initially were combined, and sometimes implicit, in the classical objects of mathematics. The study of point-sets became an attractive topic in the early 1870s, due to a revitalization of real analysis motivated by Riemann's work. Development of this theory was stimulated by its intimate connections with problematic issues in the theory of trigonometric series, integration theory, and the very notion of function. The topology of point-sets was, and remained, an important

---

[1] In 1885, Cantor wrote that pure mathematics is pure set theory (see [Grattan-Guinness 1970, 84]) and after this time he published some ideas concerning the set-theoretical foundations of number. By this time, however, he was well acquainted with Dedekind's work (chapter VII).

topic in connection with complex and real function theory (see chapters VIII and IX). But the initial hopes that topological notions would be sufficient for a general treatment of integration had to be corrected. This happened in the early 1880s, when the notion of content emerged.

Sometimes one finds the idea, expressed even by professional historians of mathematics, that point-set theory was the creation of a single man.[1] The foregoing gives plenty of evidence that this is far from the truth. In the rest of the book I shall not enter into a detailed analysis of subsequent developments in point-set theory, although in chapters VI, VIII, and IX I will make some comments in connection with other contributions of Cantor and their diffusion.[2]

---

[1] See, e.g., [Johnson 1979/81, part 1, 163].

[2] For detailed analysis, the reader may consult [Cooke 1993] and the summary given by Kanamori [1996], as well as older accounts like [Schoenflies 1900/08; Schoenflies 1913]. Some important development after 1900 are discussed in [Johnson 1979/81].

Part Two: Entering the Labyrinth – Toward Abstract Set Theory

During the last quarter of the 19th century, Dedekind and Cantor published crucial contributions to abstract set theory. Dedekind attempted to elaborate an abstract basis for the rigorous foundation of pure mathematics – arithmetic, algebra, and analysis. Meanwhile, Cantor took the radical step of beginning to explore the realm of infinite sets and in the process created what is usually called set theory in the strict sense. These contributions proved to be deeper and much more influential than contemporary ones by Peano or Frege (on which see §IX.1.2). Thus, the emergence of set theory was the result of a mostly mathematical development, although philosophical ideas played an important motivating role in it.

The transfinite realm is presupposed in classical mathematics, as long as one understands the continuum as a set of points. But it was Cantor who first realized the labyrinth behind that basic idea. In December 1873 he proved that $\mathbb{R}$ is not equipollent to $\mathbb{N}$ and saw the possibility of distinguishing 'sizes' in the infinite (chap. VI). To analyze infinite sets, he introduced a whole series of notions – power or cardinality, notions of point-set theory, and the concept of well-ordering. Cantor posed the problems and developed the concepts and basic results that would establish set theory as an independent discipline. By 1885 he had conceived of an abstract set theory based on the ideas of cardinality and ordering (chap. VIII).

Meanwhile, the abstract viewpoint had been adopted by Dedekind as early as 1872, but he pursued aims quite different from Cantor's. He focused on the foundations of the number system, establishing them on the 'logical' notions of set and mapping. With it, he was also establishing the foundations for his abstract approach to algebra and other areas of pure mathematics (chap. VII). His work had an impact on such influential authors as Hilbert, and some of his ideas – particularly the notion of chain – were taken up in the later development of set theory.

Part Two is devoted to delineating those lines of development and exploring the interaction between Cantor and Dedekind. We shall find that, unfortunately, there was a lack of collaboration and mutual reinforcement due to difficulties that emerged quite early in their relationship. But there were also several positive outcomes of their episodic correspondence and meetings.

VI   The Notion of Cardinality and the Continuum Hypothesis

> The ideas I have lately communicated to you are even for me so unex-
> pected, so new, that I will not be able to have a certain peace of mind, so
> to say, until I obtain from you, very esteemed friend, a decision con-
> cerning their correctness. So long as you have not given your approval, I
> can only say: je le vois, mais je ne le crois pas.[1]

The present chapter will discuss Cantor's first two articles on topics that would become the core of transfinite set theory – his famous work on the non-denumerability of the reals [1874] and the equipollence of continua of any number of dimensions [1878]. This was the birth of the notion of cardinality or *power* of an infinite set,[2] which Cantor presented to the public in the second paper, an epoch-making article which also contained his first version of the Continuum Hypothesis. The paper that might be regarded as the first published contribution to transfinite set theory [Cantor 1874] appeared under complex circumstances, and it was far from offering a clear idea of Cantor's actual views. The crucial idea that infinite sets have different powers had been born for him, but not for most of his readers. Apparently due to the influence of Weierstrass, his presentation failed to emphasize that point. Cantor's abstract viewpoint and the notion of power only became clear with his second paper [1878], which makes several references to the first one in order to place it in a new light [*op.cit.*, 120, 126]. Cantor even felt the need to re-formulate his 1874 proof five years later [Cantor 1879/84, part I], which under-scores the peculiar nature of the 1874 article.

The content of both papers was first discussed in Cantor's correspondence with Dedekind. Given the importance of this correspondence, and the fact that both mathematicians met a few times in the 1870s and 80s, it seems appropriate to pay

---

[1] Cantor to Dedekind, June 1877 (the French sentence means, 'I see it but I do not believe it.'). [Cantor & Dedekind 1937, 34]: "die Ihnen jüngst von mir zugegangenen Mittheilungen sind für mich selbst so unerwartet, so neu, dass ich gewissermassen nicht eher zu einer gewissen Gemüthsruhe kommen kann, als bis ich von Ihnen, sehr verehrter Freund, eine Entscheidung über die Richtigkeit derselben erhalten haben werde. Ich kann so lange Sie mir nicht zugestimmt haben, nur sagen: je le vois, mais je ne le crois pas."

[2] I shall translate Cantor's *Mächtigkeit* into 'power.' A little care and attention to the context should suffice to avoid mistaking this notion with that of power set [Potenzmenge].

close attention to their relations and exchange of ideas. The urgency to do so is all the more evident given Dedekind's grasp of foundational issues. By the early 1870s Dedekind already had developed advanced set-theoretical conceptions, and one is naturally led to wonder whether he informed Cantor of his views. The correspondence also became a source of difficulties for Dedekind and Cantor. Although most historians have assumed that the exchange of letters was only interrupted in 1882,[1] as a matter of fact trouble seems to have begun in 1874. Relations between both mathematicians were difficult after that time, although they went through brief phases of tranquility in the years 1877 and 1882. The most important consequence of these difficulties was that in the 1880s, when they came to publish their mature ideas on abstract set theory, Cantor and Dedekind turned their backs on each other, so to say. There was a lack of collaboration and mutual reinforcement, and the theories they presented suffered from this. For these reasons, I shall pay close attention to this correspondence, which will also lead us to discuss briefly some aspects of Cantor's biography.

In spite of Dedekind's collaboration, the main problems and ideas that can be found in the correspondence, those that became crucial for transfinite set theory – especially the notions of derived set and cardinality – originated with Cantor. Just as he took the initiative in the correspondence, Cantor was the force behind the development of transfinite set theory (in contradistinction to general set theory). The last three sections of this chapter discuss some crucial results that Cantor published between 1882 and 1884, all of them linked to derived sets and cardinalities. After 1878, the main motivation for his set-theoretical work was the attempt to prove the Continuum Hypothesis (CH). The study of the cardinalities of point-sets by analyzing their derived sets led to noteworthy results, particularly the Cantor–Bendixson theorem, which established CH for closed point-sets. Along the way, Cantor was led to introduce important new notions belonging to the topology of point-sets, and he offered a novel definition of continua.

1. The Relations and Correspondence Between Cantor and Dedekind

Dedekind and Cantor first came into written contact in 1872, when they exchanged copies of their papers on the theory of real numbers. Cantor must have known Dedekind as the editor of the *Vorlesungen über Zahlentheorie*, and of Riemann's papers [1854, 1854a]. In the summer of that same year they met by chance in Gersau, Switzerland, where both were spending their vacations [Cantor & Dedekind 1976, 223]. Dedekind was then about to reach 41 years of age, Cantor was 27.

This first encounter might have been of great importance, but there is no documentary evidence to inform us of any details. By then, Dedekind had developed

---

[1] See, e.g., [Grattan-Guinness 1974, 125–26] and [Purkert & Ilgauds 1987, 73–75].

very advanced foundational conceptions. As we have seen, he was in the possession of a set-theoretical formulation of algebraic number theory and algebra (§§III.2–3), detailed reflections on the set-theoretical definition of the number system (§§III.5.1 and IV.2.3), basic notions of point-set topology (§IV.4.1), and even the fundamental notions of his mature abstract theory of sets and mappings (§III.6).[1] This all suggests that by 1872 Dedekind had already come to the conclusion, presented in [1888], that all of pure mathematics – arithmetic, algebra and analysis – was based on the theory of sets and maps. One is naturally led to wonder whether, on their first meeting, he informed Cantor of his views, thereby helping to shape the future orientation of his research.

Certainly, a comparison of the published work of both mathematicians gives the impression of quite divergent research interests. Cantor worked on number theory, under the direction of Kummer, for his Ph.D. and *Habilitation*, but he quickly moved to research on trigonometric series in connection with Riemann and Weierstrass. His work up to 1884 seems to have remained within the realm of analysis and its foundations.[2] Meanwhile, Dedekind's research was in the fields of number theory and algebra, a clear contrast which might suggest that a direct, noticeable influence is implausible. Nevertheless, both mathematicians had wide-ranging interests, and a deeper look at their work reveals significant areas of overlap.

Dedekind had a lively interest in analysis, which becomes particularly clear when we consider his unpublished work on the Dirichlet principle, leading to basic notions of point-set topology (§IV.4.1), and on the analytical aspects of Riemann's differential geometry (chapter II, appendix). This helps explain why he was able to make a crucial contribution to the foundations of analysis with his work on the real numbers. On the other hand, Cantor's work on point-sets was not just related to analysis; there is abundant evidence, which we shall review, that to him it was intimately connected with the foundations of Riemannian geometry. And he also kept his interest in number theory: this is shown particularly clearly by his later concentration on transfinite number theory, but also by his lecture courses at Halle (see [Purkert & Ilgauds 1987]). To summarize, the areas of overlap included foundations of analysis, topology, geometry – all of them topics linked with the name of Riemann – and of course set theory.

Keeping this wide range of interests in mind, it is clear that many possible topics could have been touched upon in their 1872 conversations. Since both had just published and exchanged their papers on the real numbers, one may safely assume that they talked about the foundations of the number system. If so, it seems likely that Dedekind would have mentioned his conviction that sets and mappings are the basic notions of arithmetic, the basis on which it is possible to define the natural

---

[1] As the reader may recall (§III.6), the very beginning of an 1872 draft contains the extensional notion of set [System], the general notion of mapping, and Dedekind's famous definition of infinite sets.

[2] The fact that Klein gave him the assignment of refereeing Lindemann's work on the transcendence of π suggests that he was regarded as an expert in irrational and transcendental numbers.

numbers and all of their extensions [Dedekind 1872/78, 293]. The fact that Dedekind began writing his draft for the later book *Was sind und was sollen die Zahlen?* in 1872 might be related to his casual encounter with a younger colleague who was independently advancing toward set theory. It is noteworthy that, in his crucial letter of Nov. 1873 posing the denumerability problem for ℝ, Cantor introduced a set-theoretical approach without any further motivation; he seems to take it as self-explanatory, which must have been due to prior knowledge that Dedekind was perfectly accustomed to such a viewpoint.[1] But, since there is no detailed evidence concerning the 1872 conversations, we shall not speculate on the matter further. The fact is, that in 1873 both mathematicians started their well-known correspondence.

Fraenkel remarked that, although its mathematical content is limited, the correspondence offers valuable insight into Cantor's working ways and into the opposite characteristics of both minds, the 'romantic' Cantor and the 'classic' Dedekind [Cantor 1932, 456]. Indeed, as regards their approach to mathematical research, both men seem to have been almost polar opposites. Dedekind was a stickler for rigorous arguments and conceptually clean, elegant theories, while Cantor appears to have been far more of a effusive character, interested above all in obtaining advanced results by whatever means. In this connection, it is noteworthy that Cantor's early work on set theory does not seem to have been an outcome of interest in the foundations of classical mathematics. Quite the opposite, Cantor employed all kinds of traditional means of proof – taken from arithmetic and analysis – to obtain sophisticated results in the new field of research. Far from being taken as the foundation of mathematics, set theory seems to have been, in Cantor's eyes, the highest level, the upper reaches of the edifice [Medvedev 1984]. As we saw in the preceding chapter, this holds for his work up to 1884, too, and one would be tempted to say that it holds for all of his work.[2]

In Fraenkel's opinion, Dedekind's style of thought had a visible influence on Cantor's work:

Much more ... than it is visible from the letters, the differences in character between Cantor's early and later set-theoretical publications show indirectly the profound influence of Dedekind's more abstract style, which tends to proceed analytically and strives for rounded systematization, in contrast with the more constructive style of the younger Cantor, which tends to advance by single strokes.[3]

---

[1] However, this might simply have been due to Cantor's knowledge of Dedekind's work [1871; 1872].

[2] Compare the critical comments of Zermelo on Cantor's remarks [1895/97] on the foundation of the number system [Cantor 1932, 352, notes 4 and 5]. It is true that an unpublished paper of 1885 [Grattan-Guinness 1970] included remarks on set theory as the foundation of pure mathematics, but there is reason to think that these remarks were made under the influence of Dedekind's work (then still unpublished but known to Cantor since 1882, see §§VII.4 and VIII.3).

[3] [Cantor 1932, 456–57]: "Weit mehr allerdings, als es aus den Briefen ersichtlich wird, zeigen mittelbar die Verschiedenheiten in der Anlage der frühen und der späteren mengentheoretischen Veröffentlichungen Cantors den tiefgreifenden Einfluss der abstrakteren, mit Vorliebe

This seems to be a sensible way of analyzing the differences between Cantor's early work in the 1870s and, for instance, the *Beiträge* that summarized his set-theoretical work [1895/97], but it is always difficult to substantiate such claims, based on perceptions that are to some extent subjective. In support of his claim, Fraenkel cites Cantor himself acknowledging, in a letter of Aug. 1899, the "manifold stimuli and rich lessons" that he has received from his colleague's "classic" writings.[1] Due to those differences and also the significant difference in their ages, Cantor appears "by and large as the one who asks and takes,"[2] Dedekind as the critic and counselor.

The correspondence has now been published in full, although in scattered locations. Zermelo included the letters of 1899, dealing with the paradoxes, in his edition of Cantor's treatises [Cantor & Dedekind 1932]. A few years later, E. Noether and J. Cavaillès edited the mathematical portions of the correspondence, except for the 1899 letters [Cantor & Dedekind 1937]. A French translation of both sets of letters can be found in [Cavaillès 1962], and English readers may now have access to a reliable translation of that material in [Ewald 1996, vol. 2]. The remainder of the correspondence seemed lost after World War II, but it was rediscovered in the United States, as part of Emmy Noether's papers [Grattan-Guinness 1974]. Pierre Dugac included the unpublished part of the correspondence as an Appendix to his book on Dedekind [Cantor & Dedekind 1976].[3] All of these documents came from Dedekind's *Nachlass*, so that Cantor's letters are originals, while Dedekind's are his drafts [Cantor & Dedekind 1932, iv; 1937, 11; 1976, 224].

The exchange of letters we are discussing involves some of the most interesting documents of the 19th century insofar as the foundations of mathematics is concerned. But, in spite of its interest, and even though several historians have paid attention to it, it can be said that up to now it has not been fully analyzed. Most of those historians have used it as a source for studying Cantor's biography and the evolution of his ideas.[4] As a result, neither the frequency of the correspondence nor the nature of the relationship and collaboration between both mathematicians have been well understood. A partial exception is the work of Dugac [1976, 116–18], which stresses relevant facts like the tensions that aroused in 1874. Nevertheless, Dugac has not supported his narrative with further analysis, and it is even possible

---

analytisch vorgehenden Art Dedekinds, die nach abgerundeter Systematik drängt, gegenüber dem mehr konstruktiven Stil des jüngeren Cantors, der gerne zum Einzelstoss vorwärtsstürmt." In [1930, 197], Fraenkel writes "logizistischen" instead of "mit Vorliebe analytisch vorgehenden."

[1] [Cantor & Dedekind 1976, 261]: "die vielfache Anregung und reichliche Belehrung die ich aus Ihren classischen Schriften empfangen habe."

[2] [Fraenkel 1930, 196]: "dieser Unterschied, ... sowie die ... fühlbare Altersdifferenz ... lassen im grossen und ganzen Cantor als den Fragenden und Nehmenden in diesem Briefwechsel erscheinen."

[3] Dugac overlooked parts of a letter of Nov. 1877 and another of Aug. 1899; both can be found in [Grattan-Guinness 1974, 112, 129].

[4] See [Fraenkel 1930; Fraenkel in Cantor 1932], [Meschkowski 1967], [Grattan-Guinness 1971; 1974], [Dauben 1979] and [Purkert & Ilgauds 1987].

to find some errors in it.[1] Some years ago, I published a detailed analysis with the hope that it would be "balanced enough to settle the question," that is, to finally draw an adequate picture of the relations between Cantor and Dedekind [Ferreirós 1993]. Here we shall mention the most important general results, but the reader is referred to that paper for further details.

The traditional view is that, after their first encounter, Cantor and Dedekind became close friends and met each other frequently. The first biographer of Cantor, Fraenkel, speaks of frequent meetings, most of which happened in Harzburg (a town located in the Harz mountain range, not far from Braunschweig and Halle). If this were true, one would presume to find a fairly intense interchange of ideas between both mathematicians. But, in fact, the total number of meetings that we can trace from available evidence is just six in a period of 28 years [Ferreirós 1993, especially Appendix I]. The first meeting happened by mere chance, and the same seems to have been the case with the second meeting in 1874 [Cantor & Dedekind 1976, 228–29]. The other four meetings took place in 1877, 1882 and 1899.[2] After careful analysis, it seems likely that 1872 and 1882 were the only chances for significant unnoticed intellectual exchanges.

As regards the frequency of correspondence, it can be said that the exchange of letters followed a peculiar rhythm, with intense periods of contact followed by long gaps. The main exchanges occurred in 1873, 1877, 1882 and 1899, each time in connection with Cantor's current work.[3] For most of the intermediate periods, we have confirmation of the absence of any contact; we shall have occasion to mention some of the related documents.[4] It is not infrequent to find Cantor expressing regret about this state of affairs [Cantor & Dedekind 1976, 233, 258] and signs of relief when contact is renewed [*op.cit.*, 232, 259]. This pattern contrasts rather sharply with other correspondence of both mathematicians, e.g. Cantor's with Mittag-Leffler, which involved hundreds of letters over a short period of time, or Dedekind's with Lipschitz, Frobenius, and especially his sustained and constant correspondence with H. Weber.[5]

---

[1] E.g., regarding the connections between a suggestion of Dedekind and a proof of Cantor [Dugac 1976, 117], and concerning Cantor's interpretation of his theorem of 1878 [*op.cit.*, 121–22].

[2] In May 1877 Cantor visited Dedekind in Braunschweig on his way back home from the *Gaussfeier* at Göttingen [Lipschitz 1986, 88]; in September 1882 they met twice, first at Harzburg and then at the *Naturforscherversammlung* in Eisenach [Cantor & Dedekind 1937, 55; 1976, 255–56]; in September 1899 they met for the last time, presumably at Harzburg, to discuss the set-theoretical paradoxes [*op.cit.*, 260–62; Landau 1917, 54].

[3] The same is true for shorter exchanges in Jan. 1879 and Jan. 1880 [Ferreirós 1993, 356–57].

[4] See [Cantor & Dedekind 1976, 228–29 and 232–39], esp. [232, 233].

[5] Dauben [1979, 2] remarks that it is revealing of Cantor's personality that his friendships were intense but brief. But his difficulties with Dedekind were somewhat different from those experienced with Schwarz or Mittag-Leffler, in good measure due to the different personality of each partner.

2. Non-denumerability of ℝ

In 1867 Cantor defended the thesis that in mathematics the art of posing questions is of more consequence than that of solving them, and his career seems to have reflected this maxim to a good extent. The Continuum Hypothesis is, of course, his most famous problem, the one he left to posterity as an open question. Second in importance, but making that problem possible, is the question he posed in 1873: can the reals be put in one-to-one correspondence with the natural numbers? Actually, one of the most peculiar aspects of his crucial discovery of 1873 is the very fact that he came to pose the question and take it seriously.[1] As a matter of fact, Cantor's involvement with cardinality problems in the 1870s does not seem to be directly linked to open questions in real analysis, although he did hope that the notion of cardinality would come to play a role. His 1873 question was a speculative one and may seem related to his philosophical interests, in particular the infinitistic philosophy of Spinoza (see §VIII.2 and [Purkert & Ilgauds 1987]).

On the basis of documents that have since been lost, Fraenkel [1930, 199] indicated that Cantor had already considered cardinality questions in Weierstrass's seminar at Berlin, where he proved that ℚ is denumerable. In one of his letters to Dedekind from 1873, he says that he had posed the question on ℝ to himself several years before [Cantor & Dedekind 1937, 13]. According to this letter, he had always wondered whether the difficulty he encountered with it was merely subjective or inherent in the question. Many years later he remarked that he had first tried to establish an enumeration of ℝ, and only after several unsuccessful efforts did he attempt to establish that such an enumeration is impossible [Fraenkel 1930, 237].

Cantor's solution to the problem was published in a paper that, peculiarly, bore the title 'On a Property of the Collection of all Real Algebraic Numbers' [Cantor 1874]. The property in question was the denumerability of algebraic numbers, demonstrated in §1 of the paper; §2 showed the non-denumerability of ℝ and applied both results to a new proof of Liouville's theorem asserting the existence of transcendental numbers. The content of the paper had been discussed with Dedekind in letters of November and December 1873. We shall follow events in the correspondence rather closely, almost day by day, in order to clarify some aspects of the matter.[2]

2.1. Denumerability of the algebraic numbers.   On 29 Nov. 1873 Cantor posed his problem to Dedekind as follows:

---

[1] Dauben [1979, 49] tried to link it directly to the discovery that first species sets have no effect on integration, but there is no evidence to support this conjecture – rather the contrary, for Cantor said he had considered the problem years before.

[2] I should indicate that the most complete present-day edition of Cantor's correspondence [Meschkowski & Nilson 1991] only includes excerpts from the correspondence with Dedekind, which are insufficient to judge some of the events we are going to discuss. Readers who may wish to scrutinize my narrative and interpretations more closely should keep this fact in mind.

Take the collection of all positive integral individuals n and denote it by (n); consider, further, the collection of all positive real numerical magnitudes x and denote it by (x); then the question is simply, can (n) and (x) be correlated so that to each individual in one collection there corresponds one and only one in the other? At first sight one says to oneself, no it is not possible, for (n) consists of discrete parts, while (x) forms a continuum; but nothing is won with this objection, and as much as I tend to the opinion that (n) and (x) admit of no univocal correlation, I can still not find the reason, which is what interests me, but perhaps it is very simple.[1]

One tends to think, he went on, that (n) cannot be correlated one-to-one with the collection (p/q) of positive rational numbers, and yet it can be correlated not only with that collection, but also with

$$(a_{n_1, n_2, \ldots n_\nu}),$$

where the n_i are positive integers, and ν is arbitrary. By this Cantor seems to have meant what we would call the set of ν-tuples $<n_1, n_2, \ldots, n_\nu>$.

Dedekind answered immediately, acknowledging that he was not able to solve the problem. He remarked that the problem was not worthy of much effort, for it had no practical interest, but he formulated and proved in detail the theorem that the collection of all algebraic numbers was, in later terminology, denumerable. This theorem was subsequently published in §1 of Cantor's paper [1874].

At this point, I have to make clear that, as stated before, Dedekind's original letters have been lost. Moreover, there is no extant draft for any of his letters to Cantor of late 1873. All the evidence we possess for establishing the historical facts comes from Cantor's letters, and from a set of notes on the 1873 correspondence that Dedekind wrote at an unknown date. Thus, the main evidence is to be found in Cantor's letters, which are normally sufficient to establish the main facts. As regards Dedekind's notes, it is my opinion that they are quite reliable. The mere fact that he always kept them for himself, and never claimed the theorem, already gives a motive for believing them. Dedekind never had an explicit, overt interest in the matter, whereas Cantor certainly did. Likewise, the tone in which the notes are written and the way in which Dedekind underscores the importance of Cantor's contributions speak for their objectiveness.[2]

---

[1] [Cantor & Dedekind 1937, 12]: "Man nehme den Inbegriff aller positiven ganzzahligen Individuen n und bezeichne ihn mit (n); ferner denke man sich etwa den Inbegriff aller positiven reellen Zahlgrössen x und bezeichne ihn mit (x); so ist die Frage einfach die, ob sich (n) dem (x) so zuordnen lasse, dass zu jedem Individuum des einen Inbegriffes ein und nur eines des andern gehört? Auf den ersten Anblick sagt man sich, nein es ist nicht möglich, denn (n) besteht aus discreten Theilen, (x) aber bildet ein Continuum; nur ist mit diesem Einwande nichts gewonnen und so sehr ich mich auch zu der Ansicht neige, dass (n) und (x) keine eindeutige Zuordnung gestatten, kann ich doch den Grund nicht finden und um den ist es mir zu thun, vielleicht ist er ein sehr einfacher."

[2] Naturally, my impression that the notes are objective is also based on my views regarding Dedekind's personality, as a result of the general impression obtained from reading many other letters to Cantor and others, etc. It is impossible to convey this to readers of the present work, and

Cantor's answering letter of Dec.2 [Cantor & Dedekind 1937, 13] confirms that Dedekind had sent a proof of the following

Theorem 1: The field of algebraic numbers can be put in one-to-one correspondence with the set (n) of natural numbers.

The proof presented in Cantor's paper [1874, 116] goes as follows. Algebraic numbers are, by definition, roots of polynomials with coefficients in $\mathbb{Q}$ or, by a trivial transformation, in $\mathbb{Z}$. If we take irreducible polynomials with first coefficient a_0 positive, to each algebraic number ω corresponds a single polynomial $p(x)$, so that

$$p(\omega) = a_0\,\omega^n + a_1\,\omega^{n-1} + ... + a_n = 0.$$

Dedekind and Cantor now call *height* of $p(x)$ the positive integer $N = n-1+|a_0|+$ $+|a_1|+...+|a_n|$. To each algebraic number ω there corresponds one height N, to each height only finitely many polynomials, and therefore finitely many algebraic numbers. This is the fundamental fact in virtue of which one can use the height of polynomials to enumerate the algebraic numbers: one forms a sequence beginning with those numbers corresponding to polynomials of lowest height, and orders numbers of equal height according to any criterion (e.g., in the case of $\mathbb{R}$ one may use the natural order).

Cantor did not claim independent discovery of the theorem in any of his letters of the period. Nevertheless, in the letter of Dec. 2, he stated that Dedekind's proof is "more or less the same" that he used to establish the result on $<n_1, n_2, ..., n_\nu>$ mentioned previously. "I take $n_1^2 + n_2^2 + ... + n_\nu^2 = \mathfrak{N}$ and I order the elements accordingly."[1] Apparently Cantor thought that this was reason enough to use his colleague's proof in print without naming him. But both proofs are not equivalent. The main question is whether Cantor could have conjectured the result of Theorem 1.

First and foremost, at the time Cantor never claimed to have proved, nor even independently conjectured, Dedekind's result. Still, one might consider the possibility that he had seen a connection between denumerability of n-tuples and that of polynomials, from which he could have conjectured Theorem 1. Nevertheless, Cantor's proof works only when all n_i are positive, as he had explicitly required [Cantor & Dedekind 1937, 12]. To show that the set of polynomials is denumerable one must extend Cantor's result to n-tuples of integers, and to this end one needs to take into account explicitly the number ν of elements (n in Dedekind's height above). Otherwise to each $\mathfrak{N}$ there will be infinitely many n-tuples, because some n_i may be zero, and enumeration fails. This is where both proofs differ.

I must leave to other historians the task of making their own informed judgements about the issue.

[1] [Cantor & Dedekind 1937, 13]: "Der von Ihnen gelieferte Beweis, dass sich (n) den Körper aller algebraischen Zahlen eindeutig zuordnen lasse, ist ungefähr derselbe, wie ich meine Behauptung im vorigen Briefe erhärte. Ich nehme $n_1^2 + n_2^2 + ... + n_\nu^2 = \mathfrak{N}$ und ordne darnach die Elemente."

Thus, technical reasons make it implausible, too, that Cantor could even have conjectured the result. In his notes on the 1873 correspondence (published only in 1937), Dedekind says:

soon afterwards, this theorem and proof [1 above] were reproduced almost literally, including the use of the technical term *height*, in the treatise by Cantor in Crelle['s *Journal*] vol. 77, with the only divergence, maintained against my counsel, that only the collection of all *real* algebraic numbers is taken into account.[1]

Dedekind's advice of stating the proof for all (real or complex) algebraic numbers was given on Dec. 25, after receiving a letter of Cantor in which he communicated that he had written and sent to the *Journal* a short paper, in which Dedekind's "comments" and "way of expression" had been very useful.[2] The title read 'On a Property of the Collection of all Real Algebraic Numbers' [Cantor 1874; Cantor & Dedekind 1937, 17]. According to it, Dedekind's result was the main content of the paper!

Despite the objective tone that he attempted to keep in these notes, Dedekind's astonishment at his colleague's behavior is apparent. Cantor's decision to restrict the formulation to real numbers was partly linked to the circumstances in Berlin, as he explicitly acknowledged [Cantor & Dedekind 1937, 17]. We shall have occasion to consider the way he presented the matter in §3 below, but now we must turn back to early December 1873, when Cantor found his crucial result on $\mathbb{R}$.

2.2. Non-denumerability of $\mathbb{R}$.

In his notes, Dedekind wrote that his opinion had been that Cantor's original question was not worthy of much effort, but this view had been dramatically contradicted by Cantor's proof of the existence of transcendental numbers [Dedekind & Cantor 1937, 18]. Having confirmed that the difficulty was a serious one, Cantor devoted a few days to the matter, and on Dec. 7 he was able to prove rigorously that the positive real numbers $0<x<1$ *cannot* be univocally correlated with the positive integers (n). On the same day he sent the proof to Dedekind, convinced that he would be, as he said, the most indulgent critic. One day later, Dedekind sent an answer with his congratulations for the beautiful result [*op.cit.*, 14–15, 19].

Cantor proceeded by *reductio ad absurdum*. Assuming an enumeration of the reals numbers in (0,1) as given, he showed that there must be at least one such number not contained in that enumeration. The assumption that *all* real numbers in (0,1) can be enumerated thus leads to a contradiction, i.e., the reals in (0,1) cannot be mapped one-to-one to the naturals (n). The original proof of 1873 is quite differ-

---

[1] [Cantor & Dedekind 1937, 18]: "dieser Satz und Beweis ist bald darauf fast wörtlich, selbst mit dem Gebrauch des Kunstausdruckes *Höhe*, in die Abhandlung von Cantor in Crelle Bd. 77 übergegangen, nur mit der gegen meinen Rath festgehaltenen Abweichung, dass nur der Inbegriff aller *reellen* algebraischen Zahlen betrachtet wird."

[2] [Cantor & Dedekind 1937, 17]: "Dabei kamen wir, wie Sie später finden werden, Ihre, mir so werthen, Bemerkungen und Ihre Ausdrucksweise sehr zu statten."

ent from the famous, later proof by diagonalization. While this latter proof [Cantor 1892] could be regarded as purely arithmetical, the former one employed topological properties of ℝ and is quite interesting. In particular, Cantor employed the Bolzano–Weierstrass principle which, in modern terminology, states that an infinite sequence of nested closed intervals of real numbers must have a nonempty intersection (see §IV.4.2).

The assumption, then, is that we have an infinite sequence

(I) $\qquad\qquad\qquad x_1, x_2, x_3, ...$

containing all real numbers in (0,1). Cantor's proof of Dec. 7 was still quite complex, he decomposed the sequence (I) into infinitely many sequences, such that the members of each one are ordered by size. On this basis he was able to define a sequence of nested intervals, and the Bolzano–Weierstrass principle (that he assumed implicitly) yielded the existence of a real number η not in (I) [Dedekind & Cantor 1937, 14–15]. This proof admitted of a simplification that Dedekind gave in his letter of Dec. 8, by avoiding the intermediate step of decomposing (I) into further sequences. Cantor sketched the same simplification in a letter of the following day.[1]

The second theorem can be stated as follows:

Theorem 2. Given any sequence of real numbers, one can determine in *any* interval (α,β) a real number η that does not belong to that sequence [Cantor & Dedekind 1937, 16].

The proof that Cantor presented in [1874, 117] is the following. Assume given an infinite sequence of the form (I) and an interval (α,β). Call α', β' the first two numbers in the sequence (I) that fall within (α,β), so that $\alpha' < \beta'$; call α'', β'' the first two numbers in (I) that fall within (α',β'), so that $\alpha'' < \beta''$; and so on. By construction, α' must antecede α'' and β' must antecede β'' in the sequence (I), and we have $\alpha' < \alpha'' < \alpha''' < ...$ and $... \beta''' < \beta'' < \beta'$. We thus obtain a sequence of nested closed intervals $[\alpha^n, \beta^n]$; by the Bolzano–Weierstrass principle we conclude that there is a real number η that belongs to all of the $[\alpha^n, \beta^n]$. This number cannot belong to (I), for then we would have $\eta = x_p$, but x_p cannot, by construction, lie within the p-th interval (α^p, β^p), which contradicts our assumption.

As a matter of fact, the final part of Cantor's published proof is more detailed and circumvented. Instead of directly applying the Bolzano–Weierstrass principle, it considers three different cases. Either the number of intervals obtained in the above construction is finite, the last being $[\alpha^n, \beta^n]$; in this case any $\eta \in (\alpha^n, \beta^n)$ will do. Or we obtain infinitely many intervals, in which case we have a monotonically increasing, bounded sequence (α^n) with limit α^∞, and a monotonically de-

---

[1] Dedekind seems to have come to doubt the veracity of this independent simplification [Cantor & Dedekind 1937, 16, 19].

creasing, bounded sequence (β^n) with limit β^∞. Then it may be that $\alpha^\infty = \beta^\infty$, as happens when (I) is the collection of all real algebraic numbers, but $\eta = \alpha^\infty = \beta^\infty$ cannot belong to (I), for the reason given above. Or it may happen that $\alpha^\infty < \beta^\infty$, in which case any $\eta \in (\alpha^\infty, \beta^\infty)$ will do. This constitutes a proof of the Bolzano–Weierstrass principle on the basis of Dedekind's preferred proposition (see §IV.2.3) that every monotonically increasing, bounded sequence of real numbers has a limit.

Dedekind wrote later, in reference to Cantor's letter of the 7th:

> To this letter, which I received on December 8, I replied on that same day with my congratulations on the beautiful success, and at the same time I 'mirrored' the kernel of the proof (that was still quite complicated) in a greatly simplified way; this exposition was also reproduced almost literally in Cantor's treatise (Crelle['s *Journal*] vol. 77); but the phrasing that I had employed, 'according to the principle of continuity,' was avoided at the corresponding place! (p. 261, l. 10–14)[1].[2]

Here again, Dedekind betrays his astonishment: Cantor's behavior did not appear to be the result of inadvertence, for he had intentionally avoided a passage that contained a reference to Dedekind's 'Continuity and Irrational Numbers' [1872].[3] The fact that the published version contains a proof of the Bolzano–Weierstrass principle seems to support Dedekind's comments. Cantor had used the principle *implicitly* on Dec. 7, as if it were self-explanatory,[4] and was still using it as late as 1884, when he remarked that it was hardly possible to replace it by an essentially different one [Cantor 1879/84, 212]. But Dedekind, with his typical penchant for systematic development, preferred to derive it from the principle of continuity and the theorem that he took as basic for analysis in [Dedekind 1872].

Cantor's new theorem solved completely the initial question, but at the same time it opened up a vast new field of research. In one of his letters of early December he wrote:

> I conclude from that, that among the collections and sets of values there are differences in essence, that until recently I could not examine.[5]

---

[1] This corresponds to the first five lines of the last paragraph of [Cantor 1874, 117], that is, to the point where the existence of α^∞ and β^∞ is inferred.

[2] [Cantor & Dedekind 1937, 19]: "Diesen, am 8. December erhaltenen Brief beantworte ich an demselben Tage mit einem Glückwunsch zu dem schönen Erfolg, indem ich zugleich den Kern des Beweises (der noch recht compliciert war) in grosser Vereinfachung 'wiederspiegele'; diese Darstellung ist ebenfalls fast wörtlich in Cantor's Abhandlung (Crelle Bd. 77) übergegangen; freilich ist die von mir gebrauchte Wendung 'nach dem Prinzip der Stetigkeit' an der betreffenden Stelle (S. 261, Z. 10–14) vermieden!"

[3] Recall that Cantor's related paper [1872] did not mention any 'principle of continuity.'

[4] His exact words are [Cantor & Dedekind 1937, 15]: "Es lässt sich nun stets *wenigstens* eine Zahl, ich will sie η nennen, denken, welche im Innern eines jeden dieser Intervalle liegt; von dieser Zahl η, welche offenbar $^{>0}_{<1}$, sieht man rasch, dass sie in keiner unserer Reihen ... enthalten sein kann."

[5] [Cantor & Dedekind 1937, 16]: "ich schliesse daraus, dass es unter den Inbegriffen und Werthmengen Wesensverschiedenheiten giebt, die ich bis vor Kurzem nicht ergründen konnte."

The realm of infinity had turned out to be far more interesting than it had ever seemed, for it was possible to draw distinctions among infinite sets – in some sense, they were of different sizes. At this point, many new questions could be posed: are there sets of other 'sizes,' different from those of $\mathbb{N}$ and $\mathbb{R}$?; what is the precise contribution of irrational numbers to that particular difference in size?; does the new distinction allow applications in the theory of real functions? The reader should pause to consider that possibilities were wide open. For instance, it might have turned out that the set of irrational numbers is of a 'size' intermediate between those of $\mathbb{N}$ and $\mathbb{R}$, so that the 'sizes' of $\mathbb{N}$ and of the irrational numbers add up to that of $\mathbb{R}$.

Cantor considered all of these questions, and this distinguishes him from all other mathematicians of the time, including Dedekind, who never again partici-pated actively in the development of the novel ideas of his colleague. E.g., it was only in 1899 that he included again a mathematical proof in one of his letters [Cantor & Dedekind 1932, 449]. Still, there was something in common between both mathematicians, a disposition to accept the highly abstract theoretical implica-tions of results as the above. We shall see that this was far from being a common-place attitude at the time.

3. Cantor's Exposition and the 'Berlin Circumstances'

On Christmas day, 1873, Cantor informed Dedekind that he had written and sent to the *Journal für die reine und angewandte Mathematik* an article, whose title he specified, containing the results they had recently discussed in their correspondence [Cantor & Dedekind 1937, 16–17]. Initially it had not been his intention to publish, but while in Berlin he communicated those results to Weierstrass, who visited him one day later in order to get the details of the proofs; "he was of the opinion that I should publish the matter, as long as it is related to the algebraic numbers."[1] Cantor mentioned that he had made use of Dedekind's comments and ways of expression.

Weierstrass's opinion explains why Cantor chose the title 'On a Property of the Collection of all Real Algebraic Numbers;' no reference was made to the surprising property of the set of real numbers later called non-denumerability. In his answer, Dedekind recommended that he drop the word 'real' from the title, since the proof extends immediately to *all* algebraic numbers, but Cantor stuck to that restriction, partly for expository reasons, partly due to the "circumstances here," namely in Berlin [den hiesigen Verhältnissen; *op.cit.*, 17, 20].

The proof of denumerability of the algebraic numbers appears relatively trivial to us today, because we are perfectly accustomed to the notion of denumerability itself, and also to the general strategy of denumerability (and non-denumerability) proofs. Weierstrass's reaction suggests that, back in 1873, such a proof was by no

---

[1] [Cantor & Dedekind 1937, 17]: "er meinte, ich müsste die Sache, soweit sie sich auf die al-gebraischen Zahlen bezieht, veröffentlichen."

means trivial. His interest was due to the fact that the result offered him an interesting tool for the definition of special functions. Weierstrass had already employed enumerations of the rational numbers to define 'pathological' functions, and in Dec. 1874 he communicated to Schwarz an example of a continuous function $f(x)$, based on Theorem 1 above, that is differentiable for transcendent x but non-differentiable for algebraic x [Dugac 1973, 140]. Weierstrass's attitude seems to grant the conclusion that, at a time when abstract set theory was in the future, denumerability problems and proofs were anything but trivial or self-evident.

The text of the article was also adjusted to the orientation expressed in the title. The abstract result that $\mathbb{R}$ is non-denumerable was not at all emphasized. As Cantor presented it, Theorem 2 was proved in order to give an application of the main result: a new proof of Liouville's theorem that within any prescribed interval (α,β) there are infinitely many transcendent numbers (reals that are not algebraic) [Cantor 1874, 115–16]. The 'essential differences' between infinite sets, that were particularly interesting to Cantor, were only indicated as an addition or afterthought:

Besides, the theorem of §2 presents itself as the reason why the collections of real numerical magnitudes that constitute what is called a continuum (say all real numbers that are ≥ 0 and ≤ 1) cannot be univocally correlated with the collection (ν) [of all natural numbers]; thus I find the clear distinction between a continuum and a collection of the kind of the totality of all real algebraic numbers.[1]

Dugac has called attention to an interesting passage in a letter of Dec. 27 where Cantor remarks that, following Weierstrass's counsel, he had initially omitted "the comment on the essential differences between the collections," but that he could include it later as a "marginal note."[2] This he did while proof-reading the paper, and the result is the text we have quoted.

It thus seems that Weierstrass repudiated the idea that there may be essential differences among infinite sets.[3] This is a clear indication of Weierstrass's semi-constructivism, that sheds new light on the preference for the finite and the constructive evidenced in his theory of real numbers and his approach to analysis (see chapter IV). Weierstrass was not a radical constructivist, for he accepted 'classical' notions and results concerning the real numbers, but to some extent he was close to

---

[1] [Cantor 1874, 116]: "Ferner stellt sich der Satz in §.2 als der Grund dar, warum Inbegriffe reeller Zahlgrössen, die ein sogenanntes Continuum bilden (etwa die sämmtlichen reellen Zahlen, welche ≥ 0 und ≤ 1 sind) sich nicht eindeutig auf den Inbegriff (ν) beziehen lassen; so fand ich den deutlichen Unterschied zwischen einem sogennanten Continuum und einem Inbegriffe von der Art der Gesammtheit aller reellen algebraischen Zahlen."

[2] [Cantor & Dedekind 1976, 226]: "Die Bemerkung über den Wesenunterschied der Inbegriffe hätte ich sehr gut mit aufnehmen können, liess ihn auf Herrn Weierstrass Rat fort; könnte ihn aber als Randnote später, bei der Correctur, doch anbringen."

[3] In the summer of 1874 he gave a course where he stated that two "infinitely great magnitudes" are not comparable and can always be regarded as equal, and that applying the notion of equality to infinite magnitudes does not lead to any result (Hettner redaction, [Dugac 1973, 126]). This might perhaps be related to his opinions on infinite sets.

his colleague Kronecker. It is unclear when his attitude changed, but there is evidence that by the mid-1880s he was accepting the conclusion that infinite sets are of different powers (see Weierstrass's letter of 1885 in [Mittag-Leffler 1923, 194]).

A closer look at Cantor's paper indicates that his whole presentation might be called semi-constructive. By establishing an enumeration of the algebraic reals and giving a method which allowed the determination within any given interval (α,β) of numbers not contained in that enumeration, he was offering an effective procedure for the calculation of transcendent numbers. This seems to have been the expository reason behind his presentation. One might expect that kind of result would be agreeable to Weierstrass – even though his reactions and counsels suggest that this was not his initial reaction. On the other hand, the abstract non-denumerability result ran counter to traditional assumptions. Had Cantor emphasized it, as he did in the correspondence with Dedekind, there is no doubt that Kronecker and Weierstrass would have reacted negatively. To summarize, the first contribution to transfinite set theory appeared under quite complex circumstances, offering far from a clear perspective on Cantor's thoughts and interests. This explains the need he later felt to explain it again and throw new light on the crucial result of 1873.

The 'Berlin circumstances' may also have been related to Cantor's attitude toward Dedekind.[1] A milder, but somewhat similar episode of ingratitude happened a few years later, in connection with [Cantor 1878] (see §4). On other occasions, Cantor had been honest and appreciative toward his colleagues; for instance, his papers of 1870–72 on trigonometric series acknowledge debts to Heine, Schwarz, Weierstrass, and Kronecker.[2] But all of these names were closely tied to the Berlin school. Moreover, there is some evidence that Kronecker and Kummer were angered with Dedekind after the publication of his theory of algebraic numbers in 1871. Allegedly, Kronecker possessed an equivalent theory since 1858, although he published it only in 1882 (see [Edwards 1980, 329]). Frobenius wrote that Kronecker had never acknowledged Dedekind's priority, and had never forgiven him for that publication. In a letter of 1883, Cantor remarked that Kronecker would be angered by seeing his name alongside Dedekind's in connection with algebraic number theory.[3] On the occasion of a visit he paid to Kummer in the 1880s, Dedekind was received with the unfriendly salute "so you are coming to see whether I will pass away soon."[4]

These circumstances simply reflect the rather tense atmosphere of opposition between schools that reigned in Germany by the late 19th century. Cantor was in good relations with the Berlin masters throughout the 1870s. His famous confrontation with Kronecker only emerged after 1882, and there is evidence of fluid rela-

---

[1] For further details on this topic, see [Ferreirós 1993, 35–52].

[2] [Cantor 1932, 71, 82, 84]. Nevertheless, his later problems with du Bois-Reymond and Harnack (chapter V) are somewhat reminiscent of those he had with Dedekind, though milder.

[3] Frobenius to Dedekind, 1895 [Dugac 1976, 280; Ferreirós 1993, 351–52]. Cantor to Mittag-Leffler, 1883 [Meschkowski & Nilson 1991, 144].

[4] Mentioned by Dedekind in conversation with Bernstein in 1911 [Dedekind 1930/32, vol. 3, 481]: "Er sagte: Sie kommen wohl, um zu sehen, ob ich nicht bald abgehe."

tions in 1877 and 1879 [Grattan-Guinness 1974, 112; Dugac 1976, 233]. Thus, one can safely say that he never faced more opposition from Kronecker than did Weierstrass himself [Biermann 1988, 137–39]. It seems that Cantor was well aware of the feelings of Kummer and Kronecker back in 1873, and that he feared the enmity of the Berlin masters had he acknowledged Dedekind's collaboration in a published work. This would certainly have handicapped his academic career, for which Cantor entertained high hopes in the 1870s. In a word, the admiration of Weierstrass, fear of the reaction of Kronecker and Kummer, and fidelity to the Berlin school, weighed more than honesty to Dedekind, after all an isolated man without influence on university life.

Dedekind never discussed the uncomfortable events of 1873/74 openly with Cantor [Cantor & Dedekind 1937, 20]. This seems somewhat characteristic of his personality: an extremely formal man, he probably took it for granted that Cantor would have to make the first move and explain his behavior. His interest in avoiding public polemics became explicit in an article of 1901, where he employs the "important property" of the field of algebraic numbers that Cantor "had been the first" to emphasize. A footnote indicates that Dedekind had "also" found that theorem but doubted whether it was fruitful until Cantor offered his beautiful proof of the existence of transcendental numbers [Dedekind 1930/32, vol. 2, 278].

Nevertheless, from May 1874 onward Cantor's letters remained unanswered (see [Cantor & Dedekind 1937, 21; 1976, 228–29]). Both met again by chance in Switzerland around October 1874, but in the light of all the evidence available, the meeting was probably less than cordial. A letter by Cantor of October 1876 confirms that two more years passed without any news from each other [1976, 229]. In addition, many years later Cantor himself recalled the tensions; in a letter to Hilbert of 1899 he speaks about his old desire to communicate with Dedekind on the problem of the paradoxes, saying:

Only this fall did I have the opportunity to discuss it with him, because for *reasons unknown to me*, he had been angry with me for years, and *from 1874 he had almost broken off* the earlier correspondence of 1871.[1]

This letter was written during a period in which Cantor was about to begin suffering serious attacks of mental illness [Grattan-Guinness 1971, 365–68], which may explain some of its peculiarities and forces us to treat its contents carefully. Still, as regards a break in the correspondence it fits well with the rest of the evidence.

---

[1] [Purkert & Ilgauds 1987, 154; Meschkowski & Nilson 1991, 414]: "Die Gelegenheit erhielt ich von ihm *erst in diesem Herbst*, da er mir aus *mir unbekannten Gründen* jahrelang gezürnt und die alte Correspondenz von 1871 *seit* 1874 *circa abgebrochen hatte*." Italics in the original.

4. Equipollence of Continua ℝ and ℝ^n

Already in January 1874 Cantor had posed a new question, a kind of follow-up to the previous denumerability problems:

Is it possible to correlate univocally a surface (say a square with inclusion of the boundary) with a line (say a straight segment with inclusion of the extremes), so that to each point of the surface there corresponds a point of the line and, inversely, to each point of the line a point of the surface?[1]

He tended to think, no, but he was convinced that the question would be very difficult to answer. Although Cantor still sent a few letters, Dedekind did not reply, not even after Cantor repeated the question in May. Only in 1877 did Cantor find the opportunity to discuss the problem with his colleague again. The year before Dedekind had taken the initiative to send a couple of papers, his biography of Riemann and his French exposition of ideal theory. Thus he showed he was now open to reconciliation; in 1877 they discussed again on irrational numbers and finally came back to Cantor's question.

By then, Cantor had found a simple proof that it was indeed possible to correlate one-to-one domains of any finite or (denumerably) infinite number of dimensions with a segment [Cantor & Dedekind 1937, 34]. His formulation of the result in 1877 was much more abstract than it had been three years ago. As a matter of fact, Cantor followed very closely the terminology of Riemann [1854]:

A continuous manifold expanded in n dimensions can be coordinated univocally and completely with a continuous manifold of one dimension.[2]

The Riemannian aspects of Cantor's paper will be analyzed in §4.3. By this time he was also using the word "power" [Mächtigkeit] for an extremely important new notion, that would become one of the cornerstones of transfinite set theory. Thus, he was able to state his new result as follows: two continuous manifolds, one of n dimensions, the other of m dimensions, are of the same power [Cantor 1878, 122]. The 1878 paper in which he published those results bore the title 'A contribution to the theory of manifolds,' and would be of great significance.

---

[1] [Cantor & Dedekind 1937, 20]: "Lässt sich eine Fläche (etwa ein Quadrat mit Einschluss der Begrenzung) eindeutig auf eine Linie (etwa eine gerade Strecke mit Einschluss der Endpunkte) eindeutig beziehen [sic], so dass zu jedem Puncte der Fläche ein Punct der Linie und umgekehrt zu jedem Puncte der Linie ein Punct der Fläche gehört?"

[2] [Cantor 1878, 122]: "Eine nach n Dimensionen ausgedehnte stetige Mannigfaltigkeit lässt sich eindeutig und vollständig einer stetigen Mannigfaltigkeit von einer Dimension zuordnen." This is almost exactly the formulation of [Cantor & Dedekind 1937, 29].

In his letter of June 1877 Cantor indicated that, for several years, his opinion had been that such a one-to-one correlation was impossible, and so he wished to know whether Dedekind considered the proof "arithmetically rigorous" [Cantor & Dedekind 1937, 25–26]. A few days later he wrote about his anxiety to know Dedekind's reaction, for the new results were so unexpected and new that, before receiving his approval, he was only able to say: 'je le vois, mais je ne le crois pas' [I see it, but I do not believe it]. Dedekind indeed found a slight problem in that simple proof, but Cantor was in the possession of a more complex proof that his colleague acknowledged as completely rigorous. Dedekind also contributed to put the new theorem in focus by rigorously formulating the invariance of dimension under bicontinuous mappings (what we call homeomorphisms). Although these contributions were perhaps less important than the one he made to Cantor's earlier paper, once again he had the unpleasant experience of seeing his remarks included in Cantor's publication [1878] without mention of the source.

4.1. The notion of power in 1878.   Cantor's letters to Dedekind of June and October 1877 show that by then he was fully in the possession of the crucial new notion of the power or cardinality of an infinite set [Cantor & Dedekind 1937, 25, 40–41]. Interestingly, the word was taken from a lecture course by Steiner, where it was used to denote one-to-one projective coordinations [Cantor 1879/84, 151]. Cantor began to employ it in a much more general sense, and started his 1878 article with a definition of the new notion.

If two well-defined manifolds M and N can be coordinated with each other univocally and completely, element by element (which, if possible in one way, can always happen in many others), we shall employ in the sequel the expression, that those manifolds have *the same power* or, also, that they are *equivalent*.[1]

The same definition can be found in the letter of October 1877 mentioned above; today we would say *equipollent* instead of 'equivalent.' Cantor went on to present the basic idea of subset or *part* [Bestandtheil] and to indicate elementary results on powers, some of which would later be regarded as theorems whose demonstration was problematic.

While many findings in transfinite set theory have been quite surprising and contrary to presumed 'intuitions,' a few basic results appear intuitively clear. The following is one of the clearest instances. Cantor stated without proof that whenever M and N do not have the same power, either M is equivalent to a part of N (the power of M is less than that of N) or N is equivalent to a part of M (the power of M is greater than that of N) [Cantor 1878, 119]. This is the theorem of comparability of cardinals, which remained unproved until 1904 when Zermelo established it as a

---

[1] [Cantor 1878, 119]: "Wenn zwei wohldefinirte Mannigfaltigkeiten M und N sich eindeutig und vollständig, Element für Element, einander zuordnen lassen (was, wenn es auf eine Art möglich ist, immer auch noch auf viele andere Weisen geschehen kann), so möge für das Folgende die Ausdrucksweise gestattet sein, dass diese Mannigfaltigkeiten *gleiche Mächtigkeit* haben, oder auch, dass sie *äquivalent* sind."

corollary to the well-ordering theorem, by employing the Axiom of Choice (§IX.3). A decade earlier Cantor [1895/97, 285] had called attention to the need to prove carefully the result conjectured in 1878, and he was perfectly aware that well-ordering implied comparability of powers.

Coming back to the 1878 paper, Cantor went on to give some examples of powers. First we have the powers of finite manifolds, which coincide with the number of elements in the customary sense of the word. At this point, he warned the reader that infinite powers have a peculiar property: while the power of a part of a finite set M is always less than that of M itself, "this relation disappears completely in *infinite* manifolds, i.e., those that consist of infinitely many elements."[1] The fact that a manifold is a part of another, or is equivalent to a part of another, does not at all imply that the power of the first is smaller than that of the second [1878, 120]. This was traditionally regarded as paradoxical, in the strong sense of contradictory, for it went counter to the Euclidean principle that the whole is greater than the part. Ten years later, when Dedekind employed that peculiar property for a definition of infinite set, Cantor let him know that he had the priority, because he had published the above sentence back in 1878.[2] But indicating a property and using it as a definition on which to base a theory are completely different things; in this case, the first was done already by Galileo in the 17th century, and in the 19th by Bolzano, who – like Cantor – denied any contradictory character in that property of infinite sets. According to Dedekind, Cantor was astounded when in 1882 he showed him that there was indeed a simple definition of infinitude: "at times one possesses something, without judging its value and significance adequately."[3]

The series of positive integers, ℕ, constitutes the least infinite power; by this time, Cantor was not yet speaking of 'denumerable' [abzählbare] sets, an expression he started employing in 1879. But, he went on, in spite of being only the first infinite power the corresponding class of manifolds is quite rich. Among them we find the fields of algebraic numbers that Dedekind had defined in 1871,[4] and also the point-sets of the first species that Cantor had defined in 1872 (such that the derived set P^n is empty for some finite n). Also included are the set of real or complex algebraic numbers, the 'double sequences' [Doppelreihen] – sets of pairs, and the 'n-ply sequences' [n-fachen Reihen] – sets of n-tuples [Cantor 1878, 120].

Finally, Cantor presented two basic theorems that, employing terminology he introduced slightly later, can be formulated as follows:

If M is a denumerable set, then every infinite part [subset] of M is denumerable.
If M', M'', M''', ... is a finite or infinite sequence of denumerable sets, their union M is also denumerable.

---

[1] [Cantor 1878, 120]: "dieses Verhältniss hört gänzlich auf bei den *unendlichen*, d.i. aus einer unendlichen Anzahl von Elementen bestehenden Mannigfaltigkeiten."

[2] The reader will recall that Dedekind stated his definition for the first time in 1872 (§III.6).

[3] Dedekind to Weber, January 1888 [Dedekind 1930/32, vol. 3, 488]: "man besitzt bisweilen Etwas, ohne dessen Werth und Bedeutung gehörig zu würdigen."

[4] More precisely, Cantor mentions 'finite fields' [endliche Körper], i.e., finite extensions of ℚ.

In 1882 he would remark that these two theorems are the basis for any denumerability proof [Cantor 1879/84, 152].

The only other infinite cardinality that was known to Cantor at this time, was that of the continuum. His 1878 paper was devoted to an investigation of the powers of n-ply manifolds [Cantor 1878, 120] and established that $\mathbb{R}^n$ has the same power as $\mathbb{R}$, which justifies our talking of *the* power of the continuum. As a result of his investigations on infinite powers, Cantor was led to conjecturing a first version of the Continuum Hypothesis; calling "linear manifold" any infinite subset of $\mathbb{R}$, he ended the paper saying:

By an inductive reasoning, into which we shall not enter here,[1] the theorem appears plausible that the number of ... classes of linear manifolds is finite, and precisely equal to two.

According to this, the linear manifolds would consist of two classes, the first of which includes all manifolds that can be brought into the form: functio ips. ν (where ν passes through all positive integral numbers); while the second consists of all those manifolds that can be reduced to the form: functio ips. x (where x can take all real values ≥ 0 and ≤ 1).[2]

Cantor promised to return to this question in the future, and he certainly did. With this conjecture, he had stated a basic goal for his future investigations, a leading objective that would, however, remain inaccessible.

4.2. The theorem in the correspondence of 1877. As we have seen, Cantor was now in the possession of some abstract set-theoretical terminology, built around the words 'manifold' and 'power.' But, in order to formulate his questions in an arithmetically rigorous way, he preferred to phrase them in the terminology of analysis, namely speaking of variables and functions – this he called "purely arithmetical." The question whether two manifolds $\mathbb{R}$ and $\mathbb{R}^n$ can be coordinated one-to-one (see above) was rephrased as follows: consider ρ independent "variable real magnitudes" $x_1 ... x_\rho$ and another variable y, all of which take values in [0,1]; is it possible to correlate each "system of values" $(x_1, x_2, ... x_\rho)$ with a single value y, and inversely? [Cantor & Dedekind 1937, 25; Cantor 1878, 122]. As we see, this formulation is not purely set-theoretical, but employs the language of variables usual in analysis. Cantor used all kinds of means taken from traditional arithmetic and analysis in order to solve the question.

The proof that he communicated to Dedekind in June 1877 was surprisingly simple; one can infer the general strategy from the simplest case, in which we have

---

[1] For a detailed analysis of this topic, including some remarks on this 'inductive reasoning,' see [Moore 1989].

[2] [Cantor 1878, 132–33]: "Durch ein Inductionsverfahren, auf dessen Darstellung wir hier nicht näher eingehen, wird der Satz nahe gebracht, dass die Anzahl der ... Klassen linearer Mannigfaltigkeiten eine endliche und zwar, dass sie gleich zwei ist. / Darnach würden die linearen Mannigfaltigkeiten aus zwei Klassen bestehen, von denen die erste alle Mannigfaltigkeiten in sich fasst, welche sich auf die Form: functio ips. ν (wo ν alle positiven ganzen Zahlen durchläuft) bringen lassen; während die zweite Klasse alle diejenigen Mannigfaltigkeiten in sich aufnimmt, welche auf die Form: functio ips. x (wo x alle reellen Werthe ≥ 0 und ≤ 1 annehmen kann) zurückführbar sind."

only two variables x_1, x_2. Employing the decimal representation, each point in the unit square is determined by two numbers $0.\alpha_1\alpha_2...$ and $0.\beta_1\beta_2...$ It suffices to correlate those two numbers with the single one $0.\alpha_1\beta_1\alpha_2\beta_2...$ in the unit segment [Cantor & Dedekind 1937, 26]. Dedekind found the following problem: since the decimal representation is not univocal – e.g., ¼ = 0.25 = 0.2499... – , Cantor had established the convention that finite representations are excluded; as a result, some points in the segment are not correlated to any point of the square or, in modern terms, the image of the one-to-one mapping is a subset of [0,1]. (This proof and its 'faultiness' can be found in [Cantor 1878, 130–31].) Dedekind ended by saying that he did not know whether the objection was an essential one [*op.cit.*, 28]. But, of course, in some sense Cantor had proven more than he wanted, as he remarked in his reply [*ibid.*].

It seems convenient to pause at this point and consider the situation. Cantor had established a one-to-one correspondence between an *n*-dimensional continuum and a proper part of a one-dimensional continuum; one can trivially define another correspondence with the inverse effect. Was this not enough to regard the theorem as proven? None of the correspondents indicated this possibility, and it is evident that both assumed as an essential requirement that the correspondence should be exhaustive. The proof I have just sketched would rest on the Cantor–Bernstein theorem, that Cantor formulated but was not able to prove.[1] But in 1877 there was no general theorem of set theory to rely on, and the penchant for arithmetical rigor that was characteristic of both Dedekind and the Berlin school demanded the immense effort of proving every single result in full detail and in the most explicit way. The reader should take this into account in order to appreciate the nature of Cantor's early work on set theory.

A couple of days later, Cantor sent Dedekind a much more complex proof that he had found earlier than the other one. This proof is interesting in several ways; it gives us an indication of the topics that had occupied Cantor's attention in the long period from 1874 to 1877. We shall not enter into all details here.[2] Essentially, the proof rests on two lemmas: the first is a version of the above theorem, but restricted to points in ℝ and ℝ^*n* with *irrational* coordinates. Cantor takes these points as represented by continuous fractions (which completely avoids the possibility of double representation) and proves that to each irrational coordinate system of ℝ^*n* one can associate a single irrational coordinate, following the strategy indicated above, so that the correspondence is one-to-one [Cantor & Dedekind 1937, 29–30; Cantor 1878, 123]. The second lemma is a proof that the irrational numbers in [0,1] can be univocally correlated with the whole interval [0,1] [Cantor & Dedekind 1937, 30–37; Cantor 1878, 124–29].

---

[1] We shall see in §VII.4 that Dedekind was the first to prove it in 1887.

[2] See [Cantor & Dedekind 1937, 29–33] or [Cantor 1878, 122–29]. An English exposition can be found in [Dauben 1979, 60–65].

The second result was particularly difficult to establish. Although Cantor did not indicate it explicitly, the lemma establishes that the set of irrational numbers has the power of the continuum. As we see, in his research on infinite powers Cantor had considered and solved the question: what is the cardinality of the set of irrational numbers? This was extremely important in order to clarify how irrational numbers contribute to the difference in power between $\mathbb{Q}$ and $\mathbb{R}$.

In October 1877, Cantor found and sent to Dedekind a new and much easier proof that the irrational numbers have the power of the continuum. Both this [1878, 129–30] and the earlier proof found their way into the paper, because of the intrinsic interest of the lemmas involved. It will be worth presenting the second proof in exactly the same way as Cantor formulated it in the correspondence and the article, so that the reader can observe the way in which he employed variables. The theorem is:

Given a variable e that takes all irrational values $^{>0}_{<1}$, and another one x that assumes all rational and irrational values that are ≥ 0 and ≤ 1,

$$e \sim x.^1$$

$a \sim b$ means that the values of the variable a can be correlated one-to-one with those of the variable b. Denote by ϕ_v the sequence of all rational numbers in [0,1], by η_v any sequence of irrational numbers (e.g. $\eta_v = \dfrac{\sqrt{2}}{2^n}$); and consider a variable h that takes all values in [0,1] with the exception of those of ϕ_v and η_v. Then we have:

$$x \equiv \{ h, \eta_v, \phi_v \}$$
$$e \equiv \{ h, \eta_v \}.$$

Cantor employs here a peculiar notation for the union set or, more precisely, the 'union' of variables. But by a trivial transformation we can also write the last formula as follows:

$$e \equiv \{ h, \eta_{2v-1}, \eta_{2v} \}.$$

Considering that $h \sim h$, $\eta_v \sim \eta_{2v-1}$, and $\phi_v \sim \eta_{2v}$, we may conclude that

$$x \sim e.$$

The proof proceeds by decomposing both sets into disjoint sets that are pairwise equipollent, and it relies on the easy (but implicit) theorem that the unions of pairwise equipollent sets are equipollent.

As regards the above way of expression, Cantor seems to have regarded it as closer to the preferences of the Berlin school, for he kept employing it in the arti-

---

[1] [Cantor & Dedekind 1937, 41]: "Ist e eine Veränd., welche alle irrationalen Werthe $^{>0}_{<1}$ anzunehmen hat, x eine solche, welche alle rationalen und irrationalen Werthe, die ≥ 0 u. ≤ 1 sind, erhält, so ist: / $e \cong x$." See [Cantor 1878, 129–30].

cles he published in the *Journal für die reine und angewandte Mathematik,* but
abandoned it when he ceased publishing there. We see that he might have gone on
developing transfinite set theory in the language of analysis, but eventually he preferred the more abstract language, closer to the preferences of Dedekind and Riemann. At any rate, that proof may be called 'purely arithmetical' in the peculiar
sense of the time, which took some set-theoretical operations to be 'arithmetical.'
Therefore it satisfied Cantor's requirements. The same was not true of Cantor's
earliest proof of that result, which even employed the drawing of a discontinuous
curve (to be sure, it would have been easy for Cantor to define it analytically). In
that first proof, Cantor needed to show that an open interval (α,β) has the same
power as the respective closed interval $[\alpha,\beta]$. To establish this, Cantor used the
lemma that $(\alpha,\beta]$ has the same power as $[\alpha,\beta]$ which is proved with the help of the
"notable curve" that can be found below [Cantor & Dedekind 1937, 32; Cantor
1878, 127–28].

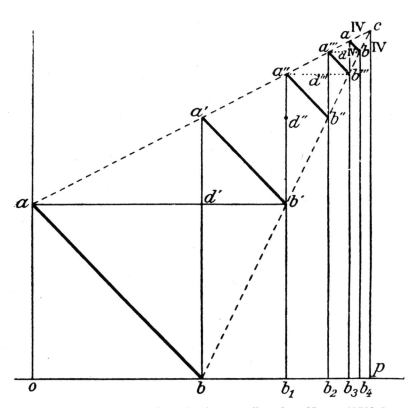

Figure 6. *Curve showing [0, 1] and (0, 1] to be equipollent, from [Cantor 1878]. It consists
of the infinitely many segments ab, a′b′,... and the isolated point c; points b, b′,... do not
belong to the curve (op = pc = 1; the a_i and b_i are obtained by halving intervals).*

All of these proofs, and the lemmas contained in them, give us some indication of the topics that Cantor studied from 1874 to 1877. His interest in the study of infinite cardinalities led him to attempt a careful analysis of the powers of sub- and super-sets of $\mathbb{R}$, including sets of irrational numbers and point-sets of the first species. It is not strange that he came to be regarded as an expert in irrational numbers: Klein gave him the assignment of refereeing Lindemann's famous 1882 paper on the transcendence of π.[1] The final picture that Cantor obtained was extremely simple: all subsets of $\mathbb{R}$, or even $\mathbb{R}^n$, seemed to be either denumerable or of the power of the continuum. Thus he formulated the restricted version of the Continuum Hypothesis; the language he used to this end was still that of variables and functions (see end of §4.1).

4.3. Riemannian issues.

Within the context of the contemporary discussion on real functions, Cantor's contribution was taken to give one more example of a 'pathological' function, a highly discontinuous $f: \mathbb{R} \to \mathbb{R}^n$. In this regard, it is still frequently mentioned together with such functions as the Peano curve. But from Cantor's viewpoint, the essential thing was its implications for the Riemannian theory of manifolds, for the discussion on geometry and the problem of space. This came out very explicitly in his paper: after presenting the notion of power and stating that the purpose of his paper was to investigate the powers of n-ply manifolds, he went on to put this in the context of recent work "on the hypotheses on which geometry is founded," started by Riemann and Helmholtz [Cantor 1878, 120–22].

In a letter to Dedekind of June 1877 Cantor goes one step further, suggesting that the whole investigation came out of his interest in the foundations of geometry. He says that, for several years [seit mehreren Jahren], he had followed with interest the efforts made, in connection with Gauss, Riemann and Helmholtz, to clarify the first assumptions of geometry [Cantor & Dedekind 1937, 33]. In this connection, he had noticed that such research depended, in its turn, on an unproved assumption: that the number of coordinates needed to determine points in an n-ply manifold is exactly n, which defines the dimension of the manifold (see §§II.6–7). He admits having shared that viewpoint, with the only qualification that he, unlike other authors, regarded that proposition as a theorem badly in need of proof. To make his viewpoint explicit, he formulated the question whether a continuous figure of n dimensions can be correlated one-to-one with a continuous segment, which he had asked several colleagues. Most of them answered, evidently not, and he too thought that a negative answer was the most plausible, until "very recently" he had found, by quite a complicated train of thought, that the question had to be answered in the affirmative [*op.cit.*, 34].

The kind of critique to which Cantor was subjecting the Riemannian assumptions on manifolds and dimension seems reminiscent of the Weierstrassian critique of analysis and function theory. He was employing his new tools for the study of infinite manifolds – one-to-one correspondences and the notion of cardinality – to analyze and clarify those assumptions. It seems that an important objective for his research at the time was to analyze basic notions of Riemann such as continuity and

---

[1] See [Meschkowski & Nilson 1991, 73ff].

dimension. As regards the notion of continuity, shortly afterward he also employed derived sets to clarify it, which led him to develop basic ideas of point-set topology and to offer an abstract definition of the continuum. All of this helps us understand why the title of his crucial papers of the period 1879–84 referred to the 'theory of manifolds' [Mannichfaltigkeitslehre].

As a consequence of his recent theorem, Cantor drew the following conclusion, with which he ended the June letter mentioned above:

Now it seems to me, that all philosophical and mathematical deductions that make use of that erroneous assumption are unacceptable. One must rather seek the distinction between figures with *different* number of dimensions in aspects completely different from the number of independent coordinates, which is taken to be characteristic.[1]

In these sentences, the words 'in aspects completely different' are quite clearly exaggerated. Dedekind warned that Cantor's words could be interpreted as casting doubt on the significance of the notion of dimension, and stated his conviction that the number of dimensions is, "now as before," the most important invariant of a continuous manifold [Cantor & Dedekind 1937, 37]. He admitted that the constancy of the number of dimensions was in need of a proof, but remarked that all previous writers on the subject had "obviously" made the implicit assumption that, in case of a change of coordinates, the new coordinates would be continuous functions of the old ones, and that which appears as continuous under one coordinate system would also be continuously linked under the second [*op.cit.*, 38]. As a result of these reflections, he believed in the following proposition:

If one succeeds in establishing a reciprocally univocal and complete correspondence between the points of a continuous manifolds *A* of *a* dimensions on the one hand, and the points of a continuous manifold *B* of *b* dimensions on the other, then, if *a* and *b* are *unequal*, this *correspondence itself* is necessarily a *completely discontinuous* one.[2]

This is a precise formulation of the theorem of dimension invariance under homeomorphic mappings, only in converse form. The idea was clearly suggested by Cantor in the introduction to his paper [1878, 121].

Dedekind went on to say that the first proof communicated to him by Cantor established what would have been a continuous mapping, had all points of the unit segment been included in the image. In the second proof, the first step involving only points of irrational coordinate is continuous, but the remaining transformations

---

[1] [Cantor & Dedekind 1937, 34]: "*Nun scheint es mir*, dass alle philosophischen oder mathematischen Deductionen, welche von jener irrthümlichen Voraussetzung Gebrauch machen, unzulässig sind. Vielmehr wird der Unterschied, welcher zwischen Gebilden von *verschiedener* Dimensionenzahl liegt, in ganz anderen Momenten gesucht werden müssen, als in der für charakteristisch gehaltenen Zahl der unabhängigen Coordinaten."

[2] [Cantor & Dedekind 1937, 38]: "Gelingt es, eine gegenseitig eindeutige und vollständige Correspondenz zwischen den Puncten einer stetigen Mannigfaltigkeit *A* von *a* Dimensionen einerseits und den Puncten einer stetigen Mannigfaltigkeit *B* von *b* Dimensionen andererseits herzustellen, so ist diese *Correspondenz selbst*, wenn *a* und *b* *ungleich* sind, nothwendig eine *durchweg unstetige*."

introduce a horrendous, vertiginous discontinuity in the correspondence, by which everything is dissolved into atoms, so that every continuously connected part of one domain, small as it may be, appears in its image completely torn and discontinuous.[1]

Dedekind ended by saying that his only intention was to avoid that Cantor enter into ungrounded open polemics against the "articles of faith" of the theory of manifolds. His correspondent was obviously very interested in those remarks, asking for more of them and saying that he would form his opinion on how to follow up the matter in their light [Cantor & Dedekind 1937, 39]. As a matter of fact, Cantor attempted to prove the invariance of dimension under bicontinuous mappings in a paper of 1879.

Cantor's suggestion [1878, 121] that dimension invariance might be a consequence of requiring the mapping to be bicontinuous provoked an immediate reaction. In that same year there were contributions by Lüroth, Thomae, Jürgens and Netto. The attempt of general proof published by Eugen Netto pointed in the promising direction of finding precise topological notions as a basis; it was well received by Cantor and Dedekind [1937, 47–48]. Nevertheless, Cantor did not find it completely satisfactory [*op.cit.*, 43] and published his own attempt in 1879. Apparently, the question was regarded as solved after these two publications, but in 1899 Jürgens called attention to the unsatisfactory character of all published proofs.[2] Only in 1911 did Brouwer finally prove the theorem of dimension invariance satisfactorily, on the basis of novel topological ideas that opened up a new era for this subject.

A few years later, Cantor published a new contribution that again suggests the influence of Riemann's geometrical thought in his mathematical work. The third installment of [Cantor 1879/84] employs the non-denumerability of the real numbers to show that there are discontinuous spaces in which continuous displacement is possible. We have seen (§II.1.2) that Riemann approached the question of physical space by means of a series of more and more restrictive hypotheses. In a similar vein, Cantor had thought that the continuity of physical space, which is not a necessity in itself, is a consequence of continuous motion [Cantor & Dedekind 1937, 52]. But this conviction vanished once Cantor noticed that in a space A, which is the result of substracting a denumerable dense set (e.g., the set of all points with algebraic coordinates) from $\mathbb{R}^n$, continuous displacement is still possible. This led him to speculate about the possibility of a modified mechanics, valid for such spaces A. This kind of physical application of his mathematical speculations is reminiscent of Riemann's work on geometry. Interestingly, that result of 1882 found application years later in function theory, but Cantor never considered this kind of use.[3]

---

[1] [Cantor & Dedekind 1937, 38]: "die Ausfüllung der Lücken zwingt Sie, eine grauenhafte, Schwindel erregende Unstetigkeit in der Correspondenz eintreten zu lassen, durch welche Alles in Atome aufgelöst wird, so dass jeder noch so kleine stetig zusammenhängende Theil des einen Gebietes in seinem Bilde als durchaus zerrissen, unstetig erscheint."

[2] For these developments, the reader may consult [Dauben 1979, 70–76] and above all Johnson's work on the history of dimension theory [Johnson 1979; 1981].

[3] Borel used it for a question of prolongation of analytical functions in his Ph.D. thesis, which also reformulated the notion of measure and gave the Heine–Borel theorem (see [Hawkins 1980]).

The notion of power, that Cantor formulated in 1878 and which he came to see as the most basic and general characteristic of sets [1879/84, 150], remained a central notion of set theory. And his lonely journey of exploration of the realm of transfinite sets continued for years to come, guided by the key objective of settling the Continuum Hypothesis. In 1895 he would introduce new numbers, the alephs, to designate the different transfinite powers (chapter VIII); with the help of the arithmetic of alephs he was able to prove the basic result of [1878] with just a few pen-strokes [1895/97, 289].

5. Cantor's Difficulties

'A Contribution to the Theory of Manifolds' [Cantor 1878] was ready for publication just seven days after the last July letter to Dedekind on the topic. During the half year that passed until the paper was published, one can notice in the letters that Dedekind had a very positive attitude toward his correspondent.[1] The good atmosphere is reflected in Dedekind's readiness to help Cantor when the latter thought there was trouble with the publication of his recent results [Cantor & Dedekind 1976, 231]. This kind of atmosphere is never again found in the correspondence, which will frequently reflect Cantor's worries that he may somehow have offended his colleague.[2] There is little doubt that Dedekind was unhappy with the way in which his suggestions were employed, without mentioning him, in Cantor's paper.

The paper was sent to Borchardt, then the editor of *Journal für die reine und angewandte Mathematik*, in July. One of Borchardt's collaborators was an old friend of Cantor, and in November Cantor knew through him that the article had not yet been composed. His friend intimated that Borchardt had plans to delay the publication, which caused him much excitement (see the letter in [Grattan-Guinness 1974, 112]). Once he got those news, Cantor started thinking about a separate publication of the work, and wrote in this connection to Dedekind, who offered his help but recommended patience. It seems noteworthy that Cantor's letter includes the following passage:

The delay on the side of the *Journal* seems to me the more inexplicable, for during my recent stay at Berlin I talked in detail with our older colleagues, who stay very close to the *Journal*, about the content of the work and I found no objection to it from any side. On the contrary, the question was new to all and they were very surprised about the result, which certainly had been unexpected for me too, but they acknowledged the absolute correction of the proof.[3]

---

[1] See [Cantor & Dedekind 1976, 230].

[2] See especially the letter of January 1880 in [Cantor & Dedekind 1976, 233].

[3] [Grattan-Guinness 1974, 112]: "Die Hinziehung von Seiten des Journals ist mir um so unerklärlicher, als ich während meines Berliner Aufenthaltes jüngst mit unseren älteren Fachgenossen, welche dem Journal sehr nahe stehen, über den Inhalt der Arbeit ausführlich gesprochen und von keiner Seite ein sachliches Bedenken dagegen gefunden habe. Im Gegentheil war allen

Less than a week later the issue had been satisfactorily solved, as Cantor told Dedekind citing further news from his Berlin friend [Cantor & Dedekind 1976, 231; Grattan-Guinness 1974, 113].[1] But the event was not without consequences: never again did Cantor send a paper to the *Journal für die reine und angewandte Mathematik*. Although no particular name is mentioned in his letters of the time, seven years later Cantor would blame Kronecker for the whole issue [Schoenflies 1927, 5].

It may be interesting to compare that episode with a similar one that involved Weber and Dedekind, in connection with their famous paper on algebraic functions. The article, sent to the *Journal* for publication in 1880, appeared only in 1882, the issue being discussed with Cantor in this last year [Cantor & Dedekind 1976, 247–54]. Kronecker regarded himself entitled to cause that delay until his related work [1882] was ready for the press. Weber and Dedekind had actually expressed their desire to see all of his investigations published, which Kronecker took as an agreement that their paper should be published after his [Cantor & Dedekind 1976, 253]. Dedekind stayed calm and showed no animosity toward his Berlin colleague; he just came to the conclusion that Kronecker was the victim of self-deception [Selbsttäuschungen].[2] Besides, he kept publishing in the *Journal*, regarding it as the leading mathematics journal. On that occasion, Cantor expressed his opinion of the "little despot" Kronecker and of the "negligence" [Bummelei] in the management of the *Journal* [*op.cit.*, 252–53]. This is actually the earliest occasion on which we have written confirmation of Cantor's growing enmity toward his former teacher.

Ever since Schoenflies published a paper on 'The Crisis in Cantor's Mathematical Creation' [1922], it has been an essential part of the folklore surrounding Cantor that the confrontation with Kronecker caused his mental illness, or at least his first nervous attack in 1884. Schoenflies based his argument on letters to Mittag-Leffler, editor of *Acta Mathematica* and a confidant of Cantor's in the mid-1880s. Shortly before his depressive crisis, Mittag-Leffler informed Cantor that Kronecker had asked him to accept, "with the same impartiality" as he was publishing Cantor's work, a paper in which he would show "that the results of the modern function theory and set theory do not possess any real significance."[3] (*Acta Mathematica* had been launched as an international journal devoted to the modern theory of functions, in close association with work of the Berlin and Paris schools.) Mittag-Leffler did not overlook that the move was as much against Weierstrass as against Cantor, but the latter could not help taking it extremely personally:

---

die Frage neu und sie waren sehr erstaunt über das, ja auch für mich unerwartet gewesene Resultat, für welches sie jedoch die völlig richtige Beweisführung anerkannten."

[1] As a matter of fact, no delay is apparent from the volume of the *Journal* in question (no. 84)—quite the opposite, for another paper written at an earlier date is printed after Cantor's.

[2] See [Dugac 1976, 253–54; Edwards 1980, 368–72].

[3] As quoted by Cantor in his answer [Schoenflies 1927, 5–6]: "dass die Ergebnisse der modernen Funktionentheorie und Mengenlehre von keiner realen Bedeutung sind ... mit derselben Unparteilichkeit in die Acta aufnehmen, wie die Untersuchungen ihres Freundes Cantor."

It is *tremendously suspicious* that he offers *you* precisely, for *your journal*, the product of the *poison* accumulated in him against function theory and set theory; I conjecture that with this he does not pursue *any other aim* but to banish me or better my articles from *Acta*, since he has attained the same perfectly well with respect to *Crelle's Journal*. The reason why for seven years I have not sent *anything there*, is none other than that I *reject* forever *any relation* with Mr. Kr[onecker].[1]

No doubt, Kronecker was a strongly opinionated person who did not hesitate to discredit the work of others, and who was in the habit of doing so mostly in personal conversations. Weierstrass too suffered from it, though of course he did not become ill (see [Biermann 1988]). However, it is doubtful whether Kronecker had bad intentions; this was at least the opinion of Dedekind and Mittag-Leffler, who regarded him as an upright man.[2]

Historians such as Grattan-Guinness [1971, 369], Purkert and Ilgauds [1987, 79–81] have carefully studied Cantor's illness, which according to their work seems to have been a manic-depressive one. They warn that Schoenflies' suggestion, that the fight on the Continuum Hypothesis was an important causal factor, is mere speculation. And they remind us that such mental problems have endogenous causes – to some extent, it was Cantor's disturbed mind and not merely Kronecker or some other external factor, which caused the trouble. On the other hand, external causes certainly contribute to the illness and the appearance of crises, and here one must take into account the impact on Cantor of the academic atmosphere, with its school fights and the presence of all-too-powerful men such as Kronecker. One should always be cautious not to accept uncritically Cantor's sometimes distorted perceptions of the episodes, and here, too, one might add or complement other factors. The difficulties with the Berlin school were partly caused by narrow-minded conceptions of what a school ought to be [Rowe 1989], but partly also by methodological disagreements regarding how mathematics ought to be conceived and practiced.[3] And it seems likely that the difficulties with Dedekind only added

---

[1] [Schoenflies 1927, 5]: "Es ist *höchst verdächtig*, dass er das Produkt des in ihm wider die Funktionentheorie und Mengenlehre angesammelten *Giftes* gerade *Ihnen* für *Ihr Journal* anbieten lässt; ich vermute, dass er hiermit *keine andere Absicht* verfolgt, als mich oder vielmehr meine Aufsätze auch aus den «Acta» zu vertreiben, da ihm dasselbe mit Bezug auf das «Crellesche Journal» durchaus gelungen ist. Der Grund warum ich seit sieben Jahren *nichts dorthin* geschickt, ist kein anderer, als dass ich für immer *jede Gemeinschaft* mit Herrn Kr. *perhorresziere*."

[2] See the letter of Mittag-Leffler in [Meschkowski 1967, 246]. When Cantor, after his depressive crisis in 1884, attempted a reconciliation with Kronecker, he treated him in a kind and respectful way [Schoenflies 1927, 9–13; Meschkowski 1967, 248–52].

[3] Cantor realized this very well; in his [1883] one finds an inspired defense of abstract mathematics, that will be considered in chapter VIII.

more tension. If Cantor tended to an abstract conception of mathematics, his dubious behavior of the 1870s cut the possibility of ties to one of its foremost exponents, whom Cantor himself regarded as the best interlocutor for the discussion of set-theoretical issues.[1]

Cantor contributed to creating tensions and difficulties, even for himself, with some of his actions and decisions [Purkert & Ilgauds 1987, 53–55, 76, 93]. For instance, as he was dissatisfied with being in Halle [Cantor & Dedekind 1976, 228], he repeatedly applied himself to the Ministry for vacant positions at other universities – a most uncommon procedure that could only make him appear as a troublemaker. Kronecker offered Mittag-Leffler a critique of function theory and set theory, but this happened after Cantor had written to the Ministry applying for a vacant position at Berlin. An oft-quoted letter from Cantor to Mittag-Leffler commenting on his application talks about the intrigues of Schwarz and Kronecker against him, and about the effect he was certain to attain – that Kronecker would feel like bitten by a scorpion and would howl as if Berlin was the African desert, with its lions, tigers and hyenas [Schoenflies 1927, 3–4]. Cantor's reaction to Kronecker's proposed critique was positive at first, but after a few days Cantor threatened to end his collaboration with *Acta* in case Mittag-Leffler would accept Kronecker's article. Such hypersensitivity indicates that his illness was in an advanced state; a few months later, in May and June 1884, he suffered his first great depressive breakdown.[2]

Although his work was finding acceptance (see §VIII.5), Cantor felt increasingly rejected by the German mathematical community. As we have seen, this was to some extent a natural consequence of the atmosphere of the period, to some degree a situation he had created for himself. In the early 1880s, when he felt definitively distanced from the Berlin school, he started looking for new allies. New possibilities opened up when Heine's position became vacant after his death in 1881, and Cantor attempted to bring Dedekind to Halle.[3] Although Dedekind gave a negative answer to his proposal, Cantor still managed to have him named first in the official list. His colleague refused to leave Braunschweig, however, for reasons such as family ties and salary conditions, but still the episode seems to have bettered their relations. Twice Dedekind expressed that he was attracted by the prospect of collaborating with Cantor [Cantor & Dedekind 1976, 239–40, 246].[4]

---

[1] The way in which Cantor treated the issue of the theorem of denumerability of algebraic numbers might reveal the advancement of his mental illness. He seems to have tended to suppress memory of the events: in January 1882 he wrote to Dedekind himself of "the serial ordering of all algebraic numbers, discovered by myself eight years ago" [Cantor & Dedekind 1976, 248] (compare [Cantor 1878, 126], where he mentions the theorem "presented by me").

[2] See his letters from 1882 to 1884 in [Meschkowski & Nilson 1991], especially [59–61, 65, 127–29, 162–73, 192–201]. As he wrote in August, probably under the influence of his physician and his wife, the crisis had not been caused by the fatigues of working, but by "frictions that I could reasonably have avoided" [Schoenflies 1927, 9].

[3] See especially [Grattan-Guinness 1974, 116–23; Dugac 1976, 126–29].

[4] This seems to conflict with my interpretations. The explanation might be found in his courtesy, or in the good sentiments that Cantor's proposal aroused; it might even be that Dedekind

During 1882, Cantor sent several letters on mathematical and institutional issues, the mathematical themes being noteworthy and constituting the most important exchange between 1877 and 1899 (see §§VII.4 and VIII.3). But by the end of the year, after a couple of personal meetings in September, the correspondence came to an end. Dedekind showed no interest in the last letters Cantor sent him, leaving them unanswered in spite of the fact that they included a detailed announcement of the introduction of transfinite numbers [Cantor & Dedekind 1976, 258; 1937, 55–59]. We do not know exactly why the correspondence ended; it may have been something that occurred in the course of the personal meetings, or perhaps some other reason.[1]

The vacant position at Halle was finally filled by A. Wangerin, the Berlin mathematicians' choice and not someone proposed from Halle. Cantor's feeling of isolation increased. He turned to foreign friends, especially Gösta Mittag-Leffler, editor of *Acta Mathematica*, a journal launched in an ambitious way and counting on the collaboration of the leading mathematicians at Paris and Berlin. After his failure to get a position at Göttingen in 1885 he came to the conclusion that he was destined to remain in Halle. Then, in that same year, Mittag-Leffler recommended that he not publish a paper that had previously been accepted by *Acta*. Cantor felt that this was merely to protect the interests of Mittag-Leffler's journal and that the Swedish mathematician was caving to the influence of Berlin. He stopped publishing in mathematical journals, feeling distanced from mathematicians.[2] Although he kept doing some research, Cantor began to devote more and more time to his philosophical, theological and literary interests, including participation in a polemic then in vogue, in which he defended the view that Francis Bacon was the real author of Shakespeare's works [Purkert & Ilgauds, 1987].

Still, the story did not have a negative ending. On the basis of the bitter experiences made in the 1870s and 80s, Cantor became a strong advocate of the creation of a German mathematical association. He hoped that this new organization would help to prevent young mathematicians from being mistreated as he thought he had been. Cantor became the visible hand during the difficult period in which the *Deutsche Mathematiker-Vereinigung* was being launched, although he was by no means the most important man behind. He was named the first president of the association and stayed in office from 1890 to 1893. Likewise, he did his best to promote international congresses of mathematics, also hoping to open impartial forums where new ideas could be freely exposed and judged.

---

had already come to the conclusion that there was something wrong in Cantor's personality.

[1] Grattan-Guinness [1971; 1974] assumed it was because Cantor felt offended by Dedekind's negative answer to the Halle offering, but as we have seen the situation was much more complex; besides, it was Dedekind who left the letters unanswered.

[2] See [Schoenflies 1927; Grattan-Guinness 1970; Purkert & Ilgauds 1987, 79–101].

6. Derived Sets and Cardinalities

In 1879 Cantor started publishing a collection of papers in *Mathematische Annalen* under the common title 'On infinite, linear point-manifolds' [Cantor 1879/84]. These papers are regarded as his masterpiece; according to Zermelo they constitute the "quintessence" of their author's work, making all other contributions appear just as precedents or complements [Cantor 1932, 246]. Although they were published at a time when the theory of point-sets was being actively studied for the purposes of real variable theory (chap. V), Cantor's articles are distinguished by the feature that they take the theory of sets as an autonomous domain of study. The six installments in this collection are of unequal length and depth, going beyond the limits suggested in the title. In fact, Cantor employed them as a means to publish his new ideas quickly; it would seem that he felt urged to establish the priority and superiority of his ideas against such potential competitors as du Bois-Reymond, Harnack, and Dedekind.

Cantor was guided, above all, by the core objective of proving the Continuum Hypothesis. To this end, he studied in detail the powers of subsets of $\mathbb{R}$ (what he called 'linear manifolds'), refining his theory of derived sets and looking for combined results on derived sets and powers. This led him to introduce some basic notions of point-set theory, which we regard today as having to do with the topology of point-sets. A note to the third installment of the series, published in 1882, contains a comment that seems worthy of attention:

Most of the difficulties of principle that are found in mathematics have their origins, it seems to me, in ignorance of the possibility of a purely arithmetical theory of magnitudes and manifolds.[1]

This suggests a way of understanding Cantor's research up to that year, a second key motivation. It is likely that, when talking about the purely arithmetical theory of magnitudes, Cantor had in mind Weiestrass's critical reworking of the foundation of analysis. The comment suggests that its author had not yet considered the possibility of reducing the theory of magnitudes (i.e., arithmetic) to set theory, so that he was still far from taking set theory as the foundation of mathematics.[2] It would seem that it was Cantor's aim to offer a detailed critique and rigorous development of basic ideas behind Riemann's theory of manifolds, on an arithmetical basis and in analogy with the work of Weierstrass. It should be added that after the introduction of transfinite numbers in 1882, Cantor's projects seem to have changed (see chap. VIII).

---

[1] [Cantor 1879/84, 156]: "Die meisten principiellen Schwierigkeiten, welche in der Mathematik gefunden werden, scheinen mir ihren Ursprung darin zu haben, dass die Möglichkeit einer rein arithmetischen Grössen und Mannichfaltigkeitslehre verkannt wird."

[2] This interpretation is consistent with the way in which he approached results in set theory, see above and §V.5.

Cantor's initial objective was to investigate 'linear point-manifolds' (subsets of $\mathbb{R}$), since in the light of [Cantor 1878] these must already present us with all possible cardinalities of subsets of a continuum. That is presented as a reason to give special attention to linear point-manifolds and to investigate their "classification" [1879/84, 139]. To this end, the first part of the series introduces again the notions of derived sets, "everywhere-dense" sets, and cardinality or "power;" each suggests a way of classifying subsets of $\mathbb{R}$. Cantor analyzes some interrelations between those notions before proceeding to a new exposition of the proof that $\mathbb{R}$ is not denumerable. He promises to employ the notion of derived set as a basis for the "simplest and most complete" definition of a continuum [Cantor 1879/84, 139].[1]

The first two articles published in 1879 and 1880 do not really contain important new results, contrasting sharply with the next three installments, which abound in new theorems. The second part was published in the *Annalen* volume that was next to the one in which du Bois-Reymond promised a paper on derived sets of an infinite order (see §V.4.2). It contains the first public presentation of an old idea of Cantor's in which the actual infinite figured prominently – that of introducing new symbols to characterize derived sets of a higher order. A footnote indicates that Cantor had come to this idea ten years earlier, i.e., around 1870.[2] But his attempted classification of linear point-manifolds ultimately led Cantor to the idea of employing derived sets for studying the powers of point-sets, and this was the source of very important results presented in parts 3 to 6. From his letters to Mittag-Leffler we know that Cantor worked "literally for years" on the proof of some theorems that can be found in these installments [Meschkowski & Nilson 1991, 88], so one may assume that around 1879 he had in mind some important results that resisted his attempts for a long time.

6.1. Derived sets of an infinite order.

In §V.4 we saw how point-sets with a derived set of an infinite order had been defined by means of the method of distributing intervals. This allowed several mathematicians to offer examples of nowhere-dense sets with positive outer content. From Cantor's viewpoint, it also gave a sense to the introduction of 'symbols of infinity' employed for a precise characterization of such derived sets. As we shall see (§VIII.3), these symbols formed the germ from which transfinite numbers grew. Although Cantor stated that he had come to the idea around 1870, it is quite unclear whether at that point he had interesting examples of point-sets with derived sets of an infinite order – that is, examples other than the rational numbers or the algebraic numbers, whose derived set is already the continuum $\mathbb{R}$ of real numbers. The available evidence (see §V.4) suggests that interesting examples were first given by Smith and du Bois-Reymond,

---

[1] The definition was actually given in 1883 (see §VI.7). In the light of the crucial role that continuous manifolds played in Riemann's work, that promise might seem to support the interpretation of Cantor's projects given above.

[2] The note did not find its way into Zermelo's edition of his treatises; see [Cantor 1984, 55] or *Mathematische Annalen* **17** (1880), 358.

who seems to have been the first to employ interval distributions. In contrast to the work of these authors, Cantor's paper is remarkable for being disconnected from more particular mathematical concerns, and for presenting the whole issue in a purely abstract way.

There is in Cantor's paper a point of interest to us, that has not been noticed previously by historians. Cantor begins by introducing notation and terminology for some of the basic relations and operations on sets, and this terminology shows a strong resemblance to that of Dedekind's algebraic number theory:

NOTION	NOTATION	TERMINOLOGY
set identity	$P \equiv Q$	
disjoint union	$P \equiv \{P_1, P_2, ...\}$	
inclusion	$(P \subset Q)$	P is a *divisor* of Q
		Q is a *multiple* of P
union	$P \equiv \mathfrak{M}(P_1, P_2, ...)$	P is the *least common multiple* of P_1, P_2, ...
Intersection	$P \equiv \mathfrak{D}(P_1, P_2, ...)$	P is the *greatest common divisor* of P_1, P_2, ...

While Cantor's notations [1879/84, 145–46] are novel,[1] his terminology for inclusion, union and intersection agree with Dedekind's [1871, 224, 252–53] (see §III.2).[2] On the other hand, Cantor was the first to suggest that one may form unions and intersections of infinite families of sets; and this was essential in the context of his paper.

After such preliminaries, Cantor remarks that the successive derived sets of a given point-set P are included in each other: ... $P^{(n)} \subset ... \subset P'' \subset P'$ [Cantor 1879/84, 146]. The process of set derivation eliminates points, but does not add new points that were not already in P'. This was a crucial realization, for it enabled him to define the first derived set of an infinite order as the intersection of all derived sets of finite order. Cantor writes:

$$P^{(\infty)} \equiv \mathfrak{D}(P', P'', ... P^{(n)}, ...) \equiv \mathfrak{D}(P^{(n)}, P^{(n+1)}, ...).$$

In general, the process of derivation will then continue with $P^{(\infty+1)}$, ... $P^{(\infty+n)}$, indeed $P^{(\infty)}$ may in its turn have a derived set of infinite order, which at this time Cantor denotes by $P^{(2\infty)}$. Going on in this way, one reaches derived sets whose

---

[1] Notation for union and intersection was at least suggested by Dedekind's work, since he employed Gothic *m* and *d* for the least common multiple and greatest common divisor of two ideals [1871, 252–53].

[2] Since this terminology makes little sense in the context of point-set theory, or of abstract set theory (Dedekind himself replaced it for this purpose in 1872), the question of its origins seems to be beyond doubt.

order is a multiple of ∞, such as $n∞+m$, and proceeds to $∞^2$, $∞^3$, ..., to polynomials like $n_0∞^m+n_1∞^{m-1}+... +n_m$, to $∞^∞$ and beyond. In the process, some derived sets are defined in the customary way $P^{(n+1)} = [P^{(n)}]'$, while others are defined by means of the intersection of infinitely many point-sets. These two different procedures are nothing less than the forerunners of the two *generation principles* that Cantor employed in 1883 for defining the transfinite ordinal numbers (§VIII.3).

Only the last paragraph gives an example of a point-set whose derived set $P^{(∞)}$ consists of a single point [Cantor 1879/84, 148]. From a mathematical viewpoint, however, this example is of the utmost importance, for without it one might fear that the whole symbolism defined by Cantor is useless – derived sets of an infinite order might always be equal to each other. Cantor stated that it is equally simple to give examples for $P^{(2∞)}$ and higher orders.

Employing a philosophical language, he wrote that "we observe a dialectical generation of concepts which leads always farther, and which in so doing remains free from any arbitrariness, necessary and consistent in itself."[1] But it is crucial to realize that the 'concepts' that Cantor is introducing relate directly to point-sets. The objective basis for the process was thus point-sets and their properties, while the 'symbols of infinity' were just notational devices which lacked any objective reference in themselves. This makes clearer the distance that separates Cantor's ideas at this stage from his later consideration of the symbols as referring to true numbers. Taking this step involved a difficult conceptual reconsideration which, as we shall see (§VIII.3–4), was only possible on the basis of the new notion of well-ordered set.

6.2. Cardinality results.[2]   Almost all of the new results that Cantor proved between 1882 and 1884 emerged from research on the process of derivation in connection with the notion of power. He studied the kinds of sets we find through derivation; when the process stops; and what information it affords about the original set, especially about its cardinality. Particularly important was the Cantor–Bendixson theorem, which establishes a unique decomposition for derived sets P' into two sets of known properties, which implied that CH is valid for this particular kind of set. (Cantor announced a weak version of it in September 1882, but he had been pursuing it for some years [Meschkowski & Nilson 1991, 88].) Attention to derived sets naturally led him to develop some topological notions of point-set theory, which he took much farther than his predecessors (but always in relation to limit points and derived sets).

In the third installment, Cantor investigated subsets of $\mathbb{R}^n$, so-called *n*-dimensional manifolds, because, as he said, they offer new viewpoints and results that are useful for the study of linear point-sets, and they are interesting in themselves and for applications [1879/84, 149]. The *n*-dimensional case allowed him to

---

[1] [Cantor 1879/84, 148]: "wir sehen hier eine dialektische Begriffserzeugung, welche immer weiter führt und dabei frei von jeglicher Willkür in sich nothwendig und consequent bleibt."

[2] It is impossible here to discuss Cantor's work of the 1880s in full detail. See [Dauben 1979, chs. 4–5] and [Hallett 1984, 81–98].

prove a key theorem for his study of cardinality and derivation: given infinitely many (closed) n-dimensional subdomains in $\mathbb{R}^n$, which are disjoint except perhaps at boundary points, the set of such subdomains is always denumerable [*op.cit.*, 152]. One easily sees that such must be the case by taking into account that any such subdomain will contain points whose coordinates are all rational, and that the set of such points in $\mathbb{R}^n$ is denumerable. Cantor did not follow this simple way, but established a one-to-one mapping between the points of $\mathbb{R}^n$ and the points in the n-dimensional sphere of unit radius in $\mathbb{R}^{n+1}$; he thus gained the possibility of enumerating the subdomains as a function of the volume of their images [*op.cit.*, 153, 157; Dauben 1979, 84–85].

That result was useful for the proof of a weak version of the Cantor–Bendixson theorem, due to Cantor alone. This weaker theorem says that whenever a point-set P is such that P' is denumerable, there exists some index α for which P^α vanishes ($= \varnothing$), and conversely, if P^α vanishes for some α then P' and also P are denumerable [Cantor 1879/84, 160; 1883, 171]. The result is remarkable not just as a first step toward the famous Cantor–Bendixson theorem, but also because of its applications in function theory. Cantor was well aware that the theorem would make it possible to give a natural conclusion to Mittag-Leffler's work on the representation of analytic functions,[1] which generalized a famous theorem of Weierstrass. Mittag-Leffler had been working on the subject from the mid-1870s; in his [1884], on the basis of Cantor's work, he was able to prove the Mittag-Leffler Theorem which solves the construction of analytic functions with an infinite set of isolated poles. Weierstrass was impressed by the work and said that the main problem of the theory, which had previously seemed a matter for the future, had found its most general solution (quoted in [Mittag-Leffler 1927, 25]). Thus, in this particular case Cantor was pursuing a result of very clear and concrete application to problems that were then regarded as central at the time. Moreover, the context of Cantor's result was intimately connected with the introduction of transfinite numbers.

In the fourth part of [1879/84] Cantor proved a restricted version of the result, for index ∞ as he wrote then (or, as he would say within a few months, the first transfinite number ω). His strategy employed the new notion of *isolated* set and the fact that isolated sets are denumerable [Cantor 1879/84, 158–59]. A point-set P is isolated if and only if none of its limit points belongs to it ($P \cap P' = \varnothing$), in which case every point in P can be enclosed in a neighborhood that does not contain any other element of the set. Taking such a collection of intervals $[\alpha,\beta]$, that are disjoint except perhaps at the extremes, Cantor could apply the theorem he had proved in the previous issue, to conclude that this collection of intervals (and therefore P) is denumerable. Cantor remarked that while the set of intervals $[\alpha,\beta]$ and the set E of the extremes α, β are both denumerable, this is not necessarily the case for the derived set of E [*op.cit.*, 154]. This remark indicates that Cantor had paid careful attention to what are now called 'Cantor sets' (in §V.4 we noticed that he was not the first to consider such sets).

---

[1] See the correspondence of both mathematicians in [Meschkowski & Nilson 1991, 88–89].

By considering isolated sets, Cantor was able to obtain a good number of results on the cardinality of point-sets. First, if a derived set P' is denumerable, then so is P – call Q the set of elements of P that do not belong to P', then P is the union of the isolated Q and a subset of P', both being denumerable. Second, writing '–' for the difference of sets and '+' for the union operation, Cantor [1879/84, 158] established the following decomposition:

$$P' \equiv (P' - P'') + (P'' - P''') + ... + (P^{\infty}).$$

Since the sets $(P^n - P^{n+1})$ are all isolated, it is easy to show that, if P^{∞} is denumerable, P' and P must be denumerable, for P' is then a denumerable union of denumerable sets. More particularly, every point-set of the first species (such that $P^n = \varnothing$ for some n) must be denumerable [*op.cit.*, 159–60].

The way to generalizing that theorem was opened by the possibility of extending the same kind of decomposition into isolated sets to any derived set of infinite order P^{α}:

$$P' \equiv (P' - P'') + (P'' - P''') + ... + (P^{\infty} - P^{\infty+1}) + ... + (P^{\alpha}).$$

But, from the fact that P^{α} is denumerable, one can only conclude that so is P' in case the union is denumerable or, what comes to the same, if one has only denumerably many indices preceding α (for the union of a non-denumerable family of nonempty sets is non-denumerable). Thus, in the context of this theorem it was natural to pay attention to sets of indices, and Cantor came to the idea of considering 'symbols of infinity' with denumerably many predecessors. This is the simplest version of the so-called *principle of limitation* that was crucial for his 1883 theory of transfinite ordinals (§VIII.3.1). When Cantor announced the theorem in September 1882 [1879/84, 171], he still lacked the basic results on the indices α that would enable him to prove it. The actual proof was only given in the sixth and last part of the series 'On infinite, linear point-manifolds' [1879/84, 220–21], after the transfinite numbers had been introduced.

To conclude this section, we shall indicate a few more results connected with the circle of ideas that we have just delineated. Reading the above theorem of Cantor's in the converse form, one has proven that if a point-set P' is not denumerable, then P^{α} is also non-denumerable, α being a natural number or one of the 'symbols of infinity' [*op.cit.*, 160]. Actually, Cantor was able to show that the derivation process only has to involve indices that are denumerable ordinals – as he said, numbers of the second number-class. The sixth part of his series of papers, published in 1884, indicated this as follows:

Actually ... the matter disposes itself so that in rigor, for point-sets within any domain G_n [$\mathbb{R}^n$], only those *derivations* play a role whose ordinal number belongs to the *first* or *second* class of numbers. For one can show the *extremely notable fact*, that for *every* point-set P *from a certain ordinal number* α on, which belongs to the *first* or *second* number-class but not to *any* higher one, the derived set $P^{(\alpha)}$ is either 0 [empty] or a *perfect* set. From which follows that derivations of a higher order than α are all identical with the derived set $P^{(\alpha)}$, and taking them into consideration is *superfluous*.[1]

The study of derived sets led Cantor to another new topological notion, that of *perfect* sets – a set P is perfect if and only if it is identical to its first derived $P' = P$. Whenever a point-set is non-denumerable, the derivation process ends up in a perfect set.

At this point, it was natural to consider decomposing any derived set P' into its perfect component and the rest. This was the idea behind the Cantor–Bendixson Theorem. Cantor showed that we always have

$$P' \equiv R + S,$$

where S is a perfect set and R is denumerable [Cantor 1879/84, 223–24]. In the *Grundlagen* [1883, 193], Cantor called R a *reducible* set, meaning that $R^\alpha = \varnothing$ for some α, but this is not true in general. The young Scandinavian mathematician Bendixson pointed out the error and showed that the correct result is as follows: there always exists an α such that $R \cap R^\alpha = \varnothing$ [Bendixson 1883]. In summary, Cantor had established that the derivation process yields only sets that are either denumerable or perfect, or composed of both. This seemed to confirm CH.

7. Cantor's Definition of the Continuum

The new notions that Cantor was introducing quickly became important for function theory and the theory of point-sets. They all emerged from the basic concepts of limit point and derived set, which makes it plainly clear why they were eventually called topological notions. In the sixth installment [Cantor 1879/84, 226, 228], the concepts of isolated and perfect sets were joined by the following new ones: a point-set is *closed* iff (if and only if) it contains all of its limit points ($P' \subset P$); a point-set is *dense in itself* iff all of its elements are also limit points of the set ($P \subset P'$); a point-set is *separated* iff none of its subsets is dense-in-itself. Cantor could now say that a point-set is perfect iff it is closed and dense-in-itself.

The notion of perfect set became the basis for quite a general definition of continua, a definition he had promised to give back in 1879 [Cantor 1879/84, 139]. Thus, a basic objective of Cantor's, clarifying the notion of the continuum, was attained. The definition was presented in §10 of the *Grundlagen*, which begins with

---

[1] [Cantor 1879/84, 218]: "In Wirklichkeit stellt sich zwar ... die Sache so, dass bei den Punktmengen innerhalb eines beliebigen Gebietes G_n strenggenommen nur diejenigen *Ableitungen* eine Rolle spielen, deren Ordnungszahl der *ersten* oder *zweiten* Zahlenclasse angehört. Es zeigt sich nämlich die *höchst merkwürdige Thatsache*, dass für *jede* Punktmenge P *von einer gewissen Ordnunszahl* α an, welche der *ersten* oder *zweiten* Zahlenclasse, jedoch *keiner* höheren Zahlenclasse angehört, die Ableitung $P^{(\alpha)}$ entweder 0 oder eine *perfecte* Menge wird; daraus folgt, dass die Ableitungen höherer Ordnung als α mit der Ableitung $P^{(\alpha)}$ sämmtlich identisch sind, ihre Inbetrachtnahme daher *überflüssig* wird."

reference to the unending philosophical discussions on the question, including the dispute between the partisans of Aristotle (the continuum is indefinitely divisible, but not a set of points) and those of Epicurus (it is composed of finite atoms). Cantor does not intend to enter into such disputes, but only to give a "sober logical" [logisch-nüchtern] definition of the notion as needed in the mathematical parts of the theory of manifolds [Cantor 1883, 191]. It seems likely that one of the motives why he does not enter into the dispute is that his whole approach always presupposed the view that the continuum is built up of infinitely many points, and defending this viewpoint against radically opposite ones brings into the picture all kinds of arduous problems.

The definition was also discussed in three letters sent to Dedekind in 1882 [Cantor & Dedekind 1937, 52–54], where we can read that a first attempt to generalize Dedekind's notion of a cut came to nothing, while the notion of a denumerable fundamental sequence complied naturally with the task. Dedekind's definition of 1872 is clearly the most important precedent of Cantor's, but it was too closely connected with the linear continuum and the assumption of a totally and densely ordered subset. Cantor wished to have a general definition that could be applied not only to continua like $\mathbb{R}$ and $\mathbb{R}^n$, but also to continuous subsets of these which might be composed of different parts, linked continuously, which might have different dimension numbers. In a word, his objective required the use of ideas of a more purely topological kind, and it is not surprising that Dedekind's standpoint did not yield good results in that direction.[1]

It is evident that continuous sets are unaltered by the derivation process: [0,1] is identical with its derived set. For this reason it was natural to require a continuum to be a perfect set. But, as we have seen, Cantor knew that perfect sets may be nowhere dense. He needed an additional notion to characterize continua, and he found it in the idea of *connectedness*, which Weierstrass [1880] had employed a few years earlier, defining it in terms of neighborhoods. According to Cantor, a subset T of $\mathbb{R}^n$ is "connected" if for every two points t, t' in T, and every real number $\delta > 0$, it is possible to determine a finite set of points t_1, t_2 ... t_n such that the distances $\overline{tt_1}, \overline{t_1 t_2}, ..., \overline{t_n t}$ are all less than δ [Cantor 1883, 194]. On this basis, Cantor defined a point-continuum within $\mathbb{R}^n$ as a *"perfect-connected set"* [*ibid.*].

Cantor's notion of connectedness is somewhat strange, since according to it the set of rational numbers is connected. Since the early 20th century a different definition of connectedness for topological spaces has become common; likewise, present definitions of continua are different from Cantor's. Nevertheless, his definition showed the way in which topology would proceed. With this and other related notions, Cantor inaugurated the study of the topology of point-sets. For himself, a different stage was reached with the general theory of order types, to which the above definition could be adapted (see §VIII.4.2).

---

[1] Dedekind's method can be applied to ordered spaces, while Cantor's is the one to be used for metric spaces. This is how Hausdorff employed them in his epoch-making handbook [1914].

8. Further Efforts on the Continuum Hypothesis

A key motivation of Cantor's after 1878 was to prove CH. This is not to say that he was unaware of possible applications of set theory here and there, as was the case with the Cantor–Bendixson Theorem. Occasionally, his papers suggested possible applications of his theory – to function theory and geometry in part three of [1879/84] (see end of §4.3), to integration theory in part four.[1] But Cantor does not seem to have pursued that kind of aim directly. Instead, he was focusing on point-sets, and later on well-ordered sets of transfinite numbers, mostly because he hoped to analyze them in sufficient detail so that the continuum problem would become solvable.

As we have seen, sometime around 1880 he surmised that a joint analysis of derivation and cardinality might be the key to the problem. Thus he obtained new results on the powers of point-sets: denumerability of first-species sets, denumerability of isolated sets, relations between the power of $P^{(\alpha)}$ and that of P (leading to the Cantor–Bendixson theorem). The process led to the transfinite ordinals (§VIII.3), but also to the important notion of a perfect set $P=P'$. The Cantor–Bendixson theorem means that derivation yields only sets that are either denumerable or perfect, or composed of both. With an eye to CH, the next obvious question was: what is the power of perfect sets? Cantor was able to solve this in the last paper of the series [1879/84]. During the mid-1880s he kept making strong efforts to prove CH; in the present section we shall consider some of these efforts.[2]

The assumption of CH implied that perfect sets have to be either denumerable or of the power of the continuum. Already for his theorems on derivation and cardinality, Cantor had needed the result that a perfect set cannot be denumerable. As we have seen, with his 1884 terminology one can say that a perfect set has to be closed ($P' \subset P$) and also dense-in-itself ($P \subset P'$); Cantor proved that a set P that is denumerable and dense-in-itself cannot be closed. To this end, he employed an argument based on the Bolzano–Weierstrass principle (§IV.4.2), and therefore his proof is reminiscent of the crucial theorem of 1874 [Cantor 1879/84, 215–18].[3] Late in 1883 Cantor was finally able to establish that perfect sets are equipollent to $\mathbb{R}$.

A perfect set P is not necessarily dense in every interval, it may, in fact, be nowhere dense. Cantor gave an example of this kind by methods that are intimately related to the interval distributions that had been employed to give examples of

---

[1] In connection with previous work by du Bois-Reymond and Harnack, he proved that if a set P is bounded and has a denumerable derived set P', then it has content zero [Cantor 1879/84, 160–61]. This is an example of the central role Cantor expected the notion of power to play in mathematics generally.

[2] On this topic, see [Hallett 1984, 74–118] and especially [Moore 1989].

[3] Since P is denumerable, it can be put into the form of a sequence (p_n); any neighbourhood of every p_n contains infinitely many points of P, since it is dense-in-itself. On this basis, Cantor defines a sequence of nested spheres that determine a limit point of P not belonging to (p_n).

nowhere dense sets with outer content zero (§V.4). The result was his famous 'ternary set' defined in an endnote to the *Grundlagen*: the set of all real numbers of the form

$$c_1/3 + c_2/3^2 + ... + c_n/3^n + ...,$$

where the coefficients c_n can only take the values 0 and 2, is a perfect set that is nowhere dense in [0,1] [Cantor 1883, 207]. The complement of the ternary set C is the union of infinitely many intervals that are densely distributed throughout [0,1], and so C is nowhere dense. Since those intervals are all open, C must contain all of its limit points, i.e., is closed. Moreover, one can prove that every number of the above form is the limit of a convergent sequence of such numbers, which means that $C \subset C'$, i.e., that C is dense-in-itself. Therefore, C is a perfect set.

In fact, all nowhere-dense perfect sets have as a complement in $\mathbb{R}$ a union of disjoint intervals that are densely distributed. And a theorem that Cantor proved in 1882 (§6.2) shows that the set of those intervals must be denumerable. Cantor was able to employ these results to prove that nowhere dense perfect sets are equipollent to [0,1]. (If P is perfect and dense in an interval $[a,b]$, it is rather trivial that it must have the power of the continuum, for in such case $[a,b]$ must be included in $P'=P$.[1]) The proof of that result is of interest to us because it confirmed CH, and because it seems to have indicated the way toward a general theory of order types. This was the subject of an unpublished paper of Cantor's in 1885, and constituted the last crucial development toward his mature abstract theory of sets (see §VIII.4). For these reasons we shall outline the core ideas of the proof, without entering into details.

Cantor proved that any subset S of [0,1] that contains 0 and 1, is perfect, and is nowhere dense, can be put into one-to-one correspondence with the full interval [1879/84, 237–41]. As we have said, the complement R of S in [0,1] is a denumerable union of open disjoint intervals (a_n,b_n), whose endpoints a_n, b_n, belong to S. The set J of the endpoints a_n, b_n, must also be denumerable, but it determines S completely; Cantor showed that $S = J'$ [*op.cit.*, 237]. Now Cantor's key idea was to establish a one-to-one correspondence between the set of intervals $\{(a_n,b_n)\}$ and the set of rational numbers in [0,1] *preserving the dense order* within each set. This must be possible, for both are denumerable sets dense in [0,1]; the matter will be taken up in §VIII.4.2.

All of the results that Cantor had obtained until 1884 contributed to making CH plausible. First, previously known sets and subsets of $\mathbb{R}$ turned out to be either denumerable or of the power $\mathfrak{C}$ of the continuum. Moreover, basic set-theoretic operations on sets fulfilling the Hypothesis yield more such sets, and for the elementary operations of transfinite arithmetic it suffices to consider those two pow-

---

[1] Nevertheless, this simple proof presupposes the then unproven Cantor–Bernstein theorem; thus Cantor had to give a rather prolix proof for the general case, see [1879/84, 241–43].

ers.[1] As regards the process of derivation, recall that the Cantor–Bendixson Theorem states that $P' = R \cup S$, where R is denumerable and S is perfect; therefore, the process yields only denumerable sets or sets of power $\mathfrak{C}$. And the fact that derivation essentially ends with transfinite ordinals of the second number-class may also suggest that CH is true.

But none of those reasons can be made into a proof of CH. As regards the derivation process, just like there are denumerable sets with a perfect derived (an obvious example is $\mathbb{Q}_{[0,1]}$) there could be point-sets whose power is different from the known two, with the same property. An essential defect of Cantor's theory of derivation and cardinalities is that it applied almost only to the derived sets; he was forced to refine the theory in order to be able to apply it more directly to the point-sets themselves. This Cantor was able to do in one particular case. Recall that derived sets always contain all of their limit points (for it is always true that $P'' \subset P'$); Cantor was led to focus on sets of that kind ($P' \subset P$), which he called *closed* sets. He was in the condition to show that every closed set is the derived set of another point-set [Cantor 1879/84, 226–27], and this, in the light of the Cantor–Bendixson Theorem, suffices to show that closed sets satisfy CH.

The sixth part of his series on linear point-manifolds ended as follows:

We thus have the following theorem:
    An infinite closed linear point-set has either the first power or the power of the linear continuum, it can thus be thought either in the form funct. (v) or in the form funct. (x) ...
    In later paragraphs it will be proven that this notable theorem also has an ulterior validity for *non-closed* linear point-sets and also for all *n*-dimensional point-sets. ...
    From this, and with the help of the theorems *proven* in [1883, 200; see §2.2] it will be concluded that the *linear continuum has the power of the second number-class (II.).*[2]

The proof that is announced here was never published. Cantor's strategy for proving CH by means of an analysis of subsets of $\mathbb{R}$ would be carried much farther by mathematicians of the 20th century. This happened in the context of descriptive set theory, and it was possible to attain stronger results,[3] but it became clear that this strategy is limited and cannot be carried until the desired end.

Nevertheless, Cantor still made some more attempts. A simplified form of the problem is the following: since non-denumerable closed sets have the power $\mathfrak{C}$, it is

---

[1] This argument can be found in an 1886 letter to Vivanti [Moore 1989, 94].

[2] [Cantor 1879/84, 244]: "Wir haben also folgenden Satz: / *Eine unendliche abgeschlossene lineare Punktmenge hat entweder die erste Mächtigkeit oder sie hat die Mächtigkeit des Linearcontinuums, sie kann also entweder in der Form Funct. (v) oder in der Form Funct. (x) gedacht werden ... /* Dass dieser merkwürdige Satz eine weitere Gültigkeit auch für *nicht abgeschlossene* lineare Punktmengen und ebenso auch für alle *n*-dimensionalen Punktmengen hat, wird in späteren Paragraphen bewiesen werden. ... / Hieraus wird mit Hülfe der in B. XXI, pag. 582 bewiesenen Sätze geschlossen werden, dass das *Linearcontinuum die Mächtigkeit der zweiten Zahlenclasse (II.) hat.*"

[3] For instance, Hausdorff and Aleksandrov established in 1916 that CH holds for the Borel sets, see [Hallett 1984, 98–118; Moore 1989; Kanamori 1996, §§2.3 and 2.5].

sufficient to find a closed set of the second cardinality. Cantor tried this path in 1884, as we know from his correspondence with Mittag-Leffler. His views changed quickly during this period. In August he believed to have found an example of closed set of the second cardinality [Schoenflies 1927, 16], but soon he realized he was mistaken. In October he communicated to Mittag-Leffler new developments that were published the next year in *Acta Mathematica* (see [Cantor 1932, 261–76]). The basis was the new notion of homogeneous set, which opened up the possibility of decomposing any point-set into disjoint parts, one isolated and two homogeneous.[1] These three parts were called *coherence, adherence* and *inherence*. Again, this train of thought did not lead to the desired end, and we shall not enter into those ideas in more detail.

In November 1884 he wrote to Mittag-Leffler with an extremely significant announcement – the refutation of CH:

And when I exerted myself again with the same purpose these days, what did I find? I found a *rigorous* proof that the continuum does *not* have the power of the second number-class and furthermore, that it does not have absolutely any of the powers that can be determined by a number.

As fatal as an error that one has sustained for so long may be, for the same reason its final elimination constitutes a much greater gain.[2]

Nevertheless, one day later he wrote again, with the news that his last proof was flawed and that CH was again on its feet [Schoenflies 1927, 18–19].

Since the publication of Schoenflies' paper it has been customary to accept his view that these unfruitful efforts, together with the annoying opposition of Kronecker, caused Cantor's mental crisis and his withdrawal from mathematics (see §VIII.5). But Cantor had been working on CH for a long time already, and he kept doing so immediately after his mental crisis in 1884; as a matter of fact, the sequence of alleged proofs and refutations that we have just reviewed came right after his recovery. Purkert and Ilgauds [1987, 79–81, 193–94] have employed the testimonies of his doctors to show that, apparently, mathematical work and particularly work on CH was among the very few things able to help Cantor come out of the low phases in his manic-depressive illness.

A new line of attack for the proof of CH emerged with the introduction of the transfinite ordinal numbers in 1882. This novel development happened within a complex context, which included Cantor's current research, particularly on the

---

[1] A homogeneous set P is characterized as being dense-in-itself, and such that sufficiently narrow neighborhoods of its elements always contain parts of P of the same power [Cantor 1932, 265]; for the definitions of coherence, adherence and inherence, see [*op.cit.*, 265, 270]. On the connection between these ideas and the CH, see [Cantor 1932, 264; Schoenflies 1927, 17].

[2] [Schoenflies 1927, 17]: "Und als ich in diesen Tagen wieder mich um denselben Zweck abmühte, da fand ich was? Ich fand einen *strengen* Beweis dafür, dass das Continuum *nicht* die Mächtigkeit der zweiten Zahlklasse und noch mehr, dass es überhaupt keine durch eine Zahl angebbare Mächtigkeit hat. / So fatal ein Irrthum, den man so lange gehegt hat, auch sei, die endgültige Beseitigung ist dafür ein um so grösserer Gewinn."

Cantor-Bendixson theorem, but also (it seems) his conversations with Dedekind in September 1882. The transfinite numbers motivated a shift in Cantor's work toward an abstract theory of transfinite sets, independent of point-sets and topological notions. In order to be able to analyze those developments in their complex context, and also for expository reasons, I shall devote the next chapter to Dedekind's theory of sets and mappings, on which he based his definition of the natural numbers and his understanding of pure mathematics. Then, in chapter VIII we shall continue our review of the development of Cantor's ideas, analyzing his work from 1883 to the end of his career.

VII   Sets and Maps as a Foundation for Mathematics

> Nothing capable of proof ought to be accepted in science without proof.[1]

> Of all the aids which the human mind has for simplifying its life, i.e., the work in which thinking consists, none is so rich in consequences and so inseparably bound up with its innermost nature as the concept of number. Arithmetic, whose sole object is this concept, is already a discipline of insurmountable breadth, and there is no doubt that there are absolutely no limits to its further development. Equally insurmountable is its field of application, for every thinking man, even if he does not clearly realize it, is a man of numbers, an arithmetician.[2]

According to Plutarch, the great philosopher Plato said: ἀεὶ ὁ θεὸς γεωμετρεῖ, God eternally geometrizes. The sentence was remembered in 19th-century Germany, and it was subject to changes that reflect the changing conceptions of mathematical rigor and pure mathematics. During the first half of the century, one of the greatest German mathematicians said, ἀεὶ ὁ θεὸς ἀριθμητίζει, 'God eternally arithmetizes;'[3] geometry had lost its privileged foundational position to arithmetic. Gauss was of the opinion that, while space has an outside reality and we cannot prescribe its laws completely *a priori*, number is merely a product of our spirit or mind [Geist; Gauss 1863/1929, vol. 8, 201]. Dedekind essentially agreed, and his most important foundational work, *Was sind und was sollen die Zahlen?* [1888], bears the motto: ἀεὶ ὁ ἄνθρωπος ἀριθμητίζει, 'man always arithmetizes.' It seems that, in Dedekind's

---

[1] [Dedekind 1888, 335]: "Was beweisbar ist, soll in der Wissenschaft nicht ohne Beweis geglaubt werden."

[2] Dedekind, undated manuscript [Dugac 1976, 315]: "Von allen Hilfsmitteln, welche der menschliche Geist zur Erleichterung seines Lebens, d.h. der Arbeit, in welcher das Denken besteht, ist keines so folgenreich und so untrennbar mit seiner innersten Natur verbunden, wie der Begriff der Zahl. Die Arithmetik, deren einziger Gegenstand dieser Begriff ist, ist schon jetzt eine Wissenschaft von unermesslicher Ausdehnung und es ist keinem Zweifel unterworfen, dass ihrer ferneren Entwicklung gar keine Schranken gesetzt sind; ebenso unermesslich ist das Feld ihrer Anwendung, weil jeder denkende Mensch, auch wenn er dies nicht deutlich fühlt, ein Zahlen-Mensch, ein Arithmetiker ist."

[3] Kronecker [1887, 252] attributes the sentence to Gauss on the basis of apparently reliable evidence. Kline [1972, 104] says it was Jacobi who coined it, and quotes Plato's *dictum* in [Kline 1980, 16].

Was sind und was sollen

die

Zahlen?

Von

Richard Dedekind,

Professor an der technischen Hochschule zu Braunschweig.

᾽Αεὶ ὁ ἄνϑρωπος ἀριϑμητίζει.

Braunschweig,

Druck und Verlag von Friedrich Vieweg und Sohn.

1 8 8 8.

Figure 7. *Title page of Dedekind's* What are numbers and what could they be? *[also: ... and what are they for?] [1888]. Notice the Greek motto: "man eternally arithmetizes."*

view, numbers are not made by God, but by men;[1] mathematics has nothing to do with a world of essences or a Platonic heaven, it is a free creation of the human mind [Dedekind 1888, 335, 360].

Dedekind was above all an expert in algebraic number theory, and everything seems to point to the conclusion that he regarded pure mathematics as the science of number. But his attempts to establish arithmetic on a sound, rigorous foundation led him to view it as built upon a set-theoretical basis, upon the notions of set and mapping. Taking into account that Dedekind regarded these notions as logical ones, the text quoted at the top reads somewhat differently – number is built directly upon the logical notions of set and map, and therefore it is indissolubly linked to the innermost nature of thought. Far from being a purely philosophical vision, the idea took the form of a detailed and precise *foundational program* in Dedekind's mind. This program was a private endeavor until the publication of his booklet [1888], to which we shall refer as *Zahlen*.[2]

The title of this work can be taken to mean 'What are numbers and what are they for?' But its second part can also be read in a subtler way, as meaning what *ought* and *could* numbers be? The prologue begins with another motto that Dedekind had coined already in 1872, 'nothing capable of proof ought to be accepted in science without proof.' In the author's view, this basic requirement had never been satisfied, "not even in the foundation of the simplest science, that part of logic which deals with the theory of numbers."[3] One of purposes of this chapter is to analyze the meaning of Dedekind's logicism and the reception it was given (§§2 and 6), but in the light of Chapter III his self-ascription to logicism should not come as a surprise.

The key ideas of *Zahlen* have to do with two main fields, general set theory and the foundations of the number system and pure mathematics. Dedekind's set-theoretical ideas proved quite influential among authors such as Hessenberg and Zermelo, in the 1900s, or Kuratowski in the 1920s. His development of the theory of ordinal numbers and mathematical induction became a model for a rigorous presentation of the transfinite ordinals. In dealing with the natural numbers, Dedekind has an eye on the ulterior rigorous development of the whole number system, beginning with the integers and rationals. Once the problem of the irrational numbers had been solved to his satisfaction, the only foundational difficulty – and a great one, to be sure – was to find a satisfactory treatment of the natural numbers, that at the same time could constitute a foundation for the whole further development.

Unfortunately, Dedekind's exposition paid too little attention to motivating his theoretical developments and suggesting their scope. Many contemporaries (in-

---

[1] Kronecker is reported to have said, in a lecture to a congress of 1886, that good God made the integers, and all the rest is the work of man [Weber 1893a, 15].

[2] Since the book is divided into numbers corresponding to definitions or theorems, we shall refer to them as follows: *Zahlen*.66 refers to proposition 66 in Dedekind's numbering.

[3] [Dedekind 1888, 335]: "selbst bei der Begründung der einfachsten Wissenschaft, nämlich desjenigen Theiles der Logik, welcher die Lehre von den Zahlen behandelt."

cluding Cantor, it seems – see §4) failed to appreciate correctly Dedekind's motives and the nature of his theory. For these reasons, I shall provide a careful analysis of the work and describe how algebra and analysis fit into Dedekind's foundational program. But, first, we have to analyze its origins, which brings us back to the beginning of our story in the 1850s.

1. Origins of Dedekind's Program for the Foundations of Arithmetic

Dedekind's *Habilitation* as a *Privatdozent* at Göttingen took place in 1854, just a few days after Riemann's. The conference he gave on that occasion has already been mentioned in §III.1.1. It bore the title 'On the Introduction of New Functions in Mathematics' [Dedekind 1854] and constitutes the earliest document of his foundational views. As examples of the introduction of new functions or operations, he examined the trigonometric functions, integration (in connection with elliptic functions), and elementary arithmetic. Dedekind presented the program of a gradual, 'genetic' development of arithmetic, departing from the natural numbers.

1.1. The program in the 1850s.   Dedekind began with the 'absolute integers,' our natural numbers, regarded primarily as ordinals. The "successive progress" from one member of the number-series to another is the first and simplest operation of arithmetic, on which all others are based. Addition is obtained by collecting "into one act" several repetitions of that elementary operation, multiplication is built in a similar way from addition, and so is elevation to powers from multiplication.

But these definitions of the fundamental operations no longer suffice for the further development of arithmetic, the reason being that it assumes the numbers with which it teaches us to operate restricted to a very narrow domain. The requirement of arithmetic, namely, to recreate again the entire existing number-domain through each of these operations, or otherwise said: the requirement of the unconditional possibility of carrying through the indirect, inverse operations of substraction, division, etc., makes it necessary to create new classes of numbers, since with the original sequence of the absolute integers that requirement cannot be satisfied.[1]

---

[1] [Dedekind 1854, 430–31]: "Aber die so gegebenen Definitionen dieser Grundoperationen genügen der weitern Entwicklung der Arithmetik nicht mehr, und zwar aus dem Grunde, weil sie die Zahlen, mit denen sie operieren lehrt, auf ein sehr kleines Gebiet beschränkt annimmt. Die Forderung der Arithmetik nämlich, durch jede dieser Operationen das gesamte vorhandene Zahlgebiet jedesmal von neuem zu erzeugen, oder mit andern Worten: die Forderung der unbedingten Ausführbarkeit der indirekten, umgekehrten Operationen, der Substraktion, Division usw., führt auf die Notwendigkeit, neue Klassen von Zahlen zu schaffen, da mit der ursprünglichen Reihe der absoluten ganzen Zahlen dieser Forderung kein Genüge geleistet werden kann."

Thus one obtains the negatives, the fractions, the irrationals, and finally the so-called imaginary numbers. Once the number domain has been extended, it is necessary to redefine the operations, until then restricted to the natural numbers, in order to be able to apply them to the "newly created" [neugeschaffenen] numbers.[1] And this extension of the definitions is not arbitrary as soon as one follows the general principle, to declare valid in general the laws that the operations obey in their restricted conception, and to derive from them, inversely, the meaning of the operations for the new number domains. This principle is analogous to Ohm's ideas on how to generalize arithmetical operations, and to the famous 'principle of permanence' formulated by Peacock around 1830 (still found in [Hankel 1867]).

To some extent, the program Dedekind presented in 1854 was one he pursued throughout his life. The above ideas resonate in §1 of his paper on irrational numbers, where Dedekind suggests how the rational numbers are defined on the basis of the naturals [1872, 317–18]. The laws of operation with the new numbers "can and must" be reduced to the operations with natural numbers [*op.cit.*, 322]. But one can find an important difference: while in 1854 he emphasized the redefinition of the operations, in 1872 Dedekind emphasized the definition of the numbers themselves. He now disliked to see the operation defined as such, preferring to determine the *result* of the operation as a certain number – one should define the sum as a number unequivocally determined by the summands, not define addition (1878 letter to Weber, [Dedekind 1930/32, vol. 3, 486]). This change is especially important in the light of the fact that, in 1872 and later, *sets* were the means that enabled Dedekind to define or 'create' new numbers. He must have begun to employ sets for this purpose before 1858 (§III.1.1), which supports the interpretation that Riemann's ideas, that he came to know in the *interim*, may have played a decisive role in the process.

In 1858 Dedekind devoted himself exclusively to the question of the foundations of the number system for a few months, until he formulated his definition of the real numbers by means of cuts ([Dedekind 1872, 316], § IV.2.3). Among his unpublished manuscripts one finds several that deal with the definitions of $\mathbb{Z}$ and $\mathbb{Q}$.[2] They present the customary definition of the integers as ordered pairs of natural numbers, and the rationals as ordered pairs of integers, including proofs of the main theorems. There is evidence that these manuscripts were written before 1872, and so it is not unlikely that they may also date back to the late 1850s. Dedekind required that the definitions of new numbers should be "purely arithmetical" and free from foreign elements [1877, 268–69]. For this reason it is noteworthy that he freely employed sets as a valid means for such definitions. The reason seems to be,

---

[1] The genetic approach, on the basis of the integers, the motivation for the expansion of the number system, and the focus on a rigorous redefinition of the operations, can all be found in Ohm (§§I.3 and IV.1). More specifically, see [Bekemeier 1987, chap. 2, particularly 103ff], or the shorter exposition in [Novy 1973, 83–89].

[2] Preserved under the signatures [Cod. Ms. Dedekind III, 2] and [III, 4]. The viewpoint is rather elementary, and the terminology is not yet that of [1872/78]; the topic is referred to in [Dedekind 1872, 317–18]. Thus, they seem to have been written before 1872

quite simply, his understanding of the notion of set as a purely logical one, and therefore admissible in any domain of thought, in connection with any 'product of our mind.'

But, if Dedekind's first contact with sets was through Riemann, it seems that he may have first encountered the genetic method of defining new numbers through William Rowan Hamilton. In 1854 Dedekind said that the extension of arithmetic to the irrational and imaginary numbers poses essential difficulties; the main difficulties of systematic arithmetic begin with the imaginaries [Dedekind 1854, 434]. Interestingly, he never came to publish, or even write a manuscript, on the topic of the complex numbers. In 1857 he read *Lectures on Quaternions*,[1] the introduction to which presented Hamilton's conception of the real numbers and his well-known definition of the complex numbers as pairs of reals [Hamilton 1853]. Thus, a few years after his Habilitation Dedekind could regard the problem of the complex numbers as satisfactorily solved, while the reals still posed a difficult problem. Hamilton's treatment of the imaginaries is the earliest example of the use of ordered couples to define new numbers, a key example of the genetic method.

The introduction to Hamilton's *Lectures*, like an earlier paper dealing with 'Algebra as the Science of Pure Time' [1837], presented quite interesting mathematical ideas under a dubious philosophical dressing.[2] For Hamilton, the essential trait of the notion of real number is the continuous, one-dimensional order of progression; for some time he doubted whether it was convenient to identify this abstract order with the intuition of pure time (see [Hendry 1984, 70–72]). Finally, under the influence of the philosophical ideas of Berkeley and Kant, he decided to do so, and this is why he regards algebra as the science of pure time [Hamilton 1853, 117]. This viewpoint had some advantages. Mathematically, reference to the continuity of intuited time eliminated the problem of the existence of irrational numbers and the need to pin down the difficult idea of the continuum. Time is treated as a given, and so many basic propositions of arithmetic follow from its properties, as happens with the law of trichotomy (given two temporal instants one has one and only one of the relations $a=b$, $a<b$, $a>b$). Philosophically, Hamilton obtained a beautiful scheme in which geometry emerged from the pure intuition of space, and algebra from that of time (the two pure forms of intuition, according to Kant). But his extreme reliance on philosophical ideas hindered the diffusion and acceptance of Hamilton's ideas.

Hamilton's theory of the real numbers was a sophisticated version of the Greek theory of ratios. He conceived of "steps" or temporal transitions that pass from one instant to another, defining the real numbers as ratios of steps [Hamilton 1853, 119–20]. (The notions of step and ratio, like that of couple below, are obviously abstract, not based on any presumed intuition of pure time). As regards the complex numbers, Hamilton writes:

---

[1] Dedekind borrowed Hamilton's book from the Göttingen Library in 1857, as evidenced by the *Ausleihregister* kept at the Niedersächsische Staats- und Universitätsbibliothek.

[2] On this topic, see [Hankins 1976; Mathews 1978, Hendry 1984]. I shall not refer to the 1837 paper, a notable piece of work, since it is very likely that Dedekind did not read it.

I thought that without going out of the same *general class* of interpretations, and especially without ceasing to refer all to the notion of *time*, explained and guarded as above, we might conceive and compare *couples of moments*; and so derive a conception of *couples of steps* (in time), on which might be founded a theory of *couples of numbers* [Hamilton 1853, 121].

This is how he presented the idea of employing ordered couples for a genetic definition of complex numbers. His manuscripts suggest that he had come to this idea independently of his general vision of algebra, under the simple form of couples of real numbers [Hendry 1984, 76–78].[1]

When Dedekind wrote, in 1888, that he regarded the number-concept as completely independent from any idea or intuition of space or time, he seems to be replying to Hamilton.[2] It is likely that, already in the 1850s, he was not attracted by the speculative idea of basing everything on the intuition of time. But he must have been attracted by the thought of defining new numbers as ordered pairs, for he employed this tool in his manuscripts on the integers and rationals. Had Dedekind come independently to this idea, he would have written at least a manuscript on the definition of the complex numbers, and he would have mentioned the question (e.g., in [1871] and [1872]). One is thus led to understand that he used the genetic method for $\mathbb{Z}$ and $\mathbb{Q}$ under the influence of Hamilton's treatment of $\mathbb{C}$, which Dedekind seems to have regarded as definitive. On the other hand, Dedekind found a more or less satisfactory frame for this kind of definition in the general theory of sets that he presented in [1888]. (Not completely satisfactory, since Dedekind failed to employ the notion of ordered pair as a primitive one, or to justify it in any other way.)

To summarize, Dedekind had arrived at the idea of a 'genetic' development of arithmetic in 1854, plausibly under the influence of Martin Ohm, but it was only later that he came to merge this with the notion of set. In his algebraic and number-theoretical work, Dedekind began to employ sets and mappings around 1856 or 1857, probably under the influence of Riemann (see above and §III.1). (The abstract conceptual methodology that he shared with Dirichlet and Riemann (§1.4) also played an important role in guiding his choices.) In 1857 he came to know Hamilton's work, which for the first time offered what Dedekind regarded as a satisfactory treatment of the complex numbers, defined as ordered pairs of reals. It seems likely that in 1858, while devoting himself to the foundations of the number system, Dedekind established definitions of the integers and rationals that were

---

[1] Interestingly, Hamilton referred to the possibility of a theory of *ordered sets* or "systems" [1853, 132], as an extension of those of ordered couples and ordered quadruples (the quaternions). The reader should take into account that Hamilton is properly talking of ordered *n-tuples*, not sets in the modern sense of the term. But one may assume that this kind of statement may have encouraged Dedekind in his tendency to introduce the abstract notion of set or "system" in arithmetic, algebra, analysis, and number theory.

[2] And to Helmholtz [1887], an article cited in [Dedekind 1888, 335]. A similar criticism can be found earlier in Cantor [1883, 191–92]; it is one of those instances in which one doubts whether they came independently to the same idea, or they discussed it in one of their few meetings.

modeled on Hamilton's. He also employed a much more complex set-theoretic notion, that of a Dedekind cut, to define the real numbers. In this way, Dedekind's genetic method came to be intimately linked with the notion of set. The general set-theoretic framework for such a genetic development of the number system is laid down in *Zahlen* [Dedekind 1888].

1.2. Arithmetical foundations. Besides the general framework for a development of the number system, the other core element of *Zahlen* is the rigorous development of elementary arithmetic on the basis of sets and maps. Until the late 19th century, the elements of arithmetic had not been rigorously treated, and there had been a cleavage between authors who dealt with the elements, normally textbook writers, and authors who contributed to number theory or mathematical research in general [Novy 1973]. The distance was narrowed by authors such as the Berlin professor Ohm, but it had not yet disappeared (Ohm was basically a teacher and textbook writer). An important step forward was accomplished by Hermann Grassmann in his *Lehrbuch der Arithmetik* [1861], which seems to be the most important precedent for Dedekind's work.[1] Grassmann based his treatment of arithmetic on the systematic application of mathematical induction, giving recursive definitions of the basic operations and rigorous proofs of the fundamental properties [Grassmann 1861, 1–10, 17–28, 73–78]. He also analyzed the introduction of rational and irrational numbers, but here he was behind Weierstrass, Dedekind and Cantor, for he did not address the key problem of the existence of irrational numbers (see Chapter IV and [Grassmann 1861, 99]). On the other hand, Grassmann's approach tended to be axiomatic rather than genetic.[2]

In comparison with Grassmann, Dedekind's final contribution in *Zahlen* is much deeper, for he gives a general, set-theoretical foundation for the recursive approach to arithmetic. Dedekind's basic set theory sufficed for *defining* the natural numbers abstractly, in such a way that he was able to justify the methods of mathematical induction and recursive definition (and even generalize them). One may say that, with his booklet, the hiatus between elementary and higher mathematics was suppressed for the first time, within the domain of arithmetic.

The lecture [1854] indicated that one should introduce the number operations on the basis of the order of succession among numbers. This viewpoint is developed in two unpublished manuscripts, entitled 'Arithmetical Foundations.'[3] The manuscripts present recursive definitions of addition and multiplication, upon which

---

[1] As late as 1876 Dedekind had not read this work [Lipschitz 1986, 74], which means that his contribution was independent.

[2] From this viewpoint, one has to say that his axiomatization of the real numbers is not adequate, for it does not include an axiom of continuity (completeness). Grassmann kept talking about 'magnitudes,' although in an extremely abstract sense; one should probably blame this reliance on a traditional approach for his inadvertence of the need to enforce a continuous number-domain.

[3] 'Arithmetische Grundlagen' [Cod. Ms. Dedekind, III, 4, II]. One should differentiate these manuscripts from the ones on $\mathbb{Z}$ and $\mathbb{Q}$ mentioned in §1.1.

Dedekind based rigorous proofs by mathematical induction of the fundamental properties (associative, commutative, and distributive laws). It seems that the manuscripts must have been written between 1854 and 1872, so one may safely take them to stem from the 1860s. The reasons why they should be dated before 1872 are the following. The ideas presented here will be published in *Zahlen*, but are not dealt with in the draft begun in 1872.[1] Right from the start, this draft adopts a much higher standpoint, which presupposes the material of 'Arithmetical Foundations.' Moreover, the last part of the manuscripts documents the emergence of the higher standpoint of *Zahlen* and the 1872/78 draft.

Both manuscripts start with the "creation" [Erschaffung] of the natural numbers, beginning with 1 and forming the "*successor*" [folgende Zahl] of any number *a* by means of the "act +1" [Cod. Ms. Dedekind III, 4, II, p. 9, 11]. One thus has the number sequence 1, $1+1=2$, $2+1=3$, $3+1=4$, ..., defined just like Leibniz did in the *Nouveaux essais*.[2] Dedekind goes on to say that, because of that definition, everything will be proven by mathematical induction. He defines addition by means of the formula "$a+(b+1) = (a+b)+1$" [*ibid.*] and proves the associative and commutative laws by induction. Similarly, he defines multiplication by $a \cdot 1 = a$ and $a \cdot (b+1) = a \cdot b + a$ [*op.cit.*, 10, 12], proving the commutative, associative, and distributive laws.

Dedekind would not be satisfied with this kind of analysis of the natural numbers, not even after it had been extended to a complete inventory of the fundamental properties of $\mathbb{N}$. A retrospective account of his investigations, in an 1890 letter to the *Gymnasium* professor Keferstein, contains the following noteworthy explanation of the problem he posed to himself:

What are the mutually independent fundamental properties of this sequence *N*, that is, those properties that are not derivable from one another but from which all others follow? And how should we divest these properties of their specifically arithmetic character, so that they are subsumed under more general notions and under activities of the understanding *without* which no thinking is possible at all but *with* which a foundation is provided for the reliability and completeness of proofs and for the construction of consistent notions and definitions?[3]

As we see, Dedekind wished to adopt a higher standpoint, which would enable him to divest the fundamental properties of $\mathbb{N}$ of their specifically arithmetic character. This he found in the 'logical' notions of set and mapping. In the next-to-last page of 'Arithmetical Foundations' he recasts the definitions of addition and multiplication in an abstract way, regarding them as functions of two arguments. Addition, for instance, becomes "$\varphi(a,d(b)) = d\varphi(a,b)$; $\varphi(a,1) = d(a)$" [Cod. Ms. Dedekind

---

[1] The recursive definitions of the operations are only mentioned expeditiously at one point in the draft [Dedekind 1872/78, 303].

[2] [Leibniz 1704, book IV, chap. 7]. But Leibniz is not necessarily the source; Ohm, e.g., did it the same way (see [Bekemeier 1987, 167]).

[3] As translated in [van Heijenoort 1967, 99–100]. The German text, and a French translation, can be found in [Sinaceur 1974, 272]. The letter was first noticed by Wang [1957].

III, 4, II, p. 12]. The "act +1" has come to be regarded as a mapping d: $\mathbb{N} \to \mathbb{N}$. Therefore, the last page of the manuscript reformulates the original idea by writing: "Sequence [Reihe] 1, $\varphi(1)=2$, $\varphi(2)=3$, $\varphi(3)=4$, ..." [*op.cit.*, 13].[1]

With this simple step, Dedekind reached an abstract, unifying standpoint. The "simplest arithmetical act," that of counting or, otherwise said, the successive creation of the infinite series of natural numbers [1872, 317], has come to be seen as a mapping φ of $\mathbb{N}$ in itself. Previously, natural numbers had been the foundation on which all of arithmetic was developed, by means of the notions of set and mapping, which had proven their usefulness in the context of algebra and number theory(§§1.1 and III.1–2). Now it seemed possible to define the natural numbers themselves on the basis of sets and maps. The implications and further development of this change in perspective would be the subject of the draft for *Zahlen*, written between 1872 and 1878. As we have seen (§III.6), this draft begins with the notions of set and mapping, of which Dedekind writes:

The concepts of system and of mapping, which will be introduced in the sequel in order to lay the foundation of the concept of number, and of cardinal number, remain indispensable for arithmetic even in case one wished to presuppose the concept of cardinal number as immediately evident ("inner intuition").[2]

The letter to Keferstein indicates that Dedekind considered the question of the independence of the basic properties he employed for defining $\mathbb{N}$, although he did not treat the problem in a formal way. He also made an effort to characterize the number sequence completely [van Heijenoort 1967, 100], and this led him (through a noteworthy reasoning that will be discussed in §2.3) to the notion of chain. This is certainly the most original notion employed in *Zahlen*, and becomes the basis for mathematical induction. But first we need to comment briefly on the general set-theoretic notions developed by Dedekind.

2. Theory of Sets, Mappings, and Chains

An important part of the contents of *Zahlen* is perfectly general, an abstract theory of sets, mappings, and chains. It has been a frequent mistake, particularly common in Dedekind's time, to think that, since the book deals with the natural numbers, its contents must be elementary, e.g., restricted to finite sets. Several remarks in the booklet suggest that Dedekind was offering an alternative foundation for the basic

---

[1] Dedekind then goes on to introduce addition, with some theorems, to define substraction, and to consider very briefly the introduction of the integers.

[2] [Dugac 1976, 293]: "Die Begriffe des Systems, der Abbildung, welche im Folgenden eingeführt werden, um den Begriff der Zahl, der Anzahl zu begründen, bleiben auch dann für die Arithmetik unentbehrlich, selbst wenn man den Begriff der Anzahl als unmittelbar evident ('innere Anschauung') voraussetzen wollte." For the meanings of 'Zahl' and 'Anzahl,' see [*op.cit.*, 300, 303].

notions of set theory, also applicable to a reconstruction of Cantorian set theory. However, Dedekind refrained from making the least explicit reference to Cantor and his work, and he did not enter into the theory of transfinite ordinals or well-ordered sets.[1] Thus, the arch giving structure to transfinite set theory is missing, and it would only be years later that Zermelo indicated how to establish a connection between both theories (§IX.4).

The most original contributions made by Dedekind in *Zahlen* are all related to the notion of mapping. This distinguishes him from Cantor, who certainly employed mappings, but never considered them explicitly nor even employed a common word for the different kinds of mappings.[2] Dedekind was the first mathematician who focused explicitly on that general notion, whose algebraic origins (§III.2) and connection with the natural numbers (§1) we have already considered. The new ideas that Dedekind gradually forged and elaborated in his draft of 1872/78 are all related to the notions of mapping and of chain (which depends on the former). Actually, the theory of chains was his most important and original contribution to abstract set theory.

2.1. Things and sets. §1 of *Zahlen* is devoted to formulating the most basic set-theoretic notions, to proving some elementary results that will be used later, and to studying the interrelations between those notions. The treatment is notable for being more succinct and systematic than anything written until then, but otherwise it is not particularly interesting. The peculiar terminology can be easily translated into the modern one: he presents notions of "thing," "system," "part" (subset) and "proper part," "compounded system" (union), and "community of systems" (intersection).[3] These are just the notions that he had fruitfully employed in the context of algebra and number theory (§III.2). It should be noted that Dedekind did not analyze more interesting operations like power set formation or Cartesian product; these would be introduced later in connection with Cantorian problems, the second by Cantor himself. An example of the results Dedekind established is the associative law for the union operation (*Zahlen*.16).

An interesting aspect of Dedekind's work was the underlying idea that all of the notions employed are purely logical. In a draft written in 1887, this section bore the title "Systems of Elements (Logic)" [Cod. Ms. Dedekind III, 1, III, p.2]. The published version eliminated the word in brackets, but as late as 1897 Dedekind still

---

[1] Actually, Dedekind eliminated a reference to Cantor that appeared in his 1887 draft of the preface (§VII.2.1 and [Cavaillès 1962, 120]). Passages that are relevant to Cantorian set theory are the footnote to [Dedekind 1888, 387] and also [*Zahlen*.34 and 63].

[2] When it comes to one-to-one mappings, Cantor speaks of correlating [beziehen] or coordinating [zuordnen] univocally and completely, or else of a correspondence (§VI.2–4); in the context of well-ordered sets, he speaks of "mappings" [Abbildungen], meaning order-isomorphisms (§VIII.4.1); and in another context he talks of "coverings" [Belegungen], which correspond to Dedekind's mappings (§VIII.7).

[3] [Dedekind 1888, 344–47]: "echter Teil," "zusammengesetzte System.," "Gemeinheit der Systeme."

referred to the "logical theory of systems" [Systemlehre der Logik; Dedekind 1930/32, vol. 2, 113]. Everything suggests that Dedekind took it for granted that a theory of sets, manifolds, or classes (synonyms that he mentions explicitly) is a logical theory. But, although he made some efforts to convince the reader that the notion of mapping is a logical one, we find no such effort with respect to sets. The reason seems to be his awareness of the circumstances described in §II.2 (which deals with issues that are essential here), regarding the state of logical theory by the mid-19th century. The logical character of the notion of set could be taken to have been established by the tradition of so-called 'formal' logic, on the basis of a connection between concepts and sets (which would lead to the principle of comprehension).

It seems that Dedekind conceived of logic, in agreement with an old tradition, as the "science" or discipline that deals with the most general laws of thought, those which are applied whenever we think about any particular subject matter. Among the notions that are always applied, we find the notion of set (linked to the very idea of concept) and the notion of mapping. In accordance with the view that the scope of logic is the broadest possible, Dedekind framed his set-theoretical ideas within the most general context. He began by defining a "*thing*" [Ding] to be any object of our thought [1888, 344], a definition that stems from 1872 at least [Dedekind 1872/78, 293]. He employed the Leibnizian definition for the equality of things: *a* and *b* are equal when everything that can be thought of *a* can also be thought of *b*, and conversely.

The notion of set was explained as follows (it is noteworthy that, in contrast to the rest of his booklet, Dedekind does not call this a "definition"):

It very frequently happens that different things *a, b, c, ...* considered for any reason under a common point of view, are collected together in the mind, and one then says that they form a *system S*; one calls the things *a, b, c, ...* the *elements* of the system *S*, they are *contained* in *S*; conversely, *S consists* of these elements. Such a system *S* (or a collection, a manifold, a totality), as an object of our thought, is likewise a thing; it is completely determined when, for every thing, it is determined whether it is an element of *S* or not.[1]

Dedekind's conception of set can be regarded as a typical example of the old 'naive' approach, although he avoided associating too closely sets and concepts. His last remark contains a clear statement of the principle of extensionality – that a set is univocally determined by its elements; at the same time, it suggests a dichotomic conception of sets, in the sense that a set is determined by partitioning the class of

---

[1] [Dedekind 1888, 344]: "Es kommt sehr häufig vor, dass verschiedene Dinge *a, b, c, ...* aus irgendeiner Veranlassung unter einem gemeinsamen Gesichtspunkte aufgefasst, im Geiste zusammengestellt werden, und man sagt dann, dass sie ein *System S* bilden; man nennt die Dinge *a, b, c, ...* die *Elemente* des Systems *S*, sie sind *enthalten* in *S*; umgekehrt *besteht S* aus dieser Elementen Ein solches System *S* (oder ein Inbegriff, eine Mannigfaltigkeit, eine Gesamtheit) ist als Gegenstand unseres Denkens ebenfalls ein Ding; es ist vollständig bestimmt, wenn von jedem Ding bestimmt ist, ob es Element von *S* ist oder nicht."

everything into two parts [Gödel 1947; Wang 1974]. Any thing may belong to a set, and every set is a thing, which opens the way to a free formation of sets of sets. Dedekind established no restriction on the possibilities of set formation, and the whole booklet (particularly *Zahlen*.66) relied on the acceptance of a universal set, "my world of thoughts, i.e., the totality *S* of all things that can be an object of my thought."[1] This is the basis for his dichotomic conception, and the door through which the set-theoretical paradoxes affected Dedekind's system.

Dedekind seems to have avoided associating sets and concepts too closely. It might have been easier to say that a set is determined by a common property shared by its elements, as he had done back in 1872 (§III.5), but Dedekind now preferred to leave undetermined what kind of "common point of view" may determine the set. He did indicate that it is sufficient that the set be determined in principle, whether or not we know a way to actually determine its elements; this made it clear that he did not share Kronecker's view that one should restrict the process of concept-formation in mathematics [1888, 345; see 338]. One can only speculate on the reasons for his decision to avoid explicit reference to concepts: it may have been that his methodological ideas, his preference for an abstract viewpoint, recommended the change; one might even wonder whether he may have been aware that a purely extensional (quasi-combinatorial) conception of sets implies the existence of sets that are not defined by any property.

At any rate, it is certain that the association between concepts and sets, the comprehension principle, guided Dedekind for a long time and determined his conceptions (see §§III.4.2 and III.5). We have seen that around 1872 he defined a "*system or collection*" [Inbegriff] of things by reference to a "common property" that they share ([1872/78, 293], §III.5). The second, 1887 draft for *Zahlen* contains the following statement: "A system can consist in *one* element (i.e., in a *single* one, in one and *only* one), it can also (contradiction) be *void* (contain no element)." What the second part of this sentence means is that a contradiction, a contradictory condition or property, determines the empty set. That is exactly the way in which Frege and Russell justified the assumption of an empty set. Likewise, while defining union-sets, Dedekind commented aside: "extension (of the concept) in contrast to restriction;" and in relation to the intersection, he wrote that it can be "void (contradiction)."[2] In *Zahlen* itself we still find a vestige of the connection between sets and concepts, when Dedekind proves a theorem of (generalized) mathematical induction, speaking of a set Σ, and then goes on to translate it into the language of properties (and of propositions) [1888, 355].

In the introduction to [Frege 1893], the great logician criticized Dedekind's views by indicating three weak points – the lack of a clear distinction between

---

[1] [1888, 357]: "Meine Gedankenwelt, d. h. die Gesamtheit *S* aller Dinge, welche Gegenstand meines Denkens sein können, its unendlich."

[2] [Cod. Ms. Dedekind, III, 1, III, p. 2]: "Ein System kann aus *einem* Element bestehen (d. h. aus einem *einzigen*, aus einem und *nur* einem), kann auch (Widerspruch) *leer* sein (kein Element enthalten). ... Erweiterung (des Begriffs) im Gegensatz zu Verengerung."

inclusion and the membership relation; the confusion between a unitary set and its single element; and the exclusion of the empty set. Frege was intent on showing that any extensional conception of sets is unsatisfactory, and that one *must* begin with the purely logical, and intensional, notion of concept [*op.cit.*, 2–3]. His first criticism certainly spotted a weak point in Dedekind's terminology and notation, a weakness that would first be remedied by Frege and by Peano, who introduced the symbol '∈.' Dedekind employed a special sign for inclusion (though I shall use the modern one), and wrote $a \subset S$ when a is an element of S. This is because he admitted the formation of a unitary set from any thing a, and denoted *both* the thing and the unitary set with the same letter (*Zahlen*.3). But he was conscious that both are quite different notions, and perfectly aware of the "danger" of contradiction hidden in his notation.[1] As regards the third point, we have seen that Dedekind defined in 1887 the empty set just as Frege did, but in the book he preferred to exclude it "for certain reasons," saying that in other contexts it might be comfortable to "imagine" [erdichten] such a set.[2] Frege was troubled by this negligent way of speaking of the empty set, for it run quite contrary to his philosophical tendencies. A manuscript written by Dedekind in the 1890s, on a new definition of finite and infinite, introduces the empty set, denoting it by '0,' and employs the notation '[a]' to distinguish the unitary set from the element a (see [1888, 342] and [Dedekind 1930/32, vol. 3, 450–60]).

2.2. Mappings.   Dedekind's exposition of the notion of mapping is quite modern, and his terminology remains to a great extent the present one. He defines a "*mapping* φ of a system S" as a "law" according to which to each element s of S there "*corresponds*" a certain thing φ(s) called the "*image*" of s.[3] Then he goes on to present the notions of the image of a part (subset) of S, the restriction of φ to a part of the domain, and the identity mapping. Later we find the notion of a mapping "*composed*" of two given ones, ψ(φ(s)) or ψφ(s) [*op.cit.*, 349]; this was quite natural for Dedekind, given the way in which he used mappings in, e.g., group theory.

One may criticize Dedekind for calling his explanation of what a mapping is, a "definition" [Erklärung]. This is objectionable, because it plays the role of a primitive notion in his work, just like the notion of set, which he did not claim to have 'defined.' To put it otherwise, establishing what a mapping is by reference to a

---

[1] See [Dedekind 1888, 345], an 1888 letter to Weber [Dugac 1976, 273], and the manuscript 'Gefahren der Systemlehre' [Sinaceur 1971], where we can read that in 1888 he planned to write an announcement of the book, where he would have discussed this delicate point, as he does in the letter to Weber.

[2] [Dedekind 1888, 345]. The reasons may have had to do with simplifying the proofs, or with his decision to begin the number sequence with 1, regarding 0 as an integer.

[3] [Dedekind 1888, 348]: "Unter einer *Abbildung* φ eines Systems S wird ein Gesetz verstanden, nach welchem zu jedem bestimmten Element s von S ein bestimmtes Ding *gehört*, welches das *Bild* von s heisst und mit φ(s) bezeichnet wird; wir sagen auch, dass φ(s) dem Element s *entspricht*."

"law" may be regarded as a useful explanation, but as a definition it seems circular. It is interesting, though, that Dedekind's definition is strongly reminiscent of Dirichlet's explanation of functions as arbitrary laws (§V.1). This supports the interpretation that his general notion of mapping is taken to be the logical background of the idea of function as used in analysis.

The German word 'Bild' can have the meaning of image or figure, but also of mental picture or idea. Thus, 'Abbildung' could be translated into "representation," which seems to capture an important part of the connotation of the term, as Dedekind employed it. The book's preface indicates that, if we consider what man does while counting, we are led to the faculty of the mind to relate things to things, to let a thing correspond to another, or "to represent" [abzubilden] one thing by another. This faculty of representing or mapping, which we always employ in counting, is also absolutely essential for thinking in general [1888, 335–36]. This is the author's characteristically brief argument for the logical nature of mappings. As we see, the very general sense of the term 'Abbildung' is coherent with the idea that it is a purely logical, completely general notion.

In §3, Dedekind introduces the notion of injective mapping with a paragraph that is a model of the modern character of his exposition. A mapping φ of S is called "*similar* (or *distinct*)" [ähnlich (oder deutlich)] when to different elements of S there correspond different images [1888, 350]. Behind the terminology employed lays the connotation of 'Abbildung' as representation, an injective mapping being a similar or distinct representation of the original (one is even reminded of Descartes' or Leibniz's clear and distinct ideas). Dedekind goes on to say that, since in this case $\varphi(s)=\varphi(t)$ implies $s=t$, we can consider the "*inverse* mapping" $\overline{\varphi}$, which is also a 'similar' mapping from $S'=\varphi(S)$ to S. It is trivial, he says, that $\overline{\varphi}(S') = S$, that φ is the inverse mapping of $\overline{\varphi}$, and that the composed mapping $\overline{\varphi}\,\varphi$ is the identity mapping on S.

This paragraph defining injective mappings ends by talking about bijective ones, which points to an ambiguity in Dedekind's terminology. One can understand it as a consequence of the fact that he regarded as trivial the restriction of the range to the image $S'=\varphi(S)$. (A later paragraph (*Zahlen*.36), devoted to defining what is meant by a mapping of a set in itself, makes it clear that Dedekind does not take the final set and the image to be always identical.)

Injective or 'similar' mappings led Dedekind to consider what he calls "similar [ähnliche] systems," i.e., equipollent sets. After proving that the relation of 'similarity' between sets is transitive, he defines equipollence classes [Dedekind 1888, 351]. Here he is considering all sets as partitioned into classes of equipollent sets, and one should take into account that he has not yet adopted any restriction as to the sets he is considering. His general notions are valid for finite and infinite sets alike, and so we arrive at his version of Cantor's notion of power or cardinality. The definition of cardinal numbers as classes of equipollent sets would be adopted by Russell [1903], after related (but intensional) ideas had been proposed by Frege and Cantor.[1]

---

[1] Frege did so in [1884], but Dedekind read this book for the first time in 1889 [Dedekind

2.3. Chains. §4 of *Zahlen*, devoted to the mapping of a system in itself, deals properly with the theory of chains and the principle of mathematical induction. The theory of chains is Dedekind's most original contribution to abstract set theory, it is more general and independent of the topic of natural numbers. This theory is built on the basis of the general notion of map, without presupposing injective maps. (Dedekind realized that this was possible while writing the second part of his early draft [1872/78, 297–99]; later he wrote that the theorem of mathematical induction does not require an injective mapping [*op.cit.*, 305–07]).

But this interesting body of theory is the best example of how, in his absolute attention to rigor, Dedekind neglected to motivate his new ideas and offer the reader a grasp of their scope. It would have been very important to explain the role of chains in connection with the definition of numbers, a topic that appears only in the letter to Keferstein. And the reception of Dedekind's work would probably have been better had he explained that results such as the Cantor–Bernstein theorem (§4) were consequences of chain theory. As he presented the matter, only a few mathematicians (especially Schröder and Zermelo, but surprisingly not Cantor) were able to appreciate his contribution adequately. Today it is easier to put everything in perspective, thanks to our knowledge of manuscripts and of the later development of set theory.

The notion of chain was obtained by generalizing the conditions that a mapping on a set must satisfy in order to make proofs by induction possible. Dedekind convinced himself of the need for such a notion through a noteworthy argument that he explains in his 1890 letter to Keferstein. Clear traces of this train of thought can be found in the very first part of his draft [Dedekind 1872/78, 295]. One is thus led to the conclusion that the notion of chain was forged in the early 1870s, most likely in 1872.

Using the concepts that we have already discussed, Dedekind was led to characterize $\mathbb{N}$ as a set with a distinguished element 1, for which a mapping φ is defined such that (*a*) φ is an injective map, (*b*) $\varphi(\mathbb{N}) \subset \mathbb{N}$, and (*c*) 1 is not an element of $\varphi(\mathbb{N})$. Under these conditions, and given his definition of infinite sets (§III.5, or §3), $\mathbb{N}$ is infinite – it is equipollent to its proper part $\varphi(\mathbb{N})$. But, considering the possible sets that might satisfy conditions (*a*)–(*c*), Dedekind concluded that one needs something else to characterize $\mathbb{N}$. The argument is model-theoretic in character,[1] and is very clearly explained in the letter to Keferstein.

The three conditions above would be valid for any system S that, besides the number sequence $\mathbb{N}$, contains an arbitrary set T, to which we can always extend the mapping φ so that it remains injective while $\varphi(T) \subset T$. Under the above definition, S would be called a number sequence, but we can define it in such a way that almost

---

1888, 342; Sinaceur 1974, 275].

[1] Model theory is a branch of mathematical logic that considers the interplay between formal axiom systems and their models (i.e., mathematical structures that satisfy the axioms). Properly speaking, it consolidated around 1950, although one can find model-theoretic arguments in the 1920s.

no theorem of arithmetic would be preserved, for proofs by induction would fail to establish the properties of the elements in T (the elements of T form, so to say, an ordering apart from the succession of numbers).

What, then, must we add to the facts above in order to cleanse our system S again of such alien intruders t as disturb every vestige of order, and to restrict it to N? This was one of the most difficult points of my analysis and its mastery required lengthy reflection. If one pre-supposes knowledge of the sequence N of natural numbers and, accordingly, allows himself the use of the language of arithmetic, then, of course, he has an easy time of it. He need only say: an element n belongs to the sequence N if and only if, starting with the element 1 and counting on and on steadfastly, that is, going through a finite number of iterations of the mapping φ (see the end of article 131 in my essay), I actually reach the element n at some time; by this procedure, however, I shall never reach an element t outside of the sequence N. But this way of characterizing the distinction between those elements t that are to be ejected from S and those elements n that alone are to remain is surely quite useless for our purpose; it would, after all, contain the most pernicious and obvious kind of vicious circle. The mere words 'finally get there at some time,' of course, will not do either; they would be of no more use than, say, the words 'karam sipo tatura,' which I invent at this instant without giving them any clearly defined meaning. Thus, how can I, without presupposing any arith-metic knowledge, give an unambiguous conceptual foundation to the distinction between the elements n and the elements t? Merely through consideration of the *chains* (articles 37 and 44 of my essay), and yet, by means of these, completely![1]

Dedekind gave two different notions that are referred to by means of the word 'chain.' Given a set S and a mapping $\varphi \colon S \to S$, a subset K of S is called a *"chain"* [Kette] if and only if $\varphi(K) \subset K$ [Dedekind 1888, 352]; this notion is essentially dependent on φ. One might say that, for any map φ, there will be subsets of the domain that are 'chains' in Dedekind's sense. A second, and crucial, notion is that of the *"chain of the system A"* [Dedekind 1888, 353], denoted by 'A_0,' where A is any subset of S. A_0 is the intersection of all those subsets K of S that are chains ($\varphi(K) \subset K$) and contain A; informally, the chain of A is the smallest chain that con-tains A, or the closure of A under φ in S. The set A_0 is univocally determined, and so we are justified in calling it *the* chain of set A; on the other hand, the notion depends completely on the basic mapping φ, and so Dedekind proposed to use the notation '$\varphi_0(A)$' if needed for the sake of clearness [*ibid.*].

Dedekind went on to prove three results that, as he remarked, suffice to charac-terize completely the notion of the chain of a set. A_0 is a subset of S such that: $A \subset A_0$ (*Zahlen*.45); $\varphi(A_0) \subset A_0$ (*Zahlen*.46); if $K \subset S$ is a chain and $A \subset K$, then $A_0 \subset K$ (*Zahlen*.47) [1888, 353]. With this new notion, it is possible to characterize $\mathbb{N}$ completely and exclude the 'intruders' t (see above) by the following condition: (*d*) $\mathbb{N}$ is the chain of $\{1\}$, which Dedekind wrote 1_0 or $\varphi_0(1)$. Otherwise said, an element n of S belongs to $\mathbb{N}$ if and only if n is an element of *every* subset K of S

---

[1] Letter to Keferstein, 1890, as translated by Wang and Bauer-Mengelberg in [van Heijenoort 1967, 100–01]. The original text can be found in [Sinaceur 1974, 273–75].

such that $1 \in K$, and $\varphi(K) \subset K$ [van Heijenoort 1967, 101]. But let us go back to the theory of chains as developed in §4 of *Zahlen*.

Dedekind investigated next some relations between the notion of the chain of a set and concepts he had introduced previously. In particular, he gave some results about the so-called "*image-chain*" [Bildkette] of a set, $A'_0 = (A_0)' = (A')_0$; the chain A_0 is the union of A and its image-chain A'_0 [1888, 354]. Finally we find the key result that underscores the importance of chains, a generalization of mathematical induction that forms the "scientific foundation" for that method of proof:

59. *Theorem of complete induction*. In order to show that the chain A_0 [$\subset S$] is a part of any system Σ – be this latter part of S or not –, it suffices to show,

    ρ. that $A \subset \Sigma$, and

    σ. that the image of any common element of A_0 and Σ is likewise an element of Σ.[1]

The proof is not difficult. Dedekind considered the intersection $G = A_0 \cap \Sigma$, which (by ρ and prop. 45 above) must contain A. Since $G \subset A_0$, we have $\varphi(G) \subset A_0$, and also (by σ) $\varphi(G) \subset \Sigma$; therefore (by definition of G) we have $\varphi(G) \subset G$, i.e., G is a chain. This, together with $A \subset G$, implies (by prop. 47 above) that $A_0 \subset G$. Thus, $G = A_0$ and A_0 is a subset of Σ, as the theorem claims.

Here, mathematical induction is not formulated for $\mathbb{N}$, but for an arbitrary chain. This constituted an important generalization of induction in two respects: $\mathbb{N}$ is the chain of a unitary set $\{1\}$, but here one considers the chain of an *arbitrary* set A; and the mapping φ of $\mathbb{N}$ is injective, while here one considers arbitrary mappings. This last point is, perhaps, less evident but more important; it seems to have been one of the main motives why Dedekind presented a general theory of mappings in *Zahlen* (see §2.2).

3. Through the Natural Numbers to Pure Mathematics

Dedekind had the "dazzling and captivating" idea of grounding the finite numbers, and all of pure mathematics, on the infinite.[2] To this end, he needed a general definition of infinite set. The fact that he was so interested in this question, while Cantor had not paid attention to it until well into the 1880s, is symptomatic of the differences in their approaches to set theory (see [Medvedev 1984]). Cantor was interested in establishing results on point-sets and transfinite sets; he took the natural numbers as given and so he could presuppose the notion of infinity (see [Cantor 1878; 1879/84]). Dedekind regarded set theory as the *foundation* of mathematics,

---

[1] [Dedekind 1888, 354–55]: "*Satz der vollständigen Induction*. Um zu beweisen, dass die Kette A_0 Teil irgendeines Systems Σ ist – mag letzteres Teil von S sein order nicht –, genügt es zu zeigen, / ρ. dass $A \ 3 \ \Sigma$, und / σ. dass das Bild jedes gemeinsamen Elementes von A_0 und Σ ebenfalls Element von Σ ist."

[2] Hilbert [1922], in [Ewald 1996, vol.2, 1121].

and wanted to employ it to define the natural numbers and to develop arithmetic rigorously; thus he could not presuppose anything from arithmetic, and needed an abstract definition of infinity.

3.1. Finite and infinite sets. The definition of finite and infinite sets that one finds in *Zahlen* is exactly the one that Dedekind formulated in 1872 (§III.5). *S* is infinite if and only if it is similar (equipollent) to a proper part of itself; otherwise, *S* is finite [Dedekind 1888, 356]. A footnote indicates that he had communicated this definition to Cantor in 1882 (see §VI.4.1) and to Schwarz and Weber some years earlier.[1]

Dedekind's step was noteworthy because, in contrast with the traditional definition, he based the notion of natural number on a general theory of finite and infinite sets. The familiar and concrete was thus explained through the unknown, abstract, and disputable. It is not easy for a modern mathematician to appreciate the difficulty of this move, since 20th-century mathematicians are schooled in the abstract approach. From a modern viewpoint, the question whether one uses one or another definition of infinity is only interesting in connection with the means of proof required to establish their equivalence, or to develop set theory (see, e.g., [Tarski 1924]). But in the present context it is important to reflect on the historical and conceptual difficulties implied in Dedekind's move. He was proposing a definition for a notion that, in Riemann's opinion, could only be characterized negatively and stood at the boundary of the representable (§II.4.2). He was defining the infinite through a property that Galileo, and even Cauchy, regarded as paradoxical, for it contradicted the Euclidean axiom 'the whole is greater than the part.'

§5 of *Zahlen*, where Dedekind gave his definition, contains a famous attempt to *prove* the existence of infinite sets (*Zahlen*.66). Much has been said about this – that proposition 66 is not worthy of the name of 'theorem,' and that its proof may be 'psychological' but not mathematical [Dugac 1976, 88–89]. One needs both historical and philosophical sensibility to judge it in its own terms and not anachronistically. The proposition was essential for Dedekind's project, since he was attempting to establish pure mathematics as a branch of logic. This was equivalent, in his view, to saying that mathematics needs no axiom (see §5), and so he needed to prove everything. It was also a question of rigor, as he explained to Keferstein:

does such a[n infinite] system *exist* at all in our world of thoughts? Without a logical proof of existence it would always remain doubtful whether the notion of such a system might not perhaps contain internal contradictions. Hence the need of such proofs (articles 66 and 72 of my essay).[2]

---

[1] As a matter of fact, it was only after the publication of Dedekind's booklet that Cantor began to give definitions of finite and infinite, without naming him [Cantor 1887/88, 414–15; 1895/97, 295].

[2] As translated in [van Heijenoort 1967, 101], though with a small change.

Dedekind is posing the problem of consistency, but in the 19th century the notion of a formal proof of consistency did not exist, and one only conceived of proving consistency by exhibiting a model (see §IV.1). Bolzano [1851, 13] was the first to propose a proof of the existence of an infinite domain, and Dedekind transformed it to suit his different philosophical ideas and his strict definition of infinity. The proof will be analyzed in §5.2, where we shall see that it is quite interesting conceptually; it tries to remain within the realm of pure logic and, precisely, to avoid reliance on psychology.

Bolzano and Dedekind have the merit of having realized for the first time the need to establish the existence of infinite sets as an explicit proposition within set theory. Their attempted proofs had to be abandoned, for they rely on the assumption of a universal set, which is dubious in the light of the paradoxes. This is why Zermelo established that proposition simply as an axiom, which he frequently called "Dedekind's axiom" (§6.2; Zermelo's axiom simply gives a particular example of a set that is Dedekind-infinite). The following commentary stems (as Landau explains) from Zermelo himself:

Instead of simply postulating it axiomatically, Dedekind wishes to establish the existence of infinite systems, on which his theory of the number sequence rests, on the foundation of our 'world of thoughts,' that is, the totality of everything that can be thought. ... But it was later shown (by Russell among others) that this world of thoughts cannot be regarded as a system in the same sense. Nevertheless, this more philosophical than mathematical foundation for his assumption is totally insignificant for the further developments [of *Zahlen*].[1]

3.2. Natural numbers. We have reviewed in §2.3, in connection with the emergence of the notion of chain, the main points of Dedekind's analysis of the natural numbers. Dedekind emphasizes the notion of ordinal number, not the cardinal numbers, and he focuses on an abstract characterization of the ordinal structure of the number sequence. Thus, his viewpoint can be labeled 'structural,' contrasting with the more 'essentialist' analyses offered by Frege [1884], Cantor [1887/88; 1895/97], and Russell [1903]. While Frege looked for concrete objects that are essentially fit to the characterization of cardinalities, Dedekind simply observed that any set isomorphic with $\mathbb{N}$ can be employed to express arithmetic theorems and to determine cardinalities univocally.[2]

§6 of *Zahlen* presents the concept of "simply infinite system" [einfach unendliches System] and defines the number sequence. A set N is simply infinite when there is an injective mapping φ of N in itself, such that N is the chain of an element

---

[1] [Landau 1917, 56]: "Die Existenz unendlicher Systeme, auf der seine Theorie der Zahlenreihe beruht, will Dedekind, anstatt sie einfach axiomatisch zu postulieren, auf das Beispiel unserer 'Gedankenwelt', d.h. die Gesamtheit alles Denkbaren, begründen. ... Aber es hat sich doch später (durch Russell u. a.) gezeigt, dass diese Gedankenwelt nicht als System im gleichen Sinne gelten kann. Doch ist diese mehr philosophische als mathematische Begründung seiner Annahme für die weiteren Entwicklungen durchaus unerheblich."

[2] Interestingly, in recent times there have been philosophical contributions that are reminiscent of Dedekind's viewpoint, see [Benacerraf 1965; Parsons 1990].

that does not belong to $\varphi(N)$. This distinguished element is called the "*base ele-ment*" and denoted '1,' and we say that N "is *ordered* by" φ. The key conditions are, therefore, (δ) that φ be injective, (α) that $\varphi(N) \subset N$, (γ) that $1 \notin \varphi(N)$, and (β) that $N = 1_0$ [Dedekind 1888, 359] (see also [1872/78, 308–09]). Here we have a slight ambiguity, that I have already discussed: one should properly say that N is the chain of a unitary set $\{1\}$ (not of an element), and write $\{1\}_0$.

Dedekind's four conditions α–δ are essentially equivalent to the axioms given by Peano in [1889]; in particular, the chain condition β is equivalent to the axiom of induction. Peano acknowledged having consulted Dedekind's book while pre-paring his own work [Peano 1889, 86; 1891, 93], but according to Kennedy [1974, 389] by then he had already arrived at his axioms, and Dedekind's work only con-firmed his results.

After proving that every infinite set contains a simply infinite one (*Zahlen*.72), Dedekind goes on to define the natural numbers 'by abstraction.' He says that any 'simply infinite' set represents the number sequence, as soon as we disregard the particular nature of its elements, considering only that they are different and that they are related by the ordering mapping φ [1888, 360]. This definition is later justified by proving that simply infinite sets are isomorphic to each other (see [*op.cit.*, 377–78]). As he put it in a letter to Weber, the natural numbers are the "abstract elements" of simply infinite sets [Dedekind 1930/32, vol. 3, 489].[1] A more modern way of saying it would be that the elements of such a set, considered in a purely formal way, are the natural numbers; or even more modern, that arith-metic studies the abstract structure $\langle N, \varphi, 1 \rangle$ of simply infinite sets. It is with re-spect to that "liberation of the elements from any other content (abstraction)," he says, that we are entitled to call the numbers "a free creation of the human mind."[2] As we see, the 'creative power' of the mind seems to be under strict limits here!

Dedekind based the theory of the order $<$ among elements of N upon the theory of chains. He shows that, if n and m are different numbers, either n belongs to the chain of m' (m'_0), or m belongs to the chain of n' (n'_0); those conditions correspond to $m < n$ and $n < m$, respectively (*Zahlen*.88–89). On this basis, one obtains all of the known laws of the less-than relation [1888, 361–68]. Dedekind introduces the no-tion of initial segment of the number sequence, denoting the set of all numbers that are not greater than n by Z_n (*Zahlen*.98); of course, $N = Z_n \cup n'_0$ [*op.cit.*, 365], i.e., initial segments are the complements of chains. Here we find, though only in a concrete setting, the notions of initial segment and remainder that Cantor would

---

[1] Weber was among the first mathematicians who showed an interest in Dedekind's work; eventually, he even published a paper on 'elementary set theory,' in connection with Dedekind [Weber 1906].

[2] [Dedekind 1888, 360]: "In Rücksicht auf diese Befreiung der Elemente von jedem anderen Inhalt (Abstraktion) kann man die Zahlen mit Recht eine freie Schöpfung des menschlichen Geistes nennen."

later formulate for well-ordered sets in general. Dedekind also proves [*op.cit.*, 364] that every subset of $\mathbb{N}$ contains one and only one least number (which shows that $\mathbb{N}$ is well-ordered).

§8 is devoted to some basic theorems on finite and infinite subsets of $\mathbb{N}$. Examples are the following: every Z_n is finite (in the sense of §3.1); Z_n and Z_m are not equipollent whenever $n \neq m$; and every subset of $\mathbb{N}$ which does not have a greater number is a simply infinite set [Dedekind 1888, 368–69]. But the most important element in the rest of his essay is the theorem of recursive definition and its applications. For the first time, a mathematician considers the question of justifying recursive definition, and Dedekind solves the problem in such a way that the extension to the transfinite case is not too complicated. His theorem reads as follows:

126. *Theorem of definition by induction.* Given any mapping θ (similar or dissimilar) of a system Ω in itself and also a certain element ω in Ω, there is one and only one mapping ψ of the number sequence N, which satisfies the conditions
  I. $\psi(N) \subset \Omega$,
  II. $\psi(1) = \omega$,
  III. $\psi(n') = \theta\psi(n)$, where n denotes any number.[1]

The theorem is restricted to $\mathbb{N}$, and not formulated for an arbitrary chain, because a chain can be a finite set endowed with a cyclic order, and in this case the conditions I–III would lead to a contradiction (*Zahlen*.130). Employing Cantor's notions, Dedekind could have generalized the theorem so that any well-ordered set plays the role of $\mathbb{N}$, which would yield transfinite recursion.

Theorem 126 is employed at three key points in the sequel. In §10, to show that all simply infinite systems are equipollent (similar) to $\mathbb{N}$, and therefore to each other [Dedekind 1888, 376]. This means that all models of the conditions α–δ above are isomorphic, i.e., that Dedekind's characterization of $\mathbb{N}$ is categorical.[2] The theorem of recursive definition is naturally also employed to justify the introduction of the arithmetic operations in §§11 to 13. These operations are defined are particular new mappings of $\mathbb{N}$ in itself, and their laws are proven by mathematical induction. And, finally, theorem 126 is used in §14 for introducing the cardinal numbers, i.e., in order to show that the ordinal numbers can be employed to express unequivocally the cardinality of a finite set.

This last result is based on theorem 159, which establishes that if Σ is an infinite set, then every Z_n (every initial segment of $\mathbb{N}$) can be mapped injectively in Σ, and conversely [Dedekind 1888, 384–86]. Dedekind remarks that the proof of the con-

---

[1] [Dedekind 1888, 371]: "*Satz der Definition durch Induktion. Ist eine beliebige* (ähnliche oder unähnliche) *Abbildung* θ *eines Systems* Ω *in sich selbst und ausserdem ein bestimmtes Element* ω *in* Ω *gegeben, so gibt es eine und nur eine Abbildung* ψ *der Zahlenreihe* N, *welche den Bedingungen* / I. ψ(N) 3 Ω, / II. ψ(1) = ω, / III. ψ(n′) = θψ(n) *genügt, wo* n *jede Zahl bedeutet.*"

[2] The notion of categoricity was first introduced by Veblen and Huntington in the 1900s.

verse, as evident as it may seem, is rather complex; given a mapping α_n into Σ for every Z_n, one has to prove that Σ is infinite. As Zermelo indicated [1908, 190], Dedekind's proof rests implicitly upon the axiom of choice.[1] On the basis of theorem 159, Dedekind proves that a set is finite or infinite according as it is equipollent to a Z_n or not. And this, together with previous results, allows him to show that, to each finite set, there is a unique equipollent Z_n. That is the basis for his definition of cardinal numbers (*Zahlen*.161). As he had written in the preface, the notion of cardinal number is actually a very complex one, at least as presented in the context of his theory.

When compared with those of Frege, Russell and Cantor, his presentation has a very important advantage. The direct definitions of the cardinal numbers given by these authors are inadmissible as a consequence of the paradoxes. Dedekind's exposition, from the ordinal to the cardinal numbers, can easily be adjusted to the frame of axiomatic set theory.

3.3. Extension of the number system. In the foregoing we have summarily analyzed the contents of *Zahlen*, but we are still far from an adequate realization of its scope. This is something that Dedekind did not indicate clearly, in spite of a few comments in the book's preface. There we find an indication that arithmetic embraces algebra and analysis [Dedekind 1888, 335], and he also reserves for himself the right to offer a joint exposition of the extension of the number concept [*op.cit.*, 338]. Actually, somewhere in the 1890s he started writing a manuscript entitled 'The Extension of the Number Concept on the Basis of the Sequence of Natural Numbers' [Cod. Ms. Dedekind, III, 2, I]. This was directly linked to *Zahlen*, and we read that Dedekind had treated there the operations on $\mathbb{N}$ in such a way that the foundation for the inverse operations was also won. The inverse operations were the motivation for the expansion of the number concept up to the complex numbers.

The first section of that manuscript deals with the introduction of zero and the negative numbers. Dedekind defines them by means of equivalence classes of pairs of natural numbers: after defining the equivalence relation on pairs, he shows that it is reflexive, symmetric and transitive, and therefore allows the introduction of equivalence classes. Next he analyzes the different kinds of equivalence classes, and establishes a correspondence between $\mathbb{N}$ and the 'positive' classes, that, he says, justifies talk of an extension of $\mathbb{N}$. Finally, before coming to the operations, Dedekind defines a mapping of $\mathbb{Z}$ in itself that establishes its linear ordering. This last might be characteristic of his views at the time, and probably was a source of difficulties as to how to proceed in the ulterior development.

The second section, that was never written, would have dealt with the rational numbers. Clearly Dedekind was convinced that the general framework delineated in

---

[1] AC is employed when he uses the mappings α_n to define new mappings ψ_n that are extensions of each other. On that basis, one can finally define a mapping $\chi\colon \mathbb{N} \to \Sigma$ by the condition $\chi(n) = \psi_n(n)$.

Zahlen was sufficient for the introduction of the different kinds of numbers, ending with $\mathbb{C}$ and its field structure. At this point, we have to take into account that, in the context of algebra, Dedekind did not operate with abstract structures, but only with number-structures, i.e., subsets of $\mathbb{C}$. His sets and mappings, which yielded the complex numbers, were thus a sufficient foundation for algebra as he conceived it. The role of particular kinds of mappings, what we call morphisms, in this context was mentioned in the 1879 version of ideal theory, and explicitly elaborated upon in the last, 1893 version.[1]

It seems that Dedekind conceived of mappings as the foundation for the functions of analysis. We have already noticed the similarities between his 'definition' of mapping and the 'definitions' of function given by Dirichlet and Riemann (§V.1). Having defined $\mathbb{R}$ and $\mathbb{C}$, real and complex functions were clearly at hand, and this is probably the way in which Dedekind understood how *Zahlen* offered a foundation for analysis [1888, 335]. Thus, we have reasons to think that the booklet was not just a detailed essay on the elementary topic of the natural numbers, but a contribution to the foundations of mathematics in general. Nevertheless, in his almost exclusive attention to the deductive structure of his theory, Dedekind neglected motivating the aims and scope of the edifice for his readers. Besides, his approach was too arduous and abstract to attract many readers as of 1890.[2]

Incidentally, there is some evidence that, in his reflections on the foundations of pure mathematics, Dedekind also took into account geometry. As is well known, projective geometry was generally taken to be the most general foundation for that branch of mathematics in the late 19th century. After his Habilitation, Dedekind gave his first course at Göttingen on projective geometry, and he took the subject up again later, most likely in the late 1870s or the 1880s. In his *Nachlass* one can find the beginnings of a paper on the "presuppositions" of projective geometry and its relations to arithmetic.[3] He presented an axiomatization of the notion of projective space, based on definitions and axioms. He conceived of that space as a point-set, accepting once again the possibility that the number of points may be finite. It is interesting that he regarded geometry as based on axioms – in the old, pre-Hilbertian meaning of the word; this contrasts with his idea that arithmetic does not depend on such axioms (the contrast will be analyzed in §5). As concerns Riemann's 'ideal' geometry, it seems likely that Dedekind may have regarded it as a part of analysis (see his manuscript in [Sinaceur 1990]).

---

[1] See [Dedekind 1893, §161, 456–57]. In this work, Dedekind presented Galois theory as dealing with groups of automorphisms.

[2] Hilbert noted down, when he visited Berlin in 1888, everybody was talking about the booklet and mostly in critical terms (see his paper of 1931 in [Ewald 1996, vol. 2, 1151]).

[3] 'Die Voraussetzungen der reinen Geometrie der Lage und deren Beziehungen zur Zahlen-Wissenschaft' [Cod. Ms. Dedekind XII, 3]. The dating (late 1870s or 1880s) is my own, on the basis of the terminology employed, which suggests that Dedekind had already developed the elements of his theory of sets in the draft [1872/78]. His use of the expression 'Geometrie der Lage' may indicate that, at this time, he was influenced by von Staudt or by Reye.

4. Dedekind and the Cantor–Bernstein Theorem

Chain theory is not limited to the purpose of characterizing the successor function on ℕ, and mathematical induction was not the only notable consequence that Dedekind established. In 1887, while finishing off the details of his booklet, he found a proof of the Cantor–Bernstein equivalence theorem. This is an important basic theorem in the theory of cardinalities, that Cantor was the first to conjecture and Bernstein the first to prove satisfactorily in print. (The theorem is sometimes given the name of Schröder–Bernstein, for Schröder presented a proof in 1896 and published it in his [1898]; nevertheless, Schröder's proof is flawed.) It reads as follows:

Cantor–Bernstein Theorem. If set A is equipollent with a proper subset of set B, and B is equipollent with a proper subset of A, then A and B are themselves equipollent.[1]

Oddly enough, Dedekind decided not to include this theorem in his booklet, thereby missing a good opportunity to show the importance of his chain theory. In an 1899 letter to Cantor, Dedekind explains how young Felix Bernstein visited him in Harzburg two years earlier, just after having proven the Theorem, "and was a bit shocked when I expressed my conviction that it is *easy* to prove with my means."[2] Zermelo wrote in 1932 that Dedekind's proof could still be regarded as "classical," and that he did not understand why neither Cantor nor Dedekind published it [Cantor 1932, 451].[3]

We know that, in the 1880s, Dedekind was aware of the importance the result had for Cantor. We have already mentioned (§VI.1, see also §VIII.3) the long meeting that Cantor and Dedekind enjoyed at Harzburg in September 1882. This was a result of the betterment in their relations, apparently due to the issue of the vacant position at Halle, and we have considered its possible connections with the introduction of transfinite numbers. The fact that Cantor and Dedekind had discussed problems in general set theory is corroborated by Cantor's letter of November 1882 where he affirms having told Dedekind about his difficulties in proving the following result:

If $M'' \subset M' \subset M$, and there is a one-to-one correspondence between M and M'', then M' has the same power as M and M'' [Cantor & Dedekind 1937, 55, 59].[4]

---

[1] The manuscript was found by Cavaillès in Dedekind's *Nachlass*, and published in [Dedekind 1930/32, vol.3, 447–48]. Cantor's conjecture is in [1883, 201] and Bernstein's proof in [Borel 1898].

[2] [Cantor & Dedekind 1976, 261] or [Dedekind 1930/32, vol. 3, 448]: "und stutzte ein wenig, als ich meine Überzeugung aussprach, dass derselbe mit meinen Mitteln (Was sind und was sollen die Zahlen?) *leicht* zu beweisen sei." Berstein's visit had been motivated by Cantor himself, see §VIII.8.

[3] Zermelo had independently found the same proof and published it in [1908, 208–09].

[4] Cantor mentions the lemma twice, the first time he regards it as solved by his theory of transfinite ordinals [*op.cit.*, 55], which is how he presented the matter in [Cantor 1883, 201]; the

This is the crucial lemma in the proof of the Cantor–Bernstein Theorem. For this theorem assumes given injective mappings $\varphi\colon A \to B$, $\psi\colon B \to A$; using Cantor's lemma with $M = A$, $M' = \psi(B)$, and $M'' = \psi\varphi(A)$, obviously M and M'' are equipollent, and so $\psi(B)$ and B are equipollent to A.

Since Cantor mentioned it both in person and in his letters, at least three times during 1882, and went on to discuss the theorem again in [1883, 201], one can hardly doubt that Dedekind was well aware of this interest. It is unlikely that when he proved the lemma and the theorem in 1887 he did not remember the matter. Nevertheless, Dedekind did not communicate the proof to Cantor. Even more mysterious: Dedekind's exposition of chain theory concluded with an obscure proposition which, as he himself remarked, was not to be used in the remainder of the book and whose proof was left to the reader. The last proposition in §4 of [Dedekind 1888] is an obscurely formulated version of the lemma above, with indication of the main points in its proof. The presentation is a bit obscure because the question is stated for arbitrary mappings. (This is coherent with the approach taken in the whole section 4, devoted to chain theory and characterized by not requiring mappings to be injective).

Given a map φ, and using the notation K' for images $\varphi(K)$, Dedekind [1888, 356] starts by assuming that $K' \subset L \subset K$. This means that K, and also L (since $L' \subset K'$) are chains (§2.3). Under these conditions, Dedekind asserts that one can always establish the following decomposition of L and K. Take $U = K \backslash L$ (the complement of L in K) and $V = K \backslash U_0$ (where U_0 is the chain of set U); then one has

$$K = U_0 \cup V \qquad \text{and} \qquad L = U'_0 \cup V,$$

where U'_0 is the image-chain of U_0 (see §2.3). The proof of this result was left to the reader (consider that U and U'_0 are disjoint, their union being U_0), and Dedekind made no further comment on its meaning. Nevertheless, it is easy to see that if the mapping φ were injective, U_0 would be equipollent to U'_0, and so L and K themselves would be equipollent. This proves Cantor's lemma precisely in the way that Dedekind proved it, and the Cantor–Bernstein theorem, in 1887 [Dedekind 1930/32, vol. 3, 447–448].

Why did Dedekind include this obscure proposition, whose purpose was unclear, and which played no role in his strictly deductive theory? It seems likely that he wanted to test the alertness of his colleague by proposing the lemma in a deliberately obscure form, in the manner of many 17th-century mathematicians. If so, Cantor apparently failed the test, for in 1895 he still considered the Theorem unproved [Cantor 1895/97, 285] and in August 1899 he accepted Dedekind's proof as new [Cantor 1932, 449–450]. Beyond showing a lack of collaboration, the incident suggests that Cantor paid little attention to Dedekind's theory. Actually, in a letter to Vivanti of 1888 he describes *Zahlen* as an "artificial system" of 172 propositions that merely "tend to the most elementary, and sometimes the most trivial," and which seems

---

second he presents it as an open problem [*op.cit.*, 59], as he would do in the 'Beiträge' [Cantor 1895/97, 285].

"more adequate to obscure the nature of numbers than to clarify it."[1] The episode confirms the lack of collaboration and mutual reinforcement between both mathematicians during the 1880s.

5. Dedekind's Theorem of Infinity, and Epistemology

Riemann and Cantor had in common a deep interest in philosophical questions. By contrast, Dedekind is normally regarded as a mathematician in the purest form. There is no indication in his mature work or letters to suggest that he paid particular attention to any philosopher. But his work presents us with philosophical motives, which, however, could be explained simply by reference to ideas transmitted through the work of contemporary mathematicians and scientists. This indicates the degree to which the German intellectual atmosphere was permeated by such ideas (see §§I.2 and 3). At the same time, it suggests that one should be cautious in ascribing direct philosophical influences to any author of the time.

Even if he had no direct interest in philosophy, Dedekind was a coherent thinker. All of his statements reflect a unitary epistemological vision, based to an important extent on ideas that appeared as 'common sense' in his time, but also on serious reflection on his intellectual activity. Not surprisingly, his basic epistemological framework seems to be that of Kantian philosophy, but he had to face the important transformation that was underway in 19th century mathematics. As a result, he rejected the *Anschaulichkeit* thesis (see §I.2) and embraced logicism. Also present in his views are traits of Leibnizianism: his early definitions of the natural numbers (§1.2), the definition of equality and the deductive organization of *Zahlen*, his position regarding actual infinity, and the logicist position itself.[2] A particularly noteworthy piece of Dedekind's work, intimately connected with his epistemological ideas, is the famous theorem of infinity that he presented in *Zahlen*.

5.1. Logicism as a philosophy. While Cantor favored Platonism, Riemann and Dedekind seemed to think that mathematics has its origins in the human mind. But Riemann believed – with Herbart – that human concepts emerge from experience and that there is no *a priori* basis for our knowledge [Scholz 1982a], while Dedekind was a logicist, who thought that number is an outgrowth of the *a priori* laws of logic. Dedekind was convinced that many mathematical notions are more complex than they appear:

---

[1] [Meschkowski & Nilson 1991, 302]: "Das künstliche System der 172 sich nur um das Elementarste und zum Theil Trivialste drehenden Dedekindschen Sätze scheint mir mehr geeignet, die Natur der Zahlen zu verdunkeln als sie aufzuhellen."

[2] Regarding deductive structure and equality, compare Leibniz's 'Non inelegans specimen' [1966, 122–30], first published in 1840. For the rest, see §I.3 and below.

most people think that everything which has become familiar to them, unconsciously and mechanically, through lifelong practice, is also simple; they do not ponder at all how long the chain of thoughts frequently is, that thereby enters into play.[1]

This was a comment apropos of ideal theory, but the same idea had previously been expressed in connection with the notion of cardinal number [1888, 337]. Many contemporaries attributed our familiarity with numbers to an 'inner intuition.' Dedekind compared it with our ability to read – both reading and arithmetic presuppose complex logical processes that a cultivated person is able to carry out unconsciously and very quickly. Dedekind's purpose was to reveal the unconscious logical process that underlies our most basic mathematical notions – to analyze them into their simplest foundations, and to present the long series of deductions on which each of our presumed 'intuitions' is based [*ibid.*].[2]

We may offer the following reconstruction of the way in which Dedekind came to his logicist convictions. As a mathematician, he tended 'naturally' to rigor; a curriculum that he wrote at age 21 emphasizes that facet of his intellect, his inability to make progress in any field without basing each principle on the preceding ones, according to a "perpetual order and ... reason."[3] Later on, he used to characterize himself as a slow mind, a "step-wise understanding" [Treppenverstand], that needed to master fully the basics of a subject in order to be able to work on it [Dugac 1976, 179, 261]. This trait is manifest in the 1854 *Habilitationsvortrag*, and a rigorously deductive organization can already be found in the manuscript on Galois theory, written about 1858 or 1860 [Dedekind 1981]. It may explain to some extent his interest in number theory, and it certainly explains his attraction for logic.

Meanwhile, a usual conception of mathematics in early 19th-century Germany was marked by the thesis of intuitiveness [Anschaulichkeit]. According to Kant, the subject's *a priori* equipment includes both the forms of intuition (space and time) and the categories (inborn concepts of the understanding). Mathematics has to do with the *a priori* 'construction' of concepts in the intuition [Kant 1787, 741]. This view seems more or less natural in the context of the traditional conception of geometry and of the idea that mathematics is the science of magnitudes (§II.1). Kant's ideas had been mirrored in the writings of authors like Hamilton and Helmholtz, who elaborated upon the topic of the intuition of time as the source of the number system.[4] But during the 19th century mathematics was more and more becoming an

---

[1] Letter to Frobenius, February 1895, in [Dugac 1976, 283]: "die Meisten halten Alles, was ihnen unbewusst durch lebenslängliche Uebung mechanisch geläufig geworden ist, auch für einfach, und sie erwägen gar nicht, wie lang oft die hierbei ins Spiel kommende Gedanken-Kette ist."

[2] Of course, this idea involves a confusion, well-known to some cognitive scientists, that is typical of the 19th century: to think that logic is somehow 'in our brains,' in our (real, psychological) thinking, and believe that the real thinking process must reproduce the steps into which the task in question can be logically analyzed.

[3] [Dugac 1976, 179]: "perpetuo ordine et certa quadam ratione sequens praeceptum antecedentibus innitatur."

[4] For this reason, we have no direct evidence for a Kantian Dedekind, and one should be care-

abstract edifice, which did not necessarily deal with 'truths,' but which was constructed along rigorously deductive lines. This was at least Dedekind's orientation,[1] and it must have impelled him to revise traditional views.

Under the influence of Riemann and Hamilton, but above all in connection with his own experiences in the fields of algebra and number theory, Dedekind's search for rigor and systematicity led him to use set-theoretical notions ever more. Having been accustomed to traditional logic since the *Gymnasium* (§II.2), it seemed clear to him that a set or class can be regarded as the extension of a concept, and that a theory of sets ought to be a part of logic. Dedekind's algebraic work, and his analysis of arithmetic, led him to focus on a second notion, that of mapping, which presupposed the idea of set. But he could convince himself that it is also a logical notion (§§2.2, 6.1). All of this suggested a way of harmonizing his mathematical tendencies, and understanding the new abstract mathematics in connection with customary epistemological ideas. Reduction of large parts of mathematics to sets and maps could thus be understood as reduction of mathematics to logic.

The tendency to abstraction and away from Kant became patent, in the clearest way, with the set-theoretical definition of the real numbers (§IV.2). The traditional definition of the reals as ratios *presupposed* continuous magnitudes, and continuity could only be an axiom, an unprovable postulate, in the realm of geometry (§IV.3). But in the context of pure arithmetic it was possible to define logically (i.e., set-theoretically) a continuous number-domain [1888, 335–36]. The notion of continuity does not have to be understood as having empirical origins, nor does it require an intuitive source in the Kantian sense. Given the usual conception of axioms as indemonstrable propositions that are evidently true, that meant that the theory of real numbers does not depend on special axioms. Which, in turn, suggested the ambitious project of a non-axiomatic development of *all* of arithmetic, beginning with the natural numbers. This was undertaken by both fathers of logicism, Dedekind and Frege.

Dedekind's logicist viewpoint comes out clearly in the first preface to *Zahlen*, where, for reasons like the above, he criticized the *Anschaulichkeit* thesis:

In calling arithmetic (algebra, analysis) just a part of logic, I declare already that I take the number-concept to be completely independent of the ideas or intuitions of space and time, that I see it as an immediate product of the pure laws of thought.[2]

---

ful with treatments like [McCarty 1995]. I myself started twice to write papers on Dedekind and Kant, and then Dedekind and Leibniz, until I became convinced that available evidence was too scanty to warrant conclusions.

[1] It was also the orientation of contemporaries like Pasch and Weierstrass, and later Hilbert.

[2] [Dedekind 1888, 335]: "Indem ich die Arithmetik (Algebra, Analysis) nur einen Teil der Logik nenne, spreche ich schon aus, dass ich den Zahlbegriff für gänzlich unabhängig von den Vorstellungen oder Anschauungen des Raumes und der Zeit, dass ich ihn vielmehr für einen unmittelbaren Ausfluss der reinen Denkgesetze halte."

This is direct criticism of Kant, but also of Hamilton [1837; 1853] and Helmholtz [1887], both of whom based the theory of real numbers on the alleged intuition of time. Dedekind believed, by contrast, that only the arithmetical notion of continuity enables us to sharpen and make precise the ideas of space and time. The meaning of 'arithmetic' is enlarged so as to include algebra and analysis, i.e., what many German mathematicians called 'pure mathematics.'[1] Also common at the time was the idea that logic is the science of the laws of thought – it can be found in such authors as Kant, De Morgan, Boole and Peirce (and the first two cases, at least, show that such conception is not necessarily associated with so-called psychologism).

But Dedekind's own reaction, like Frege's, is still framed within the overall Kantian scheme. The tendency toward abstraction refutes the thesis of intuitiveness and liberates mathematics from the bond to intuition. If we still believe, with Kant and Leibniz, that pure mathematics is *a priori*, there only remains the possibility of ascribing to it a logical origin. Thus, one falls back on Leibniz's thesis that mathematical propositions are "verités de raison" [1714, §33–35]. The doctrines of mathematics constitute purely mental, conceptual results, which, insofar as they are independent of experience, must be logical.[2]

5.2. The theorem of infinity.   This famous, or infamous, theorem is quite revealing of Dedekind's epistemological ideas, and at the same time of his analytical abilities. The mere fact that it was included in *Zahlen* (prop. 66) shows Dedekind's awareness of the logical presuppositions of his work and of set theory generally. Dedekind's definition of infinity was the core of his investigation [1888, 356], since he based his theory of natural numbers on the notion of infinite set. Even the theory of finite subsets of $\mathbb{N}$ was based on the theory of chains, which are infinite subsets. Thus, proposition 66, be it an axiom or a theorem, was strictly necessary for his approach to the natural numbers and for general set theory – a point that would be acknowledged by posterity, beginning with Zermelo (§IX.4). Moreover, as we shall see, Dedekind's logicism demanded that the existence of infinite sets should be proved, not just postulated. Dedekind also emphasized that the *existence* of an infinite set would establish the consistency of his theory (see §3.1).

Nevertheless, the origins of the theorem are rather late. It is not to be found in the draft [1872/78], but only among the papers belonging to a second draft for *Zahlen*, written in 1887 [Cod. Ms. Dedekind III, 1, II, p. 32]. We know that, in the meantime, Dedekind read Bolzano's work [1851], which he came to know through Cantor, and which included a similar proof (see [Dugac 1976, 81, 88, 256]). But there were other reasons to pay particular attention to infinite sets in the 1880s. One

---

[1] This was not uncommon, though the overall discipline was sometimes called analysis [Gauss 1801, xvii] or algebra [Hamilton 1837, 6]. Pasch [1882, 164], Kronecker [1887, 253] and Schröder [1890/95, vol. 1, 441] agree with Dedekind in calling it arithmetic.

[2] This would be supported by a free or imprecise reading of Kant's *Kritik der reinen Vernunft*, where by (transcendental) logic one understands the doctrine of the *a priori*, non-intuitive contents of the understanding [Kant 1787]. Imprecise because it would not take into account the distinction between formal and transcendental logic.

reason, of course, was the heated debate generated by Cantor's radical contributions and his defense of actual infinity. The work of Cantor and Dedekind encountered strong criticism from Kronecker, who tried to eliminate infinity from mathematics [Edwards 1989]. Kronecker regarded Cantor's ideas as nonsense or even perverted, and criticized Dedekind's ideal theory in print [Kronecker 1882]. This polemical reception of infinite sets was certainly behind Dedekind's theorem. In order to justify the actual infinite, Cantor adopted the discursive way of philosophical argument (see §VIII.1), while Dedekind tried to employ the deductive way.

Dedekind acknowledged that his theorem was similar to one of Bolzano [1851, §13]. The first paragraphs of *Paradoxien des Unendlichen* are devoted to a discussion of the notion of infinity, both from a mathematical [§§2–10] and a philosophical standpoint [§§11–12]. Having clarified the conceptual issue, Bolzano poses the problem whether this notion is also "endowed with *objectivity*" [Gegenständlichkeit; 1851, 13], whether there are things (infinite sets) to which it can be applied. This question, and Bolzano's answer, must have immediately caught Dedekind's attention. The answer was:

Already within the *domain of those things which do not have any pretension of reality, but only of possibility*, there indisputably are sets that are infinite. *The set of propositions and truths in themselves* is infinite, as one can easily see.[1]

Bolzano's standpoint may seem close to Cantor's Platonism, but it is rather subtle. He speaks of truths in themselves, but he carefully differentiates between the real and the possible, which reminds one of Leibniz and Frege. As for the proof, he considers any true proposition *A* (e.g., 'that there are truths') and forms from it a different, true proposition, '*A* is true.' He concludes that by iterating this process we obtain as many propositions as natural numbers, that is, an infinite set.

Thanks to his definition of infinity, Dedekind was in a position to perfect Bolzano's proof. But this was not enough to satisfy him: he reformulated the basic terms of the proof so that they became coherent with his epistemological ideas. Obviously Dedekind was no Platonist, but also no empiricist: mathematical objects do not exist outside of our minds, but they are no simple outcome of our perceptions or experience of the physical world. He was always convinced that mathematical objects and concepts are creations of the human mind (see §III.4.1). It is not by chance that *Zahlen* bears the motto: 'man arithmetizes always,' where man replaces the God of his predecessors. For this reason, he had some 'nominalist' bias, in the sense that he avoided reference to seemingly self-existing abstract objects, such as Bolzano's 'truths in themselves.'

---

[1] [Bolzano 1851, 13]: "Es gibt schon im *Reiche derjenigen Dinge, die keinen Anspruch auf Wirklichkeit, ja nur auf Möglichkeit machen*, unstreitig Mengen, die unendlich sind. *Die Menge der Sätze und Wahrheiten an sich* ist, wie sich sehr leicht einsehen lässt, unendlich."

Coherently with that, Dedekind does not consider Bolzano's set of truths, but "my realm of thoughts" [meine Gedankenwelt; 1888, 357]. It is not a matter of an objective world of propositions, independent of the mind, but of the totality of things that can be the object of human thought. Now, according to his definition, a set S is infinite iff there is a proper part of S equivalent with S; we just need to show an adequate mapping. To any element s of the 'realm of thoughts' S, we correlate the following image $\varphi(s)$: 's can be an object of my thought.' Under these conditions, $\varphi(s)$ is a new possible thought, therefore an element of S. And the mapping φ is injective, for if a and b are different thoughts, then $\varphi(a)$ and $\varphi(b)$ are also different – they express different propositions, for although the predicate is the same, the subject is not.[1] It remains to show that $\varphi(S)$ is a *proper* part of S, which Dedekind does by pointing to an example like "my own ego" [mein eigenes Ich], which is a thing that can be an object of my thought (or so do Occidentals tend to think), but is not a proposition of the form $\varphi(s)$.[2]

Similarly to Bolzano, Dedekind does not consider actual, real thoughts, but *possible* thoughts. (This dispenses with the most obvious criticisms, such as those that Russell presents in [1919].) In some sense, the realm of logic is the realm of possibility, a very traditional idea indeed. Zermelo would write [1930, 43] that mathematical existence is ideal existence. Bolzano and Dedekind are not trying to show that actual infinity exists in the real world. They simply try to establish that the notion of infinity is valid within the realm of thought. But Dedekind's proof fails because one cannot safely assume that 'my realm of thoughts' is a set (§VIII.8). In any event, these attempts have to be counted among the few serious ones to conclusively establish the consistency of the assumption of infinite sets.

5.3. Dedekind's deductive method.   Dedekind influenced in important ways the axiomatization of the natural numbers, the real numbers, and set theory. But the approach he took in his works on the foundations of the number system is rather peculiar and calls for explanation. In particular, the question is why Dedekind's strong deductivism did not lead to an axiomatic approach. Clarifying this point is important in order to judge his influence on Hilbert and his school (§6.2).

Zahlen is notable for its deductive structure and, to some extent, all of Dedekind's work is a natural precedent for the modern axiomatic method. The attempt to derive step by step, in a completely rigorous way, *all* of the propositions that are needed is one of the characteristic traits of the book. This is certainly a modern trait, which makes Dedekind come close to such authors as Frege, Russell, or Hilbert. His theories intended to satisfy the following requirement (see full quotation in §IV.2.3):

---

[1] Connections between theorem 66 and traditional logic, particularly the analysis of propositions into the form subject–predicate, are emphasized in 'Über den Begriff des Unendlichen,' an unpublished article in response to Keferstein that can be found in [Sinaceur 1974].

[2] If we call 'my own ego' v, the chain $\varphi_0(v)$ is a simply infinite set and could be identified with $\mathbb{N}$, on the basis of *Zahlen*.73.

in replacing all technical expressions by newly invented arbitrary words (that until then lacked any meaning), the edifice, if well constructed, should not collapse, and for instance I affirm that my theory of the real numbers would bear that test. [Dedekind 1930/32, vol. 3, 479]

If Dedekind said this about his theory of the real numbers, it must be true all the more of his 1888 booklet, which exhibits a rigid deductive structure and establishes the general framework in which the definition of ℝ can be set up. In 1890 he stated that his theory of sets and the natural numbers seemed to him an "edifice built according to the canons of the art, perfectly compact in all its parts, unshakable."[1]

But in other respects Dedekind's deductive method seems rather strange, and could even be called anti-axiomatic. Against Euclid's classical model, later followed by Hilbert and his followers, in *Zahlen* we find no postulate or axiom, only definitions and theorems. The whole theory is derived exclusively from the basic notions of set and mapping, together with definitions of union, intersection, chain, etc. In my opinion, this peculiarity is intimately related to Dedekind's logicist convictions and his conception of logic. His exposition, although completely *abstract*, is not *formal*: the notions of formal inference and formal proof, which Frege [1879; 1893] was beginning to use, are absent from his work. The underlying elementary logic – although transparently employed – is not made explicit, and above all arithmetic is understood as requiring no axiom. All of this places Dedekind's contribution in a peculiar historical position, as an intermediate step that would quickly be abandoned (or, if you wish, superseded). But we shall see that it was not extraordinary in the context of his time.

The traditional conception of axioms regarded them as evident, true propositions that do not admit of a proof. We have seen (§3.3) that in the context of geometry Dedekind admitted that axioms play an essential role, but not so in arithmetic. At this point, it is convenient to recall the development of 19th century geometry and compare it with the development of work on the number system. Euclid's parallel axiom was certainly regarded as an acceptable assumption, but it had become clear that there were logical alternatives. By contrast, Dedekind 'discovered' that a continuous number-domain could be *defined* starting with the natural numbers, and that the natural numbers, in turn, could be *defined* through sets and mappings. In his eyes, this meant that in arithmetic there are no axioms – everything is just immediate consequences of logical notions and definitions. In this light, it would seem that he wanted to distinguish necessary propositions from possible assumptions; the first were immediate consequences of logic, while the second were axioms properly.

It may be that, in this issue, Dedekind adhered to the influential Kantian epistemology. Kant defined axioms to be "synthetic *a priori* principles," which are "immediately true" [Kant 1787, 760]. Since logic was conceived as a purely analytical science, no synthetic principle ought to play a role in it: logic was radically foreign to the use of axioms. This Kantian conception can still be found in the work of the German mathematical logicians Schröder [1890/95, vol. 1, 441] and Frege. It was

---

[1] [Sinaceur 1974, 259–260]: "ein[em], wie ich glaube, kunstvoll gefügten, in allen seinen Theilen fest geschlossenen, unerschütterlichen Gebäude."

apparently for that reason that Frege [1893; 1903] did not talk about arithmetical 'axioms,' but about the "fundamental laws" [Grundgesetze] of arithmetic. Similarly, Dedekind seems to have thought that the logicist program demanded the development of arithmetic in a rigorously deductive way, without any recourse to axioms.

The idea that a deductive theory can be based on definitions alone was not completely unknown in this period. In Schröder's first logical work, devoted to the Boolean calculus, we can read that all theorems of logic are intuitive, for as soon as they are brought to conscience, they become immediately evident. And for this reason the basic statements (that Schröder introduces as axioms) could also justifiably be presented as consequences that are immediately given with the definitions [Schröder 1877, 4]. Schröder preferred to base the logical calculus on axioms, perhaps due to the strong influence that Grassmann exerted on him,[1] while Dedekind's was the opposite choice. Anyhow, Dedekind made a conscious effort to derive all immediate consequences of his definitions that would be needed later on.

In a word, it seems that for Dedekind a theory can only be judged strictly logical when its propositions follow from basic logical *notions*, like those of set and mapping, without the use of any axioms. In this way, his theory can be considered strictly deductive but non-axiomatic – definitional instead. From a modern standpoint, this has to be considered a drawback, since axiomatic treatment allows better control of the theory. But one might add that Dedekind's [1888] seems very easy to axiomatize. The only necessary principle that certainly escaped his attention was the Axiom of Choice (see §3.2).

6. Reception of Dedekind's Ideas

In considering the diffusion of Dedekind's ideas, one should emphasize that the way he chose to present them was probably inconvenient. Instead of calling attention to the set-theoretical contents of his booklet, or perhaps to its logicist program, the title of the book was the question: what are numbers, and what should they be (resp., are they for)? This could certainly be appealing for a general reader, but not for a professional mathematician. Even worse was the way in which he presented the contents. Although Dedekind was convinced that the book could be understood by anybody who possessed "sane common sense" [1888, 336], this is probably far from the truth, and he made no effort to simplify the task for his readers through side-comments. Instead, he developed in great precision a rather limited project, without making programmatic statements; and he left out results that could have enlightened readers, like the Cantor–Bernstein Theorem. Few readers were able to grasp properly the scope and possibilities of his theory – it was not well understood even by Cantor and Bernstein!

---

[1] This position was more Leibnizian (see, e.g., [Leibniz 1704, book IV, chap. 7]), while Dedekind's would appear to be Kantian.

The only thing that could have compensated these defects was the name of the author. A book on the natural numbers, written by one of the foremost specialists in number theory, certainly deserved attention. But most readers must have been surprised by the great complexity with which a seemingly simple topic was being treated. Dedekind anticipated that his readers would scarcely recognize their familiar numbers in the "shadowy forms" [schattenhaften Gestalten] that he was bringing before them, and that they would be frightened by the long series of simple inferences, proving truths that are supposed to be known by inner intuition [Dedekind 1888, 336].[1] In good humor, he wrote to Felix Klein, who had not read *Zahlen*, that he felt pity for those who might feel obliged to read it, and continued:

What will the forbearing reader say at the end? That the author, in a squandering of indescribable work, has happily managed to surround the clearest ideas in a disturbing obscurity![2]

In fact, the reaction of some mathematicians was quite negative: du Bois-Reymond said that the work was "horrendous" [grässlich; Dugac 1976, 203]. But the overall impact was not so negative – a number of authors received this contribution positively, and in time its effect became greater.

6.1. Reception among mathematical logicians. One of the most interesting aspects in a study of the reception of *Zahlen* is to consider the reaction of contemporary logicians. Dedekind was a 'pure' mathematician, who had never published anything on logic, but in 1888 he dared to present a presumed logical theory as a foundation for arithmetic, and even for the whole of mathematics.[3] His pioneering contribution is certainly surprising when viewed from the standpoint of 20th-century logic: it is nothing but a general theory of sets and maps, which of course does not contribute to quantification theory, and which contains no mention even of propositional logic! Of course, Dedekind's logic is not so strange when compared with the then-reigning algebra of logic, a logical calculus that had as its primary interpretation the operations on classes. Even if we consider Russell's conception of logic as of 1903, Dedekind's work can indeed indeed be regarded as logic – for logic has three parts, the calculus of propositions, the calculus of classes, and the calculus of relations [Russell 1903, §13], and Dedekind's contribution is related with the last two.

---

[1] By contrast, the possibility of forming such deductive chains was for Dedekind convincing proof that those truths are *not* gained by inner intuition.

[2] Dedekind to Klein, April 1888 [Dugac 1976, 189]: "Und was wird der geduldige Leser am Schlusse sagen? Dass der Verfasser mit einem Aufwande von unsäglicher Arbeit es glücklich erreicht hat, die klarsten Vorstellungen in ein unheimliches Dunkel zu hüllen!"

[3] Although he published four years after Frege [1884], Dedekind worked independently and did not even know that book until 1889 (see [Dedekind 1888, 342–43; van Heijenoort 1967, 101]). Thus, Frege's work could not make him feel secure or motivate him to publish (he was motivated by papers of Kronecker and Helmholtz, see [1888, 335]).

Among the diverse 19th-century conceptions of logic (see §II.2), we are interested in mathematical logic. Up to 1900, the principal names were Frege, Peano, Peirce, and Schröder, and we shall consider how these logicians reacted to *Zahlen*. There is substantial evidence that Dedekind's logic (sometimes with more or less important modifications) was taken as a part of logic by all of these contemporaries. But, first of all, one should emphasize that Dedekind devoted practically no space to philosophical reflections. This contrasts not only with Frege and Russell, but also with Schröder, and it seems related to the main shortcomings in his presentation. The precise logical underpinnings of the crucial notions of set and map were not spelled out. As we have seen, Dedekind relied on the idea that every concept determines a set, but he chose to de-emphasize this point, and consequently he did not formulate the principle of comprehension. His approach to sets remained as vague as those we find in Cantor [1895/97] or Schröder [1890/95, vol. 1]. Regarding mappings, he had the promising insight that they could be regarded as purely logical, but again he failed to clarify how exactly. Thus, his work made a strong case for the technical plausibility of the logicist program, but left crucial issues in the foundations of logic open.

Dedekind did not feel a need to argue for the logical character of the notion of set, since this seemed to be sufficiently availed by traditional views and by recent work of logicians like Schröder.[1] But his first preface to *Zahlen* showed a desire to convince the reader that the notion of mapping is indeed purely logical. According to him, mappings [Abbildungen] are an expression of the faculty of the mind to relate things to things, to let a thing correspond to another, or to represent [abbilden] a thing by another thing [Dedekind 1888, 336]. This faculty, he says, is an indispensable ingredient of thought. He explicitly proposed to regard mappings as a kind of relation; at the same time, with his reference to *abbilden* he may have had in mind the role of words, 'ideas,' or any other form of representation of external objects, in the thinking process.

The proposal came exactly at the right time and was very well received. For that was the time of development of a general theory of relations in the hands of Peirce and Schröder (and later Russell). Charles S. Peirce did not undertake a logicist position, but in 1901, writing on 'Logic' for a dictionary of philosophy and psychology, he pointed to Dedekind's work as showing that the borderline between logic and pure mathematics is almost evanescent [Peirce 1931/60, vol. 2, 124–25]. In 1911 he would say that Dedekind's conception of mappings was "an early and significant acknowledgment that the so-called 'logic of relatives' is an integral part of logic" [Peirce 1931/60, vol. 3, 389]. On the other hand, one should mention that Peirce did not appreciate Dedekind's contribution very much. He wrongly thought that Dedekind had not contributed a single result that Peirce had not proven before, and he even accused him of plagiarism for the definition of infinity.[2]

---

[1] There are references to [Schröder 1877] in the last pages of [Dedekind 1872/78].

[2] See [Gana 1985]. Peirce's accusation stumbles upon the fact that Dedekind formulated his definition in 1872.

Peano was soon acquainted with Dedekind's booklet, acknowledging that he employed it while giving final form to his famous axiomatization of arithmetic [Peano 1889, 86; 1891, 93]. He had already made contributions connected with set theory, and mathematicians related to his school followed up the work of Cantor and Dedekind. Bettazzi and Burali-Forti were the ones who employed Dedekind's work the most, publishing several articles in the years 1896/97 that dealt with the definition of infinity, in which context they discussed the Axiom of Choice [Moore 1982, 26– 30]. Peano's logical work was in the tradition of Boole in that a single formal language was given a dual interpretation, both in terms of propositions and of classes (§IX.1.2). It is clear that he took the notion of set or class to be logical, presupposing the principle of comprehension, but he was ambivalent toward the logicist viewpoint.[1]

By and large, Schröder was the logician who received Dedekind's foundational work more positively. He worked on the subsumption of his theories under the algebra of relatives, and it seems that Dedekind's contribution brought his conversion to logicism. In 1890 he seemed to leave open the question whether one should join "those who, with Dedekind, consider arithmetic as a branch of logic" [Schröder 1890/95, vol. 1, 441], but in 1898 he said that pure mathematics seemed to him just a branch of general logic [Schröder 1898, 149].[2] Schröder devoted several pages of his major work, *Vorlesungen über die Algebra der Logik* [1890/95, vol. 3, 346–52], to offer a careful review of *Zahlen*. He stated that one of the "most important objectives" of his work was to incorporate all the essential parts of Dedekind's book into the edifice of general logic that he was setting up [*op.cit.*, 346].

Schröder underlined the importance of Dedekind's "epoch-making" contributions, emphasizing that he had acutely filled in a great gap, that until then could be found in all handbooks of arithmetic and algebra. He stressed how much the calculus of logic had to advance in its development, in order to make it possible to establish the lost connection in a really conclusive way [Schröder 1890/95, vol. 3, 349]. Two out of 12 lectures in volume three of the *Vorlesungen* are devoted to an examination of Dedekind's theories: chapter 9 deals with chain theory,[3] and chapter 12 with the theory of mappings, which of course Schröder conceived as relations of a particular kind. There is little doubt that his treatise was an instrument for the diffusion of Dedekind's ideas around the turn of the century, especially outside Germany (Dedekind's booklet was probably more widely read within Germany).

---

[1] On the Peano school, see [Borga, Freguglia & Palladino 1985] and also [Rodríguez-Consuegra 1991] for a discussion of its influence on Russell.

[2] The sense given to 'arithmetic' by Schröder is the same general sense in which Dedekind used the word [Schröder 1890/95, vol.1, 441 footnote]. Schröder's logicism has been analyzed in detail by Peckhaus [1991; 1993], though without emphasizing the role of Dedekind's work in his conversion to logicism.

[3] Schröder generalized the theory of chains further, observing that it does not strictly require mappings, but can be applied generally to binary relations.

6.2. Frege, Dedekind, and pre-Russellian logicism. Frege wrote that Dedekind's book was the most complete work on the foundations of mathematics which had come to his knowledge lately [Frege 1893, vii]. Nevertheless, he was the most critical of all contemporary logicians and expressed openly his dissatisfaction with the work of the pioneers of set theory (see his more particular criticisms of Dedekind in §2.1). He thought that contemporary notions of set were not really abstract notions, nor of course logical notions: they were just generalizations of the naive idea of a grouping of things. In his view, it was necessary to give primacy to the intensional viewpoint: sets or classes are no satisfactory foundation, and everything should be stated, instead, in terms of concepts.[1] With these criticisms, Frege was rightly underscoring the need for more careful work on the foundations of logic, and the peculiar nature of the notion of set (Russell [1903] would rightly emphasize that it was half extensional, half intensional). But with his excessive emphasis on intensionality, Frege's standpoint risked deviating from the mathematical theory of sets (see, e.g., the work of Weyl in chapter X).

At any rate, his remarks did not mean that Frege wanted to dispense with set-theoretical ideas. For instance, he thought it possible to restate Cantor's results in a more rigorous, purely logical way. Indeed, for his logicist reduction of arithmetic Frege needed an analogue of sets or classes, which he introduced as 'extensions of concepts,' preserving the old terminology of German logicians. It is sometimes said that Frege was the first to formulate in a precise form the principle of comprehension, with the basic law V of the *Grundgesetze*. However, as we shall see (§IX.1.2), that law was nothing but a principle of extensionality, while comprehension was implicit in the very notation employed by Frege. Thus, it seems that nobody pinned down the crucial principle of comprehension before the emergence of the paradoxes.

Frege subjected Dedekind's mappings to objections which are similar to those he presented against sets. Mappings are not purely logical tools; instead, one should speak in the intensional tongue, namely about relations. However, as we have seen, Dedekind himself had pointed out that mappings are a kind of relation, saying that the ability of mapping is the ability of relating one thing to another. At any rate, Frege proposed to replace Dedekind's systems by concepts, and his mappings by relations; to end his discussion of Dedekind's ideas, he wrote: "Concept and relation are the basic stones on which I erect my edifice," namely the *Grundgesetze der Arithmetik* [Frege 1893, 3]. Notice how other elements in Frege's logical theory, like propositional connectives and quantifiers, are not mentioned in this sentence. This is because, although necessary for a careful development of his foundational program, those elements were not "basic stones:" they were not crucial for the logicist project. Meanwhile, Frege's definition of natural numbers cannot be stated without referring to concepts, extensions of concepts, and relations.

---

[1] Frege criticized Dedekind's notion of set in the introduction to his *Grundgesetze* [Frege 1893], and Cantor's theory in reviews (see [Dauben 1979, 220–28]; [Frege 1895] criticizes Schröder in a similar vein). A lengthy and sophisticated defense of the extensional viewpoint can be found in [Schröder 1890/95, vol. 1, 83–101].

In spite of Frege's criticisms, there is a remarkable parallelism between his pair of basic notions, concept and relation, and Dedekind's basic ideas of set and mapping. If we ignore the choices made by Frege on the basis of his preference for the purely intensional, this is just confirmation that Dedekind's was indeed a logical theory. Most important, it indicates that *the versions of logicism presented by Dedekind and Frege rested essentially on the same basis*. The parallelism underscores how the logicist program needed set theory, or an equivalent device, in order to subsume arithmetic under pure logic. Logicism depended essentially on the notions of set and relation, conceived extensionally or otherwise. The lower levels of logical theory (propositional and quantification theory) did not play such a crucial role – which seems to undermine the argument that Dedekind was not really a logicist because he did not use a modern theory of formal logic. Set theory was an indispensable ingredient of the logicist's logic, which is why the first wave of logicism was shaken by the paradoxes. For, as we shall see, the paradoxes undermined the traditional justification of the logical character of sets (the concept–set connection, the principle of comprehension) and necessitated radical changes in the received picture of logic and its demarcation.

Logicism can be considered to have been in the air since modern mathematics in general, and set theory in particular, began to develop. But it was after the 1880s, when Frege and Dedekind published their independently developed views, that a growing logicist trend emerged within the mathematical community. In the 1890s and early 1900s, followers of logicism appeared in all the main European countries: Dedekind, Frege, and Schröder in Germany, Jourdain and Russell in Great Britain, and Couturat in France are the outstanding names. Throughout this period, Frege's work was paid very little attention, and so one is led to think that Dedekind was – in the public's eye – the main proponent of logicism until Russell came into the scene. It is significant that Schröder and Peirce associate the logicist position always with Dedekind, and never with Frege. The early form of logicism, however, has been mostly forgotten since from the 1900s Russell's conceptions became enormously influential. But Russell and the *Principia Mathematica* [Whitehead & Russell 1910/13] represent a second phase in the history of logicism, forced by the effect of the paradoxes.[1]

6.3. Hilbert and his school. Among the mathematicians who showed an early interest in Dedekind's work, the most interesting example is David Hilbert. Dedekind's influence is not only visible in the notions Hilbert employed in the context of algebra and algebraic number theory – like those of field and ideal [Hilbert 1897] – but also in his set-theoretical terminology and his approach to set theory. It is not a coincidence that Hilbert began his classic *Grundlagen der Geometrie* with the words: "We conceive of three systems of things: the things of the first system we call *points*." The terms 'system' and 'thing' are taken from Dedekind's *Zahlen*, a

---

[1] A more detailed and focused discussion of these issues can be found in [Ferreirós 1996], and an analysis of the fall of logicism in [Ferreirós 1997].

book that attracted Hilbert's attention as soon as it was published.[1] Also in the famous address 'Mathematische Probleme' we find Dedekind's terminology, precisely where Hilbert explains the Cantorian problems of the power of the continuum and of well-ordering; he even seems to prefer the word 'System' for the abstract notion of set [Hilbert 1900, 298–99; also 301]. And the same applies to his
axiomatization of the real numbers [1900a, 1094–95].

The high esteem in which Hilbert held Dedekind's foundational work is clearly
visible from several of his papers, and indirectly from the work of his students. In
1904 he wrote that Dedekind had clearly acknowledged the mathematical difficulties which the founding of ℕ poses, and had offered an "extremely sagacious"
construction of the theory of natural numbers [Hilbert 1904, 130–31]. In his famous
lecture on the infinite, he called *Zahlen* an epoch-making work [1926, 375] and in
[1922] he labeled Dedekind's thought of grounding the finite numbers on the infinite "dazzling and captivating," although he wrote that it led to a dead end. This last
comment points to the main problem we face while trying to ascertain the influence
of Dedekind on Hilbert. All of his recorded comments on foundational matters
come from 1900 or later, by which time Cantor had made him aware of the set-
theoretical paradoxes, which created enormous difficulties for Dedekind's theory
(§IX.1). Thus, even the strong assumption that Hilbert was a convinced logicist in
Dedekindian style for a decade (1888–98) would be compatible with the historical
record.

It is noteworthy that Hilbert's Göttingen was the place where Dedekind's heritage could more clearly be felt (also in the work of Emmy Noether). Indeed, some
examples taken from the work of Hilbert's students suggest that he may have
stressed the importance of Dedekind's work. In *Das Kontinuum*, Hermann Weyl
presents a rather radical alternative to classical mathematics, criticizing traditional
ways of proof and definition. He proposed an altered (predicative) conception of
sets, and in this connection indicated that Dedekind's set-theoretical treatment of
the natural numbers in no way constitutes a 'reduction' of them to pure set theory.[2]
A footnote, however, stated that it was not his intention to deny [anzutasten] the
great historical significance of *Zahlen* for the development of mathematical thought
[Weyl 1918, 16]. Weyl linked the abstract, set-theoretical conception of mathematics, in which he was schooled at Göttingen, with the names of Dedekind and Cantor:

---

[1] [Hilbert 1930, 4]: "Wir denken drei verschiedene Systeme von Dingen: die Dinge des ersten
Systems nennen wir *Punkte*." For his reaction in 1888, see [Dugac 1976, 93; Ewald 1996, vol. 2,
1151].

[2] For set theory has to be axiomatized, and then its formal develoment must presuppose the
'intuition of iteration' and the number sequence [Weyl 1918, 116–17]; see §X.1.

Confined by tradition in that complex of thoughts which is linked above all to the names of Dedekind and Cantor, and which today certainly enjoys absolute dominance in mathematics, I found for myself and traversed the way leading out of this circle that I have here staked off.[1]

The first part of this sentence should be read as a testimony of the atmosphere reigning, during the 1900s and early 1910s, at Hilbert's and Minkowski's Göttingen.

My second example comes from a very different direction, which makes the argument stronger. According to his own recollections, Zermelo came to work on set theory and foundations under the influence of Hilbert [Moore 1982, 89]. His earliest work on set theory was exclusively related to Cantor's contributions, but later, espacially from 1905, he payed close attention to Dedekind's [Peckhaus 1990, 89–90]. He devoted a crucial paper to show how the entire theory "created by Cantor and Dedekind" can be reduced to a few definitions and seven axioms [Zermelo 1908, 200]. Zermelo studied carefully the works of both predecessors in order to establish the basic postulates involved in set theory, and it has frequently been mentioned that Dedekind's booklet may have suggested some of his axioms.[2] In one particular case, there is no doubt: he called the axiom of infinity "Dedekind's axiom" [Zermelo 1908, 204; 1909, 186]. Moreover, Zermelo used chain theory for some proofs presented in his papers and employed a transfinite generalization of Dedekind's chains for his second proof of the Well-Ordering theorem (see §IX.4). The latter is, in my view, the best example of how his work synthesized and intertwined the ideas of Dedekind and Cantor.

In the 1920s, Hilbert was in the habit of referring to the work of Dedekind, Cantor and Frege as the origins of modern mathematics and foundational research.[3] In a lecture course of 1920, he said that Minkowski and him had been the first in the younger generation of German mathematicians to take "Cantor's side," the side of abstract set theory [Ewald 1996, vol. 2, 946]. As we have seen, there are reasons to think that by 1900 or even 1910 Hilbert identified the set-theoretical approach also with Dedekind. But it is not by chance that he later preferred to mention Cantor, since Dedekind's approach, in its concentration upon systematics and foundations, must have seemed somewhat one-sided to Hilbert. Characteristically, he decided to associate the name of set theory with the man who posed completely new questions in this area, opening up a pure, abstract paradise. This is, after all, the spirit of 'Mathematische Probleme.' And Hilbert's decision has deeply influenced later perceptions of the emergence of the set-theoretical approach.

---

[1] [Weyl 1918, 35; see also 36]: "Durch Tradition eingesponnen in jenen ja heut in der Mathematik zur unbedingten Herrschaft gelangten Gedankenkomplex, der vor allem an die Namen Dedekind und Cantor anknüpft, habe ich für mich den aus diesem Kreise herausführenden Weg gefunden und durchmessen, den ich hier abgesteckt habe."

[2] See Noether's comments in [Dedekind 1930/32, vol. 3, 390–391], and [Moore 1978; 1982].

[3] See, e.g., his articles in [van Heijenoort 1967, 375] and [Ewald 1996, vol. 2, 1119, 1121, 1151]; or the appendixes to [Hilbert 1930].

VIII The Transfinite Ordinals and Cantor's Mature Theory

The *essence* of *mathematics* lies precisely in its *freedom*.[1]

Just since our recent meetings in Harzburg and Eisenach [Sept. 1882], God Almighty saw to it that I attained the most remarkable and unexpected results in the theory of manifolds and the theory of numbers, or rather that I found what fermented in me for years and what I have long been searching for.[2]

On the whole, one may differentiate four phases in the development of Cantor's research on sets. The first, from about 1870 to 1872, was devoted to the study of point-sets through their derived sets for the purposes of the theory of trigonometric series. The second stretched from 1873 to 1878 and focused above all on the study of infinite cardinalities (chap. VI). The third period, 1879 to 1884, was guided by the core objective of proving the Continuum Hypothesis (CH). Cantor studied in detail the powers of subsets of $\mathbb{R}$, looking for combined results on derived sets and powers, which led him to introduce basic notions of the topology of point-sets. Up to this point, however, he had not distilled an abstract conception of set theory, dissociated from topological properties. With the introduction of transfinite ordinal numbers, in 1883, he found a way of defining an increasing sequence of consecutive powers or cardinalities. His interests thereafter shifted from the theory of point-sets to that of ordered sets, and by 1885 he had conceived of a general theory of order types (i.e., types of totally ordered sets). Thus he finally arrived at a general, abstract analysis of sets based on the notions of cardinality and order. This fourth period went from 1885 to the end of his career.

The crucial shift to abstract set theory occurred while Cantor was publishing 'On infinite, linear point-manifolds' [1879/84], a collection of six articles of unequal length and depth. In the third installment, published in 1882, Cantor began to

---

[1] [Cantor 1883, 34]: "das *Wesen* der *Mathematik* liegt gerade in ihrer *Freiheit*."

[2] Cantor, November 5, 1882 [Cantor & Dedekind 1937, 55]: "... gerade seit unserm jüngsten Zusammensein in Harzburg und Eisenach hat es Gott der Allmächtige geschickt, dass ich zu den merkwürdigsten, unerwartetesten Aufschlüssen in der Mannigfaltigkeitslehre und in der Zahlenlehre gelangt bin oder vielmehr dasjenige gefunden habe, was in mir seit Jahren gegährt hat, wonach ich lange gesucht habe."

GRUNDLAGEN

EINER

ALLGEMEINEN

MANNICHFALTIGKEITSLEHRE.

———

EIN

MATHEMATISCH-PHILOSOPHISCHER VERSUCH

IN DER

LEHRE DES UNENDLICHEN.

VON

Dr. GEORG CANTOR,

ORDENTLICHER PROFESSOR A. D. UNIVERSITÄT HALLE-WITTENBERG.

———◆———

LEIPZIG,

COMMISSIONS-VERLAG VON B. G. TEUBNER.

1883.

Figure 8. *Title page of Cantor's* Foundations of a general theory of manifolds *[1883]. The subtitle reads: a "mathematico-philosophical attempt" to contribute to the "theory of infinity."*

discuss foundational questions and the scope of set theory, and this tendency reached a peak with the fifth installment [Cantor 1883], a veritable tour-de-force that dealt with both mathematical and philosophical questions. It was here that Cantor presented the transfinite numbers, which had an enormous importance in the development of his work. Transfinite numbers mark the crucial point at which Cantor turned to an abstract theory of sets. The importance he attached to them is reflected in the fact that he published this part of the series as a separate book entitled *Foundations of a General Theory of Manifolds* [Cantor 1883].

Cantor was conscious that he was plunging into a direct confrontation with widely held views regarding mathematical infinity and the notion of number. He decided to take the implications of his move seriously, entering into a detailed discussion of the philosophical and theological implications, which also allowed him to unbridle his speculative interests. For the first time, the public came to know aspects of his viewpoints and personality that had previously been hidden. The *Gundlagen* [Cantor 1883] is thus an admixture of mathematical, foundational and philosophical considerations, one of the most extraordinary articles in the history of modern mathematics and, to be sure, one of great interest.

In the present chapter, we shall outline the development of Cantor's mature theory of sets, from the ordinals to the paradoxes. Starting from his foundational reflections and his notion of set as of 1882, we proceed to the transfinite numbers and the related interest in ordered sets. This marks a clear shift in the evolution of his views, differentiated neatly from the early work discussed in chapter VI (related to derivation and point-sets). Still, we shall consider the many ways in which Cantor's theory of point-sets suggested viewpoints and results for his theory of ordered sets. After a section devoted to the reception of his views and Cantor's withdrawal from mathematical publications in the 1880s, the final sections deal with his work of the 1890s: Cantor's Theorem [1892], the synthetic presentation of transfinite set theory in the *Beiträge* [1895/97], and the discovery of the paradoxes around 1896/97.

1. "Free Mathematics"

The turn toward an abstract conception of mathematics and its methodology, implicit in Cantor's research from the early 1870s and in his publications from 1879, became explicit in the *Grundlagen*. Cantor thought it necessary to confront the views of those mathematicians, like Kronecker, who objected to the free introduction of new notions (particularly those having to do with the infinite) and the application of traditional methods of proof to them. As we have seen (§IV.4.2), already in the early 1870s Kronecker objected to the proof of the Bolzano-Weierstrass theorem as an obvious sophism. Apparently, he thought that a pure existence proof was not enough to guarantee the existence of a number having the desired properties. Kronecker's position amounted to a rejection of the *tertium non datur* as applied to infinite sets.

1.1. Against finitism.[1] §4 of the *Grundlagen* describes a standpoint of radical arithmetization of mathematics associated with Kronecker, although Cantor did not name him directly [1883, 172–73]. Kronecker planned to produce a constructive development of all branches of pure mathematics on the basis of the natural numbers and the literal calculus [Buchstabenrechnung] of algebra [Kronecker 1887; Edwards 1989]. He rejected the various definitions of the real numbers (§IV.2) since these involved the actual infinite and could not be reduced to notions defined algorithmically on the basis of the natural numbers. Cantor conceded that this kind of approach may help to eliminate errors and to attain some methodological advantages, but he stated that it represents a kind of excess of zeal. Though in general he maintained a moderate tone, at times Cantor used somewhat derogatory expressions, as when he said that in this way a definite, but "rather banal and obvious" [ziemlich nüchternes und naheliegendes] principle is recommended to all as a guideline; or later, that no real progress has ever been due to such an approach and that it would hold back mathematics and confine it into the narrowest bounds [Cantor 1883, 173].[2]

Since Kronecker's objections to the new developments in mathematics reminded Cantor of the Greek skeptics, and given his inclination to philosophy, it is no wonder that he turned his reply into a discussion of the most important philosophical arguments against actual infinity.[3] Cantor distinguished the *"proper infinite"* [*Eigentlich-Unendliches*], which corresponds to what is commonly called actual infinity, from the *"improper infinite"* [*Uneigentlich-Unendliches*] of potential infinity [Cantor 1883, 165–66]. He quoted above all the arguments of Aristotle, which led to the scholastic sentence 'infinitum actu non datur,' but he also gave references to philosophers like Locke, Descartes, and even Spinoza and Leibniz [*op.cit.*, 173–79]. In essence, his argument was that the classical objections against infinite numbers were always based on a *petitio principii* – they assume or require that those numbers must possess all properties of the finite numbers, including some which contradict the nature of the transfinite [*op.cit.*, 166, 177–78]. For instance, if one stipulates that the result of an enumeration must always be the same, independent of the way the elements are ordered, then the notion of a transfinite 'number' would be contradictory. In and of themselves, however, transfinite numbers do not entail the least contradiction or inconsistency. Transfinite numbers constitute a completely new kind of number, whose laws and properties depend "on

---

[1] A very detailed, though perhaps over-systematic, discussion of Cantor's views can be found in [Hallett 1984, ch.1]. The reader will find there an interesting analysis of Cantor's "finitism," in the sense of his quasi-finitistic or quasi-combinatorial conception of infinite sets. Here I employ the term 'finitism' in the usual meaning.

[2] It is not surprising that this kind of open polemics would bother Kronecker, who had never published his ideas. In personal conversations he went so far as to call Cantor a corrupter of the youth [Schoenflies 1927, 2].

[3] See [Cantor 1879/84, 212–13; Schoenflies 1927, 12]. Cantor's interest in philosophy dates back to his student times, when he studied carefully the views of Spinoza [Purkert & Ilgauds 1987, 183]; see also [Cantor 1932, 62].

the nature of things" and ought to be the subject of careful research, not of our prejudices or arbitrariness [Cantor 1932, 371–72].

In the *Grundlagen*, Cantor wrote:

To the thought of considering the infinitely great not merely in the form of what grows without limits – and in the closely related form of the convergent infinite series first introduced in the seventeenth century – , but also fixing it mathematically by numbers in the determinate form of the completed-infinite, I have been logically compelled in the course of scientific exertions and attempts which have lasted many years, almost against my will, for it contradicts traditions which had become precious to me; and therefore I believe that no arguments can be made good against it which I would not know how to meet.[1]

He went on to clarify that, in speaking of traditions, he did not only refer to what he had personally experienced (e.g., the viewpoints of the Berlin school), but also to traditions going back to the founders of modern science and philosophy.

Along the way, particularly important in his endnotes, Cantor discussed all kinds of philosophical and theological questions. The notion of God or the Absolute played an important part in his thought, also in connection with set theory.[2] Among philosophers it had been customary to identify actual infinity with the Absolute, and to deny the possibility of determining the actual infinite by numbers, since this would have amounted to the impious idea that God can be determined by human reason. Cantor took it as his duty to defend his theory of the transfinite against theological objections, and he entered into correspondence with many theologians, particularly Catholics. He wished to convince philosophers and theologians that between the finite and the Absolute there is still an "unlimited hierarchy" of concepts, the transfinite numbers [Cantor 1883, 176], and that his theory did not in the least imply that it is possible to determine the Absolute [*op.cit.*, 175–77, 205]. He proposed the following maxim: "Omnia seu finita seu infinita *definita* sunt et excepto Deo ab intellectu determinari possunt" – all things finite or infinite are *definite* and, God excepted, can be determined by the intellect [*op.cit.*, 176].

In an endnote [1883, 205] Cantor stated that the Absolute can only be acknowledged, not known, not even approximately known. And he went on to suggest that the absolutely infinite sequence of the transfinite numbers seems to be an adequate "symbol" for the Absolute [*op.cit.*, 205]. This served as an important background

---

[1] [Cantor 1883, 175]: "Zu dem Gedanken, das Unendlichgrosse nicht bloss in der Form des unbegrenzt Wachsenden und in der hiermit eng zusammenhängenden Form der im siebenzehnten Jahrhundert zuerst eingeführten convergenten unendlichen Reihen zu betrachten, sondern es auch in der bestimmten Form des Vollendetunendlichen mathematisch durch Zahlen zu fixiren, bin ich fast wider meinen Willen, weil im Gegensatz zu mir werthgewordenen Traditionen, durch den Verlauf vieljähriger wissenschaftlicher Bemühungen und Versuche logisch gezwungen worden und ich glaube daher auch nicht, dass Gründe sich dagegen werden geltend machen lassen, denen ich nicht zu begegnen wüsste."

[2] This topic has been studied by all biographers, see [Meschkowski, ch. 8; Dauben 1979, ch. 6 and 10; Purkert & Ilgauds 1987]. For its mathematical implications, see especially [Hallett 1984, 40–48, 165–76] and [Jané 1995].

for his reaction to the set-theoretical paradoxes, since it suggested that there is no set of all ordinals, because the Absolute cannot be determined. If the totality of ordinals represents the Absolute, it should also be beyond what can be mathematically determined, beyond the transfinite. There would be no set, no collection into a whole, of all ordinals (see §VIII.8).[1]

1.2. Mathematical existence.   Cantor's critique of Kronecker's views reappeared from a different standpoint in §8. Here Cantor explained his philosophical conception of mathematics and discussed the methodological requirements that are essential in this discipline, comparing the work of the mathematician with that of the natural scientist. His basic idea was a distinction between two senses in which one may treat the question of existence concerning any concept or idea, and in particular the existence of the transfinite numbers. First, one may judge the *"intrasubjective* or *immanent reality,"* which only depends on the concept being well defined, free from contradiction, and entering into fixed relations with previously available and accredited concepts [1883, 181–82]. A different question is that of the *"transsubjective* or *transient reality,"* which is ascribed to a given notion insofar as it represents processes or relations in the external world [*op.cit.*, 181].

With this distinction in mind, Cantor affirmed that mathematics is distinguished from all other scientific disciplines in that, while scientists are busy above all with transient reality,[2] mathematics "has to take into account *only* and *exclusively* the *immanent* reality of its concepts."[3] In mathematics, one freely introduces new notions, which however will be abandoned whenever they turn out to be unfruitful or inconvenient. Otherwise said, the essence of mathematics lies precisely in its freedom, and restrictions on this freedom to form consistent notions are very dangerous. For this reason, Cantor would replace the usual expression "pure mathematics" by the more pregnant name *"free mathematics"* [1883, 182].

As justification for his viewpoint, Cantor mentioned first some philosophical ideas, and then the recent development of mathematics. Regarding the latter, the impressive development of function theory, of the theory of differential equations, and of algebraic number theory would never have come about without the above-mentioned freedom [*op.cit.*, 183]. But, to believe his words, Cantor was led to his views on mathematics first and foremost by the conviction that both kinds of reality always correspond to each other – that every concept that exists in the immanent sense possesses a transient reality too. This is due to the essential unity of reality, the unity of the whole to which we ourselves belong [*op.cit.*, 181–82]. We find here the source of Cantor's Platonism, of his conviction that transfinite sets exist in a world of ideas or in God's intellect, and at the same time also in Nature (see end of §2).

---

[1] The point seems to have been first realized by Purkert; see [Purkert & Ilgauds 1987].

[2] Cantor went so far as to say that these disciplines are *"metaphysical"* in both their foundation and goals [1883, 183]; the idea would reappear in the following years.

[3] [Cantor 1883, 182]: "dass nämlich [die Mathematik] bei der Ausbildung ihres Ideenmaterials *einzig* und *allein* auf die *immanente* Realität ihrer Begriffe Rücksicht zu nehmen ... hat."

Cantor's inspired defense of the abstract approach to mathematics was certainly one of the first. No wonder that Hilbert used to name him, alongside Dedekind, as one of the founders of the modern approach to mathematics (see §VII.6.2). If we undress his ideas from the philosophical language in which he couched them, and from his metaphysical convictions, Cantor's opinion that one can only require mathematical notions to be well defined and consistent is a forerunner of Hilbert's famous view – that we are entitled to assume the mathematical existence of an object as soon as the corresponding axiom system is consistent. Transfinite numbers have certainly passed Cantor's test, in the sense that they are generally acknowledged to be well-defined, intuitively consistent, and fruitful.

2. Cantor's Notion of Set in the Early 1880s

The third and fifth installments of Cantor's series on linear point-manifolds [1879/84] contained, for the first time, general reflections on the basic notions of set theory. Before discussing them, it will be worthwhile to try clarifying Cantor's linguistic usage. He employed preferentially the words manifold [Mannichfaltigkeit] and set [Menge], but these terms seem to have different connotations. Manifold is the more general term, 'Menge' is used in particular for sets of points or sets of numbers, i.e., the common *mathematical* examples of manifolds.[1] Cantor regarded the theory of manifolds as going beyond mathematics [1879/84, 152], including or being intimately connected with logic and epistemology [1883, 181].

Some aspects of Cantor's thought about sets remained basically constant over the years. Like all other authors in the early period of the history of sets, he tended to think of them as given by a concept or a law, indeed he emphasized this aspect more than other contemporaries. To give some examples, in his definition of the real numbers he required the fundamental sequence to be "given by a law" [Cantor 1872, 92–93], and while defining derived sets he said that P' is "conceptually given" together with P [*op.cit.*, 98]. The proof of non-denumerability of $\mathbb{R}$ begins by considering a sequence of real numbers given by any law [1874, 117], and throughout the paper Cantor always writes 'Inbegriff' instead of set.[2] As late as 1882/83 we find similar statements. He explained a linear point-set by saying that it is a manifold of points belonging to the line which is "given by a law" [gesetzmässig gegeben; 1879/84, 149]. Most important, his first explicit definition of set (see below) requires that the elements be joined into a whole by a law [1883, 204].

Other elements of his conception were incorporated slowly. Starting in 1878, Cantor insisted that manifolds or sets have to be "well-defined;" for instance, his very definition of cardinality assumes two given well-defined manifolds [1878,

---

[1] See, e.g., the first sentence of endnote 1 in [1883, 204].

[2] The term 'Inbegriff,' meaning class, is etymologically related to 'Begriff,' or concept. In my translations I have rendered it as 'collection' for lack of a better choice.

119; also 124]. Similarly, the condition of being well-defined appears in the definition of well-ordered set, and the "law" of well-ordering is taken to be valid for well-defined sets [1883, 168, 169]. In 1882 he explained what he meant:

> I say that a manifold (a collection, a set) of elements that belong to any conceptual sphere is *well-defined*, when on the basis of its definition and as a consequence of the logical principle of excluded middle it must be regarded as *internally determined*, *both* whether an object pertaining to the same conceptual sphere belongs or not as an element to the manifold, *and* whether two objects belonging to the set are equal to each other or not, despite formal differences in the ways of determination.[1]

Cantor went on to say that one should not require that it be possible to actually take the relevant decisions with available means, it is only essential that the set be determined internally (or in principle). Here we find a precedent for the abstract methodology that was explained a year later by means of the idea of 'immanent reality' (§3.2).

The above passage points quite clearly in the direction of the principle of extensionality – that a set is completely determined by its elements, irrespective of the way it may be given through defining properties, etc. It also underscores a key aspect of set theory that would be criticized by Brouwer and his followers: the application of the *tertium non datur* to infinite totalities. Finally, it emphasizes a characteristic of Cantor's notion of set that differentiates him from, say, Dedekind or Frege. For him, the elements of a set must pertain to a certain "conceptual sphere" [Begriffssphäre], be it the domain of arithmetic, of function theory, of geometry, or even those of logic or of epistemology [1879/84, 141, 150; 1883, 181].[2] In this connection, it is noteworthy that Cantor never considered sets having elements of unequal kinds – he studied sets of numbers, of points, or even of functions, but never mixed sets, and he seems to have refrained from considering sets of sets. It seems that he implicitly required the elements of a set to be homogeneous and to be apparent individuals. This may be linked to the above requirement of restriction to a certain conceptual sphere. It might explain some surprising characteristics of his views, for instance, why he did not formulate Cantor's Theorem in terms of the power set (§6).[3]

---

[1] [Cantor 1879/84, 150]: "Eine Mannichfaltigkeit (ein Inbegriff, eine Menge) von Elementen, die irgend welcher Begriffsphäre angehören, nenne ich *wohldefinirt*, wenn auf Grund ihrer Definition und in Folge des logischen Princips vom ausgeschlossenen Dritten es als *intern bestimmt* angesehen werden muss, *sowohl* ob irgend ein derselben Begriffsphäre angehöriges Object zu der gedachten Mannichfaltigkeit als Element gehört oder nicht, *wie auch* ob zwei zur Menge gehörige Objecte, trotz formaler Unterschiede in der Art des Gegebenseins einander gleich sind oder nicht."

[2] This is reminiscent of the conditions for significativity that Weyl [1918] imposes on logic and therefore on his predicative set theory. Weyl, in turn, seems to have been influenced by Husserl.

[3] It may also explain why he never employed equivalence classes in the context of his definition of real numbers, and why he distinguished higher kinds of real numbers in 1872 and 1883

Disregarding other conceptual spheres, the theory of manifolds embraces all of pure mathematics: arithmetic, algebra, function theory and geometry. In Cantor's opinion, it embraces these domains and brings them to a higher unity on the basis of the notion of power, under which both the discontinuous and the continuous fall [1879/84, 151]. The notion of cardinality is the most basic and important notion in set theory, for the power of a set is an invariant attribute of any well-defined manifold and should be regarded as "the most general genuine aspect for manifolds."[1] Moreover, the concept of cardinality includes the notion of integer, "this foundation of the theory of magnitudes," as a special case [*op.cit.*, 150]. Even after the introduction of the transfinite ordinals, Cantor kept insisting on the view that cardinality is the simplest and primary notion of set theory. This viewpoint is implicit in the organization of the *Beiträge*, where the ordinals are discussed only after the finite and infinite cardinals, and where Cantor regards the notion of power as the "most natural, brief, and rigorous foundation" of finite numbers [1895/97, 289]. I wish to emphasize this point, because it has recently been suggested that Cantor was immersed in an ordinal conception of sets, and an attempt has been made to explain his views systematically on this basis [Lavine 1994, 77–86]. That reconstruction, sharp and interesting in itself, meets with a lot of contrary evidence.[2]

Only in 1883 did Cantor offer for the first time an explicit definition of set:

By a manifold or a set I understand in general every Many that can be thought of as One, i.e., every collection of determinate elements which can be bound up into a whole through a law, and with this I believe to define something that is akin to the Platonic εἶδος or ἰδέα.[3]

Once again, Cantor employs here the word 'Inbegriff' [collection] and emphasizes that it must be possible to turn the collection, the Many, into a whole or One, which is done by a law. Here, like a year earlier [1879/84, 149], it seems that one should understand by 'law,' in general, a conceptual condition;[4] this reading is supported by references to Plato's ideas and to Spinoza's adequate ideas at several other places (endnotes 3, 5, 6 [Cantor 1883, 205–07]). The peculiar language of Many and One reflects the usage of ancient Greek philosophers, particularly Plato.

(see §IV.2).

[1] [Cantor 1879/84, 150]: "das allgemeinste genuine Moment bei Mannichfaltigkeiten."

[2] It seems clear, as a historical fact, that the reason why Cantor attributed great importance to order types was because only well-ordered sets allowed him to establish a satisfactory theory of transfinite powers. He regarded denumerability by ordinals as a distinguishing feature of transfinite sets [Hallett 1984, 146–50], but this does not imply that he had an ordinal conception of sets.

[3] [Cantor 1883, 204]: "Unter einer Mannichfaltigkeit oder Menge verstehe ich nämlich allgemein jedes Viele, welches sich als Eines denken lässt, d.h. jeden Inbegriff bestimmter Elemente, welcher durch ein Gesetz zu einem Ganzen verbunden werden kann und ich glaube hiermit etwas zu definiren, was verwandt ist mit dem Platonischen εἶδος oder ἰδέα."

[4] Perhaps Cantor employs 'law' instead of 'concept' because the condition may logically be more complex than a property. Lavine's interpretation [1994, 85] seems forced and contrary to texts of a year earlier. Cantor's reference (immediately after the text I have quoted) to the "μικτόν" seems to be simply related to his idea that transfinite sets are infinite but at the same time determinate.

With his reference to the relation between sets and Platonic ideas, Cantor introduced for the first time an element that would become characteristic of his philosophy of mathematics – extreme realism with respect to the objects of mathematics.[1] It appears that, up to the early 1880s, Cantor essentially accepted the views of his teacher Weierstrass, and of Dedekind too, according to which pure mathematics is related to pure thought. References to logic in connection with set theory abound in these years – to the principle of excluded middle in 1882 [1879/84, 150], to Well-Ordering as a law of thought [1883, 169], to the notion of the continuum as a *"mathematico-logical"* one and his definition of it as soberly logical [1883, 191], or to the principles of generation and limitation of ordinal numbers as "logical functions" [1883, 196, 199]. But, since Cantor had always been interested in the philosophical views of Spinoza and others, it seems likely that throughout this period there was an unresolved tension between that logical approach to mathematics and his philosophical convictions regarding mathematical objects.

The tension began to be resolved with the speculative principle that 'immanent reality' and 'transient reality' are always coincident. The ultimate reason for this notable convergence of idealism and realism is "the *unity* of the *all to which we ourselves belong.*"[2] An endnote relates that principle to the philosophies of Plato, Leibniz, and Spinoza; the latter, for instance, wrote the famous sentence that the order and connection of ideas is the same as the order and connection of things [1883, 206–07]. Cantor was convinced that his theory of sets opened the way to a satisfactory development of the metaphysical and scientific views of Spinoza and Leibniz, leading up to an "*organic* explanation of nature" that would complement or even substitute the one-sided mechanical explanation [*op.cit.*, 177]. At several places in the *Grundlagen* he stated his conviction that the transfinite is present and real in the physical world and in the mental world. Cantor was certainly a romantic thinker, but the strong presence of these ideas, so close to the *Naturphilosophie*, is noteworthy.[3]

Later in his life, Cantor distanced himself from the logical approach to mathematics, and particularly from Dedekind's logicism.[4] The paradoxes gave him a powerful argument against the views of Dedekind and other logicists, showing that one can at times define a Many that cannot be thought of as a One. His old philosophical and theological convictions aided him in accepting this situation and reacting to it positively (see §VIII.8). At that time, he affirmed that he had reached

---

[1] See, e.g., the second quotation in [1895/97, 282], taken from Francis Bacon. Cantor compares his mathematical activity with the work of a "faithful scribe" transcribing the revelations of Nature.

[2] [Cantor 1883, 181–82]: "[die] *Einheit* des *Alls, zu welchem wir selbst mitgehören.*"

[3] See [Cantor 1883, 177, 199, 204–07; Cantor 1932, 374–75], his published views of 1885 in [Cantor 1932, 261–76], and also the letters to Mittag-Leffler and Hermite in [Meschkowski 258–59, 275].

[4] However, in 1888 he still praised Dedekind's tendency to give a purely logical foundation for arithmetic, although he criticized his work for other reasons (see his letter to Vivanti in [Meschkowski & Nilson 1991, 302]).

this standpoint already in the *Grundlagen*, but the evidence suggests that this may have been an overstatement, and that only gradually (especially in the 1890s) did he come to reject logicism and realize its limitations.

3. The Transfinite (Ordinal) Numbers

We have seen (§VI.6) that the proof of a result connected with the Cantor–Bendixson theorem necessitated focusing on the 'symbols of infinity' and considering the set of all such symbols that have denumerably many predecessors. This was the motivation for a crucial turning point in Cantor's work on transfinite sets – the introduction of transfinite numbers. Cantor was perfectly aware of the importance of this revolutionary step, and obtained from Klein, the editor of *Mathematische Annalen*, permission to publish the article separately as a book. The title underscored the importance of its contents and revealed the new philosophical turn that Cantor was giving to his work: *Foundations of a General Theory of Manifolds. A philosophico-mathematical attempt in the theory of infinity* [Cantor 1883]. These are the introductory words of the paper:

The foregoing account of my researches in the theory of manifolds has reached a point where further progress depends on extending the concept of true integral number beyond the previous boundaries; this extension lies in a direction which, to my knowledge, no one has yet attempted to explore.

My dependence on this extension of number concept is so great, that without it I should be unable to take freely the smallest step further in the theory of sets.[1]

The developments analyzed in §VI.6 indicate that this was not mere rhetoric, but a true depiction of the situation in his research. Cantor went on to say that, although he was going to propose a daring extension of the number system "beyond the infinite," he was firmly convinced that it would someday be seen as a simple, adequate and natural step.

3.1. Origins in point-set theory. Let us first refresh our memory of how the 'discovery' of transfinite numbers was motivated by Cantor's attempts to prove part of the famous Cantor–Bendixson theorem. His aim was to show that, whenever a derived set P^α is denumerable, for any 'symbol of infinity' α, so is P', and con-

---

[1] [Cantor 1883, 165]: "Die bisherige Darstellung meiner Untersuchungen in der Mannichfaltigkeitslehre ist an einen Punkt gelangt, wo ihre Fortführung von einer Erweiterung des realen ganzen Zahlbegriffs über die bisherigen Grenzen hinaus abhängig wird, und zwar fällt diese Erweiterung in eine Richtung, in welcher sie meines Wissens bisher von Niemandem gesucht worden ist. / Die Abhängigkeit, in welche ich mich von dieser Ausdehnung des Zahlbegriffs versetzt sehe, ist eine so grosse, dass es mir ohne letzere kaum möglich sein würde, zwanglos den kleinsten Schritt weiter vorwärts in der Mengenlehre auszuführen."

versely. This was to be based on the following decomposition of P' into isolated sets and P^{α}:

$$P' \equiv (P' - P'') + (P'' - P''') + ... + (P^{\infty} - P^{\infty+1}) + ... + (P^{\alpha}).$$

Since isolated sets are denumerable, and so is P^{α} by hypothesis, one may conclude that P' is denumerable, but only in case the union is denumerable or, what comes to the same, if there are only denumerably many indices preceding α (for the union of a non-denumerable family of nonempty sets is non-denumerable). Thus, in that context it was natural to pay attention to sets of indices, and Cantor came to the idea of considering 'symbols of infinity' with denumerably many predecessors. This is the simplest version of the so-called *principle of limitation* (see §3.3).

The foregoing considerations apply to proving the 'only if' part of the theorem, but the 'if' part was more problematic: given a denumerable P', Cantor wished to establish the *existence* of an α such that $P^{\alpha} = \varnothing$. This in fact required considering the class of all denumerable ordinals as a whole, and introducing the first non-denumerable ordinal. One can now appreciate that Cantor's theorem necessitated the development of a theory of the 'symbols of infinity' and suggested a move toward considering them as mathematical objects. Since only this might justify, from his standpoint, that we consider sets of them and talk about existence results.

It was easy for Cantor to see that all of the symbols he had previously introduced – all that can be expressed as equations in ∞ or, as he now wrote, ω – satisfied the denumerability condition. Turning this fact into a principle, Cantor considered the class of all indices α that have denumerably many predecessors; this became the *second number-class*, the first number-class being that of the finite or natural numbers [Cantor 1883, 197]. Once he began considering the second number-class, Cantor discovered that it was an example, not only of a transfinite power, but of the power *immediately greater* than that of denumerable sets [*op.cit.*, 197–200]. This opened the way for a long-desired development of the theory of cardinalities, since until then he lacked a natural definition of the *higher* powers [*op.cit.*, 167]. The number-classes of "true integral, determinate-infinite numbers" turned out to be the natural and simple representatives of the regular succession of increasing cardinalities.

Cantor must have expected from this a significant advancement with regard to the Continuum Hypothesis. The transfinite number-classes made it possible to define a 'scale' of growing, consecutive cardinalities, against which it should in principle be possible to 'measure' the power of the continuum. Thus, from his standpoint, the transfinite numbers far outstripped the Cantor–Bendixson theorem in importance. This transpires in Cantor's letters communicating the great novelty to his colleagues Dedekind, Weierstrass, Mittag-Leffler and Kronecker. To Dedekind he wrote that God Almighty had seen to it that he attained the most remarkable and unexpected results in the theory of sets and numbers, that he found "what fermented in me for years and what I have long been searching for" (see opening quotation).

3.2. From symbols to true numbers.   Thus far the 'symbols of infinity' had just been indices in the derivation process, lacking any reality independent of the point-sets that carried them. One might compare them to operators like ∇, which were treated formally. The very name employed by Cantor, in its "modesty" [Cantor & Dedekind 1937, 57], suggests this subsidiary character. Cantor was not able to think about the transfinite ordinals and the new number-classes until he convinced himself that it was possible to regard the indices as true numbers, as objects. His concern about the issue is clearly visible in the *Grundlagen* [1883, 165–70] and in the letters he sent to his colleagues. To Kronecker he explained how they "have to be conceived as *numbers*" based on the facts that it is possible to determine their arithmetical relations, and that they can be conceived under a common viewpoint with the familiar finite numbers [Meschkowski 1967, 240]. More concretely, it was possible to define precisely a transfinite arithmetic (see §3.3) and it was possible to analyze the cardinal and ordinal aspects of the new numbers, in such a way that the old ones appear as specializations in the finite case [Cantor 1883, 168, 181].

Above all, the shift in Cantor's viewpoint depended upon the explicit consideration of ordered sets, particularly well-ordered ones, for the first time in Cantor's career. As Cantor himself remarked, all of this happened just after his meetings with Dedekind in mid-September of 1882 [Cantor & Dedekind 1937, 55]. We know from letters to Mittag-Leffler that in mid-October he was already in the possession of the new idea. The process thus occurred rather abruptly, within a month. I have been led to conjecture that his contacts with Dedekind may have played an important heuristic role in the discovery,[1] since in September Cantor read at leisure, over more than a week, Dedekind's draft for the later book [Dedekind 1888]. Let us briefly explore this issue, considering how knowledge of Dedekind's ideas may have suggested to Cantor the introduction of ordered sets and the idea of regarding the 'symbols of infinity' as ordinal numbers.

We have seen (§§VII.1 and 3) that Dedekind definitely favored the view that the ordinal aspect of numbers is the primitive one. He defined the natural numbers as the elements of an ordered set of a special kind, while their use as cardinal numbers was the result of much subsequent development. The whole theory of numbers and sets that he established was based on so-called *chain theory*, which in fact formed the core of Dedekind's draft [1872/78]. Chains were the main tool for defining $\mathbb{N}$, for rigorously establishing the properties of numbers, for proving results about finite and infinite subsets of $\mathbb{N}$, and for defining the cardinal numbers. In this context, to any number n there corresponds a chain n_0, which is simply the ordered set of all numbers $m \geq n$ (in Cantor's terminology, a well-ordered set of type ω). Dedekind introduced and used extensively notions of initial section and remainder, like the ones Cantor would later employ [1895/97, 314] in his general theory of well-ordered sets. Thus, his still unpublished work emphasized ordinal considerations, including extensive use of the simplest well-ordered sets (though of course without a general notion of well-ordering).

---

[1] This conjecture is discussed in section 2 of [Ferreirós 1995].

Assuming that Cantor and Dedekind discussed the latter's draft at all in 1882, it is almost certain that the question which aspect of numbers is the primary one, cardinality or ordinality, must have been touched upon. For Cantor always held the opposite view: that numbers are primarily cardinal numbers. In 1882 he stated that the notion of cardinality is the most basic and important notion in set theory, since cardinality is an invariant attribute of any well-defined manifold [Cantor 1879/84, 150]. Moreover, the concept of cardinality includes the notion of integer, "this foundation of the theory of magnitudes," as a special case [*ibid.*]. Even after the introduction of the transfinite ordinals, Cantor kept insisting on the view that cardinality is the simplest and earliest idea, the "matrix notion" of set theory, as he wrote in 1885.[1] The same viewpoint is implicit in the organization of the *Beiträge*, where Cantor regards the notion of power as the "most natural, brief, and rigorous foundation" of finite numbers [1895/97, 289].[2]

Dedekind was well aware that this was Cantor's position as early as 1887. In a second draft for his book, written that year, we find a version of its preface in which, after stating that cardinality is really a very complicated notion and not a simple one, he writes: "contrary to Cantor" [Cavaillès 1962, 120; *Cod. Ms. Dedekind* III, 1, II, 41]. This seems to evidence that they actually discussed the matter in 1882, since that had been the last time they met personally. In fact, Cantor's letters to Dedekind after the Harzburg meeting reveal an increased awareness of orderings [Cantor & Dedekind 1937, 52–54]. Here we find, for the first time, the idea that a set can be given many different orderings, and that some of its properties will depend on the assumed ordering. Cantor emphasized that it is only relative to some particular ordering that a set can be called a continuum [*ibid.*]. Thus, it is likely that Dedekind's views stimulated Cantor to turn to ordinal considerations and reconsider his previous problems in this new setting, conceiving of the 'symbols of infinity' as ordinal numbers [Ferreirós 1995, 37–41].

The former 'symbols of infinity,' if considered as elements of sets, offered examples of *well-ordered* sets, i.e., totally ordered sets such that every subset has a least element. Cantor was thus able to abstract and define the extremely important notion of well-ordering. Since, quite apart from the derivation process, the symbols are apt to represent the different types of well-ordered sets, "the reality ... of [the number] is established, even in those cases that it is definitely infinite."[3] Only by dissociating the symbols from the context of point-sets and derived sets, and by linking them to well-ordered sets, was Cantor in the position to regard them as true numbers, ordinal numbers.

---

[1] Unpublished paper of 1885 [Grattan-Guinness 1970, 86]: "Sie erscheint mir daher als der *ursprünglichste*, sowohl *psychologisch*, wie auch *methodologisch einfachste Stammbegriff.*"

[2] In [1895/97], Cantor discussed the ordinals only after the finite and infinite cardinals. Indeed, this is responsible for some methodological weaknesses of his exposition (see Zermelo's editorial comments in [Cantor 1932]).

[3] [Cantor 1883, 168]: "so ist durch diesen Zusammenhang ... die von mir betonte Realität der letzteren auch in den Fällen, dass sie bestimmt-unendlich ist, erwiesen."

There were still other reasons why Cantor thought he was perfectly entitled to regard the transfinite ordinals as true numbers. One can rigorously define the arithmetical operations on the new numbers and investigate their properties. The transfinite and the finite numbers can be compared among themselves, obeying fixed laws. And the transfinite numbers entail no contradiction except if one unduly requires them to share all properties of the finite numbers – a *petitio principii* (see §1.1). Lastly, the principle of limitation (see §3.3) makes it possible to establish a definite connection between the ordinal numbers and the transfinite cardinalities. If one now descends from the transfinite to the finite, the ordinal and cardinal properties of numbers come down to the well-known properties of finite numbers [Cantor 1883, 181]. To emphasize his new standpoint and the idea that the novel numbers represent definite steps within the properly infinite, Cantor ceased employing the equivocal symbol ∞ and chose instead ω to represent the first transfinite number [*op.cit.*, 195].[1]

3.3. Cantor's definitions. The presentation of the transfinite ordinals in *Grundlagen* is "purely constructive," as Zermelo said [Cantor 1932, 209], or even purely intuitive. Cantor does not make explicit the assumptions behind his new notion – he just accepts actual infinities unrestrictedly, taking for granted the existence of ever greater sets. This is presupposed in the possibility of unending application of the two basic principles that define the ordinals. The "*first generating principle*" [*erste Erzeugungsprinzip*] consists in adding a unit to a previously given number; this is assumed to be always possible. The "*second generating principle*" [*zweite Erzeugungsprinzip*] is more complex: given a sequence of transfinite numbers without a greatest element, it allows the "creation" of a new number that will be regarded as the "*limit* of those numbers, i.e., will be defined as the number immediately greater than all of them" [Cantor 1883, 196]. Both principles are built by analogy with the two different ways of defining derived sets, either as sets of limit points of a previously given set, or as intersections of a whole family of sets (§VI.6.1).[2] The second principle justifies the introduction of numbers such as ω itself, $\omega 2$ (which follows $\omega+n$ for all n), ω^2 (which follows all ωn), and ω^ω (which follows all ω^n). The first justifies the introduction of numbers such as $\omega+1$, $\omega 2+n$ or $\omega^m n_0 + \omega^{m-1} n_1 + \ldots + \omega n_{m-1} + n_m$.[3]

The two processes can be employed repeatedly, and lead to transfinite numbers that go beyond every boundary. It seems, as Cantor said, that we may risk getting lost in the unlimited [1883, 196–97]. To avoid this, he introduced a third principle that plays an essential role, for it establishes a well-defined connection between the

---

[1] Cantor takes ∞, as it figures in expressions such as $\lim\limits_{x \to \infty} f(x)$, to be a prototype of the improper infinite.

[2] But this process, unlike the 'generation' of transfinite numbers, had $\mathbb{R}$ or $\mathbb{R}^n$ as a clear basis.

[3] Instead of following the conventions of [Cantor 1883], where he wrote 2ω, we are following the definitive notational convention introduced by Cantor in [1895/97] (as we shall see, addition and product of ordinals is not commutative).

transfinite ordinal numbers and the powers or cardinalities. This is the *"principle of restriction or limitation"* that, as we have seen, was suggested by Cantor's strategy for proving the Cantor–Bendixson theorem (§3.1).

If we now notice that all of the numbers previously obtained and their next successors fulfill a certain condition, [that the set of their predecessors is denumerable,] then this condition offers itself, *if it is imposed as a requirement on all numbers to be formed next*, as a new *third* principle ... which I shall call *principle of restriction or limitation* and which, as I shall show, yields the result that the second number-class (II) defined with its assistance not only has a higher power than [the first number-class] (I), but precisely the *next higher*, that is, the *second power*.[1]

The condition is, more generally, that one shall "create" new numbers in accordance with the generating principles only as long as the corresponding set of predecessors has the power of a previously defined number-class [1883, 199]. The first number-class is simply $\mathbb{N}$, the second number-class is the set of all denumerable ordinals. Employing the aleph-notation that Cantor introduced in [1895/97, 293], the cardinality of $\mathbb{N}$ is $\aleph_0$, whereas the cardinality of the second number-class is $\aleph_1$; the third number-class (III), consisting of ordinal numbers with $\aleph_1$ predecessors, has cardinality $\aleph_2$; and so on.

In the process, number-classes are defined by a cardinality condition embodied in the principle of limitation, and the cardinalities are defined through the number-classes. The outstanding process that Cantor was thus able to contrive is not circular but, one might say, helical. The connection between ordinals and cardinalities made the transfinite numbers invaluable for Cantor. Only with their aid was he in a position to define the higher powers in an orderly and satisfying way, thereby laying a cornerstone for his mature theory of transfinite sets.

3.4. Number-classes and powers.   Cantor was able to prove rigorously that the second number-class (II) has a power that is immediately greater than that of $\mathbb{N}$ [Cantor 1883, 198–200]. First, he showed that (II) is not denumerable by *reductio ad absurdum*, employing an argument reminiscent of the 1874 proof about $\mathbb{R}$. Essentially it went as follows. Suppose we are given a denumerable sequence (α_n) of numbers of the second class. If a certain number β in (α_n) is greater than all others, then $\beta+1$ is a number of (II) that is not in the sequence. Otherwise the set of all numbers smaller or equal than those in (α_n) has no greater element but must be denumerable; therefore, the second and third principles (§3.2) allow for the 'creation' of a new number γ that belongs in (II) and is not in the sequence.

---

[1] [Cantor 1883, 197]: "Bemerken wir nun aber, dass alle bisher erhaltenen Zahlen und die zunächst auf sie folgenden eine gewisse Bedingung erfüllen, so erweist sich diese Bedingung, *wenn sie als Forderung an alle zunächst zu bildenden Zahlen gestellt wird*, als ein neues, zu jenen beiden hinzutretendes *drittes* Princip, welches von mir *Hemmungs- oder Beschränkungsprincip* genannt wird und das, wie ich zeigen werde, bewirkt, dass die mit seiner Hinzuziehung definirte zweite Zahlenclasse (II) nicht nur eine höhere Mächtigkeit erhält als (I), sondern sogar genau die *nächst höhere*, also *zweite Mächtigkeit*."

Next, Cantor proved that the power of (II) immediately follows that of $\mathbb{N}$. This reduces to showing that any subset T of (II) is either finite, or denumerable, or of the power of (II), "quartum non datur." Since it is always possible to order the elements of T and index them with numbers of (II), either we exhaust the second number-class and T has the same power as (II), or the indexation ends with a number α of that number-class. And the predecessors of such an α must form at most a denumerable set. With these results, it is justified to call the cardinality of (II) the *second* transfinite power, and to denote it by $\aleph_1$, as Cantor did in 1895. The Continuum Hypothesis now reads: [0,1] has the same power as (II) [Cantor 1883, 192].

Cantor was confident that his three principles for the generation of transfinite numbers and the limitation of number-classes sufficed to prove that there is an unending succession of number-classes, and therefore of powers. He wrote:

> by observing these three principles one can with the greatest certainty and evidence attain ever newer number-classes and with them all the different, successively ascending powers occurring in corporal and mental nature.[1]

As one can see, Cantor was convinced that higher powers would be found in Nature, a theme that resonates through other papers of the 1880s. His absolute faith in both the mathematical and metaphysical existence of actual infinities (see §1) seems to have made him unable to anticipate possible objections and present his readers with stronger arguments for his standpoint.

In an endnote he made it clear that we never reach a limit that cannot be trespassed, and that to every transfinite ordinal there is a corresponding number-class and power [1883, 205]. But of course, anybody who did not share Cantor's absolute faith in the existence of actual infinities would not be convinced. What is the justification for the extreme possibilities of continuation that are posited by the two generating principles? Cantor would later (but not in [1883]) emphasize that by reordering the elements of $\mathbb{N}$ we obtain all possible types of denumerable well-ordered sets; with this we seem to gain a secure foothold for the second number-class, since the example of $\mathbb{R}$ shows that there is at least one cardinality greater than that of $\mathbb{N}$.[2] But what about the first transfinite number of the class whose power is immediately greater than that of $\mathbb{R}$? Are we really entitled to assume its existence? Years later, Cantor would make a definite step forward in relation to these questions (see §7).

With the transfinite numbers, and particularly the second number-class, Cantor gained a powerful new tool for trying to settle the cardinality of the continuum. The Continuum Hypothesis was now reformulated as follows: the continuum is equi-

---

[1] [Cantor 1883, 199]: "mit Beobachtung dieser drei Principe kann man mit der grössten Sicherheit und Evidenz zu immer neuen Zahlenclassen und mit ihnen zu allen in der körperlichen und geistigen Natur vorkommenden, verschiedenen, successive aufsteigenden Mächtigkeiten gelangen."

[2] Otherwise, we would have to explain why (or postulate that) we are entitled to collect all denumerable ordinals α, or all possible orderings of $\mathbb{N}$, into a set.

pollent to (II), it has the power of the second number-class ($\aleph_1$ in the notation of 1895). Cantor immediately promised to give a proof of CH with the new means at his disposal [1883, 171]. Already in 1882, just after introducing the transfinite numbers, he had seen the possibility of this new approach; he wrote to Dedekind:

So far as I can see, our finite irrational numbers can be relatively easily determined with the help of the numbers α [of the second number-class], which I intend to pursue farther.[1]

The idea was to make the limit ordinals correspond to irrational numbers. To this end, one would need to introduce analytic tools that could be applied jointly to the transfinite numbers of the second number-class and to the real numbers (see §§4.2 and VI.8). Cantor worked on this new attempt until at least February 1885. That train of thought led him to consider $\mathbb{R}$ from a more abstract viewpoint, as a totally ordered set, but without taking into account its metric or topological properties. This suggested the development of a general theory of order types (§4.2). With this new theory, his set-theoretical ideas reached their maturity .

4. Ordered Sets

Cantor emphasized in the *Grundlagen* that a second key attainment due to the transfinite numbers was the notion of the "*number*" [*Anzahl*] of elements of a "*well-ordered*" [*wohlgeordneten*] infinite manifold [Cantor 1883, 168]. The introduction of transfinite numbers led him to focus on the new notion of well-ordered set, which he came to view as fundamental for the general theory of manifolds [1883, 169].

4.1. Well-ordered sets. With the word 'Anzahl,' Cantor was making reference to the ordinal properties of the set, to the ordinal number associated with it. He distinguished the number [Zahl] from the cardinality or power [Mächtigkeit] on the one hand, and the ordinality [Anzahl] on the other. A well-ordered set is commonly defined as a totally ordered set such that every subset has a *least* element. Cantor actually proved that subsets of the second number-class (II) always have a least element [1883, 200], but his definition of well-ordering was more complex, although equivalent:

---

[1] [Cantor & Dedekind 1937, 59]: "So viel ich sehen kann, lassen sich unsere endlichen Irrationalzahlen verhältnismässig einfach unter Zuhülfenahme der Zahlen α bestimmen, was ich noch weiter verfolgen will."

By a *well-ordered* set one should understand any well-defined set in which the elements are bound to one another by a precisely given succession, according to which there is a *first* element of the set and both to every single element (provided it is not the last in the succession) there follows a certain other, as also to every finite or infinite set of elements there corresponds a certain element, which is the *next* following element of all them in the succession (unless there is absolutely no follower to them all in the succession).[1]

Just as the cardinality of a set is invariant under one-to-one mappings, its ordinal number [Anzahl] is invariant under one-to-one correspondences that preserve the orderings, so that $a < b$ iff $f(a) < f(b)$. In an unpublished paper of 1885 Cantor called these particular kinds of correspondences 'Abbildungen' [Grattan-Guinness 1970, 86–87], a denomination that he kept employing in [1895/97].[2] As we have seen, later on it became customary to employ 'Abbildung' for the general notion of mapping.

Another theorem that is valid for well-ordered sets in general, but which Cantor formulated here for numbers of the second number-class (II), is the following. Given a sequence (a_i) of elements of a well-ordered set, such that the elements diminish gradually in magnitude (if a_α precedes a_β then $a_\alpha > a_\beta$), the sequence is finite [Cantor 1883, 200]. The result may be surprising, but is easily established: the set of all elements in the sequence (a_i) that have finite index is a subset of a well-ordered set and will therefore have a least element a_λ; since (a_i) is a diminishing sequence and a_λ is the least element with finite index, a_λ can have no successor; therefore the sequence is finite. The results formulated in *Grundlagen* indicate that Cantor was quickly in the possession of detailed knowledge of the theory of well-ordered sets. But, as he ceased publishing mathematical work in 1885, the theory only received a detailed treatment in the second part of his *Beiträge* [Cantor 1895/97].

On the basis of well-ordered sets Cantor defined the operations of addition and product on the transfinite ordinal numbers. This was of the utmost importance to him, since the possibility of defining rigorously the basic operations among transfinite and finite numbers justified calling them 'numbers.' Given two disjoint well-ordered sets A and B, with ordinals α and β respectively, the union of A and B defines the ordinal $\alpha+\beta$ as soon as we stipulate that elements of A always precede those of B, and that the ordering within each set is preserved. The ordinal $\alpha \cdot \beta$ corresponds to the well-ordered set that would result from substituting disjoint copies of A for every element of B, preserving the ordering in B between elements

---

[1] [Cantor 1883, 168]: "Unter einer *wohlgeordneten* Menge ist jede wohldefinirte Menge zu verstehen, bei welcher die Elemente durch eine bestimmt vorgegebene Succession mit einander verbunden sind, welcher gemäss es ein *erstes* Element der Menge giebt und sowohl auf jedes einzelne Element (falls es nicht das letzte in der Succession ist) ein bestimmtes anderes folgt, wie auch zu jeder beliebigen endlichen oder unendlichen Menge von Elementen ein bestimmtes Element gehört, welches das ihnen allen *nächst* folgende Element in der Succession ist (es sei denn, dass es ein ihnen allen in der Succession folgendes überhaupt nicht giebt)."

[2] Although it conflicted with Dedekind's usage. Likewise, he called order-isomorphic sets 'similar,' the same term that Dedekind [1888] employed for equipollent sets.

of the different A-copies [Cantor 1883, 170]. As above, the definitions which we have just given follow the notational conventions of [1895/97], not those of [1883].

With those definitions Cantor established the existence of general laws for the elementary operations on transfinite numbers [1883, 170, 201]. The operations are non-commutative, and so the order of factors is essential. It is easy to see that $1+\omega$ is different from $\omega+1$: we may take as representatives for both sets $<a, 1, 2, 3, ...>$ and $<1, 2, 3, ... a>$, the first corresponding to the ordinal $\omega(1+\omega=\omega)$, while the second has a last element and thus corresponds to an ordinal different from ω. Likewise it is easy to see that $\omega 2 = \omega + \omega$ while $2\omega = \omega$: take two infinite sequences (a_n) and (b_n), then $\omega 2$ corresponds to $<a_1, a_2, ..., b_1, b_2, ...>$, while 2ω corresponds to $<a_1, b_1, a_2, b_2, ...>$. On the other hand, both operations are associative, and they satisfy the distributive law, but only under the form $\gamma(\alpha+\beta) = \gamma\alpha+\gamma\beta$ [1883, 201].

An important distinction that Cantor introduced is that between transfinite numbers with an immediate predecessor, and those without such a predecessor (presently called limit ordinals) [1883, 202]. We shall not pursue in detail further results of transfinite number theory, whose position has been as isolated as that of elementary number theory up to the 19th century. It suffices to mention a few more points. Cantor considered the issue of inverse operations, which of course is quite complicated due to the lack of commutativity [op.cit., 201–02]. He gave a definition of *prime* transfinite number as a number α such that $\alpha=\beta\gamma$ is only valid in case $\gamma=1$ or $\gamma=\alpha$ [1883, 170]. Prime transfinite numbers may or may not be limit ordinals (the first two limit primes are ω and ω^ω); Cantor mentions that the sets of denumerable prime numbers of the first and of the second kind both have the same power as (II) [op.cit., 202–03]. Finally he considered decomposition into prime factors and promised to show that, under certain conditions, it is essentially unique [op.cit., 170, 204]. These topics would only receive detailed treatment in the *Beiträge* [1895/97].

From the standpoint of well-ordered sets it was possible to give a new definition of the transfinite numbers, one that was both more elegant and more convincing. We find this clearly and concisely explained in a letter to Kronecker written in August 1884, immediately after their reconciliation (which did not last long).[1] Cantor wished "especially" [besonders gern] to come to an agreement with Kronecker concerning the transfinite numbers of the second number-class. He said that these are "concepts, resp. signs or characters," which he needed *indispensably* for the *characterization* of point-sets. (It seems likely that his reference to signs was meant to reflect the standpoint of Kronecker, who preferred to regard numbers simply as symbols [Kronecker 1887].) Cantor went on to describe a foundation for these numbers which is somewhat different from the one given in his papers, and which he hoped would be more agreeable to Kronecker:

---

[1] Cantor had already presented the matter this way during the *Naturforscherversammlung* of September 1883, and thought about publishing it [Meschkowski & Nilson 1991, 130, 136–38]. He indicates the idea also in [1879/84, 213–14].

I depart from the concept of a "well-ordered set" and call well-ordered sets of the *same type* (or the same [ordinal] number) those which can be related to each other in a reciprocally univocal way, *preserving the rank-order in both sides*, and now I understand by a number the sign or the concept for a *certain type* of well-ordered sets. By limiting oneself to the *finite* sets, one obtains in this way the finite integers. But if one goes on to overview all of the types of well-ordered sets of the *first* power, one necessarily arrives at the transfinite numbers of the second number-class, and through these to the *second* power.[1]

From this standpoint, the objectionable 'generating principles' that Cantor had employed in [1883] became superfluous.

But to complement this approach, and actually the whole elementary theory of transfinite numbers, a key element was missing – the Well-Ordering Theorem. In *Grundlagen* Cantor made a comment that shows he was aware of the problem to some extent:

The concept of *well-ordered set* turns out to be fundamental for the entire theory of manifolds. It is always possible to bring any *well-defined* set into the *form* of a *well-ordered* set; this seems to me a fundamental and momentous law of thought, especially remarkable because of its general validity, to which I shall come back in a later treatise.[2]

The presumed 'law of thought' would guarantee that the growing cardinalities of the number-classes are *all* of the transfinite powers, that is, it would warrant the comparability of cardinalities. Over the years Cantor came to think that it was necessary to give a detailed proof of the Well-Ordering Theorem and of cardinal comparability. But the problem remained open, and in the *Beiträge* [1895/97] he only mentioned cardinal comparability, failing to emphasize again the importance of well-ordering. Actually he wanted to devote a third part of the *Beiträge* to this issue, but for reasons that will be analyzed in §VIII.8 he never published it. Although it was one of the most important gaps in the Cantorian theory of transfinite sets, the question remained little known except for specialists until its importance

---

[1] [Meschkowski 1967, 251–52]: "Es sind dies Begriffe resp. Zeichen oder Charactere, welche ich zur *Characteristik* von Punctmengen *unentbehrlich* brauche. ... / Ich gehe von dem Begriff einer "wohlgeordneten Menge" aus, nenne wohlgeordnete Mengen von *gleichem Typus* (oder gleicher Anzahl) solche, die sich unter *Wahrung der beiderseitigen Rangfolge* ihrer Elemente gegenseitig eindeutig aufeinander beziehen lassen und verstehe nun unter Zahl das Zeichen oder den Begriff für einen *bestimmten Typus* wohlgeordneter Mengen. Beschränkt man sich auf die *endlichen* Mengen, so erhält man auf diese Weise die endlichen ganzen Zahlen. Geht man aber dazu über, die sämtlichen Typen wohlgeordneter Mengen der *ersten* Mächtigkeit zu übersehen, so kommt man mit Nothwendigkeit zu den transfiniten Zahlen der zweiten Zahlenclasse und durch diese zur *zweiten* Mächtigkeit." Cantor does not forget to flatter the Berlin master by mentioning his "mathematical superiority" [mathematische Ueberlegenheit], that would be needed to solve open problems in transfinite number theory.

[2] [Cantor 1883, 169]: "Der Begriff der *wohlgeordneten Menge* weist sich als fundamental für die ganze Mannichfaltigkeitslehre aus. Dass es immer möglich ist, jede *wohldefinirte* Menge in die *Form* einer *wohlgeordneten* Menge zu bringen, auf dieses, wie mir scheint, grundlegende und folgenreiche, durch seine Allgemeingültigkeit besonders merkwürdige Denkgesetz werde ich in einer späteren Abhandlung zurückkommen."

was brought forward by Hilbert in his famous address 'Mathematische Probleme' [1900] and subsequently became the topic of Zermelo's polemical papers [1904, 1908a], where it was proved by using the Axiom of Choice.

4.2. Elements of a theory of order types.

Until 1970 it seemed that the theory of order types had been developed by Cantor in the 1890s, since it was first published in the *Beiträge*. But Grattan-Guinness [1970] discovered and published a manuscript, entitled 'Principles of a Theory of Order Types,' that was ready for publication and even partly typeset in 1885. This paper outlined the last great development in Cantor's theory of sets, which ended up being structured around a theory of powers or cardinalities, and a theory of order types, linked with each other through the types of well-ordered sets.(It also became clear that a disagreement with Mittag-Leffler, in connection with this paper, motivated Cantor's conscious decision to stop publishing in mathematical journals; see §5.2.)

Late in 1883, Cantor proved that perfect sets have the power of the continuum (see §VI.8). His key idea was to establish a one-to-one correspondence between the complement of a perfect set S, i.e., a set of intervals $\{(a_n,b_n)\}$, and the set of rational numbers $\mathbb{Q}_{[0,1]}$. In modern terminology, he showed that a set of disjoint open intervals is order-isomorphic with $\mathbb{Q}$. Cantor assumed both sets given in the form of sequences (m_n) and (q_n) and gave a procedure for reordering (m_n) in such a way that the correspondence φ between elements of the same index n preserves the dense order in both sides.[1] As we see, the proof depended upon establishing an order-isomorphism between two sets that intuitively are very different – a set of rational numbers and a set of intervals.

By that time, Cantor had also done the step to considering the transfinite numbers as invariants associated with order-preserving correspondences between well-ordered sets, that is, as the *order types* of well-ordered sets. From this abstract viewpoint, the correspondence between $\{(a_n,b_n)\}$ and $\mathbb{Q}_{[0,1]}$ could be understood as showing that both sets have the same order type. Besides, Cantor was considering the possibility of correlating limit ordinals with irrational numbers, which suggested looking at the continuum from an abstract, ordinal point of view (ignoring metric relations). This seems to have been the way in which he came to the idea of studying not only the types of well-ordered sets, but also those of "simply ordered" sets.[2]

According to Cantor, two simply ordered sets are 'similar' [ähnlich] to each other if there exists a one-to-one correspondence which preserves the rank-order of the elements on both sides. The *order type* of a set is defined by Cantor as the "*general concept*" under which all simply ordered sets fall, that are similar to it [Grattan-Guinness 1970, 87]. The notion of power is defined in a similar way, as a 'gen-

---

[1] Having done that, it was possible to expand φ into a one-to-one correspondence between S and $[0,1]$, since both the elements of L and the irrational numbers in $[0,1]$ can be defined through fundamental sequences of elements of, respectively, $\{(a_n,b_n)\}$ and $\mathbb{Q}_{[0,1]}$.

[2] What we now call totally ordered sets, for which an order relation $<$ is given such that, whenever a and b are different elements, then either $a<b$ or $b<a$ is the case.

eral concept' associated with a class of equipollent sets [*op.cit.*, 85].[1] Of the notion of power he said that it has

emerged by *abstraction* from all of the *peculiarities* that a set of a *determinate* [equipollence] *class* may offer, either in relation to the *constitution* of its *elements*, or with respect to the *relations* and *orderings* in which the *elements* may be, *be it among themselves* or with *things that lie outside of the set.*[2]

Cantor seriously meant that every natural and transfinite number is a concept, since he criticized other authors for not defining them this way [Meschkowski & Nilson 1991, 302].

The examples of order types given by Cantor were rather obvious – the types 1, 2, 3 ... of finite ordered sets, the type ω of the set of natural numbers, the type η of the set of rational numbers, and the type θ of the set of real numbers. These order types are the ones that he studied in more detail in the *Beiträge* [Cantor 1895/97, §§7–11, 296–311]. In 1885 he paid special attention to η, proving a theorem that characterized this order type in purely set-theoretical terms. A simply ordered set M is of order type η if and only if it is denumerable, possesses neither a least nor a greatest element, and is such that between *any two* different elements m, n there exist always, under the given ordering, infinitely many other elements [Grattan-Guinness 1970, 87–88]. The proof, Cantor said, is exactly the same that he had employed to show that perfect sets have the power of the continuum – take M and $\mathbb{Q}_{[0,1]}$ in the form of sequences and reorder the first to obtain the desired correspondence.[3] This seems to confirm the key role that this result played in the new turn of Cantor's ideas.

In his *Beiträge* Cantor would complement this theory by studying the type θ of the linear continuum. He showed that a simply ordered set X is of type θ if and only if it is perfect, and there is a denumerable subset S such that between any two elements of X there are elements of S [Cantor 1895/97, 310–11]. Once again, the proof has an important precedent in the 1880s, the analysis of perfect sets and the definition of continua (§VI.7). As we see, within the new context Cantor replaced the condition of connectedness for that of having a denumerable dense subset.

The unpublished paper presented also the operations of addition, multiplication, and inversion of types [Grattan-Guinness 1970, §§4 and 6]. The inverse of type α is denoted by α^*, and it is easy to see that $\eta = \eta^*$, just like $(1+\theta+1) = (1+\theta+1)^*$; but in general an order type is different from its inverse. The definitions of addition and

---

[1] This model was also followed in the *Beiträge*, see [Cantor 1895/97, 282–83, 296–98]. Such definitions had already been published in Cantor's 1885 review of [Frege 1884], see [Cantor 1932, 441].

[2] [Grattan-Guinness 1970, 86]: "entstanden durch *Abstraction* von allen *Besonderheiten*, die eine Menge von *bestimmter Classe* darbieten kann, sowohl in Ansehung der *Beschaffenheit* ihrer *Elemente*, wie auch hinsichtlich der *Beziehungen* und *Anordnungen*, in welchen die *Elemente sei es untereinander* oder zu *ausserhalb der Menge liegenden Dinge* stehen können."

[3] This theorem is the main result in §9 of *Beiträge* [1895/97, 304–06].

multiplication are rather direct extensions of those given previously in the *Grundlagen*. Just like in that paper, we find a definition of prime order type, and with it new problems that are similar to those of number theory, including the issue of decomposition of an order type into prime types. This kind of question is properly what Cantor meant by 'theory of order types' [*op.cit.*, 84]. The theory of order types also allowed some clarifications regarding the special properties that differentiate the transfinite ordinals from other order types. While in general order types admit of many automorphisms that preserve the order, the transfinite ordinals can only be 'similar' to themselves in one way [*op.cit.*, 89–90]. On the other hand, if α is a finite ordinal, it is always true that $\alpha = \alpha^*$, a condition that is *never* valid for transfinite ordinals.

From the foregoing one might be tempted to conclude that Cantor's 1885 work on order types was nothing but a rather trivial generalization from well-known sets and previous results of him. But, in fact, there is evidence that his main objective was to create tools that could be applied simultaneously to point-sets and ordered sets, finding a new way of approaching the proof of CH. It has to be said, however, that the paper of 1885 did not advance significantly toward a solution of the problem. Cantor established in §7 an ordinal analogue of the notion of limit point, suited to the abstract context of totally ordered sets. This was the notion of "principal element," on which a generalization of the notion of fundamental sequence was based; both can also be found in §10 of the *Beiträge*.

> *e* being an element of *A*, it *can* present us with the following *phenomenon*; if ′*e* denotes *any* element of *A* that appears *before e* according to the rank ... and if *e*′ denotes *any* element of *A* that appears *after e* according to the rank, ... then between ′*e* and *e*′ (according to the rank) there lie always infinitely many elements of *A*; if *e satisfies* this *condition*, we shall call it a *principal element* of *A*.[1]

This corresponds to the notion of limit point, but stripped from its properly topological aspects and reduced to purely ordinal terms; from the viewpoint of order types, distance conditions have no sense. Similarly, the generalized 'fundamental sequences' have nothing to do with the Cauchy condition. A denumerable sequence of elements in *A* is nothing but a well-ordered subset of *A* of type ω (or of the inverse type ω^*, which Cantor defined in this paper). If there is in *A* a least element *s* the rank of which is greater than that of any element in the sequence, then we may say that *s* is a "limit" of the sequence in *A*.[2]

---

[1] [Grattan-Guinness 1970, 92]: "Sei *e* ein Element von *A*, so *kann* dasselbe folgendes *Vorkommniss* darbieten; wird mit ′*e irgend* ein dem Range nach *früher* als *e* vorkommendes Element von *A* bezeichnet ... bezeichnen wir ferner mit *e*′ *irgend* ein dem Range nach *später* als *e* vorkommendes Element von *A*, ... dann fallen zwischen ′*e* und *e*′ (dem Range nach) stets unendlich viele Elemente von *A*; *erfüllt e* diese *Bedingung*, so wollen wir *e* ein *Hauptelement* von *A* nennen."

[2] Zermelo gives a simple example, the set {0, ½, ...1 − 1/*n*, ..., 2} of type $\omega + 1$, which possesses a principal element, 2, that is the limit of an ω-sequence [Cantor 1932, 354]; if we regard it as a point-set in $\mathbb{R}$, obviously the set does not contain its only limit point, 1.

On that basis, Cantor introduced a number of notions that paralleled strictly those he was using at the time in the theory of point-sets. He formulated for ordered sets notions of dense-in-itself, isolated, closed, perfect, coherence, and adherence [Grattan-Guinness 1970, 93–95].[1] Quite clearly he was attempting to elaborate an abstract approach to both theories, with the hope that the parallel analysis would make it possible, e.g., to define a one-to-one mapping from the well-ordered second number-class (II) to $\mathbb{R}$. We lack any information on his final conclusions regarding this attempt, in case there were any, but the fact is that in later years he did not follow this path again.

In contrast with the *Grundlagen*, the new paper was of a purely mathematical character, due to Cantor's fear that otherwise the public would only see the philosophical aspects of his writings.[2] Nevertheless, the work contained some remarkable statements on the theory of sets and its applications. As regards set theory, for the first time Cantor distinguished the theory of point-sets from general set theory. Point-sets were regarded as the subject of a branch of applied mathematics, by side of function theory(!) and mathematical physics [Grattan-Guinness 1970, 84]. Pure set theory was regarded as structured around the basic notions of power [*op.cit.*, §3] and order type [*op.cit.*, §4]. This conception and scheme are exactly the ones that determine the organization of the *Beiträge* [1895/97].

After explaining what he understood by a theory of order types, Cantor went on to say:

It constitutes a large and important part of the *pure theory of sets* (Théorie des ensembles), and therefore of *pure mathematics*, for the latter is in my opinion nothing but *pure set theory*.[3]

This was a decisive step – the theory of sets or manifolds was no longer seen as one more branch of mathematics, but as the very foundation of the discipline. At the beginning of this chapter we saw that in the early 1880s Cantor seems to have regarded pure mathematics as based on arithmetic, and divided into two general disciplines built upon it, the Weierstrassian theory of magnitudes and the theory of manifolds that Riemann was the first to suggest. By 1885 his conception had changed, and we find the clearest statement he ever made that set theory is the foundation of mathematics. It seems likely that Dedekind's ideas, particularly the set-theoretical foundation of the number system that Cantor came to know in detail

---

[1] On coherence and adherence, see §VI.8; theis connection with CH is indicated in [Cantor 1932, 264; Schoenflies 1927, 17]. Most of the ordinal notions that Cantor elaborated in these years reappear in the *Beiträge* [1895/97], but not so these concepts.

[2] He took this impression from a review written by Tannery, and probably from the counsels of Mittag-Leffler [Grattan-Guinness 1970, 84]. Even so, §1 discussed briefly the relations between mathematics and metaphysics.

[3] [Grattan-Guinness 1970, 84]: "Sie bildet einen wichtigen und grossen Theil der *reinen Mengenlehre* (Théorie des ensembles), also auch der *reinen Mathematik*, denn letztere ist nach meiner Auffassung nichts Anderes als *reine Mengenlehre*."

in 1882 (§VII.4), played a decisive part in this change. This might explain why Cantor did not present that vision again in his *Beiträge*.

5. The Reception in the Early 1880s

It has frequently been said that Cantor's theories were not well received in his time, but this is clearly an overstatement. Schoenflies [1922, 99–100] wrote that the triumphal march of his ideas can be dated back to publications of Mittag-Leffler and Poincaré, in the mid-1880s, which showed their great importance for function theory. But there are many more examples of quite a good reception, which is particularly noteworthy given the radical and sometimes speculative nature of Cantor's theories. Even so, Cantor felt rejected by German mathematicians and finally stopped publishing in mathematical journals.

5.1. Reception.   We have previously seen that the notions of derived set and set of the first species were almost immediately well received by du Bois-Reymond, Dini and Harnack (chap. V). The proof that $\mathbb{R}$ and $\mathbb{R}^n$ are equipollent provoked an immediate reaction, with a host of papers by Thomae, Lüroth, Jürgens and Netto (§VI.4.3). Neither did journals fail to support Cantor. *Mathematische Annalen* was open to the publication of his crucial series [1879/84] and the editor Klein was in good relations with him and showed interest in his work. Similarly, volume two of *Acta Mathematica* (1883) included the most important part of Cantor's research, translated into French by disciples of Hermite, among them Poincaré. The new topological notions that he developed around this time were also immediately taken over by authors such as Poincaré, Mittag-Leffler, and even Weierstrass.[1] The work of Mittag-Leffler [1884] on the representation of analytic functions employed extensively sets of the first and second species, and the transfinite numbers of the second number-class. (Students of Mittag-Leffler such as Bendixson and Phragmén also worked on the theory of point-sets.) The work of Poincaré dealt with automorphic functions and employed tools from the theory of point-sets, for instance Cantor's results on nowhere-dense perfect sets in [Poincaré 1885].

Nevertheless, this is not to say that all of the implications that Cantor saw in his work, particularly those of a more radical and abstract nature, were well accepted. His cherished ideas on the general notion of power and the transfinite numbers stirred up doubts in many circles, and were particularly unappealing for older mathematicians. It was possible to accept the details of his 1878 proof that $\mathbb{R}$ and $\mathbb{R}^n$ can be put into one-to-one correspondence, and still reject his definition of power as nonsensical. The proof would then show the need for a refinement of the

---

[1] For Klein, see the letters in [Purkert & Ilgauds 1987, 186, 190–91] or [Meschkowski & Nilson 1991]. As regards *Acta*, see [Mittag-Leffler 1927, 26], which ought to be corrected in the light of [Dugac 1984, 70–71]. For Weierstrass, see [Mittag-Leffler 1923, 195; Dugac 1973, 141].

notion of dimension, but not the radical result that $\mathbb{R}$ and $\mathbb{R}^n$ are 'equal' infinities in some sense.

A typical reaction to Cantor's work may be that of Hermite. When Mittag-Leffler obtained his collaboration for translating Cantor, he was confident that the French would soon accepts his ideas [Meschkowski 1967, 242–43]. But as the translation proceeded, Hermite's opinion worsened, and in April 1883 he wrote:

The impression that Cantor's memoirs produce on us is disastrous. Reading them seems to us a complete torture. ... Even acknowledging that he has opened a new field of research, none of us is tempted to follow him. It has been impossible for us to find, among the results that can be understood, just one that possesses a *real and present interest*. The correspondence between points in the line and in the surface leaves us completely indifferent, and we think that this remark, insofar as nobody has inferred anything from it, proceeds from such arbitrary methods that the author would have done better retaining it and waiting.[1]

Apparently, Appell and Picard joined Hermite in his opinions, and only Poincaré judged his work more positively. According to the latter, the problem with Cantor's papers was that they lacked examples (letter to Mittag-Leffler, March 1883 [Dugac 1984, 70–71]). Thus, the numbers of the second, and especially of the third number-class had the appearance of form without matter, which was repugnant to the French spirit. But in his opinion this was just a defect in the exposition, that obstructed the understanding of this "beautiful work," the *Grundlagen*.

It seems likely that Hermite's comments correspond to the views that most contemporary mathematicians held. Cantor was strongly interested in the issue of transfinite cardinalities, and convinced that the notion of power would come to play a decisive role in analysis, but his approach to those questions was speculative and had little to do with research problems of a 'real and present interest.' To some extent Hermite revealed his ignorance of German work, like the papers on integration written by Harnack and du Bois-Reymond,[2] or perhaps, more than that, the differences in style and methodology between the French style and the more abstract one of German mathematicians. But it is quite clear that, in talking about sets of *any* power and ordinal number, Cantor was only supported by a strong confidence that he had, based more on philosophical and theological motives than in mathematical ones. It is not surprising that other mathematicians did not share his expectations and looked with distrust at the development of a daring theory almost on the air.

Even Dedekind and Weierstrass did not express an interest in Cantor's transfinite numbers. This indifference, and above all the negative reaction of members of the Berlin school such as Kronecker and Schwarz, affected Cantor strongly. One may conclude that he was hypersensitive and too anxious to see his ideas acknowl-

---

[1] [Dugac 1984, 209], as translated in [Moore 1989, 96].

[2] A similar ignorance is found in Borel, when many years later he wrote that Camille Jordan "rehabilitated" set theory by showing its usefulness for integration theory (see the quotation in [Hawkins 1970, 96]).

edged and admired. Another reason for his reaction was that, either consciously or unconsciously, he saw not only his ideas, but also his academic career threatened. Although Cantor felt rejected by German mathematicians, the young generation acknowledged his originality and worked in connection with his investigations. Du Bois-Reymond included the theorem of non-denumerability of $\mathbb{R}$ in his handbook [1882, 191–99].[1] Harnack emphasized the importance of the notion of power [1885, 245], and employed Cantorian terminology and the transfinite ordinals. Scheffer, who died prematurely, applied Cantor's ideas in function theory.[2] Similarly, young Swedish mathematicians contributed to the theory of point-sets, following in Cantor's footsteps.

To summarize, it can be said that, in spite of the speculative aims of his research, Cantor was able to attract the attention and recognition of a great many mathematicians. There seems to have been a generation gap, for members of the young generation were much more easily attracted. But their work, especially in cases such as those of Mittag-Leffler and Poincaré, helped convince the older generation of the meaningfulness of work in set theory. Cantor's perception of the situation, as he expressed it in his letters, was far from being objective.

5.2. Withdrawal from mathematics.   In March 1885, Mittag-Leffler suggested to Cantor that he should temporarily withdraw the paper that was then being typeset, since the publication seemed premature to him [Grattan-Guinness 1970]. This caused their correspondence to stop, and seems to have motivated Cantor's retirement from mathematical publication over the following decade. But one should add that the circumstances were complex.

In May-June 1884 Cantor suffered his first mental crisis, following a period of great irritability, particularly in connection with Kronecker. After Cantor's critique [1883] of Kronecker's standpoint, and his later decision to apply directly for a vacant position at Berlin, Kronecker counterattacked in 1884 with his proposal of a paper for publication in *Acta Mathematica* (§VI.5). These frictions seem to have figured prominently among the factors leading to the crisis. One year later, taking into account the skepticism toward Cantor's speculative ideas that reigned in Paris and Berlin, Mittag-Leffler wrote to warn him of the dangers of premature publication:

It could well be that you and your theories are not done justice in the time of our lives. Then they would be discovered again by somebody after 100 years or more, and it would be subsequently noticed that you had everything already, and finally there would be justice, but in that way you would not have exerted any important influence in the development of our discipline. And of course you wish to exert such an influence, as any other who devotes himself to science.[3]

---

[1] Du Bois-Reymond acknowledged Cantor's originality but also claimed his part in some results and ideas [1882, 178–79].

[2] Cantor himself wrote a necrological note [1932, 368–69].

[3] [Grattan-Guinness 1970, 102]: "Ja es kann wohl sein dass man Ihnen und Ihre Theorien nie

As we know from letters of 1896, Cantor understood this paragraph to mean that Mittag-Leffler regarded his theory as premature by a hundred years [Grattan-Guinness 1970, 104–05]. No doubt, Mittag-Leffler was acting with the interests of his journal in mind, but he does not seem to have been dishonest. At any rate, his advice was not bad, since Cantor was as far as before from a proof of CH or any other important result.

Cantor considered that *Acta* was from then on closed to him, just as he saw the Berlin *Journal* since 1878. In his irritability, he was convinced that his former partner had abandoned him under pressure from his enemies, the Berlin mandarins. Besides, in 1885 Cantor failed to get a vacant position at Göttingen (which went to Klein) and he finally abandoned any hopes of ever coming out of Halle. Cantor felt rejected by the German mathematicians. He did not stop doing research, but he decided to quit publishing in mathematics journals. Later in the decade he published a couple of papers in the *Zeitschrift für Philosophie und philosophische Kritik*; these appeared in 1890 as a separate volume. He started devoting more and more time to interests of a philosophical, theological, and literary kind. Philosophers and above all theologians became his preferred interlocutors, and he even approached rather dubious personalities, such as the social reformer Julius Langbehn.[1] In 1891 he wrote to Langbehn that, in spite of taking part in the *Naturforscherversammlung* [Congress of natural scientists and mathematicians], he would find the time to see him daily, for "I must confess that, among the hundreds of colleagues that will gather around me this month, there is none who understands me as well as you do."[2]

It was not long before Cantor distanced himself from Langbehn, and in 1895 he decided to publish again in *Mathematische Annalen*. The outcome was an attempt to summarize and systematize his contributions to the theory of transfinite sets. Even before, in 1892, he had published a short but noteworthy piece in the first transactions of the *Deutsche Mathematiker-Vereinigung*. We shall devote the last two sections of the present chapter to these contributions of the 1890s.

---

in unserer Lebenszeit Gerechtigkeit zu Theil kommen lässt. So werden die Theorien wieder einmal nach 100 Jahren oder mehr von Jemand entdeckt und dann findet man wohl nachträglich aus, dass Sie doch schon das alles hatten und dann thut man Ihnen zuletzt Gerechtigkeit, aber auf diese Weise werden Sie keinen bedeutenden Einfluss auf die Entwicklung unserer Wissenschaft ausgeübt haben. Und einen solchen Einfluss auszuüben das wünschen Sie natürlich wie jeder Anderer der die Wissenschaft treibt."

[1] See [Meschkowski 1967; Dauben 1979; Purkert & Ilgauds 1987] and Cantor's letters in [Meschkowski & Nilson 1991]. In particular, on his turn to a deeper religiousness see [Dauben 1979, 140–48], on the Bacon–Shakespeare polemics see [Purkert & Ilgauds 1987, 82–92].

[2] [Purkert & Ilgauds 1987, 100]: "muss ich bekennen, dass unter den hunderten von Collegen, die sich in diesem Monate hier um mich vereinigen werden, keiner ist, der mich so gut versteht wie Sie." Langbehn wrote a best-selling book and is regarded as an intellectual forerunner of Nazism. Cantor broke with him and his racist ideas, but their correspondence is an interesting document of his opinions (see [*op.cit.*, 95–101]).

6. Cantor's Theorem

Cantor's communication to the first, 1891 congress of the *Deutsche Mathematiker-Vereinigung* bore the title 'On an Elementary Question in the Theory of Manifolds' [Cantor 1892]. The contents of this paper would constitute a keystone of transfinite set theory, more particularly of the theory of cardinalities. Here we find the proof of a result that is presently formulated as follows: the power set (set of all subsets) of a given set L has a greater cardinality than L itself. But, although it complements his set-theoretical work in an extremely important way, Cantor's Theorem was not taken into account in his summarizing papers [1895/97]. It was only in the 1900s that other mathematicians, above all Russell and later Zermelo, reformulated Cantor's Theorem in a purely set-theoretical way on the basis of the Power Set Axiom, and placed it in the cornerstone position that it merits. Moreover, Russell was led to his famous paradox by reflecting on that theorem as applied to some alleged sets, like the set of all sets (see §IX.2).

In the 1870s, Cantor had satisfactorily proved that, once we accept as given the set of real numbers, there are at least two different cardinalities of infinite sets. Since all known examples of sets turned out to belong to one of these cardinalities, Cantor conjectured in the Continuum Hypothesis that they were the first and second transfinite powers. Later on, with the absolutely infinite sequence of transfinite ordinals, Cantor thought to have established in the *Grundlagen* that there is no maximal transfinite cardinality. In one of the endnotes [1883, 205] he wrote that the sequence of ordinals trespasses any possible limit and, furthermore, that to any transfinite number γ there is a corresponding γth cardinality. This alleged proof was completely dependent on the dubious second generating principle (§3) that Cantor introduced as a basis for his definition of the ordinals. If that principle were valid, it would follow that there is a non-denumerable ordinal, etc., but this was precisely the weakest point in his whole presentation. The principles on which the 1883 'proof' rested were by no means as clear as the Power Set Axiom, which can be taken to be the basic principle behind the 1891 theorem.

The notion of the ordinals had even led Cantor to start generalizing CH: the first suggestion of the Generalized Continuum Hypothesis can be found in an endnote [Cantor 1883, 207] which indicates that the set of all real functions has the power of the third number-class. This remark presupposes that the cardinality of the set F of real functions is greater than that of $\mathbb{R}$, which suggests that Cantor may already at this early time have been able to prove Cantor's Theorem. The Generalized CH is then contained in the precision that this higher power of F is exactly the third transfinite cardinality ($\aleph_2$). Unfortunately, Cantor did not enter into further details in the *Grundlagen*. The question remained open whether there exist even greater infinite cardinals.

[Cantor 1892] presents for the first time the well-known method of diagonalization, which is employed in the usual proof that $\mathbb{R}$ is non-denumerable. Incidentally, a proof procedure that is intimately related to diagonalization had already been used by du Bois-Reymond in the 1870s [Wang 1974, 570], but Cantor did not acknowledge this precedent. A new proof that $\mathbb{R}$ is non-denumerable constitutes the first part of his paper. Actually, Cantor considered the more general case of the set of all infinite sequences formed out of two different "characters" m and w; this set is not denumerable [Cantor 1892, 278]. (That result can be applied to $\mathbb{R}$ by assuming the real numbers in $[0,1]$ given in dual representation, and excluding cases of double representation, as in §VI.4.2.) From any enumeration r_i of such sequences, one can define a new sequence s not belonging to the enumeration. It suffices to make the first decimal cipher of s different from the first decimal cipher in r_1 (first sequence in the enumeration), its second decimal cipher different from the second cipher in r_2, its n-th cipher different from the n-th one in r_n. Cantor goes on:

This proof seems remarkable not only because of its great simplicity, but especially because the principle followed therein can be extended immediately to the general theorem that the powers of well-defined manifolds have no maximum, or, what is the same, that on the side of any given manifold L one can always put another M whose power is greater than that of L.[1]

This is called Cantor's Theorem, and is normally formulated as follows: the set of all subsets of L has a greater power than L itself.

In his paper, Cantor took L to be a linear continuum, and M to be the set of all functions $f\colon L \to \{0,1\}$. Such an M is essentially equivalent to the set of all subsets of L, since each f can be regarded as the characteristic function of a subset of L (f takes the value 1 for elements of L that belong to the subset). Apparently, Cantor never saw it that way: L is a set of numbers while M is a set of functions, and a function was, in his eyes, something different from a set (we shall take this matter up in chapter IX). That may have been the reason why he did not include the notable theorem of 1891 in his *Beiträge* [Cantor 1895/97].

The proof of Cantor's Theorem can be given briefly. Let us consider an arbitrary set L and the corresponding set of functions M, defined as above. First, it is clear that the power of M is not less than that of L, for the functions which take the value 1 for just one argument $l \in L$ each, form a subset of M which is obviously equipollent to L. Assuming the law of trichotomy for transfinite powers – either two powers are equal or one is greater than the other – , it now suffices to show that the power of M is not equal to that of L. Were both sets equipollent, one could

---

[1] [Cantor 1892, 279]: "Dieser Beweis erscheint nicht nur wegen seiner grossen Einfachheit, sondern namentlich auch aus dem Grunde bemerkenswert, weil das darin befolgte Prinzip sich ohne weiteres auf den allgemeinen Satz ausdehnen lässt, dass die Mächtigkeiten wohldefinierter Mannigfaltigkeiten kein Maximum haben oder, was dasselbe ist, dass jeder gegebenen Mannigfaltigkeit L eine andere M an die Seite gestellt werden kann, welche von stärkerer Mächtigkeit ist als L."

index the functions in M with elements of L, so that each function would appear under the form f_l. But such an assumption is contradictory; it is possible to define a new function $g: L \to \{0,1\}$ that by construction cannot be any of the f_l. Just apply the above principle of diagonalization and take $g(l) \neq f_l(l)$ for all $l \in L$, by making $g(l)=1$ if $f_l(l)=0$, and $g(l)=0$ if $f_l(l)=1$ [Cantor 1892, 280].

7. The *Beiträge zur Begründung der transfiniten Mengenlehre*

After ten years of almost complete silence, Cantor published his 'Contributions to the Founding of Transfinite Set Theory' [Cantor 1895/97] in *Mathematische Annalen*, the journal that had accepted his crucial work of 1879/84. These two papers summarized his previous work, offering an overview of transfinite set theory and constituting an important instrument for its diffusion. As I have argued, there were few essential novelties in the *Beiträge*, but many of the ideas had remained unpublished or were accessible only in philosophy journals. The presentation was innovative, since Cantor made a serious attempt to present his ideas systematically, paying great attention to methodology.

He started with his famous definition of 'set' (see §8.1) and a discussion of the theory of cardinal numbers. He defined the operations on cardinal numbers, addition as the cardinal number that corresponds to the disjoint union of two sets of given cardinals, and product as the cardinality of the set of pairs $M \times N$ [Cantor 1895/97, 285–87]. He was thus the first author to consider the Cartesian product of two sets, which he called the *"connection-set"* of M and N [Verbindungsmenge], and denoted $(M \cdot N)$.[1] A novelty of the 1890s was Cantor's definition of the "exponentiation" [Potenzierung] of cardinals. To this end, he defined a *"covering"* [Belegung] of a set N with elements of the set M (which is simply a mapping of N in M, a function from N to M),[2] and considered the *"covering-set of N with M,"* that is, the set of all such mappings. If N has cardinality $\mathfrak{b}$ and M has cardinality $\mathfrak{a}$, then $\mathfrak{a}^{\mathfrak{b}}$ is the cardinal number of the set of all mappings from N to M [1895/97, 287–88]. Cantor showed how the basic properties of exponentiation made it possible to derive in just a few lines the whole content of [Cantor 1878] (see §VI.4.2).

As one can see, the theory of transfinite sets led Cantor to consider set-operations that are rather strong, defining his 'connection-set' and 'covering-set.' Nevertheless, the *Beiträge* did not introduce the operation of power set formation nor discuss the Cantor Theorem. This is quite surprising, because it could have been done with the notions and operations introduced in the first part of the work,

---

[1] The notion of cardinal number, together with definitions of addition and multiplication, had already been given in the 'philosophical' paper [1887/88, 411–14], although at that point Cantor was not yet using the Cartesian product.

[2] [Cantor 1895/97, 287]. He wrote that "in a certain sense" a "covering" is "a univocal function" of elements of N.

and since it was such an important complement to his theory of cardinals. This failure might perhaps be explained as a result of Cantor's doubts regarding the connection between functions and 'coverings.'

An important novelty in Cantor's presentation was his emphasis on the idea that a few basic results on cardinalities had yet to be proved [Cantor 1895/97, 285]. These included the law of trichotomy for cardinals – if $\mathfrak{a}$ and $\mathfrak{b}$ are cardinal numbers, then one and only one of the following relations holds: $\mathfrak{a} = \mathfrak{b}$ or $\mathfrak{a} < \mathfrak{b}$ or $\mathfrak{a} > \mathfrak{b}$ –, and the Cantor–Bernstein theorem (see §VII.4). Cantor promised to prove these results after having analyzed the "growing succession of the transfinite cardinal numbers" and their interrelations, but this material was not covered in the published papers. It would have been natural, at this point, to mention the Well-Ordering theorem too, since this yielded those basic results. The reason why Cantor refrained from doing so was, probably, that he regarded the notion of cardinal as more basic and prior to that of ordinal.

The first part of the *Beiträge* concluded with the foundations of the theory of finite cardinals, and of $\aleph_0$. The presentation was not exempt from methodological flaws, as Zermelo remarked in his edition of Cantor's works [1932, 352–55], particularly because Cantor lacked an adequate definition of finite set.[1] It was here that Cantor first employed the Hebrew letter aleph for the cardinal numbers. $\aleph_0$ is the cardinality of denumerable infinite sets, $\aleph_1$ the next infinite cardinal, which is that of the class of denumerable ordinals, and so on. It is easy to prove that $2^{\aleph_0}$ is equipollent with $\mathbb{R}$, and so the Continuum Hypothesis could now have read:

$$2^{\aleph_0} = \aleph_1,$$

though Cantor did not write this formula in the *Beiträge*. Similarly, the Cantor Theorem could have been formulated as follows: for any cardinal number $\mathfrak{a}$ it holds that $2^{\mathfrak{a}} > \mathfrak{a}$ (see [Cantor & Dedekind 1932, 448]).

Two years later, Cantor published a detailed presentation of the theory of linearly ordered sets, well-ordered sets, and the transfinite numbers of the second class (denumerable ordinals). This included, in five sections, the material that we have already reviewed in connection with the unpublished paper of 1885 (§4.2): order types and their operations, the particular cases of the orders η and θ, and the expanded notion of fundamental sequence. Then followed a detailed and careful presentation of the material first treated in the 1883 *Grundlagen*: three sections presenting the theory of well-ordered sets and introducing the transfinite ordinals, and six sections having to do with the theory of denumerable ordinals. Particularly noteworthy is the satisfactory treatment of well-ordered sets, leading up to a result that is the basis for the comparability of ordinal numbers [Cantor 1895/97, 319, 321]. As for denumerable ordinals, Cantor showed again that the cardinality of the

---

[1] A rigorous introduction of finite cardinals and $\aleph_0$ would have required the previous development of the theory of well-ordered sets, but Cantor thought this ran contrary to the natural order.

set of denumerable ordinals is the cardinal that immediately follows $\aleph_0$, and thus the "natural *representative*" of $\aleph_1$ [*op.cit.*, 332–33]. He gave a definition by transfinite recursion for the exponentiation of ordinals, and then analyzed particular questions such as "normal forms" for denumerable ordinals, and the ε-numbers (ordinals ε such that $\varepsilon=\omega^\varepsilon$).

The *Beiträge* represent a mature contribution in which Cantor presented abstract set theory as such, independent of the real numbers and point-sets. They served the purpose of making transfinite set theory known and transmitting its basic results, and some of its open problems, to a new generation of cultivators.[1] The first part of the work was immediately translated into Italian and French, and a full French translation was published in 1899; the English version had to wait until 1915. However, Cantor's lifework remained incomplete in several respects. The *Beiträge* did not address the most important problem that had guided his research, the Continuum Problem. In fact, their scope was considerably narrower than his previous work: Cantor did not go beyond the transfinite ordinals of the second number-class, which define the second transfinite cardinal. This may have partly been for expository reasons, since it was clear that his methods could readily be extended to a wider scope. But, above all, he did not come to publish his most original ideas of the period, related to the set-theoretic paradoxes and Well-Ordering, although he had planned to do so in a third part of the *Beiträge*. In the absence of a proof of Well-Ordering the work remained incomplete, lacking the key element that should connect the theories of ordinals and cardinals and show that every transfinite cardinality is an aleph. Although he did not mention the problem in print, Cantor was well aware of this shortcoming.

8. Cantor and the Paradoxes

Fortunately, Cantor's original ideas concerning the paradoxes and Well-Ordering were discussed with Dedekind, with Bernstein (a student of Cantor), and with Hilbert. Through them, particularly through Hilbert and Bernstein, they came to the knowledge of a few central figures in the history of set theory. In line with the approach I have followed in Part Two, we shall study the paradoxes mainly through Cantor and his letters to Dedekind and Hilbert, relegating to §IX.2 a short discussion of their public emergence.

By early 1897 Cantor had discovered the paradox of the 'set' of all alephs, and realized that it contradicted the usual conception of sets as concept-extensions (i.e., the principle of comprehension). His reaction to the paradoxes was not at all despairing. In fact, this finding seemed to support his Platonistic conception of sets as opposed to the logicistic views of Dedekind and Frege. Furthermore, he realized

---

[1] The most important problems that had not been discussed here were highlighted by Hilbert in his famous address on mathematical problems [1900]. See §IX.1.

that the paradox could be turned into a proof of Well-Ordering, a crowning achievement that would constitute an adequate conclusion for his new series of papers. Indeed, it seems likely that he may have found the paradoxes by reflecting about the Well-Ordering theorem.

8.1. Paradoxes and theology. Cantor decided to approach Dedekind again, in order to discuss with him this delicate material which required a sound judgement in foundational questions. Since Hilbert was now acting as editor of *Mathematische Annalen*, he also wrote to him regarding his purpose. Judging from his letters to Hilbert [Purkert & Ilgauds 1987, 224–27], in 1897 Cantor formulated the paradox of all alephs as follows. If the totality $\sqcap$ of all alephs is a transfinite set, it will have a certain cardinal number $\mathfrak{a}$, and the Cantor Theorem will imply that there is another aleph greater than $\mathfrak{a}$. But then, this new aleph would both belong and not belong to $\sqcap$, a contradiction [Meschkowski & Nilson 1991, 388].

The argument is not detailed enough to be completely convincing. Two years later he sent a much more careful proof to Dedekind [Cantor & Dedekind 1932, 448]: if $\sqcap$ is a set, so is $T = \cup M_\mathfrak{a}$ for all $\mathfrak{a} \in \sqcap$, where $M_\mathfrak{a}$ is any set of cardinality $\mathfrak{a}$ (this step involves both the Axioms of Union and Replacement). Calling $\mathfrak{b}$ the cardinal of T, by the Cantor Theorem there is $\mathfrak{b}' > \mathfrak{b}$; but then T would contain a subset of greater cardinality than T itself, a contradiction. From these arguments, Cantor inferred that one has to differentiate two kinds of well-defined sets. In his letters to Hilbert of September 1897 he said that a set can be *"finished"* [fertig] or not, and only in the first case it is a transfinite set (on the notion of 'finished,' see below). Since the assumption that $\sqcap$ is a transfinite set leads to a contradiction, one has to conclude that it is not a finished set.

The reason why Cantor was not shocked by the contradiction is that, since 1883 at least, he had differentiated sharply between the transfinite and the absolutely infinite. When Cantor introduced the transfinite ordinals on the basis of the two 'generating principles' he was guided by the intuitive idea of an absolutely limitless sequence. This was linked with theological questions, for the infinite had traditionally been identified with God's Absolute (see §1). Cantor wrote that, although he was certain that the transfinite numbers take us always farther, never reaching an insurmountable limit, he was equally convinced that in that way we never come to exhaust even approximately the Absolute. "The Absolute can only be acknowledged, but never known, not even approximately known;"[1] that is to say, it cannot be mathematically determined. Thus the "absolutely infinite number-sequence" [absolut unendliche Zahlenfolge] seemed to him an adequate symbol for the Absolute. The contradictory character of $\sqcap$ merely suggested that it is an absolutely infinite collection, and thus beyond our thinking abilities.

---

[1] [Cantor 1883, 205]: "Das Absolute kann nur anerkannt, aber nie erkannt, auch nicht annähernd erkannt werden." The triadic distinction between finite, transfinite and absolute becomes even sharper in the philosophical papers (e.g., [Cantor 1887/88, 378]). See [Hallett 1984, 32–48; Jané 1995].

The letters to Hilbert suggest that Cantor had indicated the existence of the paradoxes obscurely in his papers of 1883 and 1895, but as regards the first this seems more than doubtful.[1] On the other hand, it seems clear that the definition of set which opens the *Beiträge* was actually meant to suggest the above viewpoint, that is, to prevent absolutely infinite collections from being called 'sets' (see letter to Hilbert in [Purkert & Ilgauds 1987, 227]), although it did so in an obscure and inefficient way:

By a 'set' we understand any reunion M to a whole of definite, well-differentiated objects m of our intuition or our thought (which are called the 'elements' of M).[2]

The collection of all alephs cannot be made a whole, for this entails a contradiction, and so it is not a 'set.' The above definition has time and again been presented as a perfect example of the naive standpoint in set theory, which is not quite true.[3] The naive standpoint can be found explicitly in Frege and Russell, implicitly in Riemann and Dedekind. But the frequent confusion is understandable, for Cantor's subtle distinction had not been sufficiently clarified at all.

The first step that Cantor took, in order to clarify his new conception of the paradox and of 'finished sets,' was to write Dedekind and ask his student Bernstein to visit him.[4] Bernstein says that Cantor had immediately realized that the contradiction affected the 'system of all things' that underlied Dedekind's theorem of infinity. In his correspondence of 1899, Cantor formulates the paradoxes using Dedekind's terminology of systems, and he says explicitly that they affect the collection of everything thinkable [Cantor & Dedekind 1932, 443]. Dedekind had stated that every system is a thing [1888, 344] but this now became untenable, and his theorem that there is an infinite set vanished, bringing into question his whole logicistic project. Otherwise said, Cantor had shown that Dedekind's logicistic notion of set is not sufficient as a basis for set theory, for it allows both 'finished' and 'unfinished' collections. According to Bernstein, in 1897 Dedekind had not arrived at a definite opinion, but in his reflections he had almost come to doubt whether human thought is completely rational.

As regards Hilbert, his first reaction to Cantor's argument was a perfect example of the traditional logical standpoint (Cantor quotes his words literally in the next letter). He could not accept that the set of all alephs is contradictory:

---

[1] As late as 1888 Cantor praised Dedekind's work and did not indicate any inconsistency in it (see below). Thus, Purkert's interpretation [1989] that the paradoxes were the *motivation* that led Cantor to write the *Grundlagen* seems to me over-optimistic.

[2] [Cantor 1895/97, 282]: "Unter einer 'Menge' verstehen wir jede Zusammenfassung M von bestimmten wohlunterschiedenen Objekten m unsrer Anschauung oder unseres Denkens (welche die 'Elemente' von M gennant werden) zu einem Ganzen."

[3] This was first realized by Purkert (see [Purkert & Ilgauds 1987, 150–59]).

[4] Bernstein's recollections, in [Dedekind 1930/32, vol. 3, 449].

The collection of alephs can be conceived as a definite well-defined set, for certainly if any thing is given it must always be possible to determine whether this thing is an aleph or not; and nothing more belongs to a well-defined set.[1]

This might as well have been Dedekind's first reaction, even in the terminology employed. Underlying this approach is the conception of the universal class, the collection of all possible things; a set is determined as a well-defined subcollection of this universal class. But, precisely, Cantor was implying that the universal class is not a 'finished' thing, a set.

In 1899 Cantor had occasion to discuss the matter again by letter with Dedekind; it was the last episode in their relations. He sent some very interesting letters including a detailed discussion of his attempt to prove the Well-Ordering theorem. First of all, Cantor tried to clarify his distinction between two kinds of collections. Now he employed a different terminology, calling a well-defined collection of things a 'multitude' [Vielheit] or a 'system' [Cantor & Dedekind 1932, 443], implying that Dedekind's basic notion was inconsistent or at least insufficient. The key new idea is the following:

A multitude can be constituted in such a way that the assumption that *all* of its elements 'are together' leads to a contradiction, so that it is impossible to conceive the multitude as a unity, as 'one finished thing.' Such multitudes I call *absolutely infinite* or *inconsistent multitudes*.[2]

When a multitude can be collected to "*one* thing" without contradiction, it is called a '*consistent multitude*' or a set [Menge]. This was an interesting immediate reaction to the problem of contradictory 'sets,' that is reminiscent of von Neumann's distinction between sets and classes some twenty years later. But in the absence of any independent criterion for 'consistency,' it was not satisfactory.[3] Dedekind's only remark was that he did not understand what Cantor meant by a 'being together' of the elements [Cantor & Dedekind 1976, 261], again a clear example of their different styles of thought, intuitive and logical.

---

[1] [Meschkowski & Nilson 1991, 390]: "Der Inbegriff der Alefs lässt sich als eine bestimmte wohldefinirte Menge auffassen, da doch wenn irgend ein Ding gegeben wird allemal muss entschieden werden können, ob dieses Ding ein Alef sei oder nicht; mehr aber gehört doch nicht zu einer wohldefinierten Menge."

[2] [Cantor & Dedekind 1932, 443]: "Eine Vielheit kann nämlich so beschaffen sein, dass die Annahme eines 'Zusammenseins' *aller* ihre Elemente auf einen Widerspruch führt, so dass es unmöglich ist, die Vielheit als eine Einheit, als 'ein fertiges Ding' aufzufassen. Solche Vielheiten nenne ich *absolut unendliche* oder *inkonsistente Vielheiten*."

[3] For instance, ℕ might be inconsistent and we have simply not yet found the contradiction. See [*op.cit.*, 447–48], where Cantor proposes to introduce two 'axioms'—that finite sets are consistent, "'*the axiom* of arithmetic' (in the old sense of the word);" and that each aleph is consistent.

Cantor reacted to the paradoxes through subtle philosophical (or, if you wish, purely verbal) distinctions. What was needed was clear criteria for sethood, not a discussion of the triad finite/transfinite/absolute, however useful the latter may have been heuristically. The distinction between consistent and inconsistent systems, as it is, was far from solving the difficulties posed by the paradoxes. In his reflections on these matters, Cantor was actually led to formulate conditional set-existence principles that anticipate some of the Zermelo–Fraenkel axioms. His August 1899 letter to Dedekind contains the following: every part of a set is a set (somewhat related to the Separation Axiom); two equipollent systems are either both sets or both inconsistent (close to the Axiom of Replacement and to von Neumann's axiom – see §XI.3); given a set of sets, the union of its elements is also a set (Axiom of Union). This suggested a satisfactory way of approaching the issue, but neither Cantor nor Dedekind were any more in the position of carrying it through.

8.2. Attempted proof of the Well-Ordering theorem.   By 1899 Cantor had refined his arguments for the existence of set-theoretic paradoxes. He may have come to think that the paradox of the 'set' of all ordinals was more cogently formulated, for he based his proof of the Well-Ordering theorem on it.

Consider the system of all transfinite ordinal numbers, which Cantor denoted by Ω. In the *Beiträge* Cantor had shown that if two ordinals α and β are different, then either $\alpha < \beta$ or $\beta < \alpha$; likewise, he had shown that the relation $<$ is transitive. This means that Ω is linearly ordered by $<$. Moreover, his theorems on well-ordered sets implied that every part of Ω has a *least* element, so

The system Ω in its natural order of magnitude constitutes a 'sequence' [a well-ordered multitude].[1]

Now, if we regard 0 as belonging to Ω, as the first ordinal, it can be said in general that the ordinal number of the set of ordinals $\{\beta: \beta < \alpha \}$ is precisely α; otherwise said, α is the *type* of the set of its predecessors. Since Ω is well-ordered, if it were a set it would have an ordinal number δ greater than all of the numbers in Ω. But, by definition of Ω, δ would also belong to Ω; thus we obtain $\delta < \delta$, a contradiction. Hence, Ω is an inconsistent, absolutely infinite collection.[2]

Now, the problem is to show that every transfinite power, every cardinality of a set, is an aleph. The basic idea of the proof was as follows: if there were a collection V whose cardinality is not an aleph, the whole system Ω would be "projectible" into V; we would thus obtain a subsystem V' equipollent to Ω; thus V' is inconsistent, and so is V. Therefore, every consistent multitude, every transfinite set, has a

---

[1] [Cantor & Dedekind 1932, 444–45]: *"Das System Ω bildet daher in seiner natürliche Grössenordnung eine 'Folge'."* A "Vielheit" is called a "Folge" in case it is "wohlgeordnet."

[2] Next, Cantor appealed to the existence of a one-to-one mapping between the ordinals and the alephs (see [Cantor 1883, 205]) in order to derive from the paradox of ordinals the paradoxical character of the system of all alephs [Cantor & Dedekind 1932, 446].

definite ordinal number and an aleph as its cardinal number. (Note that the proof relies on the three set-existence principles mentioned at the end of §1.2; it requires not just Replacement, but the principle that if Ω can be mapped one-to-one onto a system, then this system is 'inconsistent.')

The weak point in the proof is the first step, in which Cantor claims that it must be possible to 'project' or map Ω into V. He probably reasoned that one can 'pick' elements of V successively, making them correspond to the ordinals in their natural order, and that the only way in which V can have no aleph as cardinality is by being so large that we employ all of the ordinals in the process. This is a very good intuitive starting point, but in a rigorous proof one would need to replace the intuitive idea of 'successively picking' by an abstract principle. Zermelo remarked, in Kantian spirit, that Cantor applies the intuition of time to a process that goes beyond all intuition; he criticized Cantor for positing a fictitious entity of which it is assumed that it could make *successive* arbitrary choices [Cantor 1932, 451]. The needed abstract principle is precisely the Axiom of Choice, "which postulates the possibility of a *simultaneous* choice" [*ibid.*]. Zermelo also remarked that the employment of 'inconsistent multitudes' in the proof might throw doubts on its validity, for which reason he had avoided them in his own proofs of the Well-Ordering theorem.

Had Cantor published the attempt, it would certainly have constituted an extremely important contribution. It would have simplified the complex story of the discovery of the paradoxes, but those surprising arguments only came to the knowledge of a few mathematicians, having to be reelaborated independently by others. It might have started a public debate, and it is clear that an analysis of its underlying assumptions would have led to the Axiom of Choice. But the reason why Cantor did not publish the promised third part does not seem to be related to the problems that Zermelo indicated. The distinction between consistent and inconsistent collections posed the general problem, which collections are sets. Cantor's response was to look for 'axioms' of transfinite set theory and arithmetic like the following: there exist things; if V is a consistent multitude and δ a thing not in V, then $V \cup \{\delta\}$ is also consistent. He would have also needed an Axiom of Infinity, but we do not know how he tried to formulate it (perhaps: ω is a consistent system). He was still busy with this matter in 1900, and he never completed the paper (see letters to Dedekind and Hilbert, [Cantor 1932, 447–48; Meschkowski & Nilson 1991, 405–13, 425ff]).

As regards Dedekind, his reaction to the paradoxes shows that he was no longer able to find a way out. In September 1899 Cantor visited him to discuss the matter personally.[1] It seems that Dedekind's only suggestion was that the problem might be related to the contradictions that arise when one does not differentiate between belonging and inclusion. He presented his colleague a detailed proof of one such

---

[1] In response to a mathematical almanach announcing that he had died the very same date of Cantor's visit, Dedekind wrote the editor that day and month might be correct, but not the year. He had spent that day in very stimulating conversation with Cantor, who dealt a death-blow not to himself, but rather to one of his errors [Landau 1917, 54].

contradiction [Cod. Ms. Dedekind III,1,V, p. 83]. In 1903 Dedekind refused to let his *Zahlen* be reprinted because of the paradoxes. His last reaction to the problem can be found in the preface to the new reprint of 1911. He admits the significance and partial legitimacy of doubts about the "security" of important foundations of his views, but expresses his "trust [Vertrauen] in the inner harmony of our logic." He thinks that a detailed investigation of the "creative power" [Schöpferkraft] of the mind will lead to an unobjectionable formulation of the work. That creative power is identified, very narrowly, with our ability to create from determinate elements a new determinate object, their set [System], which is necessarily different from those elements [Dedekind 1888, 343].

Two aspects of this suggestion are noteworthy. Since Zermelo had sent his axiomatization [1908] to Dedekind, and the latter does not even mention it, there is reason to think that he could not accept it as a satisfactory way out. It must have seemed to him a compromise solution that, by resorting to axioms, run contrary to his logicistic convictions. Second, in emphasizing the idea that sets are formed out of their elements, Dedekind departed from the traditional idea of concept-extension and prefigured the iterative conception. In so doing, he was simply abandoning a more or less philosophical conception of sets, and resorting to a regular trait of his mathematical practice, embodied in the definitions of the different kinds of numbers, of ideals, etc. The iterative conception is presently taken to be, by many authors, a satisfactory intuitive picture underlying axiomatic set theory. Prefigurations of it can also be found in other early authors, like Cantor himself (see [Wang 1974]) and Hadamard, although it properly stems from work of the 1930s and 40s by Zermelo and Gödel (see §XI.2).

Part Three: In Search of an Axiom System

Born in the 1850s with Riemann and Dedekind, the set-theoretical approach was championed from about 1900 by Hilbert, who used his influence to foster the axiomatization of mathematical theories on the basis on set theory (§IX.1). Meanwhile, abstract set theory came of age with the contributions of Zermelo and Hausdorff in the 1900s. The years up to 1914 were thus a crucial period of diffusion and recognition for set theory in all its aspects – as a basic mathematical language, as a possible foundation for mathematics, as an independent branch of the discipline.

But the 1900s and 1910s were also the high time of Russell and his collaborator Whitehead. This should be enough to remind us that, while the period was one of recognition, it was also a time of ambivalence and confusion regarding prospects for the young theory. Russell heralded the contradictions or paradoxes that affected the foundations of set theory, calling for deep reform (§IX.2). Even more important, the 1900s saw a heated foundational debate, stimulated above all by Zermelo's introduction of the Axiom of Choice and his proof of the Well-Ordering theorem: the acceptability of abstract mathematics was in question (§IX.3). Given this situation, various solutions and approaches were offered, of which the most influential were Zermelo's axiomatic system (§IX.4) and Russell's theory of types (§IX.5). Just at the time when set theory was winning more and more adepts, a bifurcation occurred that would mark the following three decades of development.

When World War I ended, there was not one single system of set theory. The situation had actually worsened, since Brouwer, Weyl, and others began to propose highly deviant, constructivist systems as alternatives to 'classical' set theory. The ensuing foundational debate, the so-called 'crisis' (§X.1 and 5), established the atmosphere in which the final steps toward a satisfactory axiom system were to be taken. While the contrast between constructivists and 'classicists' consolidated, axiomatic set theory and the theory of types entered a noteworthy process of convergence. Type theory was given a so-called 'simple' formulation, closer in spirit to set theory (§X.3–4). And the latter became a logistic system, supplemented by new axioms that eventually suggested an intuitive motivation which is close to a key guiding principle of type theory (§§XI.2 and 5). The convergence between both systems was also behind some of the most important developments of the 1920s and 30s, including Zermelo's late work and Gödel's relative consistency results (§XI.4). We shall study them in detail, together with the contributions of Skolem, Fraenkel, von Neumann, and Bernays that led the now usual first-order axiomatic systems (§§X.1–2, 5).

Figure 9. *The Göttingen Mathematics Society in 1902. Sitting at the table we find Klein and Hilbert, at the extreme right Zermelo; standing in the second row are E. Schmidt (behind Klein) and Bernstein (behind Zermelo). Courtesy Niedersächsische Staats- und Universitätsbibliohek Göttingen.*

IX   Diffusion, Crisis, and Bifurcation: 1890 to 1914

> That the word 'set' is being used indiscriminately for completely differ-
> ent notions and that this is the source of the apparent paradoxes of this
> young branch of science, that, moreover, set theory itself can no more
> dispense with axiomatic assumptions than can any other exact science
> and that these assumptions, just as in other disciplines, are subject to a
> certain arbitrariness, even if they lie much deeper here – I do not want to
> represent any of this as something new.[1]

The years up to 1914 were a crucial period of diffusion and recognition for set
theory. During the 1890s the new vision of mathematics and the Cantorian ideas
spread out, while the 1900s saw fundamental new contributions in the hands of a
new generation of cultivators – Zermelo and Hausdorff above all. But this was also
a period of heated debates surrounding the notion of arbitrary set and its expressions, the Axiom of Choice and the Well-Ordering Theorem. It was also the time in
which the paradoxes emerged, heralded by Russell. Thus, the diffusion of set theory
was accompanied by much ambivalence and confusion. The acceptability of abstract mathematics was in question, as were the relations between logic and set
theory.

Zermelo's axiomatization and Russell's theory of types turned out to be the
most important and ambitious attempts at rebuilding set theory on adequate foundations. Zermelo treated set theory in the style of Hilbert and took a decided step
toward its full extensionalization, while Russell tried to rescue as much as possible
of the old naive (and intensional) approach based on the principle of comprehension and the conception of sets as a part of logic. With their work, the future began
to look better for the young discipline, but at the same time there was much unclar-
ity since, as they were formulated in the 1900s, both approaches seemed irreconcil-
able. Only in the 1920s and 30s a more harmonic picture emerged in a gradual
process that will be the topic of the last chapter.

In Part Two (§VIII.8) we studied the paradoxes through Cantor and his letters to
Dedekind and Hilbert, here (§2) we shall offer a short discussion of their public
emergence. §1 analyzes the diffusion of set theory up to the early 1900s, and §3
discusses the early foundational debate surrounding the increasing recognition of
the peculiarities of abstract mathematics. Then, in §§4 and 5 we shall consider the

---

[1] Julius König [1905] as translated in [van Heijenoort 1967, 145].

systems proposed by Zermelo and Russell. The last section reviews some other pre-War developments briefly.

1. Spreading Set Theory

In retrospect, and disregarding [Cantor 1892], the 1890s appear as a period in which no central contributions to set theory were published, a period of stagnation. The most original work of the period with some connection to set theory came from the field of logic. But the panorama is not at all a negative one, since during that decade set theory enjoyed unprecedented diffusion and found new cultivators. An extremely important instrument for making abstract set theory known was Cantor's *Beiträge*, but the theory attracted interest mostly through its applications. This means, above all, applications in analysis and function theory.

An increasing number of new handbooks of analysis emphasized the importance of the theory of point-sets. The process had already begun in the 1870s with Dini's *Fondamenti* for the theory of real functions [1878]. It continued in Italy with the work of Peano, particularly the great collective work he directed, *Formulaire de mathématiques* [1908] (five editions from 1895). This was an encyclopedic attempt to translate all major mathematical results into an unambiguous symbolic language, in which the notion of class played a primary role. In France, the influential *Cours d'analyse* by Camille Jordan [1893] showed plainly the important role that set theory was to play. Among other things, Jordan dealt with the problem of measuring areas and sets, refining the notion of content in order to treat adequately the integration of functions of two or more real variables. Borel would write that Jordan 'rehabilitated' set theory by showing that it was a useful branch of mathematics [Hawkins 1970]. Subsequently, the French school of function theory made intensive use of set-theoretic notions. One has to mention here Borel's *Leçons* [1898], which started with basic notions of point-set theory and introduced his definition of measure. The book included Bernstein's proof of the Cantor–Bernstein theorem, the first correct published proof (a previous attempt by Schröder failed, see §VII.4).

The theory of point-sets and its applications was also made known in Germany by Schoenflies, in an exhaustive report on the development of the theory of point-manifolds [1900/08], and in England by the Youngs with their [1906]. The diffusion was, thus, simultaneous in all the main scientific languages of that time.

A kind of public breakthrough came with the First International Congress of Mathematicians in 1897, where two keynote speakers emphasized the importance of set theory in analysis. The French Hadamard spoke on possible applications of set theory, and the German Hurwitz on the theory of analytic functions. Developing work of Mittag-Leffler (§VI.6.2), Hurwitz suggested a classification of analytic functions based on the corresponding set of singularities, where Cantor's notions of denumerable, closed and perfect played an important role (see [Purkert & Ilgauds 1987, 144]). He also indicated the interest of investigating the topology of closed sets, a question that would be taken up by Schoenflies.

1.1. Hilbert and his circle.   Another path along which set theory spread out was the various activities of Hilbert and his circle. Surprisingly, this has been less emphasized by historians, probably because they have tended to focus narrowly on analysis. Hilbert's work in algebra and algebraic number theory employed freely the set-theoretical approach. His decided support of the set-theoretical approach and the work of Cantor came out very clearly in his 1900 address to the Second International Congress of Mathematicians. As is well known, the first mathematical problem he posed for future research was Cantor's Continuum Problem,[1] and he also emphasized the importance of proving the Well-Ordering theorem. He was thus calling attention to the most important open problems in the field of abstract set theory, in a way that Cantor himself had not done in his last papers (of course, his correspondence with Cantor must have played a decisive role here). Hilbert's presentation is not without interest: the Continuum Hypothesis is formulated in the weak form that every subset of $\mathbb{R}$ is either denumerable or equipollent to $\mathbb{R}$; concerning the other question, Hilbert does not ask for a general proof, but for a definable well-ordering of $\mathbb{R}$ [1900, 298–99]. The set-theoretic terminology he employed mixes those of Cantor and Dedekind (§VII.6.2).

Less noticed has been the fact that Hilbert's second problem, the consistency of the axioms for the real numbers, is also related to the issue of set theory. Hilbert himself suggested this in [1900, 301], saying that the "existence" of Cantor's "higher classes of numbers and cardinal numbers" can be established by a proof of consistency – just like the existence of the system of all real numbers. And he added: "unlike the system of *all* cardinal numbers or of *all* alephs," for which no consistent axiom system can be set up. This is actually the first published mention of the paradoxes in Cantorian set theory – without making any fuss of it.

But what is most important is to realize that the kind of axiomatization which Hilbert was proposing at the time (still today the most frequent among mathematicians) had set theory as its basis. I have already remarked that in *Grundlagen der Geometrie* he gives axioms for the elements [Dinge] of three sets [Systeme], and his axiom system for the reals is similar: one starts with a "system" of "things" and defines axiomatically relations and operations between them [Hilbert 1900, 300–01; 1900a]. Most of the axioms give conditions on the elements, and therefore can be formalized in first-order logic. But Hilbert felt free to formulate axioms dealing with sets of elements; to formalize them, one needs to quantify over sets of elements or, in modern terminology, one needs second-order logic.[2] The conspicuous example of this is the famous Axiom of "Completeness" [Vollständigkeit] that Hilbert included first in his axiomatization of the reals [1900a], and then in subsequent editions of the *Grundlagen der Geometrie* (starting with the French and English translations).

---

[1] Hilbert tried to solve it twice unsuccessfully, in [Hilbert 1926] and two years later.

[2] The same happens, e.g., with Dedekind's chain-condition in his definition of the natural numbers (§VII.3.2).

In the case of the reals, Hilbert's axiom of completeness says, essentially, that numbers form a set of elements which is maximal, that is, which is capable of no further extension as long as all of the other axioms hold [1900, 300; 1900a]. A possible way of formalizing it would be the following: the set S of real numbers is such that, whenever $S \subset T$ and T satisfies the remaining axioms (I–IV.1), then $S = T$. Sometimes it is said that Hilbert's complicated axiom is metamathematical, because in its formulation he referred to the satisfaction relation between axioms and models;[1] sometimes it is remarked that the axiom is second-order. In fact, he was just reasoning in a way that was becoming customary in algebra: one defines a certain kind of structure (Archimedian ordered field, in this case) and thinks about all possible sets that are realizations of the structure; then, a maximality condition is enough to characterize univocally the set of real numbers. Completeness, in the usual sense, is a by-product of maximality.

An interesting way of understanding what Hilbert did is to think that set theory belongs to the underlying logic in which the axiom system is formulated. This is likely to have been Hilbert's own viewpoint by the late 1890s, as we have seen. The point is that, if talk of sets and elements (or systems and things) is just logical language, there is no essential difference between conditions affecting the elements and conditions affecting sets. For whenever we have a realm of things, reasoning about sets of such things is just logical reasoning. Of course, later Hilbert learnt that in foundational work one must be more careful, developing logic and mathematics simultaneously [1904], and even later, in the 1920s, he came to use *formal* axiom systems, which is a completely different way of working axiomatically. Set theory, however, kept playing a background role in regular mathematical work, which may help the reader understand the key importance that Hilbert ascribed to it.

Hilbert contributed indirectly to the development of set theory by stimulating students and collaborators to work on it. Zermelo says that he started work on set theory under the influence of Hilbert, and that he realized its fundamental importance thanks to the joint work of the Göttingen mathematicians [Moore 1980, 130]. Although he had been introduced to set theory by Cantor himself, Bernstein did under Hilbert his doctoral work, in which he offered some abstract results and worked on generalizing the decomposition that Cantor had established for closed sets of reals (§VI.8), which implied that the Continuum Hypothesis holds for them. Schoenflies was also close to Hilbert's circle; he applied notions of point-set theory to simple closed curves in the plane, in contributions that led to Brouwer's work on topology (§6). Thus, a good number of the most important German contributors to abstract and topological set theory were inspired by him. The list becomes almost a who is who if we take into account that Hausdorff's work on order types began under the influence of Bernstein.

---

[1] But one could take axioms I–IV.1 to define a set-predicate, and in this way avoid relying on the satisfaction relation.

1.2. Contributions in logic. Already in 1884, Frege proposed to base arithmetic upon the notion of cardinal number, which he explained in a way that had some common points with Cantor. As we have seen (§VII.6.2), Frege criticized set theorists for employing an extensional notion as basic, being convinced that the only way to found arithmetic upon logic was to take the *intensional* ideas of concept and relation as a basis. In his view, assigning a number is making an assertion about a concept; if I say 'Venus has 0 moons,' that means there is no object falling under the concept 'moon of Venus' [1884, §46]. Frege defined a notion that plays the role of equipollence: concept F is "equinumerical" [gleichzahlig] with concept G if there is a one-to-one relation between the objects that fall under F and those that fall under G. Notice that this notion of 'equinumerical' does not presuppose number. On the contrary, now Frege defines the [cardinal] number that corresponds to the concept F as the "extension [Umfang] of the concept: equinumerical with the concept F" [Frege 1884, §§68, 72]. In case $\aleph_0$ (or 3) objects fall under F – i.e., have property F – we would have defined the first transfinite cardinal (or the number three).

Frege's definition has been frequently construed as if his concept-extensions were nothing but classes. If so, each cardinal number would have been defined as a class of equipollent classes. This was actually Russell's proposal [1903, §111], but it is not faithful to Frege's thought.[1] Frege developed his approach in full detail, on the basis of a formal (but interpreted) system of logic, in his *Grundgesetze der Arithmetik* [1893; 1903]. It is well known that his most important innovation was a very detailed system of second-order logic that is extremely close to 20th-century systems of mathematical logic.[2] In the *Grundgesetze* he employed freely the notion of "course-of-values" of a (logical) function, which is a generalization of concept-extensions [Frege 1893, §3]. If the function is what Frege calls a concept, say $F(x)$, its course-of-values agrees with what had been called its extension [Umfang]; but the function can also be a relation, etc., and in these cases we can still speak of the corresponding course-of-values. Naturally, one is tempted to interpret Frege as taking a class to be the course-of-values of $F(x)$ and a class of n-tuples as the course-of-values of an n-ary relation. But all he requires is that the course-of-values be an object, and that the same object correspond to concepts which apply to exactly the same things. His stipulations make it plainly clear that one can take any object whatsoever as the extension of a concept (see [Frege 1893, §§9–10]).

Frege regarded the introduction of courses-of-values in his system as one of the most important innovations he had made. He remarked that not only the cardinal numbers, but also the negative, irrational, and in short all numbers have to be defined as concept-extensions [1893, 14]. Without courses-of-values, then, it would be impossible to develop his project [*op.cit.*, ix–x]. Frege denoted by '$\acute{\varepsilon}\Phi(\varepsilon)$' the

---

[1] That can also be presented as a version of Cantor's definition, according to which a cardinal number is a general concept under which equipollent classes fall, but again not faithfully. We can here observe how by 1900 set theory was not yet completely extensionalized.

[2] This can be found already in [Frege 1879], but one of the novelties in the meantime was precisely the introduction of 'courses-of-values.'

course-of-values of any given function $\Phi(x)$, and his symbolism allowed *always* the formation of courses-of-values (with only some formal restrictions [*op.cit.*, §9]). Thus, the very symbolism incorporated the principle of comprehension, for it put no restrictions on the formation of concept-extensions. Frege's basic law V [1893, 36, 240] stated that two courses-of-values are identical when the corresponding concepts apply to exactly the same objects. In modern notation (and without respecting some of Frege's conventions) we can render it as follows:

$$[\grave{\varepsilon}\Phi(\varepsilon) = \grave{\alpha}\Psi(\alpha)] \leftrightarrow \forall x[\Phi(x) = \Psi(x)].$$

In words, a generalized equality can always be transformed into an equality of courses-of-values, and conversely; this is an analogue of extensionality. Frege anticipated that some authors might object his basic law V because it had not previously been used by logicians, but he was convinced of its purely logical character [1893, vii]. His preface expressed great confidence in his viewpoint as the only rigorous one; he even said that nobody would be able to show that his principles lead to plainly false conclusions [*op.cit.*, ix, xxvi]. Little did he anticipate the coming of Russell and his paradox; as evident as the above ideas and assumptions seemed at the time, Russell's paradox showed plainly that they were untenable (§2).

The theory of sets, under some form or another, constituted an integral part of all the important works on mathematical logic at the time. The first volume of Schröder's *Vorlesungen* [1890/95] was a prolix presentation of the Boolean calculus of classes, on the basis of an auxiliary discipline that he regarded as purely mathematical [*op.cit.*, vol. 1, 157]: the "identical calculus with domains in a manifold" [identischer Kalkul mit Gebieten einer Mannigfaltigkeit]. In fact, this is essentially the Boolean algebra of subsets of a given set, but Schröder did not clarify what he understood by a 'manifold.'[1] The calculus of propositions [*op.cit.*, vol. 2] was also presented as a particular case of the identical calculus. Schröder's work was very influential, not only through direct followers of note like Löwenheim and later Skolem, but also because it was employed by Russell, Peano and Zermelo among others.[2] It has been emphasized that Schröder's tradition is responsible for the emergence of a metatheoretical approach to logic, and particularly of model theory [Goldfarb 1979; Moore 1987].

Peano spent much effort in creating a precise and concise formal language suited for expressing mathematical propositions. The logical language that he elaborated from 1888 onwards was a refinement of Boole's calculus, in which the notion of class had a primary role. His careful choice of an adaptable symbolism led him to introduce what is essentially the modern logical notation, although his treatment of quantification and of relations was imperfect. One of his most impor-

---

[1] He did establish two interesting requirements on manifolds in order that the identical calculus become applicable. They must be "consistent" and "pure" [rein; *op.cit.*, vol. 1, 212–13, 248, 342].

[2] See [Russell 1903; Zermelo 1908]. His work led him to some ideas that played a role in the history of lattice theory, and stimulated Dedekind to important but not very influential contributions in this area [Mehrtens 1979].

tant contributions was the clear distinction between membership and inclusion, a merit he shares with Frege. Peano employed symbols like 'ε,' '$\supset$,' '$\cap$,' '$\cup$,' '$-$,' among others [Peano 1891; 1908, 3–10], but historians of logic frequently tend to forget that, like Boole and many others, he gave a double interpretation of the symbolism, propositional and class-theoretical. '$\supset$' was read both as the conditional and as inclusion, '$\cap$' as the conjunction and as intersection, 'ε' was ambiguous between a sign of predication and class-membership. Peano accepted the principle of comprehension, responsible for the emergence of contradictions, although he only relied on it implicitly [1908, 4–5, 9].

Of the systematic work done by Peano and his collaborators in *Formulaire de mathématiques*, that on the foundations of arithmetic and geometry was the most influential.[1] In the present context, it is particularly important to mention that his famous axiomatization of the theory of natural numbers [Peano 1889; 1891] is based on the notion of class. In this respect it is exactly like Hilbert's early axiom systems (see §1.1). That is particularly visible in the axiom of induction, which uses a sign K for class; translated to modern symbolism, it postulates:

$$[k \in K \wedge 1 \in k \wedge ((x \in N \wedge x \in k) \to x+1 \in k)] \to N \subset k.^2$$

It is significant that some of Peano's associates (Vivanti, Gerbaldi, Bettazzi, Burali-Forti) published contributions to set theory during the 1890s.

But soon the most influential logician was Russell, who combined elements from the traditions of Peano and Frege with innovations of his own (see [Rodríguez-Consuegra 1991]). Russell's *The Principles of Mathematics* [1903] was an ambitious review of much previous work on the foundations of mathematics, particularly of analysis, geometry, and logic.[3] The book included many original viewpoints, of which the most important is, of course, his work on the paradoxes – or contradictions, as he wrote. Russell mixed freely philosophical and mathematical considerations, a trait that was to characterize all of his related work. He became the herald of the logicistic viewpoint.

In the *Principles*, Russell spoke very favorably of the doctrines of Cantor and Dedekind. To give a couple of examples, he accepted Dedekind's theorem of infinity and devoted a chapter to discussing his approach to arithmetic, although Russell was not sympathetic to Dedekind's ordinal and structural conception of numbers [Russell 1903, §§234–43 and 338–39]. He also endorsed Cantor's theories of cardinal and ordinal transfinite numbers [*op.cit.*, §§283–98]. As a matter of fact, in the preface Russell said that his main debts in mathematical issues were to Cantor and Peano.[4]

---

[1] On Peano and his school see [Kennedy 1974; Borga, Freguglia & Palladino 1985]. [Rodríguez-Consuegra 1991] discusses carefully his influence on Russell.

[2] $k \in K$ means that k is a class, or belongs to the class of all classes; N is the set of natural numbers.

[3] With particular attention to authors such as Weierstrass, Cantor, Dedekind, von Staudt, Pasch, Pieri, Peano, Frege and Schröder.

[4] By the time he wrote this, he was just falling under Frege's influence; Frege's work was

It has to be noted that it was Russell, not Cantor in his published work, who focused on the Cantor Theorem as a central result of great importance. He seems to have been the first mathematician who presented it as showing that the set of all subsets of S has always a greater cardinality than S itself [*op.cit.*, §§346–47]. Thus, it was Russell who formulated it for the first time as a purely set-theoretical result. (In Cantor's version it showed that, given a set S, a certain set of *functions* has greater cardinality, and functions were *not* taken to be sets.) In the process, Russell was the first to emphasize something like the Power Set Axiom. All of this was in itself an important contribution, for until then the theorem lay rather forgotten in the first annual report of the *DMV* and its significance had not been clearly grasped. Moreover, the Cantor Theorem led Russell to the discovery of his paradox. Thus Russell's early reformulations of previous work were important in the process of extensionalization of set theory, although he was never partisan of a purely extensional conception of classes (see [1903, §§66–79]), and he gradually became more and more a Fregean on this account.

2. The Complex Emergence of the Paradoxes

The arguments that Cantor found between 1896 and 1899 showed the inadequacy of the logical conception of sets defended by Dedekind, Frege, Peano and others. Unfortunately he did not publish, and the arguments had to be rediscovered by others in a complex and convoluted process. It is plainly false that, as some have written, the paradoxes immediately created a stir and attracted the attention of mathematicians after the publication of a paper by Burali-Forti [1897]. It was only in 1903 that it became clear for the mathematical community at large that there was trouble with the very notion of set – Russell's work, *The Principles of Mathematics*, heralded the news. Here I shall just give a schematic account of the process, which has been very well studied by other authors.[1]

In [1897], Burali-Forti published an argument that is formally close to the paradox of the class Ω of all ordinal numbers (see §VIII.8). Nevertheless, Burali-Forti had misunderstood Cantor's notion of well-ordered set, and so he did not realize that the assumption that Ω is a set leads, by his argument, to a contradiction. Instead, he thought the argument applied to a different kind of ordered sets, what he called 'perfectly ordered classes,' and showed that such sets are not well-ordered [Moore & Garciadiego 1981]. The Italian mathematician had been close to the discovery of the paradox of the largest ordinal, but he missed it due to conceptual unclarities. Even when he realized that he had misconstrued Cantor's definition, he saw no contradiction between Cantorian set theory and his work. It would be Rus-

---

studied in an appendix.

[1] On this topic see [Grattan-Guinness 1978; Coffa 1979; Moore & Garciadiego 1981; Garciadiego 1985, 1986, 1992; Moore 1988]. For short summaries see [Garciadiego 1994; Moore, forthcoming].

sell [1903, 323] who reformulated the argument of Burali-Forti as a contradiction and gave it its present name.

Similarly, Russell himself published in 1901 an argument that is close to the paradox of the class Π of all alephs (see §VIII.8). But he did not interpret it as showing that Π cannot be assumed to be a set – at that time he rather thought he had shown that there is a very subtle fallacy in Cantor's proof of the Cantor Theorem. Actually, Russell had originally been quite skeptical towards Cantor's work, and it was only gradually that he came to accept it (see [Garciadiego 1992; Moore 1993]). By 1902 he regarded the Cantor Theorem as a correct result and thus came to think that the notion of set or class had to be essentially refined. He reformulated the previous argument as the paradox of the largest cardinal.

As we see, the earliest publications that are related to the set-theoretic paradoxes were far from transmitting the idea that set theory is inconsistent. Moreover, they did not cause any great impact, not even when Hilbert indicated, without making fuss of it, that it is possible to show that the system of all cardinalities, or the system of all alephs, are not "mathematically existing concept[s]" [Hilbert 1900, 301]. By 1900 doubts regarding the foundations of set theory were beginning to raise, but most authors hoped to be able to find alternative explanations that would leave the notion of set unaltered. The reason may be in the fact that the paradoxes of the largest ordinal and the largest cardinal elaborated on rather complex notions of transfinite set theory. One could thus hope that their source would be found in technical details of this particular theory, not in general set theory.

The situation changed substantially after·Russell hit on the paradox that bears his name. This had a distinctive character, for it did not employ sophisticated notions of Cantorian set theory – it was based on very simple notions that were then generally regarded as basic, *purely logical* ones: set, the membership relation, all, and not. The path which led Russell to his argument is noteworthy, and once more it shows how convoluted the whole issue was and how many obscurities surrounded the notion of set at the time. To begin with, in the 1890s Russell had been an adherent of idealist philosophy, convinced that in mathematics one always finds contradictions. Before formulating his famous paradox, he toyed with several others, including the Leibnizian paradox of the largest number [Moore 1988; 1993]. In the gradual process of abandoning idealism and embracing Cantorism, Russell came to be shocked by an apparent paradox. The Cantor Theorem had to be false, for it is plainly clear that there is a greatest infinite number, the number of all things (remember that classes were taken to be things). Russell was convinced that there must be a universal class, a class of everything, and regarded this as a commonsense assumption.[1] This is how he began to look for a fallacy in the Cantor Theorem.

---

[1] To Couturat he wrote that if one grants that there is a contradiction in the concept of a class of all classes, then the infinite always remains contradictory [Moore & Garciadiego 1981, 327].

In the first half of 1901, Russell engaged in a more detailed analysis of the Cantor Theorem. Hoping to find the fallacy, he applied Cantor's proof to the class of all classes, and the method of diagonalization led him to consider the class of all classes which are not members of themselves [Coffa 1979]. Let us call this class R; it is defined by $R = \{x: x \notin x\}$, and there is no doubt that the principle of comprehension warrants its existence. (In semi-Fregean notation, $R = \acute{\varepsilon}(\varepsilon \notin \varepsilon)$.[1]) By considering the definition of the Russell set, it is easy to see that

$$R \in R \text{ if and only if } R \notin R.$$

Using only basic notions and the principle of comprehension, Russell had found what we might call an elementary contradiction in the logical theory of classes. This went against all expectations that set theory had a secure place in logical theory, and therefore was consistent.

Even so, Russell must have been unclear what the real significance of his argument was, for he kept it to himself during a whole year. In June 1902 he finally decided to write the masters of logic, Peano and Frege, in order to know their reaction. Frege's reply was extremely clear: the argument had a fundamental importance, it cast doubt on the notion of course-of-values (or of concept-extension) and showed the inadequacy of his basic law V. By implication it cast doubt on the logicistic program as a whole [van Heijenoort 1967, 124–28]. This was spelled out clearly in the Appendix to volume 2 of his *Grundgesetze*:

I cannot see how arithmetic could be given a scientific foundation, how numbers could be conceived as logical objects and introduced, if it is not allowed – at least conditionally – to go from a concept over to its extension. Can I always speak of the extension of a concept, of a class? And if not, how can I know the exceptions? Can we always conclude, from the fact that the extension of a concept coincides with that of a second one, that every object that falls under the first concept also falls under the second?[2]

Frege immediately began to look for solutions, but it was not long before he came to conclude that all such attempts are unnatural, and that the logicistic project had failed. By the end of his life he asserted that the paradoxes had "destroyed" set theory, and he looked for a geometrical foundation of arithmetic [Frege 1969, 298–302]. Russell, on the other hand, would be much more optimistic about prospects to save Frege's program (§5).

---

[1] Frege did not employ a relation of membership.

[2] [Frege 1903, 253]: "Und noch jetzt sehe ich nicht ein, wie die Arithmetik wissenschaftlich begründet werden könne, wie die Zahlen als logische Gegenstände gefasst und in die Betrachtung eingeführt werden können, wenn es nicht – bedingungsweise wenigstens – erlaubt ist, von einem Begriffe zu seinem Umfange überzugehn. Darf ich immer von dem Umfange eines Begriffes, von einer Klasse sprechen? Und wenn nicht, woran erkennt man die Ausnahmefälle? Kann man daraus, dass der Umfang eines Begriffes mit dem eines zweiten zusammenfällt, immer schliessen, dass jeder unter den ersten Begriff fallende Gegenstand auch unter den zweiten falle?"

After the publications of Frege and Russell, the public became aware of the existence of difficulties in the foundations of set theory. The reaction was of course different among different groups, as can be seen in the very names given to the paradoxes.[1] Russell [1903; 1908] preferred the straightforward but severe name 'contradictions,' which is also the one Frege used. It is a purely logical expression with no other connotations, but in fact Russell was of the opinion the 'contradiction' that bears his name springs directly from common-sense [1903, §104]. If so, there is something intrinsically wrong in our commonsense assumptions, in our logic, as several other authors have also suggested [Quine 1941; Gödel 1944]. That viewpoint is aptly conveyed by the name 'antinomy,' which, as Kant used the word [1787], indicates an unavoidable contradiction to which our thought leads. It seems that Poincaré was the first to use 'antinomy' in the 1900s [1905/06]; later on the term was employed by Zermelo [1908] and popularized in Germany by Fraenkel [1928]. The softest name is 'paradox,' meaning an *apparent* contradiction (see König's quotation at the beginning of this chapter), and suggesting that it is only a incorrect formulation of set theory that leads to trouble. Not by chance, it is presently the most common term.

Russell thought that, since the contradictions spring from common sense, a deep reform of logic would be needed. He was not just interested in safe systems for mathematics, but (like Frege) in logic as the universal language [van Heijenoort 1967a]. Others would think that the problem was a purely mathematical one, that the paradoxes were another symptom that recent work had been going along unacceptable lines. In this connection, the paradoxes were just one more element of the foundational debate – though, certainly, a powerful one (see §3). Still others thought that the paradoxes called for a simultaneous reform of logic and set theory; this was the case of Hilbert.

Hilbert and some members (at least) of his circle had long been prepared for the emergence of the paradoxes, since he and Bernstein had first-hand information on the topic from Cantor, starting six years earlier. Zermelo had even found the Russell paradox before Russell himself, but he does not seem to have regarded it as a menace to set theory, and he did not even care to publish.[2] In 1904 Hilbert stated his opinion that paradoxes like Russell's

show, it seems to me, that the conceptions and means of investigation prevalent in logic, taken in the traditional sense, do not measure up to the rigorous demands that set theory imposes. ... a partly simultaneous development of the laws of logic and of arithmetic is required if paradoxes are to be avoided.[3]

---

[1] I thank Alejandro Garciadiego for calling my attention to this topic.

[2] The fact that he found it by 1900 is well-established by statements of his own, of Hilbert and of Husserl: see [Rang & Thomas 1981], also [Zermelo 1908a, 191] and [Peckhaus 1990, 25–26].

[3] [Hilbert 1904], as translated in [van Heijenoort 1967, 131].

The trouble was more in logic than in set theory, taken as a mathematical theory. But, at any rate, it was necessary to clarify and systematize the foundations of set theory. One needed to rethink it thoroughly, for, unless one wished to embrace the full Cantorian Platonism, the only previously available alternative had been to base the notion of set upon logic. The task was taken upon himself by Zermelo, who solved it in an elegant and quite complete way (§4).

But the reform of logic, that most people thought necessary, was complicated even more by the emergence of new paradoxes after 1904. In 1905, Richard and König published articles with paradoxes in which the notion of definability is present. The simplest of these was discovered by Berry and communicated to Russell: "the least ordinal not definable in a finite number of words" has just been defined in a finite number of words [Moore, forthcoming]. The Richard paradox [1905] is interesting in that it uses Cantor's diagonal procedure to present what Richard himself regarded as a merely apparent contradiction. As regards König, in 1904 he tried to show that the cardinality of the continuum could not be an aleph (§3), and the year after he presented an argument intended to show that it was not well-orderable. Assume it is well-ordered, and consider the set of definable ordinals; this must be denumerable (since definitions are finite combinations of a finite alphabet), and so there are undefinable ordinals. But then, by Well-Ordering, there is "the least un-definable ordinal," a contradiction in terms [König 1905].

As we see, paradoxes of different kinds emerged during a period of about ten years. Russell [1908] went on to compile a list of them, including the millenary paradox of the liar (or of Epimenides) and indicating that these are only a few out of an indefinite number of possible contradictions. He looked for a new system of logic in which *all* of these paradoxes would be solved in a more or less natural way (§5). A completely different line of attack was suggested by Peano [1906], who was of the opinion that the Richard paradox, and by implication those of Berry and König, belong to "linguistics," not to mathematics or logic. König's paradox, for instance, can be interpreted as showing that the notion of definability is imprecise and relative; it only becomes precise when we assume a well-established formal language, with a fixed set of symbols, and then the contradiction disappears. Thus, König's paradox does not show that the assumption of a well-ordered continuum is contradictory.

Peano's long-standing interest in formal languages and awareness of the ambiguities of natural language offered him a vantage point from which to judge the whole situation and simplify it. His proposal, however, only became common property after Ramsey made it again in a paper [1926] that proposed a simplification of Russell's type theory (see §X.3.2). Other apt mathematicians and good thinkers tried to revise logic without relying on formal languages, which led them into marshy terrain.[1] The variety of assumptions, confusions, and suggestions that were

---

[1] See, e.g., [König 1914], which tried to establish a consistent theory of arithmetic and set theory on the basis of a "synthetic logic" based on immediate intuition and with psychologistic overtones [*op.cit.*, iii–iv]. Still, König was deeply acquainted with Cantor's work and influenced by Dedekind and Hilbert. He distinguished many different senses of the word 'set,' rejecting

made in connection with the paradoxes helps us understand the long and confuse debate that followed. As a consequence of this, and of the foundational debate in general, a generalized feeling of insecurity prevailed among authors interested in foundational questions, lasting up to the 1930s (§X.1). In the end, the paradoxes of set theory have probably been the most important argument for the generalized use of formal axiom systems and formal logic in mathematics. It was only gradually that other advantages of formalization, particularly for metamathematics, became clear.

3. The Axiom of Choice and the Early Foundational Debate

After the turn of the century the field of set theory was full of activity, as is re-flected in the fact that since 1905 it was mentioned in chapter 2 of the review jour-nal *Jahrbuch über die Fortschritte der Mathematik,* under the rubric 'philosophy, set theory and pedagogy' (after 1916 it got a separate chapter, see [Purkert & Il-gauds 1987, 145]). But the decade of 1900 would also be full of polemics and criti-cism of set theory. Fortunately, although the polemics created confusion, it did not endanger the future of the young branch of mathematics. In any event, all of this shows that the mathematical community had reached a high degree of maturity regarding set-theoretic questions.

The so-called foundational crisis is, of course, a very famous episode in the history of mathematics. But this is not to say that the related historical facts are well-known or properly appreciated. The 'crisis' is normally associated with the first third of the 20th century, and it is usually taken to have been caused by the set-theoretic paradoxes. Both points can be disputed. The paradoxes were an integral part of the polemics, but the great excitement and discussion that surrounded set theory in the 1900s was not due exclusively or even primarily to them. If we look for the most central and enduring topic of the discussion, we shall find that the debate was above all about the acceptability of abstract mathematics. The trend to abstraction that we have seen unfolding in Part One reached a peak with Zermelo's [1904] proof of Well-Ordering on the basis of the Axiom of Choice. Since many found the idea of a well-ordering of $\mathbb{R}$ particularly implausible, Zermelo's proof started a heated debate.[1] But if we interpret the foundational debate as having to do primarily with abstract mathematics, its earliest expression can be found around 1870, in Kronecker's objections to Weierstrass (see §§I.5 and IV.4.2).

---

extensionality, and advanced toward a notion of 'Cantorian set' that allowed him to derive the classical theory.

[1] This conception of the debate was first emphasized by Moore [1978; see also 1982].

As we have seen, Hilbert revitalized the question of Well-Ordering by mentioning it in the context of the first problem he posed in the 1900 address before the Second International Congress of Mathematicians. Four years later, the Third Congress heard the Hungarian Julius König deliver a lecture in which he claimed to show that the power of the continuum is not an aleph. Cantor himself attended the lecture and it is said that, deeply moved, he thanked God for having allowed him to see this refutation of his error.[1] The alleged proof was quite impressive technically and made extensive use of cardinal arithmetic (see [Moore 1982, 86–88]). But König relied on a proposition that Bernstein had established in his dissertation:

$$\aleph_\alpha^{\aleph_0} = \aleph_\alpha \cdot 2^{\aleph_0}, \text{ for every ordinal } \alpha.$$

On the basis of this lemma and results of his own, König showed that the assumption that the cardinality of $\mathbb{R}$ is an aleph (i.e., that it can be well-ordered) leads to contradiction. After the Congress ended, a few mathematicians met to discuss König's argument; the group included Cantor, Hilbert, Schoenflies and Hausdorff (see [Schoenflies 1922, 100–01]). Later that year Hausdorff published a paper [1904] casting doubt on Bernstein's lemma and establishing a correct related result. The lemma turns out to be inadequate when α is a limit ordinal, which is the crucial case, and so König's refutation of Cantor's Continuum Hypothesis fails.[2] Even so, König's work was not useless, for his proof can be turned into the result that $2^{\aleph_0} \neq \aleph_\beta$ for any limit ordinal β cofinal with ω.

At any rate, the episode at the International Congress had again focused attention on the Continuum Hypothesis and Well-Ordering. A month and a half after the Congress, Zermelo sent a letter to Hilbert for publication in *Mathematische Annalen*, which presented his proof of the Well-Ordering Theorem. This theorem was a crowning achievement, an essential complement to the elementary theory of transfinite cardinals: all infinite cardinalities (the continuum in particular) are alephs, all powers are comparable. This, of course, plainly contradicted König, who, ironically, had inadvertently employed in his lecture the Axiom of Choice (AC), which was the basis for Zermelo's proof [Moore 1982, 86].

Zermelo was actually the first to present clearly AC and to claim that it is a mathematical axiom:

The preceding proof rests on the assumption that in general there exist coverings γ [see below], that is, on the principle that even for an infinite totality of sets there always exist correlations by which to each set corresponds one of its elements, or formally expressed, that

---

[1] [Kowalewski 1950, 202], who also reports that the newspapers mentioned the news of König's lecture.

[2] Some authors tell the episode differently, following Kowalewski, who in 1950 wrote that Zermelo found the error just the day after the lecture [Moore 1982, 87]. But this is contradicted by another witness who was close to Hilbert and Zermelo, Schoenflies [1922, 100]. Since the only documentary evidence from that early time, Hausdorff's paper, does not mention Zermelo, it may well be that Kowalewski misremembered.

the product of an infinite totality of sets, each of which contains at least one element, is different from zero [the empty set]. Indeed, this logical principle cannot be reduced to a still simpler one, but it is unconsciously used in numerous mathematical deductions. So for example the general validity of the theorem that the number of parts into which a set is divided is less than or equal to the number of its elements, cannot be demonstrated otherwise than by thinking that each one of the parts in question is coordinated with one of its elements.[1]

By the "product" of an infinite totality Zermelo may have meant the Cartesian product, a notion that had not been clearly formulated yet. The idea of employing the axiom for the proof was not original of Zermelo; from the beginning [1904, 139, 141] he acknowledged that it had been suggested to him by a disciple of Hilbert, the analyst Erhard Schmidt.

AC is a prototype of abstract mathematics. It asserts that, given certain sets, another exists which in general we are not at all in a position to define explicitly (otherwise AC would be avoidable). It is a purely existential postulate of the kind that has no role in constructive mathematics. But, in fact, Zermelo was quite right in claiming that the principle had been unconsciously used in many mathematical deductions. Even future critics of AC had previously used it.[2] The earliest cases of implicit but essential use found by Moore [1982, 14–16] are in a theorem of Cantor on sequential continuity of functions [Heine 1872, 183], and a result of Dedekind on modules [1877, 20–21]; one instance belongs to analysis, the other to algebraic number theory. The casual use of sequences of arbitrary choices was very widespread around 1900 in the field of analysis. To name a couple of examples, Borel's proof of the so-called Heine–Borel theorem made implicit (but avoidable) use of AC; and Lebesgue's proof that his measure is countably additive relied essentially on the axiom [Moore 1982, 65, 69–70].

Likewise, several basic results of set theory presuppose the axiom. Such are Cantor's claim that the union of a denumerable family of denumerable sets is denumerable, and his theorem that every infinite set has a denumerable subset; both require the denumerable form of AC [Moore 1982, 9]. Actually, Cantor's set-theoretical work is full of implicit uses of AC [*op.cit.*, 31ff]. A more explicit case was that of Dedekind's theorem in *Zahlen* [1888, 384–86] that if there is an injective mapping from Z_n to Σ for all n, then Σ is infinite.

Dedekind's proof revealed quite clearly one way in which the axiom enters it, and it almost led to the first public discussion of the axiom [Moore 1982, 22–30]. His implicit use of AC was pointed out by a colleague of Peano in Turin, Rodolfo Bettazzi, in 1896; Bettazzi questioned as ill-advised the idea of accepting such a postulate. In this he had been preceded, and probably influenced, by Peano himself, who in 1890 wrote – in the context of a paper on differential equations – that one cannot apply infinitely many times an *arbitrary* rule by which one assigns to a class an individual of this class [*op.cit.*, 76]. But another associate of Peano and Turin mathematician, Burali-Forti, presented a new postulate on which he based a proof

---

[1] [Zermelo 1904, 141], as translated by Bauer-Mengelberg, with some changes.

[2] A detailed analysis of implicit uses prior to 1904 is given in [Moore 1982, chap. 1].

of Dedekind's theorem. The new postulate was a consequence of the denumerable form of AC, but it convinced Bettazzi of the correctness of Dedekind's result and to close this short, early discussion.

The situation was quite different in 1904, because now the axiom had been used to establish a result that many thought implausible. Zermelo's proof implied that there is a well-ordering of $\mathbb{R}$, but most mathematicians were convinced that it is impossible to determine *effectively* such a well-ordering. In the ensuing discussion there was a good measure of confusion as to the role of existence results in mathematics. Actually, this discussion and the whole foundational debate have been a fundamental contribution to the clarification of the difference between abstract and constructive mathematics. While Zermelo's short contribution only occupied three pages, the next volume of *Mathematische Annalen* (vol. 60, 1905) included four papers that polemized against it, and two others that touched on the issue.[1] The polemic papers were signed by Borel, Jourdain, Bernstein and Schoenflies. It was a clear reflection of the controversy that extended throughout Europe, giving rise to heated debates in Germany, England, and France [Moore 1982, chap. 2].

Two noted followers of Cantor objected to Zermelo's proof because they thought it employed principles that led to paradoxes, in particular to the Burali-Forti paradox. The proof applied AC to the power set of any set M in order to get a well-ordering of M. AC amounts to the existence of a "covering" [Belegung] γ that to each non-empty subset $M' \subset M$ assigns a distinguished element $\gamma(M')$. Considering certain well-ordered subsets of M that he called γ-sets, Zermelo defined a well-ordered set L_γ as the union of all γ-sets.[2] Finally he showed that $L_\gamma = M$: he reasoned that L_γ is clearly a subset of M, and, if they were not equal, $M \setminus L_\gamma$ would have a distinguished element m', so that $L_\gamma \cup \{m'\}$ would be a γ-set but not included in L_γ, which is absurd [Zermelo 1904]. The final step, in which L_γ is extended by a new element, raised the suspicions of Bernstein and Schoenflies, since the addition of a new element to Ω had given rise to Burali-Forti's paradox. But the new element in Zermelo's reasoning is already in M.[3] At any rate, in 1908 Zermelo presented a new proof of Well-Ordering that dispensed with that method (see §4.1). Jourdain claimed to have proved the result earlier and in a simpler way. He had used certain principles to establish that every set has an aleph as its cardinality. Zermelo would later [1908a] analyze obscure points in that proof and indicate that Jourdain's principles, which allow Ω, are not sufficient to show that $\mathbb{R}$ is a set.

---

[1] König's final version of the lecture given the year before, and a paper by Georg Hamel on real functions. Only Hamel took the position of openly accepting the axiom.

[2] Zermelo thus relied on the existence of implicitly defined well-ordered subsets of M. This peculiar approach may be quite consistent with Cantor's conception of sets as given with an ordering (he regarded pure sets as obtained by abstraction from the nature *and ordering* of the elements of a given set).

[3] Their wrong appreciation of the situation was due to the fact that Bernstein and Schoenflies wished to accept the class Ω of all ordinals as a set, so they had to restrict the 'extension' of sets.

Finally, Borel objected to the axiom itself. His paper had been requested by Hilbert as editor of the *Annalen*, and it now appears as the most important of the four. According to Borel, Zermelo had shown the equivalence of the problem of Well-Ordering and the problem of choosing a distinguished element from each subset of *M*. But he had not advanced a single step toward solving the second question, which seemed to him a most difficult one in cases such as that of the continuum [Moore 1982, 93]. In his view, any argument that assumes uncountably many arbitrary choices was outside the domain of mathematics [*op.cit.*, 85]. His skepticism was due to the fact that Borel was asking for an effective definition of a well-ordering of $\mathbb{R}$. His paper circulated among several first-rate French mathematicians and originated a very interesting discussion. The letters between Hadamard, Borel, Baire and Lebesgue were published in that same year [Hadamard *et al.* 1905].

Borel, Baire and Lebesgue were distinguished French mathematicians who worked all on real functions. In their work they made extensive use of notions of point-set theory, along the lines of Jordan (§1), and they relied implicitly on AC or used results that were dependent on AC [Moore 1982, 64–70]. But when it came to confront the axiom directly, their affinity to constructivism became clear and they objected to it.[1] The key idea is that a mathematical notion (e.g., a function or a set) does not truly *exist* unless it has been finitely defined by means of characteristic properties, in the sense that one has determined a *rule* which allows its explicit construction. When the notion embraces only finitely many cases, one can sidestep this requirement and think that it is *in principle* possible to comply with it, but the situation changes essentially in the infinite case. As we see, the three French mathematicians were close to the viewpoint of Kronecker, as comes out explicitly in the following letter of Lebesgue:

if we wish to regard Zermelo's argument as completely general, it must be granted that we are speaking about an infinity of choices whose power may be very large; furthermore, no law is given for this infinity, no law for any of the choices. We do not know if it is possible to name a rule defining a set of choices having the power of the set of the [subsets] *M*'; we do not know if it is possible, given an *M*', to name a [distinguished element] *m*'.

In sum, when I scrutinize Zermelo's argument, I find it, like many other general arguments about sets, too little Kroneckerian to have meaning (of course, only as an existence theorem ...).[2]

Lebesgue himself noted that in his doctoral thesis he had proved the existence of a measurable set that is not Borel-measurable, although he continued to doubt that any such set can be named. Under these conditions, it was not legitimate to base an argument on the assumption that such a set is given.

---

[1] French mathematics had been less prone to abstraction than German mathematics throughout the 19th century.

[2] [Hadamard *et al.* 1905, 267], as translated in [Moore 1982, appendix, 316] with minor changes.

Of all them, Hadamard was the only decided partisan of abstract mathematics and set theory. It seemed totally correct to him, and useful, to speak of the *existence* of an object without being able to name it explicitly. He wrote:

there are two conceptions of mathematics, two mentalities, in evidence. After all that has been said up to this point, I do not see any reason for changing mine. I do not mean to impose it. ...
   I believe that in essence the debate is the same as the one which arose between Riemann and his predecessors over the notion of function. The *rule* that Lebesgue demands appears to me to resemble closely the analytic expression on which Riemann's adversaries insisted so strongly. And even an analytic expression that is not too unusual. [Footnote:] It seems to me that the truly essential progress in mathematics, from the very invention of the infinitesimal Calculus, has resulted from successively annexing notions which, some for the Greeks, some for the Renaissance geometers or the predecessors of Riemann, were 'outside mathematics' because it was impossible to describe them.[1]

In later years, the abstract approach won the field, not least due to the influence of Göttingen and Hilbert's authority, in spite of strong criticism on the side of intuitionists and constructivists, which led to frictions and even ruptures in the 1920s [Mehrtens 1990; van Dalen 1995].
   To Hilbert, 'existence' simply meant non-contradictoriness: whenever an axiom system is consistent, we are entitled to regard the set of objects it describes as existing. One is tempted to translate this into the slogan – mathematical existence is nothing but logical possibility. In the end, the mathematical community has come to acknowledge, more or less consciously, that the two approaches – abstract and constructivist – offer valuable results and constitute important parts of mathematics. Hadamard's 'two mentalities' have come to coexist peacefully and complement each other.
   One must add that the foundational debate in the 1900s and later was considerably confuse and involved, because several different issues coalesced around it. All kinds of arguments and positions were exposed by the participants. Poincaré, for example, was primarily intent on refuting logicism and showing that some kind of intuition is an essential element in mathematics [Poincaré 1905/06, Goldfarb 1988]; he made a key point of the paradoxes or 'antinomies' (§2). The reader should take into account that many other secondary figures entered the debate, making it considerably difficult for anybody to reach a conclusion and adopt a coherent and well-argued standpoint. The issues under debate included the paradoxes and the proper conception of logic, the role of natural vs. formal language, and the proper conception of mathematics, including the role of existence results and constructive methods in it. Different authors assigned quite different weights to each of them. Zermelo came to the conclusion that it was urgent to axiomatize set theory in order to clarify the situation.

---

[1] [Hadamard *et al.* 1905, 270], as translated in [Moore 1982, appendix, 318] with minor changes.

4. The Early Work of Zermelo

Although some elements of modern axiomatic set theory only emerged in the 1920s and 1930s, one can safely say that Zermelo's work represents the coming of age of the theory. The elements of the Cantorian edifice are rounded off, the viewpoints of Cantor and Dedekind are intertwined, and a rather precise system of axioms is proposed that allows recovery of their main results without contradictions. A good starting point for overcoming the foundational crisis is thus found.

Ernst Zermelo went to Göttingen in 1897, after studying mainly at Berlin and working three years as an assistant to Planck at the Institute of Theoretical Physics.[1] His early reputation was as an expert in applied mathematics and theoretical physics, but then he fell under the influence of Hilbert, who became a most important support for his career and brought changes in his field of activity:

Thirty years ago, when I was a *Privatdozent* at Göttingen, I came under the influence of D. Hilbert, to whom I am certainly the most indebted for my scientific development, and I began to occupy myself with the foundational questions of mathematics, especially with the fundamental problems of Cantorian *set theory*, whose full significance became conscious to me only then, through the extremely fruitful collaboration of the Göttingen mathematicians.[2]

In 1900/01 Zermelo gave a lecture course on set theory, following closely Cantor's *Beiträge*, by which time he independently discovered the Russell paradox. He seems to have viewed it as showing merely that any set which contains all of its subsets as elements is self-contradictory; the set of all sets is an example [Rang & Thomas 1981]. In 1902 he published a paper on the addition of transfinite cardinals, and the following year he discussed Frege's theory of number, in comparison with those of Dedekind and Cantor, before the Göttingen Mathematical Society [Moore 1982, 89–90]. But it seems clear that the main motivation for his later concentration on foundational issues was the controversy generated by his proof of the Well-Ordering theorem.

To confront the critics, Zermelo thought it necessary to make explicit an axiomatic framework that would be sufficient to rescue Cantorian set theory and derive his own theorem, and that at the same time avoided the known paradoxes. From about 1905 he focused on this kind of question and quickly became convinced that he could overcome all criticisms [Peckhaus 1990, 29–30]. Until then, Zermelo had based his work mainly on Cantor's, as is clearly visible from his 1904 proof (§3), e.g. in his use of the notion of 'covering.' After 1905 he developed an interest in

---

[1] For biographical data see [Peckhaus 1990, 77ff].

[2] [Moore 1980, 130; Peckhaus 1990, 82]: "Schon vor 30 Jahren, als ich Privatdozent in Göttingen war, begann ich unter dem Einflusse D. Hilberts, dem ich überhaupt das meiste in meiner wissenschaftlichen Entwickelung zu verdanken habe, mich mit den Grundlagenfragen der Mathematik zu beschäftigen, insbesondere aber mit den grundlegenden Problemen der Cantorschen *Mengenlehre*, die mir in der damals so fruchtbaren Zusammenarbeit der Göttinger Mathematiker erst in ihrer vollen Bedeutung zum Bewusstsein kamen."

the foundations of number and studied in detail Dedekind's work.[1] This is clearly visible in his papers of 1908, which make explicit reference to Dedekind and employ his notions and results. The papers were written in 1907, with just a few days distance, and they are closely linked to each other [Moore 1978]. As Moore has emphasized, it is quite clear that the paradoxes were not his main motivation for undertaking the work, although of course he had to confront them. Zermelo's axiomatization was a rather decisive solution to the problems, but it took a long time before this came to be generally acknowledged.

4.1. Defense of the Well-Ordering theorem.   Zermelo's theorem was a very important result that complemented the elementary theory of transfinite numbers, showing that all infinite cardinalities are alephs. But it found a negative reception, at least among the authors who decided to publish their views on the subject. Only Hadamard clearly favored it, and it was certainly strong support since Hadamard was one of the leaders of French mathematics. But his most influential compatriot, Poincaré [1905/06], criticized the proof in a subtle way. Russell, who found a form of AC independently in 1904 [Grattan-Guinness 1972, 107], remained skeptical of it. Fortunately for Zermelo, Hilbert himself stood on his side on the matter, but only privately or in letters.[2] It was thus imperative to come up again with a defense of the axiom and the proof, as Zermelo did in 1908.

To well-order a set M by means of AC, as Erhardt Schmidt suggested, one needs to apply the axiom to the power set $\wp(M)$. This step is employed in both of Zermelo's proofs, but the second was simpler in that it avoided bringing into the picture well-ordered subsets of M. It defined explicitly a well-ordering of M on the basis of AC and $\wp(M)$, using only simple set-operations. Moreover, its axiomatic assumptions were clearly laid out.

The Axiom of Choice implies that, for each set M, there exists a choice function, an arbitrary (non-injective) mapping Θ: $\wp(M) \to M$, such that $\Theta(S) \in S$ for non-empty subsets S of M. *Intuitively*, it is easy to see that, on the basis of that mapping, one should be able to define a well-ordering of M. We start with $\Theta(M) = m_0$; now, we consider $S_1 = M \setminus \{m_0\}$, and take $\Theta(S_1) = m_1$; we proceed to $S_2 = S_1 \setminus \{m_1\}$, taking $\Theta(S_2) = m_2$; and so on until we exhaust M. Since by definition $\Theta(S) \in S$, the elements m_0, m_1, ... are different from each other and we obtain a well-ordering. But the difficulty lies in avoiding the imprecise traits of this sketch of a proof.

First, it is unclear whether we can *successively* 'choose' the required elements; Zermelo replaces that imprecise notion by an abstract postulate that implies the existence of a simultaneous choice for $\wp(M)$. The word 'choice' itself is employed merely because it suggests the above intuitive notion, but it does not really convey what is going on. Second, it is necessary to define abstractly all of the subsets of M

---

[1] For further details on this point, see [Peckhaus 1991, 90–97], which offers numerous quotations from Zermelo's manuscripts and letters to Hilbert.

[2] See his 1905 letter to Hurwitz, in [Dugac 1976, 271] or [Moore 1982, 109].

that will be needed, without making appeal to the transfinite ordinals. Otherwise one might have to resort to the class of ordinals Ω, as Cantor had done, which would throw doubts regarding the reliability of the proof. Zermelo's proof avoids employing 'inconsistent sets' or any other dubious notion.

Zermelo found the key for solving the second point in Dedekind's notion of chain, which he generalized to the transfinite case by using a customary method of definition. This kind of approach had already been taken by Hessenberg in an important paper on the 'Basic Notions of Set Theory' [1906] that included the first careful development of the theory of ordered sets. Zermelo wrote that his procedure was modeled upon Dedekind's theory of chains, and for the rest was customary in set theory [Zermelo 1908a, 190]. He could not use Dedekind's chain theory directly, since this was restricted to sets of type ω. (The well-ordering of a transfinite set M cannot be described by means of a mapping $\varphi: M \to M$. Some elements of M will play the role of limit ordinals (ω, $\omega + \omega$, ...) and they will lack an immediate predecessor; thus such 'limit elements' cannot be characterized as the images of a preceding element.)

Zermelo sidestepped the problem by means of a clever reconceptualization of the notion of order. Determining a total order among the elements of a set amounts to the same as associating to each one of the elements m a remainder R_m, i.e., the set of all its successors. Now, the order among the elements m can be characterized in terms of the generalized chain of their remainders R_m, which enables one to apply the viewpoint directly to the transfinite case by using the following definition [Zermelo 1908a, 184–85].[1] A subset K of $\wp(M)$ (i.e., a set of subsets of M) is a "Θ-chain" if and only if

(a) if $S \in K$, then $S' = S \setminus \Theta(S)$ also belongs to K, and
(b) for every subset $A = \{S, T, ...\} \subset K$, its intersection $\cap A \in K$.

Condition (a) guarantees the step from S to S', so that we reach any S^n but not S^ω; condition (b) is the key to ensure such transfinite steps to 'limit elements.' Thus, Zermelo's generalized chains not only resemble Dedekind's theory, but their definition employs the characteristic idea that Cantor used for defining, e.g., derived sets of limit order. This is a graphic example of how the theories of both mathematicians were synthesized by Zermelo.

Analogously to Dedekind's definition of the chain of a set, Zermelo now considers the intersection of all Θ-chains K to which M belongs, which is a new Θ-chain **M** ($= \Theta_0(M)$, to use Dedekind's notation). **M** is the set that we needed to define a well-ordering of M. Without getting too much into details, it will be clear that, in virtue of the definition of **M**, we have a one-to-one mapping $\Theta: \mathbf{M} \to M$.

---

[1] I introduce a minor modification in Zermelo's definition to make it strictly parallel to Dedekind's. This makes it more general, since Zermelo included right away a condition (c) that $M \in K$.

Every element $m \in M$ is the image of a set in **M** that can be regarded as its remainder R_m. And **M** is well-ordered by reverse inclusion, $\supset$, its first element being M. Thus the one-to-one mapping Θ induces a well-ordering on M.

Zermelo's new proof was regarded as classical. A clear exposition of it can be found in Hausdorff's great handbook [1914, 136–38]. Hausdorff had actually been one of the few mathematicians who employed AC for obtaining original results in his work on order types (see §6).[1] Yet the most important early application occurred in Steinitz' pathbreaking work on abstract field theory [1910]. Steinitz' work was of fundamental importance for future research on 'modern algebra,' and he made explicit use of AC in order to prove, among other things, that any commutative field has a unique algebraic closure (up to isomorphism). Discussing this in the introduction to his article, he wrote:

Many mathematicians sill stand opposed to the Axiom of Choice. With the increasing recognition *that there are questions in mathematics which cannot be decided without this axiom*, the resistance to it must increasingly disappear. On the other hand, in the interest of purity of method it seems expedient to avoid the above-named axiom in so far as the nature of the question does not require its use.[2]

Here he proposed an attitude that was also being followed by Russell and Zermelo himself: the axiom was avoided whenever possible, and the dependence of other results on it was carefully investigated and clearly stated. This kind of approach was carried on with particular interest by Sierpiński, who published in 1918 a lengthy survey on the role of AC in set theory and analysis [Moore 1982, chap 4]. With his work, mathematicians started to become aware of how deeply analysis, and particularly the work of the French school, depended on the axiom.

4.2. Axiomatization of "the theory created by Cantor and Dedekind." By 1906 Zermelo had established the general plan for his axiomatization, which was published in 1908. The programmatic statement that opens his paper is noteworthy:

In the present paper I intend to show that the whole theory created by G. Cantor and R. Dedekind can be reduced to a few definitions and seven 'principles' or 'axioms,' apparently independent among themselves.[3]

It turned out to be a fundamental contribution, the basis for what has been the generally accepted approach to set theory since the 1920s, particularly in the second half of the 20th century. The axiom system was built with the aims of reconstructing Zermelo's own proof of Well-Ordering, Dedekind's theory of finite sets and natural numbers, and Cantor's theory of transfinite sets, their cardinalities and order

---

[1] For other cases in different areas, like Hamel and Vitali, see [Moore 1982, 100–01, 112].

[2] [Steinitz 1910, 170–71], as translated in [Moore 1982, 172].

[3] [Zermelo 1908], as translated by Stefan Bauer-Mengelberg in [van Heijenoort 1967, 200].

types (see [1908, 201; 1908a, 189]). In the years 1908 and 1909 Zermelo published articles that showed how the system was well adapted for these purposes. His contributions synthesized the work of Cantor and Dedekind both at the level of goals and methods, and started developing their heritage.

[Zermelo 1908] starts by presenting set theory as the branch of mathematics that investigates the fundamental notions of number, order, and function, thus developing the "logical foundations" of all of arithmetic and analysis.[1] But the existence of this discipline seems threatened by certain contradictions that "can be derived from its principles – principles necessarily governing our thinking, it seems." The "Russell antinomy" of the set of all sets that do not belong to themselves has the consequence that

it no longer seems admissible today to assign to an arbitrary logically definable notion a set, or class, as its extension.[2]

The author is thus very explicit in locating the key responsible for the emergence of contradictions in the traditional connection between concepts and sets – the principle of comprehension, that played such an important role in the first 50 years of development of the set-theoretic viewpoint.

For this reason, Zermelo goes on, Cantor's definition of set has to be restricted, but it has not been possible to replace it by another definition that is as simple.[3]

Under these circumstances there is at this point nothing left for us to do but to proceed in the opposite direction and, starting from set theory as it is historically given, to seek out the principles required for establishing the foundations of this mathematical discipline. [1908, 200]

Axiomatization was assuming a key methodological role in mathematics, for instance with Peano, and most influentially with Hilbert and his circle. Taking this lead, Zermelo looked for a system of principles sufficiently restricted to exclude all contradictions and, on the other hand, sufficiently wide to retain all that is valuable in set theory. His paper intends to show that the entire theory of Cantor and Dedekind "can be reduced to a few definitions and seven principles, or axioms."

In analogy with chapter I of Hilbert's *Grundlagen* [1930] (first edn. 1899, second 1903), Zermelo postulates a "*domain* $\mathfrak{B}$ of individuals" among which are the sets and also urelements (non-set individuals). There is only one "*fundamental relation*" that is specific of set theory: membership, denoted by 'ε.' An object b of the domain is called a set if and (except for the empty set) only if it has an element

---

[1] The expression 'logical foundations' is ambivalent, since it can be taken (or not) to mean that the foundations of mathematics are purely logical.

[2] [Zermelo 1908], as translated in [van Heijenoort 1967, 200].

[3] As we have seen (§VIII.8) Cantor's intention had been to exclude 'inconsistent sets' by emphasizing the 'collection *into a whole*' of the elements. But in the absence of a more detailed explanation and development, his attempt was not even noticed by Zermelo and others.

a, so that *a* ε *b* [1908, 201]. The axioms postulate some conditions that must be valid for the fundamental relations among objects of the domain:[1]

I. Axiom of "Determinacy" [Bestimmtheit]. If every element of *M* is an element of *N* and vice versa, then *M* = *N*. Every set is fully determined by its elements.[2]

II. Axiom of Elementary Sets [Elementarmengen]. There is a (fictitious) set, the empty set, which Zermelo denotes '0.' Given any two objects of the domain *a*, *b*, there exist the sets {*a*} and {*a*, *b*}.

III. Axiom of Separation [Aussonderung]. "Whenever the class-statement[3] $\mathfrak{E}(x)$ is definite for all elements of a set *M*, *M* possesses a subset $M_{\mathfrak{E}}$ containing as elements precisely those elements *x* of *M* for which $\mathfrak{E}(x)$ is true."

IV. Axiom of the Power Set [Potenzmenge]. To every set *T* there corresponds another set $\mathfrak{U}T$ [$\wp(T)$], called the power set of *T*, that contains as elements precisely all subsets of *T*.

V. Axiom of Union [Vereinigung]. To every set *T* there corresponds another set $\mathfrak{S}T$ [$\cup T$], called the union of *T*, whose elements are precisely all elements of elements of *T*.

VI. Axiom of Choice [Auswahl].[4] "If *T* is a set whose elements all are sets that are different from 0 [∅] and mutually disjoint, its union $\mathfrak{S}T$ includes at least one subset S_1 having one and only one element in common with each element of *T*."

VII. Axiom of Infinity [des Unendlichen].[5] There is in the domain at least one set *Z* such that ∅ ε *Z* and that is so constituted that if *a* ε *Z*, then {*a*} ε *Z*.

A particular difficulty is posed by Axiom III, which Zermelo presents as "in a sense" furnishing an adequate substitute for the general definition of set, by which he probably means the principle of comprehension. It embodies Zermelo's response to the different paradoxes, where, following Hessenberg [1906, chap. 23 and 24], he distinguishes the purely set-theoretic ["ultrafinite"] paradoxes from those that have to do with definability and the like [Zermelo 1908, 202]. The set-theoretic paradoxes are avoided by the expedient that Axiom III can never be used to define a set *independently*; it only serves to define a subset of a previously given set. This

---

[1] [Zermelo 1908, 201–04], as translated by S. Bauer-Mengelberg in [van Heijenoort 1967].

[2] This is the principle of extensionality, which had been indicated by Dedekind and, perhaps not so clearly, by Cantor.

[3] By 'class-statement' Zermelo means a logical condition in one variable, i.e., what Russell was calling a propositional function.

[4] The axiom is formulated for a family of disjoint sets in order to make it simple and more intuitive.

[5] Zermelo indicates that this axiom is esentially due to Dedekind. Indeed, he simply postulates a set that is infinite according to Dedekind's definition, the relevant mapping being *a* → {*a*}.

excludes the 'set of all sets' and the 'set of all ordinal numbers.' In fact, the principle of comprehension can be interpreted as Axiom III applied to the assumption of a 'universal set.' But Zermelo turns the Russell paradox into proof by *reductio ad absurdum* that such a set does not exist. This means that "*the domain* $\mathfrak{B}$ *is not itself a set*," which disposes of the "Russell antinomy" as far as Zermelo is concerned [1908, 203].

The solution to the second kind of paradoxes depends on the introduction of the notion of a "definite" assertion or class-statement. He writes:

A question or assertion $\mathfrak{E}$ is said to be *definite* if the fundamental relations of the domain, by means of the axioms and the universally valid laws of logic, determine without arbitrariness whether it holds or not. Likewise a "class-statement" $\mathfrak{E}(x)$, in which the variable term x ranges over all individuals of a class $\mathfrak{K}$, is said to be definite if it is definite for *each single* individual x of the class $\mathfrak{K}$.[1]

There is a certain amount of ambiguity in this definition, which enters when Zermelo refers to the 'universally valid laws of logic' and with his reference to a 'class.' As he would remark years later [1929, 340], at the time there was no generally accepted system of logic, so he could not base this part of the system on readily available work. Nor could he have made an explicit proposal of his own without entering into a full discussion of mathematical logic. His solution was pragmatic, but in my opinion a very clever and adequate one. The path to be followed was clearly indicated: he suggests that acceptable statements are those built from notions of mathematical logic and the fundamental relation 'ε' (or at the most new notions defined from these); and he calls such statements "definite" if and only if logical laws and the set-theoretic axioms, taken together, determine without arbitrariness whether the statement is true or false. The issue would be further clarified with the subsequent development of logical theory (see §§X.5, XI.1 and XI.5).

As Hessenberg noted [1909, 90], Axioms II and VII establish the simplest cases of finite and infinite sets, while IV and V allow us to ascend to all of the finite and infinite cardinalities.[2] Axioms IV, VI, and VII have a purely existential character that is noteworthy. The Axiom of Infinity did not give rise to polemics (except for strict constructivists) because it was so deeply ingrained in the traditional orientation of mathematics, for instance in analysis. The Power Set Axiom is an extremely powerful instrument, but only Baire took the step of denouncing it [Hadamard *et al.* 1905, 264; Moore 1982, 313]. This is probably because it seems intuitive enough as an assumption about sets.

While presenting the axioms, Zermelo proceeded to establish some simple consequences of them, and then he went on to develop in full detail the "theory of equivalence" [1908, 205–15]. Axiomatization normally implies some measure of

---

[1] [Zermelo 1908], as translated in [van Heijenoort 1967, 201], with a small change to accommodate the word 'Klassenaussage.'

[2] This is not quite true, for $\aleph_\omega$ (not to mention large cardinals) cannot yet be reached, see §XI.1.

artificiality. Zermelo was not able to define in a natural way the notion of mapping, nor Cantor's cardinal numbers. He decided to dispense with Cantor's numbers completely and to work directly with abstract sets. This was a high price to pay for axiomatic security, and in the 1920s von Neumann showed how to recover a certain degree of naturalness in the axiomatic setting.[1] As regards the other notion, Zermelo lacked ordered pairs, since means for defining the ordered pair within pure set theory were not yet known. But he was able to define a [Cartesian] "product" of a family of sets [1908, 204] and to find a partial substitute for mappings, one that worked only with disjoint sets [*op.cit.*, 205].[2]

Despite these inconveniences, his treatment of cardinal equivalence was masterly. He established the Cantor–Bernstein equivalence theorem, essentially as Dedekind had done twenty years earlier [*op.cit.*, 208–09], proved the Cantor Theorem that every set is of lower cardinality than its power set [*op.cit.*, 211–12], and demonstrated a very general theorem on cardinalities to which Fraenkel gave Zermelo's name.[3] Finally, he proved that the number sequence, which he had previously defined, is infinite, and (on the basis of AC) that every infinite set contains a denumerably infinite subset [*op.cit.*, 214–15].

Zermelo planned to publish a sequel to his axiomatization, developing the theory of well-ordered sets and its application to finite sets and the principles of arithmetic [1908, 201]. He did not come to publish this work, except for the part that had to do with finite sets and mathematical induction. He treated the topic in [1909] polemizing with Poincaré; this constituted his version of Dedekind's theory. As for well-ordered sets, his second proof [1908a] of Well-Ordering contained enough to suggest how he would have treated the subject, and there was also the previous work of Hessenberg [1906]. Thus, he could be more than reasonably confident that his axiom system was sufficient for a development of all the essentials of abstract set theory.

It is certainly true that Zermelo's motivation for axiomatizing set theory came mainly from the polemics surrounding his 1904 theorem, and that his axiom system was in good measure the outcome of analyzing the postulates needed to frame his proof of Well-Ordering [Moore 1982, 142–60]. But one should not overemphasize that, at least not to the extent of forgetting that Zermelo analyzed extensively the work of Cantor and Dedekind (at least since 1900 and 1905, respectively). His axiom system was not a hodgepodge of principles extracted from a single proof and gathered without any clear underlying conception. They could well involve some measure of arbitrariness, but they were the result of careful analysis of "set theory at it [was] historically given." Thus it is not surprising that they have fared so well in the subsequent development of the theory.

---

[1] Around 1915, Zermelo himself was also working on an axiomatic definition of the ordinal numbers, see §XI.2.

[2] He was thus forced to prove that, given sets M and N, there exists another set M' equivalent to M and disjoint from N [*op.cit.*, 206].

[3] This [*op.cit.*, 212] was a generalization of König's inequality, as Zermelo went on to notice. Of course, he formulated it for abstract sets, not for cardinal numbers.

5. Russell's Theory of Types

Russell had been one of the authors who formulated most clearly the 'naive' theory of sets or classes, i.e., the theory based on the principle of comprehension. This remained for him an inevitable component of any possible theory of classes, as we shall see. Likewise, he remained convinced that common sense leads us to assume the existence of a universal class, and of a complement for any existing class.[1] But the paradoxes had shown that common sense is contradictory – "common sense is bankrupt, for it wound up in contradiction," wrote Quine [1941, 153]. It was obligatory to give up some common-sense logical hypothesis [Russell 1903, §105]. Even so, Russell remained optimistic as to the prospects of logicism, and so he looked for a solution that would save as much as possible from the Fregean approach (see [Whitehead & Russell 1910, viii]). In essence, his solution was to cling to the principle of comprehension, restricting it by the severe conditions of type theory.

5.1. The way to type theory.   Appendix B to Russell's *Principles* [1903] contained the first exposition of the doctrine of types, as an attempt to solve the contradictions. As a matter of fact, one can find here *simple* type theory, but in a very rough version. As Russell presents the idea, every propositional function $\varphi(x)$ has a range of meaning, i.e., a "type," understood as the class of all objects for which $\varphi(x)$ is a meaningful (true or false) proposition [1903, §497]. Thus we have a type of individuals, a type of classes of individuals, a type of classes of classes of individuals, and so on. But several traits differentiate this early version from the later, mature one. First, and most important, we find here a conflation of types with classes, which is dangerous for the theory. As a matter of fact, Russell was willing to accept a "range of all ranges" or type of all types, but this brings new paradoxes into the picture [*op.cit.*, §§498–500]. Apparently, it was only much later that he finally gave up the idea of universal class.[2]

Second, the theory is complicated because Russell thinks that a progression of types starts with each different kind of object [1903, §§497–98], and back in 1903 he regarded individuals, ordered pairs, propositions, and numbers (at least) as different kinds of objects. Even if we reduce this proliferation of objects to the usual ones among logicians, individuals, classes and relations, the theory remains complex. From ordered pairs we proceed to classes of relations (extensional relations), but also to relations of relations, etc. Therefore, instead of a simple hierarchy, we have a branched tree of types. This problem was to remain so long as relations were not reduced to sets (§X.3.1); it can still be found in *Principia Mathematica*. But the

---

[1] Thus, Boolean algebra should apply to the whole universe of classes, not just to the subsets of a given set.

[2] Perhaps his unwillingness to do so was one of the reasons why he abandoned the theory of types for a few years.

theory in *PM* is much more complex yet, for after 1906 Russell adopted the vicious circle principle as his basic guiding principle in response to the paradoxes of all kinds. This led to *ramified* type theory.

In [1906], written late in the previous year, Russell returned to the problem of the paradoxes, from which he concluded that a propositional function of one variable does not determine a class. The problem was now to determine when such a function determines a class, when, as he said, it is "predicative." He proposed three different possible approaches, none of which coincides with type theory. First, what he called a "zig-zag theory," which would admit as classes the extensions of some "fairly simple" propositional functions, but not of "complicated and recondite" ones. This would preserve two traits of the naive theory of classes that Russell regarded as natural: there would be a class of all classes, and each class would have a complement. But he found it difficult to implement this idea, in particular to characterize the 'sufficiently simple' propositional functions.[1] A second approach would be a theory of "limitation of size," which would ban those classes that are 'too big,' particularly the "class of all entities." Russell mentioned a previous attempt in this direction by Phillip Jourdain. This would make it possible to preserve much of Cantor's work, but Russell found it imprecise, since it was unclear where exactly to put the limit, how far up the series of ordinals it is legitimate to go. (Zermelo set theory has frequently been called a theory of limitation of size, even by important set-theorists like Fraenkel and Bernays; in my opinion, however, that characterization is not very apt.[2]) Finally, there was a third approach, the most radical: a "no-classes theory," where classes and relations are "banished altogether" and one operates directly with propositional functions. Russell believed that this would require abandoning much of Cantor, but he preferred it because it seemed the most secure way out of the paradoxes [1906, 45–57].

In September 1906 Russell returned again to the theory of types, which he combined with elements from the no-classes view, but particularly with a new basic idea that he regarded as the key to solving the paradoxes – the vicious circle principle. This novelty seems to have been an outcome of Poincaré's debate with Couturat and Russell, among others, on the foundations of mathematics and the paradoxes. Poincaré [1905/06] was above all interested in questioning the logicist program, affirming the key role of intuition in mathematics, and vindicating Kant's philosophy of mathematics.[3] But in the course of his work he proposed notions that would play an extremely important technical role in logic. This happens in the third part of his 'Mathematics and Logic,' published in 1906, precisely in response to

---

[1] Some authors have argued that Quine's systems *NF* and *ML* are closest to being a zig-zag theory (see [Fraenkel, Bar-Hillel & Levy 1973; Ullian 1986; Wang 1986]. For a discussion of Quine's systems in the historical context of set theory and logicism, see [Ferreirós 1997].

[2] The *ZF* system seems to be compatible with postulating sizes as big as one wishes. The view that the system implements a 'limitation of size' will only appeal, it seems, to those who regard the universal class as natural, and who are foreign to the notion of the cumulative hierarchy.

[3] On this topic, see [Goldfarb 1988].

Russell's paper [1906] mentioned above. Poincaré focuses on the "Cantorian antinomies," of which he emphasizes those of Burali-Forti, Zermelo–König, and Richard. After discussing Russell's three ways out of the paradoxes, he proposes that the Richard paradox suggests the solution to all of the problems [Poincaré 1905/06, 1063]. As Richard himself had indicated, the paradox was due to a vicious circle: once we take as given the class E of finitely definable decimals, it becomes possible to define a new decimal finitely. But this new number should not be taken to belong to E; in later terminology, it is of a different order than the elements of E. Thus, he concludes, "*the definitions which ought to be regarded as non-predicative are those which contain a vicious circle*" [*ibid.*]. Poincaré means, in Russell's terminology, that a propositional function which contains a vicious circle does not determine a class. He immediately goes on to apply the principle to block the Burali-Forti paradox, but also the logicist definition of finite number and Zermelo's proof of Well-Ordering [*op.cit.*, 1063–69].[1]

In a reply article, published in French that same year, Russell accepted Poincaré's diagnosis [1906a]. He went on to reform logical theory accordingly, and the new type-theoretic system was presented in [1908]. The paradoxes are *all* solved by noting that they share the characteristic of self-reference or reflexiveness. In each contradiction something is said about all cases of some kind, and from what is said a new case seems to be generated, which engenders a contradiction because it both is and is not a member of the totality. But the vicious circle principle postulates that the assumption of a class is illegitimate if it automatically leads to new members defined in terms of itself:

'Whatever involves all of a collection must not be one of the collection,' or conversely: 'If, provided a certain collection had a total, it would have members only definable in terms of that total, then the said collection has no total.' [Russell 1908, 155]

5.2. Types and orders. Russell's formulation of the vicious circle principle is somewhat loose, but the principle becomes clearer later, when Russell describes his logical system. He explains that the type of an expression is determined by the variables contained in it: the expression must be of a higher type than the possible values of those variables [1908, 163]. There must be no propositions of the form $\phi(x)$, in which x has a value which involves ϕ [Whitehead & Russell 1910, 40]; an expression like '$\phi(\{x : \phi(x)\})$' is always meaningless.

Russell and Whitehead were thus led to speak of different *types* associated to propositional functions. They obtained not just a hierarchy of types, but a tree with infinitely many branches at each level. This is because the type of a propositional function $\phi(x)$ depends both on the type of its arguments – the possible values of x –,

---

[1] The latter is due to the fact that the last step in Zermelo's proof considered γ-sets defined in reference to L_γ, a vicious circle (see §3). The 1908 proof uses an impredicative definition, too.

and the type of the objects in the range of its bound variables [Russell 1908, 164–65; Whitehead & Russell 1910, 48–53]. To give an example, the propositional function

$$\exists \varphi \; \forall y \; F(\varphi(y), x)$$

has only the free variable x, an individual variable. We might thus think that it is of a second type, next above that of individuals, but the vicious circle principle forces us to pay attention to quantified variables too. In this way we notice that there are two bound variables, φ and y, of which φ ranges over propositional functions. Thus, the example in question must be at least a propositional function of the *second order* in Russell's terminology.

The theory was presented in a way that is somewhat unclear, for the connection between types and orders could have been explained more straightforwardly [Whitehead & Russell 1910, 37–55, 133, 161–67]. But we can think of it as two superimposed hierarchies, the hierarchy of types and the hierarchy of orders, the latter being the one that generates the 'ramification' (in later terminology). A *type* is defined as the range of significance of some function. Russell assumes that there are always values of x for which $\phi(x)$ is not just false, but meaningless. The type of ϕ is formed by the arguments with which $\phi(x)$ becomes meaningful and has values. Types must either coincide or be mutually exclusive [*op.cit.*, 161]. On the other hand, a function is of the *first order* if it involves no variables except individual variables; it is of the $(n+1)th$ order if it has at least one argument or bound variable of order n, and none of a higher order [*op.cit.*, 167]. One can only quantify over variables of some specified type and order, and if we quantify over variables of order m, the resulting expression is of order $m+1$.

Thus, we can never legitimately speak of 'all properties of a,' which creates great difficulties in connection with mathematics. To give a couple of examples, we may define the 'real numbers' as objects of a certain type and order m, and consider a set S of them. If we now go on to, e.g., the greatest lower bound of S, it will be given as an object of order $m+1$, which does not belong to the 'real numbers' as previously defined. Classical analysis will become impossible. The same happens even with mathematical induction, for it can only be formulated for number-properties of a certain order, and thus induction will not be valid for properties of a higher order [Russell 1908, 167]. Clearly one must either abandon classical mathematics[1] or introduce some new assumption that makes it possible to recover the lost ground. This was the purpose of the infamous Axiom of Reducibility, which could only seem puzzling to readers.

Russell calls a propositional function *predicative* if it is of the lowest order compatible with that of its argument [Whitehead & Russell 1910, 53]. Thus, a function of individuals that has no bound variables, except perhaps individual variables, is a predicative function. Similarly, a function of functions of the mth order that has bound variables of at most the mth order, is a predicative function. Sym-

---

[1] As Weyl did in *Das Kontinuum* [1918], developing a predicative alternative to set theory.

bolically, one uses '$\phi(x)$' for propositional functions whose order has not been specified, and '$\phi!(x)$' for a *predicative* function of x; it is only allowed to quantify over predicative functions, which involves no loss of generality [*op.cit.*, 165]. The Axiom of Reducibility reads:

every propositional function is equivalent, for all its values, to some predicative function of the same argument or arguments. [Whitehead & Russell 1910, 166]

Symbolically and slightly modernized, $\exists f\ \forall x\ (\phi x \leftrightarrow f!x)$, and the corresponding form for relations [*op.cit.*, 167]. Now it suffices to postulate mathematical induction for predicative functions, since the axiom automatically extends it to propositional functions of any order; and similarly for the theory of real numbers. Talk of 'all predicative functions of a' becomes an efficient replacement for talk of 'all functions of a.' As Ramsey and others noticed, the Axiom of Reducibility has the effect of abolishing the hierarchy of orders and letting us to back simply to the hierarchy of types.

It is particularly interesting to note that, in Russell's eyes, the assumption of Reducibility was similar but essentially weaker to the assumption of classes [1908, 167; Whitehead & Russell 1910, 58, 166]. This shows how deeply the principle of comprehension was ingrained in his mind: he could not imagine a theory of classes except as based on that principle. Thus, given a function ϕ of any order, the assumption of classes would warrant that '$\phi(x)$' is equivalent to '$x \in \alpha$,' where α is the corresponding class [*op.cit.*, 58]. Since α would be an individual, the second expression is a predicative function of x. Thus, in the context of Russell's logical theory, the principle of comprehension appears as a means to reduce the order of expressions. But,

there is no advantage in assuming that there really are such things as classes, and the contradiction about the classes which are not members of themselves shows that, if there are classes, they must be something radically different from individuals. It would seem that the sole purpose which classes serve, and one main reason which makes them linguistically convenient, is that they provide a method of reducing the order of a propositional function. We shall, therefore, not assume anything of what may seem to be involved in the common-sense admission of classes, except this, [the Axiom of Reducibility above] [Whitehead & Russell 1910, 166]

Russell thinks he is retaining as much of classes as there is any use for, and little enough to avoid the contradictions which a "less grudging" admission of classes would entail [1908, 168; Whitehead & Russell 1910, 167].

Classes and relations are introduced in *PM* as abbreviations, as so-called "incomplete symbols" which merely serve to formulate shortly other expressions in which there is no reference to classes, but only to propositional functions [Whitehead & Russell 1910, 71–84]. Russell regarded this as a great advantage of his viewpoint, for it involved the elimination of apparent but superfluous entities. Nevertheless, as Quine emphasized again and again, he had only reduced one Platonistic abstract entity – classes – to another – abstract properties or attributes. In Quine's opinion, there is no call even to distinguish properties from classes, except that classes are identical when their members are the same, while properties may still differ. Thus, Russell's definition "rests the clearer on the obscurer, and the more economical on the less" [Quine 1941, 147–48]. To put it simply, Russell avoids postulating a realm of sets, as Zermelo had done, but in order to do so he resorts to postulating a realm of properties, for which he is even forced to assume strong existential postulates like the Axiom of Reducibility. This is one of the reasons why Weyl complained that *PM* did not reduce mathematics to logic, but to

a sort of logician's paradise, a universe endowed with an 'ultimate furniture' of rather complex structure and governed by quite a number of sweeping axioms of closure [Weyl 1944, 272].

Even if we disregard the hierarchy of orders and the uncomfortable assumption of Reducibility, Russell's proposal still had some unpleasant traits. There is no longer one empty class, but an infinite series of empty classes, one for each type (the same applies to quasi-universal classes). Even worse, the same happens with numbers defined in the Frege–Russell style (§1.2): one finds new but different 'copies' of 0, 1, ... for each type [Quine 1941, 152]. Several symbols that Russell uses, like 'Λ' (empty class), 'V' (universal class), *cls* (class of classes), and even 'ε' (membership), are ambiguous because they only have a definite meaning when restricted to a specified type [1908, 174].

5.3. *Principia Mathematica.*   Whitehead & Russell's *PM* is a work that has received the highest compliments, being called one of the greatest intellectual monuments of all time [Quine 1941, 139] and the most representative work of modern logic [Tarski 1964, 229]. This is no doubt due to the ambitious project that the authors set to themselves – to build a complete system of logic and to carry in detail the derivation of mathematics from it – and to the thorough and exhaustive way in which they developed it. *PM* conveyed the impression of a completely rounded off and extremely difficult work, that culminated the development of logical theory.[1] Its sheer extension, more than 2,000 pages in three volumes, certainly did much to impress the scientific world. Logicians such as Hilbert, Skolem, Carnap, Quine, Gödel and Tarski, essentially all who began to work in the period 1915–1930, studied it carefully; thus it has been epoch-making for the influence it exerted [Tarski 1941, 229]. But it has to be acknowledged that, in some crucial points of detail, the work was less definitive and polished than it seemed.

PM was a thorough treatment of logical theory in the whole extent that the word had as of 1910. It is not just an axiomatic development of modern logic, not even a system of higher-order logic: *PM* should be regarded as a detailed treatise of set

---

[1] It was a great contribution to the axiomatization of logic, corroborated by the painfully detailed, explicit derivation of hundreds of propositions. This seems to be the main reason why Hilbert, for instance, greatly admired the book.

theory too. To justify this statement, it suffices to review the contents summarily. Part I deals with mathematical logic as the theory of propositional connectives, quantifiers, classes and relations, based on Russell's theory of types. Parts II and III develop the arithmetic of finite and infinite cardinals, complemented in part V on the basis of the theory of ordinals.[1] Finally, part VI introduces and studies the integers, the rationals and the real numbers, ending with an analysis of the logical basis of measurement. According to a knowledgeable opinion, the portions of *PM* that present the theory of cardinals and ordinals remained the authoritative work on the topic 25 years later, in point of rigor and comprehensiveness [Quine 1941, 158].

The work did not, by far, emphasize the logicist program as much as Russell's *Principles*, perhaps because Whitehead was not so convinced, particularly in view of certain difficulties. The fact that "Mathematics is just Symbolic Logic" was one of the most important discoveries of his time, wrote Russell [1903, §4], and even in 1937 he did not see any convincing reason to modify that view.[2] But no comparable statement can be found in *PM*. More surprising, not even the difficulties that such a viewpoint encounters are properly mentioned in *Principia*. The reduction of mathematics to logic was only possible on the conditional assumption of the Axiom of Infinity (for type one, i.e., for individuals) and the Multiplicative Axiom, a form of Choice. These assumptions are introduced along the way (e.g., [Whitehead & Russell 1910, 388, 481, 536–37]) but they are never mentioned in the introduction, not even in the new introduction of 1925. At any rate, it seems clear that, properly construed, it was only logicism in a pickwickian sense: mathematics had just been reduced to the non-logical Axioms of Infinity and Choice (not to mention the contentious Axiom of Reducibility).

But, surprisingly, the weakest elements in this monumental book are all located in part I, the essential basis for the rest. The presentation of the logical system is marred by several faults; in Gödel's severe judgement:

> It is to be regretted that this first comprehensive and thorough-going presentation of a mathematical logic and the derivation of mathematics from it is so greatly lacking in formal precision in the foundations (contained in *1–*21 of *Principia*) that it presents in this respect a considerable step backwards as compared with Frege. What is missing, above all, is a precise statement of the syntax of the formalism. [Gödel 1944, 120]

Whitehead and Russell did not differentiate clearly between propositions of the system and rules of inference,[3] so that the second were poorly presented. Here, as with differentiation between symbolic expressions and their referents, Frege's work was clearly ahead. Actually, the problem is not so much an imprecise statement of the syntax, as Gödel thought. Russell and Whitehead did not handle the distinction between syntax and semantics at all, which was the source of complications and

---

[1] Part IV presented a generalization of ordinal arithmetic, the so-called theory of 'relation numbers.'

[2] Introduction to the second edition of [1903].

[3] A noteworthy example can be found in [Russell 1908, 170].

obscurities (see [Quine 1941, 140–42]). It has to be said, on their behalf, that logicians only started to get clear about that distinction after World War I, thanks to the pioneering work of authors like Skolem and Hilbert.

Russell's peculiar conflation of syntax and semantics has the effect that his work is dealing with philosophical logic, and even metaphysics, throughout.[1] With these characteristics in mind, it is easier to understand why he looked for a solution to all of the paradoxes, not just the set-theoretic ones, and why he thought that the solution required scrutiny and deep revision of the fundamental logical ideas [Whitehead & Russell 1910, 60]. When the distinction between syntax and semantics became more and more customary in the 1920s, it was possible to envisage a great simplification of the theory of types.[2] This was first proposed by Ramsey in 1926, on the basis of the distinction between two kinds of paradoxes, the logical ones and those that have to do with the notions of truth, definition, and the like. The second kind of paradox was blocked simply by the adoption of a perfectly specified formal language, and thus they could be taken to belong to 'linguistics' or semantics, as Peano had suggested.

The theory of types in *PM* is also obscure and even contradictory in its motivation. The vicious circle principle becomes the basis for a severe revision of logic that leads to ramified type theory. But then an axiom is introduced, simply on pragmatic reasons, that contradicts that Principle – the Axiom of Reducibility. The least one can say is that this axiom is self-effacing, for, in case it is true, the ramification was pointless to begin with (Quine in [van Heijenoort 1967, 152]). The worst is that the axiom contradicts the basic Principle that Russell regarded as the necessary element for solving the paradoxes. It decrees the existence of a "logician's paradise" where one can do what one wished to [Weyl 1944, 272]. No wonder that the authors denied the axiom any "self-evidence," claiming only "inductive evidence" for it, since its consequences appear to be indubitable and nothing "probably false" can be deduced [Russell & Whitehead 1910, 59]. Caught between the vicious circle principle and the desire to justify classical mathematics, Russell was certainly honest in presenting the matter as he did. But, of course, his system could only puzzle attentive readers.

Ramsey decided simply to accept several traits of classical mathematics as embodied in systems of set theory. In particular, he extensionalized the theory of types and accepted impredicative definitions, i.e., he rejected the vicious circle principle. In his view, this principle was only used in *PM* for solving the paradoxes, but the semantic paradoxes were blocked by formal languages, and the 'logical' ones by the hierarchy of types. In this way, he was able to go back to the simple theory of types, that was then adopted and refined by Gödel and Tarski (see §X.4). With this

---

[1] The strange features of the famous *Tractatus* by his student Wittgenstein [1921] are thus more a symptom than a deviation.

[2] Likewise, systems of propositional and predicate logic started to be presented in the now customary way during the 1920s. Elements of the distinction between syntax and semantics can be found in Peano and Schröder, and also in Frege, insofar as he differentiates clearly between a name and its referent, between use and mention of an expression

modification, much of *PM* (especially from Part II) could remain unaltered, since the complications of ramified type theory were quickly lost of sight. Whitehead & Russell had already indicated that hardly anything in the book would be changed by the adoption of a different theory of types [1910, vii]. But if the letter remained valid, the spirit of the theory changed significantly.

6. Other Developments in Set Theory

After 1900, set theory was gradually enriched with many new results in different directions, which would merit an independent chapter. Here I can only indicate some of the relevant developments summarily.[1]

Cantor proved for closed sets a decomposition $P = R \cup S$ into a denumerable and a perfect subset, which implies that the Continuum Hypothesis (CH) holds in this case (§VI.8). In his dissertation of 1901, Bernstein was among the first to work on generalizing that result. This kind of work studied sets of reals that are definable in different ways, and led to descriptive set theory. The contributions of the French analysts Borel, Baire and Lebesgue, in their study of real functions and integration, merged with that line of development. Thus, for his theory of measure Borel considered sets obtainable from intervals by complementation and countable union. These were called the *Borel sets*, and studied by several authors. Hausdorff [1914] defined a now classic hierarchy of Borel sets, and in [1916], simultaneously with Aleksandrov [1916], established the pathbreaking result that any uncountable Borel set of reals has a perfect subset. This satisfactory generalization expanded the domain of validity of CH.

Meanwhile, the early study of consequences of AC affected that line of development, for in 1905 Vitali proved that there is a set of reals that is not Lebesgue measurable, and in 1908 Bernstein showed the existence of a set of reals with no perfect subset. Both of them built on the assumption of a well-ordering of ℝ. These results suggested difficulties for the above line of attack to the Continuum Problem, which in fact encountered peculiar difficulties from the 1920s. Descriptive set theory emerged from about 1915 with the work of the Moscow school headed by Luzin, to which Aleksandrov and Suslin belonged. They defined and investigated the analytic sets, and in the 1920s the projective sets [Kanamori 1995].

During the 1900s Hausdorff did extensive work on uncountable order types (linear orderings not restricted to well-orderings), refining Cantor's ideas and advancing further in the exploration of the transfinite [Hausdorff 1908]. He was thus led to the notion of cofinality, to the Generalized CH, and to the notion of large cardinal. In this work, Hausdorff employed AC freely, and in his handbook he explored aspects of the Well-Ordering theorem which led him to advance a maximality principle closely related to Zorn's Lemma of 1935 [Hausdorff 1914, 140ff].

---

[1] For a brief review, written from a modern standpoint, see [Kanamori 1996].

He also formulated the Hausdorff paradox [*op.cit.*, 469ff], a forerunner of the better known Banach–Tarski paradox, showing the surprising consequences to which AC leads in the classical mathematical framework.

Hausdorff [1908] considered the possibility of "exorbitant numbers" that are nowadays called the *weakly inaccessible* cardinals, but he was of the opinion that they would hardly ever come into consideration in set theory [Hausdorff 1914, 131]. Axioms of *strong infinity* asserting the existence of weakly inaccessible cardinals were formulated in the early 1910s by Mahlo, in his pathbreaking work investigating hierarchies of such cardinals. The topic of large cardinals would expand slowly at first, to become one of the major areas of axiomatic set theory in the second half of the century.[1] In this process, Alfred Tarski and his Berkeley school would be one of the main driving forces. Tarski and Sierpiński [Tarski 1986, vol. 1, 289–97] introduced the notion of (strongly) *inaccessible* cardinal in 1930, simultaneously with Zermelo [1930], and Tarski [1938] went on to present axioms of strong infinity for (strongly) inaccessible cardinals.

Coming back to the 1910s, Hausdorff's handbook presented a purely mathematical approach to set theory, developed informally in the style of Cantor, although more sophisticated. He spoke favorably of Zermelo's axiomatization, but refrained from advancing a clear opinion on the issue and preferred to work naively in a book for beginners.[2] There was no mistrust of the axiomatic viewpoint on his side: he presented an "axiomatization of point-set theory" [1914, vi] based on the notion of neighborhood, that is justly renowned (see below). All in all, it was a very interesting introduction to set theory, full of new viewpoints throughout, but particularly oriented toward point-sets and their applications: more than half of the book (over 200 pages) is devoted to point-sets, topology, real functions and measure theory. In contrast to Zermelo, Hausdorff was not acquainted with Dedekind's work,[3] but his approach favored a set-theoretic conception of mathematics. As Kanamori [1996, 17–19] has emphasized, he presented a purely set-theoretic notion of function contrasting sharply with the viewpoints of Cantor or Russell. He offered a reduction of ordered pairs to sets (although one that is neither very elegant nor convenient in an axiomatic framework) and he proceeded to define functions as sets of ordered pairs [1914, 32ff, 70ff]. He also pointed out the correlation between sets and their characteristic functions [*op.cit.*, 37], which served to emphasize the connections between the Cantor Theorem and the power set, as Russell had done earlier (see §1.2).

A question of particular interest in the context of the present work is the emancipation of topology, since it marks, by contrast, the public recognition of abstract set theory as an independent subject. It can be said that this last differentiation was prepared in the early 1910s, although its wide adoption should probably be dated in

---

[1] On this topic, see [Kanamori 1994].

[2] [Hausdorff 1914, 1–2]. His confidence in the Zermelo system is even clearer in the second, very abridged edition [1927, 34].

[3] The bibliography only cites [Dedekind 1872]; see also chapters 2 and 9, where he fails to mention Dedekind in connection with functions and mappings.

the 1920s.[1] Early in the century, Schoenflies applied notions of point-set theory to simple closed curves in the plane. His work was employed by Brouwer, who showed [1910] that most of Schoenflies' results were incorrect. Brouwer undertook the study of *n*-dimensional manifolds, introducing a new level of precision in the study of topology and opening a new era with his [1911] proof of dimension invariance. His work began to combine the notions of point-set theory with the combinatorial methods that originated in Poincaré (see [Johnson 1979; 1981]), thus sowing the seeds for the spectacular flowering of topology after the War. But mathematicians still lacked an adequate definition of topological spaces that could establish a general framework for the subject. Here Hausdorff's contribution was very influential.

Not surprisingly, there were precedents for Hausdorff's work. From about 1900, several mathematicians applied notions of point-set theory to new domains, such as sets of curves and sets of functions. This called for an abstract approach; Hilbert proposed in 1902 to define the topological notion of 2-manifold abstractly by means of neighborhood conditions (see [Hilbert 1930, appendix IV]). Independently, Fréchet proposed in 1906 that metric spaces could be axiomatized on the basis of a generalized notion of limit. In his handbook, Hausdorff analyzed different possible ways of founding the notion of a topological space – either on the basis of distance, of neighborhoods, or of limits – establishing a hierarchy of increasing generality in the order just given. He preferred to employ neighborhoods because in a sense they are more basic, since limits presuppose sequences, i.e., denumerable sets. Chapter 7 of his [1914] defined a "topological space" by means of four well-known axioms [*op.cit.*, 213]. Hausdorff's axiom system was happily selected: it is well-adapted to applications, precise, and sufficiently general, although he included the separation axiom that is characteristic of Hausdorff spaces. His exposition became a model of axiomatic development and made possible the emancipation of topology.[2]

Nevertheless, Hausdorff himself regarded topology simply as a part of set theory, and even presented a theoretic argument to justify that viewpoint [1914, 209–10].[3] Similarly, Schoenflies and the Youngs regarded the theory of point-sets and measure theory as parts of general set theory. Meanwhile, a minority of authors had begun to handle the distinction between abstract set theory and point-set theory or topology. This is notably the case with Dedekind's *Zahlen* [1888] and Cantor's

---

[1] The emergence of topology was a very complex, many-sided process. Here we pay attention to developments in set-theoretic topology leading up to the fundamental notion of topological space; for further details see [Manheim 1964, chap. 6; Johnson 1979; 1981]. Aspects of the rise of combinatorial and algebraic topology are studied in [Bollinger 1972; vanden Eynde 1992; Epple 1995].

[2] Weyl had made a similar contribution previously [1913], also stimulated by Hilbert's proposal. Hausdorff claims that his work was independent and that he presented it in 1912 at the University of Bonn [1914, 456–57].

[3] The argument emphasized the similarity between the theory of order types and that of topological spaces: an order relation can be taken to be a two-valued function of two arguments, and topology can be developed on the basis of a distance function.

Beiträge [1895/97]; also with Zermelo's work, that deals exclusively with abstract set theory. In general, however, it was only after World War I that abstract set theory found an important number of cultivators (§XI.1–3). Topology came to be seen as an independent branch of mathematics extensively studied by authors of the North-American school, by Aleksandrov, Kuratowski, and others.

X   Logic and Type Theory in the Interwar Period

> Today there exits no single general set theory – but a naive, an intuitionistic, as well as several formalistic-axiomatic systems.[1]

> The simplified theory of types or the Zermelo set theory ... are at the present time the safest cities of refuge for the classicist in mathematics.[2]

The 1920s atmosphere, among experts in foundational studies, was one of great insecurity. Most of them – intuitionists excluded – were looking for a symbolic, formal system that might provide a framework for all of mathematics. It *had* to be possible to reconstruct mathematics on a completely secure basis, to find a system maximally immune to rational doubt.[3] But, in carrying out that project, caution was the keyword. Reminders were the by-then legendary paradoxes, which ruined the work of Frege and imperiled that of Cantor, the intuitionists' indictment against modern mathematics, and the proliferation of divergent systems. During the interwar period there was a great level of experimentation in the area of foundations, not infrequently leading to systems that turned out to be contradictory.[4] Even those who were convinced of the final vindication of the 'classical' viewpoint of Cantor and Dedekind, like Hilbert, had to look for very careful ways of proceeding if they wanted to solve satisfactorily all of the problems posed.

Until the 1920s few authors adopted Zermelo's axiom system explicitly, and even among those who preferred a formal axiomatic standpoint, many kept favoring the theory of types. The reason was that it seemed to offer a safer framework. A noteworthy example is van der Waerden's *Moderne Algebra*, the first textbook to advance the structural conception of its subject. The work begins by explaining the

---

[1] [von Neumann 1928, 321]: "...dass es heute keine einheitliche allgemeine Mengenlehre gibt: sondern eine naive, eine intuitionistische, sowie mehrere formalistisch-axiomatische Systeme."

[2] [Church 1937, 95].

[3] This viewpoint, which dominated work in logic and foundations until World War II, has been called *foundationalism* [Shapiro 1991, 25].

[4] Examples are systems presented by Church in the early 30s (see [Kleene & Rosser 1935]) and Quine's *ML* in its original [1940] presentation (see [Rosser 1942] and [Ullian 1986; Wang 1986; Ferreirós 1997]).

rudiments of set theory building on the principle of comprehension and the doctrine of types [van der Waerden 1930, 4ff]. According to the great logician Alonzo Church, in spite of superficial differences, these two "widely accepted symbolic systems," Zermelo set theory and simple type theory, are "in their currently accepted forms essentially similar" [1939, 69–70]. But Church argued that there is no convincing basis for a belief in the consistency of either system "even as probable" [*ibid.*]. This was certainly a strong judgement, but at any rate it is quite representative of the period and the reigning 'atmosphere of foundational insecurity,' as I shall call it.

By emphasizing the similarity between type theory (TT) and the Zermelo–Fraenkel system, Church gives expression to the fact that, during the interwar period, the formulation of type theory was affected by influences coming from its competitor. In the hands of Ramsey, Gödel and Tarski, it became a system much closer to axiomatic set theory in its setup and spirit: an impredicative, Platonistic system, formulated in the way of Hilbertian axiomatics. Likewise, the Zermelo–Fraenkel system underwent a process of refinement in which it incorporated traits of simple TT (see §XI.2). This convergence of viewpoints, following the pre-War bifurcation, is one of the most interesting aspects of developments in this period, but up to now it has been paid little attention.

The interwar atmosphere played also an important role as the context within which first-order logic emerged as the indispensable basic logical system. Its emergence cannot be understood apart from its natural environment, the foundations of mathematics. Serious doubts regarding the proper conception of mathematics, and desire of intellectual security, placed strong requirements on systems of logic. Above all, they should as far as possible avoid assumptions that were the topic of discussion in current foundational work. In its final form (§XI.5) the axiomatization of set theory would be framed within first-order logic. For all of these reasons we need to consider the foundational debate in some detail; then, we shall outline the developments in logic and its assumed province, type theory, that would affect strongly the evolution of axiomatic set theory.

1. An Atmosphere of Insecurity: Weyl, Brouwer, Hilbert

Debate over abstract mathematics can be traced back to around 1870, when Kronecker started making objections to the procedures of Weierstrass, Dedekind, and Cantor (§§I.5, IV.4). Constructivist thinking surfaced again in a more polemical setting with the French school of Baire, Borel and Lebesgue, after Zermelo's 1904 proof of Well-Ordering. Deserving special mention are the influential but somewhat peculiar views of Poincaré, who offered some key new arguments but did not elaborate them into a coherent philosophy of mathematics (§IX.5.1). The polemics entered a new stage after the great War, reaching an unprecedented height due to the radical proposals of the intuitionists. The entrée of Weyl and Brouwer from 1918, and above all a series of conferences by Weyl in 1920, which he himself described as a "propaganda pamphlet" for intuitionism [van Dalen 1995], started a heated debate in which Hilbert undertook to defend abstract mathematics. This was the start of a polemical discussion in which conceptual, professional, and even political issues coalesced, leading up to the 'war of the frog and mice,' as Einstein was to call it.[1]

Hermann Weyl, one of the most universal mathematicians in his generation, has been regarded as Hilbert's most gifted student. According to his own account, until the early 1910s he was a convinced follower of the abstract, set-theoretic mathematics that he identified with the names of Dedekind and Cantor [Weyl 1918, 35–36]. But around 1913 he started to question his earlier beliefs, partly under the influence of the philosophy and logical investigations of Husserl,[2] partly as a result of his attempt to make fully precise Zermelo's axiom system. On the occasion of his Habilitation at Göttingen in 1909, he gave a lecture 'On the definition of the basic concepts of mathematics' [Weyl 1910]. There he tried to perfect Zermelo's system, particularly the notion of a 'definite statement' (§IX.4.2) by making explicit some logical principles of definition (see §5.1). Subsequently he tried to build set theory on this precise basis, but without presupposing the notion of the natural numbers [Weyl 1918, 36]: like Zermelo, he was trying to preserve the Dedekindian project of establishing set theory as a firm foundation for *all* of mathematics, and especially for numbers. This led to complexities that he could not overcome.

Only in connection with general philosophical insights, to which I was led by the rejection of conventionalism, did I realize that I was wrestling with a scholastic pseudo-problem, and I became firmly convinced (in accordance with Poincaré, as little as I agree with his philosophical position) that the *conception of iteration, of the natural number-sequence, is an ultimate foundation of mathematical thought* – in spite of Dedekind's "theory of chains."[3]

Furthermore, Weyl [1918, iii] was led to the conviction that classical analysis was an edifice built on sand, based on a vicious circle, and undertook a thorough revision of set theory and analysis. Thereafter, he tended to prefer constructive approaches to mathematical theories, a clear indication being that, allegedly, in his papers he was careful never to use the Axiom of Choice.[4]

In his 1918 book *Das Kontinuum*, Weyl proposed a modified, predicative version of analysis. Although his approach was not as radical as Brouwer's, his com-

---

[1] These wider aspects, including academic politics and outright political issues, have been explored by Mehrtens [1990] and van Dalen [1995].

[2] He had married a disciple of Husserl; see [Weyl 1918, 35–37] [Weyl 1954].

[3] [Weyl 1918, 36–37]: "Erst im Zusammenhang mit allgemeinen philosophischen Erkenntnissen, zu denen ich mich durch die Abkehr vom Konventionalismus durchrang, gelangte ich zur Klarheit darüber, dass ich hier einem scholastischen Scheinproblem nachjagte, und gewann die feste Überzeugung (in Übereinstimmung mit Poincaré, so wenig ich dessen philosophische Stellung im übrigen teile), dass die *Vorstellung der Iteration, der natürlichen Zahlenreihe, ein letztes Fundament des mathematischen Denkens ist* – trotz der Dedekindschen 'Kettentheorie'."

[4] Dieudonné in [Gillispie 1983, vol.13, 285]; as Dieudonné indicates, fortunately (or intentionally?) he dealt with theories in which he could do so with impunity.

ments on modern mathematics and set theory are noteworthy and instructive. Classical analysis is guilty of making two untenable moves. First, it abandons the realm of well-defined notions and works with completely vague, nebulous conceptions – those of arbitrary set and arbitrary function in the sense of Dirichlet [Weyl 1918, 15]. According to Weyl, it is the essence of infinity to be inexhaustible. For this reason, one can never deal with infinite sets combinatorially, as if one could select infinitely many elements in an arbitrary way and collect them into a set. An infinite set can only be determined by explicitly giving properties that characterize its elements [Weyl 1918, 13–15, 32–33]. But classical analysis forgets that properties are logically prior to sets.

Second, there is a logical vicious circle in that one feels entitled to quantify over all sets, or all real numbers, as if there were a well-defined realm of properties corresponding to both lawful and arbitrary sets, or a realm of numbers corresponding to both definable and undefinable real numbers [Weyl 1918, 19–23].[1] If we define the reals as Dedekind cuts on the rational numbers, that can only be done in reference to properties of rational numbers. The first section of a cut is made up of all rational numbers that possess a given property and, in the absence of 'arbitrary properties,' there is no way of proving that there always exists a least upper bound for any bounded set of real numbers [Weyl 1919, 45].

As we see, Weyl stuck to the traditional conception of sets as concept-extensions. His solution for the paradoxes consisted in strict adherence to the vicious circle principle, i.e., to *predicativism*. His motives were similar to Russell's, but he was more radical in carrying through the program, partly because he was not guided by the goal of reducing mathematics to logic.[2] In mathematics as in other fields, one deals with previously given objects, and previously given properties and relations among them. To avoid vicious circles, one must first build up predicatively, from those primitives, a well-defined realm of properties and relations. Only after this process has been completed can we, in a second step, proceed to extensionalize. And it is evident that the new objects, the sets, are completely different from the primitive ones; they belong in a wholly different sphere of existence [1918, 15]. Thus, we obtain a realm of one- and multi-dimensional sets (sets and relations, as we now say), corresponding to the constructed properties and relations [*op.cit.*, 31–32]. We shall later (§5.1) analyze in more detail the logical basis of Weyl's 1918 contribution. Suffice it to say that he was able to establish a constructive theory of the real numbers where only those real numbers that are explicitly definable from the rationals are obtained:

---

[1] More precisely, one makes the predicative levels (see below) collapse by quantifying over numbers or sets of any level whatsoever.

[2] To believe his own account [1918, 35], Weyl developed his ideas before coming to know of the related ones of Frege and Russell.

Through this conceptual restriction, a heap of points, so to say, is picked out from the fluid mass of the continuum. The continuum is pulverized into isolated elements ... Therefore I speak of an *atomistic conception of the continuum*.[1]

He acknowledged [1918, 65ff] that the outcome was a far cry from the intuitive idea of the continuum, but at least it was an irreproachable, clear and sufficiently powerful mathematical replacement. In this constructive version of analysis, Cauchy's criterion of convergence for sequences of real numbers is retained.[2]

In that same year of 1918, Brouwer started the systematic development of intuitionistic mathematics. It is well known that he had been considering a reform of pure mathematics since the time of his 1907 doctoral dissertation, in which he attacked set theory, e.g., the Well-Ordering Theorem. Although his ideas first came to the light in 1912, it was only after the Great War that he set to a careful development. This is reflected in his paper 'Foundations of Set Theory, without Use of the Logical Principle of Excluded Middle' [Brouwer 1918]. This is not the place to enter into a discussion of intuitionism, but it is important to give some basic indications of his proposals. Brouwer explored in detail the consequences of adopting a very traditional position, the rejection of actual infinity. He elaborated on ideas suggested already by Kronecker, but he developed them in full detail with the aim of building an alternative to classical analysis. As Weyl wrote later [1946, 275], Brouwer sees the sequence of numbers as "a manifold of possibilities open towards infinity ... [which] remains forever in the status of creation." Given the openness of the realm of numbers thus conceived, the principle of excluded middle cannot be applied to properties of natural numbers. It makes no sense to ask whether there *exists* a number of a given property, with the expectation that either such a number exists or every one has the opposite property. The existential statement only makes sense in case we have adequate procedures for the *construction* of a number with the property; and the negative general statement only when we have a construction procedure that, given any number, shows that it does not have the property.

Brouwer developed an intuitionistic theory of sets that deviates enormously from Cantorian set theory. On this basis he established an intuitionistic topology of point-sets and intuitionistic analysis. Actually, what he called "Menge" in the 1920s was later termed "spread," in order to avoid confusion. Spreads became the basis for an extremely original theory of the continuum. A spread is "a law" which associates certain actions to natural numbers; given a number, the spread generates either a certain sign, or nothing, or the stopping of the process [Brouwer 1918, 3; see 1925, 244–45]. Brouwer requires that, after each non-stopped sequence of $n-1$

---

[1] [Weyl 1921, 149]: "Durch diese Begriffseinschränkung wird aus dem fliessenden Brei des Kontinuums sozusagen ein Haufen einzelner Punkte herausgepickt. Das Kontinuum wird in isolierte Elemente zerschlagen ... Ich spreche daher von einer *atomistischen Auffassung des Kontinuums*."

[2] Also retained is a version of Cantor's proof that the continuum is not denumerable, even though one can enumerate all possible sets of natural numbers [Weyl 1918, 25]. For further remarks on how the set-theoretical ideas of Dedekind and Cantor are affected by his proposal, see [*op.cit.*, 16, 19, 37].

choices, one may always choose a new n-th numeral that does not stop the process. Each sequence of signs, be it finite or not, is an element of the spread, but, of course, the elements and the spread "cannot be presented as finished" [*ibid.*]. Since the 'signs' of Brouwer may always be represented by numbers, a spread may be thought of as a tree with natural numbers at its nodes, where some particular nodes are forbidden – the process is 'inhibited' at those points. Infinite paths down the tree of a spread are Brouwer's "choice sequences," sequences which are becoming or emerging through choice acts. They were the key new element that allowed a novel theory of the continuum. The continuum arises as a "medium of free becoming," as Weyl said [1921, 151]; the single real numbers fit in it, but it is not conceived as an actually infinite set of real numbers.[1]

In 1921, Weyl published his lectures of the year before, a "pamphlet" that described his own and Brouwer's approaches, supporting the latter. He employed an evocative language in order to "rouse the sleepers" [van Dalen 1995]. Making free use of a political metaphor, he compared the situation in mathematics with the great political and economic instability in Germany at the time. He began describing the explanations given to the paradoxes by important mathematicians as belonging to "that sort of half to three-quarters honest attempts at self-deception" that one frequently finds in politics and philosophy [Weyl 1921, 143]. Explaining Brouwer's viewpoint, he used the following metaphor [*op.cit.*, 156–57]: if mathematical knowledge is a treasure, a purely existential result is just paper, a promissory note indicating that somewhere there ought to be a treasure. The existential theorem lacks value: only the concrete construction which provides a real example is valuable. Thus, he was led to compare modern mathematics with the "paper economy" that reigned in his native country, and he cried: "*this order cannot be maintained in itself* ... and Brouwer – that is the Revolution!" [*op.cit.*, 158].[2]

The proposals of Brouwer and Weyl were noteworthy, even if few mathematicians were open to such strong changes. One could hardly dismiss as unimportant the fact that several great mathematicians, from Kronecker and Poincaré to Weyl and Brouwer, doubted the soundness of the new trend of abstract mathematics. Witness of the importance that the intuitionist proposals attained is the fact that Fraenkel included a discussion of them in his textbooks on set theory [1923; 1928]. During the 1920s, the issue of the foundations of set theory could not be dealt with completely apart from the so-called 'crisis.' Von Neumann wrote in 1928 that at the time there was no single general set theory, but several systems – one naive, one intuitionistic, and several formalistic-axiomatic.[3] The importance of intuitionism

---

[1] Brouwer introduced also another notion that corresponds to a different aspect of the classical notion of set – the notion of a "species," roughly meaning a class of mathematical entities corresponding to a given property [Brouwer 1925, 245–46]. On Brouwer's theory of spreads and species, see [Heyting 1956; van Heijenoort 1967, 446–63; Troelstra & van Dalen 1988].

[2] Years later, he wrote that only with hesitation he could acknowledge those lectures, which in their "bombastic" style reflected the mood of the excited times after World War I [*op.cit.*, 179].

[3] The latter included, beyond ZFC and von Neumann's system, the weaker ones of Russell and of Hilbert and his school [*op.cit.*, 321–22].

was even enhanced by the way in which Hilbert reacted to it, and in the process adopted some of its ideas.

Since the beginning of the century, Hilbert had been the champion of the axiomatic method and the leading spokesman for abstract mathematics. When his admired pupil Weyl published his propaganda pamphlet for Brouwer's intuitionism, Hilbert took it as his duty to demolish these critics and defend modern mathematics. He felt, it seems, that his own reputation was in question, and he put in the service of the cause not only his time and efforts, but also his best rhetoric abilities and his whole influence within the mathematics community. The contrast between Hilbert's position before and after the entrée of Weyl and Brouwer can be measured from a comparison of his lectures [Hilbert 1918] and [Hilbert 1922].

At first, he had been interested in axiomatics as the best method for the conscientious development of any theory, mathematical or not, for the investigation of dependence and independence relations among its propositions, and for a deeper analysis of its foundations. It was mainly in this spirit that he studied the foundations of geometry in the famous 1899 work (with several reeditions, some of them substantial; see [Hilbert 1930]). After the publication of the set-theoretic paradoxes he began to explore the possibilities of the axiomatic method as a means for securing once and for all the consistency of mathematical theories [Hilbert 1904]. Even so, before 1920 both aims, the study of the inner structure of theories and the proof of consistency, were presented in a balanced way [1918, 148].

But in 'New Founding of Mathematics' [1922] the aim of establishing consistency, in order to provide a final justification for classical analysis and set theory, was the only one in his mind.[1] Hilbert was determined to show the falsity of the viewpoints of Weyl and Brouwer, which he compared with a dictatorship *à la* Kronecker [Hilbert 1922, 159]:

I believe that, just as Kronecker in his day was unable to get rid of the irrational numbers ... so today Weyl and Brouwer will be unable to push their programme through. No: Brouwer is not, as Weyl believes, the revolution, but only a repetition, with the old tools, of an attempted coup that, in its day, was undertaken with more dash [viel schneidiger] but nevertheless failed completely; and now that the power of the state has been so well armed and strengthened by Frege, Dedekind and Cantor, this coup is doomed to failure.[2]

Hilbert presents Cantor, Dedekind and Frege as the founding fathers of the abstract mathematics which he champions. They inaugurated the critique of analysis which led to deep axiomatic theories, particularly those of Zermelo and Russell, and to the logical calculus [*op.cit.*, 162]. Hilbert's famous statement about Cantor's paradise, quoted at the beginning of the next chapter, would follow three years later [1926, 375–76].

---

[1] [Hilbert 1922, 159, 161–62, 174–75, 176–77].

[2] [Hilbert 1922, 160] as translated in [Ewald 1996, 1119] with minor changes.

Hilbert planned to dispel doubts concerning the foundations of mathematics by axiomatizing fully mathematics *and* logic, i.e., by going one step beyond customary axiomatics. The fact that notions and proofs can, indeed, be completely formalized by means of a finite system of axioms and rules had been empirically established by Whitehead & Russell [1910/13]. Hilbert's new insight was that, having formalized a theory completely, the question of possible proofs in it becomes a combinatorial question that can be studied from a *finite standpoint*. One should thus turn the very notion of mathematical proof into the topic of a detailed mathematical investigation [Hilbert 1918, 155]. This was the origin of his *Beweistheorie* [proof theory], aimed not only at establishing consistency but, among other things, at finding decision procedures for mathematical problems and establishing Hilbert's credo that all mathematical questions are solvable [*op.cit.*, 153].

To carry out this project, he drew Paul Bernays to Göttingen as an assistant, and they started serious work in 1920 [Bernays 1935, 202]. There is little doubt that the aging Hilbert profited very much from the exceptional abilities of his young collaborator. They proceeded step by step, beginning with simple subsystems of arithmetic for which consistency was proved [Hilbert 1922, 170–72]. The systems were progressively enlarged until Hilbert proved, or thought to have proved, the consistency of the full system of Peano arithmetic [*op.cit.*, 176]. The essential traits of his *Beweistheorie*, as far as the goal of consistency was concerned, had been established by 1923.[1] The key point was to show the consistency of applying the principle of excluded middle to propositions which make free use of 'all' and 'there is' [Hilbert 1923, 181]. Hilbert sketched a formal system for analysis and described a way of proceeding in the consistency proof [*op.cit.*, 188–90]. This would also establish the bridge to set theory. Since the proof of consistency would be finitary, Hilbert's approach would have rehabilitated modern mathematics by the use of means acceptable to all parties in the foundational debate. Although Brouwer was not satisfied with a mere proof of consistency, there is little doubt that a success of the Hilbertian program would have sufficed to convince practically all mathematicians.

Actually, Hilbert tended to present the matter as if it had already been solved.[2] At first it seemed that the consistency proof for analysis, i.e., for the theory of real numbers, worked [Bernays 1935, 210–11]. Hilbert's approach was developed by Ackermann [1924], who also thought to have solved the question. Later, von Neumann [1927] clarified some difficulties in Ackermann's work and presented a consistency proof for a more restricted system. All of the participants thought that the investigations of Ackermann and von Neumann had actually established the consistency of arithmetic, and that the expansion to analysis was a matter of details;[3] even critics like Weyl accepted this diagnosis [1927]. Only after Gödel established his incompleteness results [1931] did it become clear that the above-mentioned

---

[1] A more detailed presentation of the underlying ideas was given in [Hilbert 1926], and further details concerning the system in [1927].

[2] See [Hilbert 1922, 176–77; 1923, 178; 1926, 383; 1927, 479].

[3] See [Bernays 1935, 211] and also, e.g., [Hilbert 1929] and [Bernays 1930, 58].

proofs were not completely general. For Gödel indicated how to demonstrate that a formal system for mathematics satisfying very general conditions cannot codify a proof of its own consistency.[1] The prevailing optimistic state of opinion by 1930 makes it even more surprising that Gödel was able to tackle the problem as he did, looking for proofs of incompleteness and unprovability of consistency. His brilliant and surprising metatheoretical results, established by methods which complied with the strictest constructivist restrictions, contradicted some of Hilbert's key assumptions. They increasingly convinced mathematicians that Hilbert's project of a finitary consistency proof could not be carried out.[2]

Thus, even though it seemed for a while that Hilbert's project was going to succeed, insecurity was again the outcome of the process. But this is not the place to discuss further the history of Hilbert's program or its reception. Now we turn to considering the impact of the foundational debate on logic.

2. Diverging Conceptions of Logic

The whole debate on logic and the paradoxes in the first decade of the century had the effect of making quite unclear the notion of logic itself, and the scope of logical theory. Up to that point, the theory of classes or sets had been an undisputed component of logic, indeed its very core (§§II.2, VII.6, IX.1.2). But in the 1900s there was much confusion regarding what to do in response to the paradoxes. One could either try, with Russell, to preserve a strongly modified class theory as the province of logic, or treat the theory of sets as a properly mathematical one, like Zermelo.

Even worse, during the first third of the 20th century there was not a single tradition in logical theory, but different traditions endorsing diverse conceptions, none of which coincides with the first-order logic that has been regarded as the central core of logic since the 1950s.[3] A well-informed person living in the 1920s would have taken notice of the following main traditions (leaving aside the proponents of traditional logic): the tradition of *Principia*, i.e., of Frege, Peano and Russell; the algebraic tradition of Schröder, then in a weak position but presenting powerful results due to Löwenheim and Skolem; an emerging formalistic tradition,

---

[1] [Gödel 1931] only contains a sketch of the proof of this second incompleteness theorem, based on the idea that some reasonings in his first incompleteness theorem could be strictly formalized. He planned to devote a second paper to the issue, but did not come to do so. See [Gödel 1986, 137–38].

[2] See, e.g., [von Neumann 1947] and the 'Nachtrag' to [Bernays 1930, 60–61]. Hilbert himself did not think so [Hilbert & Bernays 1934/39]. For detailed analysis of Hilbert's program, see [Kreisel 1978; Detlefsen 1986; Hallett 1991].

[3] This historiographical point was prepared by van Heijenoort [1967] and forcefully established by Goldfarb [1979] and Moore [1988]. Moore [forthcoming] lists five main approaches around the turn of the century.

guided by the needs of axiomatics and led by Hilbert; and, quite importantly, the deviant proposals of the critics of modern mathematics, Brouwer, Weyl, and also Skolem.

The reigning confusion can be clearly grasped from prevailing opinions regarding the relations between logic and mathematics. Until 1930, *Principia Mathematica* marked the conception of mathematical logic more than any other work. As we shall see, a simplified form of type theory was generally acknowledged as the natural system of mathematical logic around that time.[1] Similarly, many logicians of the period evidenced the influence of *Principia* in their adherence to logicism [Ferreirós 1997], but this philosophical viewpoint proved to be much more controversial than simple type theory.

Hilbert [1904, 131] and others believed that logic and mathematics need each other and have to be treated side by side, none of them being the foundation of the other discipline.[2] A logic independent of mathematics would be a purely intuitive theory, but when it is formalized it starts depending on mathematics. Formalized logic presupposes the idea of finite iteration, which is basic for defining what a proof and a well-formed expression are. It would thus make little sense to attempt an ultimate foundation of number on the basis of formal logic [Skolem 1928, 517]. On the other hand, it is clear that mathematics needs logic for the conduct of proofs. Axiomatic mathematical theories, beginning with Peano arithmetic, have to rely on an underlying logic. Thus, the foundations for both disciplines must be laid simultaneously and in an interrelated way.[3]

To add to the confusion, Brouwer viewed mathematics as a primary, original activity of man, and thought that classical logic had been derived from that activity. Our customary logic evolved from experience with finite sets, but in the course of time it came to be applied, with no justification, to infinite sets. As Weyl liked to say [1946, 276], "this is the Fall and original sin of set theory," punished with the paradoxes. Thus, intuitionism is situated at the antipodes of logicism – logic is based on the primary activity of mathematics, rather than the latter being based on the universal 'conceptual language' of logic.

Modern logic can be seen to have arisen from the confluence of three of the main traditions indicated above – that of the logicists Frege and Russell, that of the algebraists of logic Peirce, Schröder and Löwenheim, and that of Hilbert's *Beweistheorie*. Frege and Russell introduced the crucial idea of formal proof and all the formal machinery employed by modern logic, although their systems embraced higher-order logic. They pursued the dream of a grand logic (*logica magna*), the one true logic that speaks of the Universe in a fixed way; in their eyes the logical system is an *interpreted* system, not a purely formal one. Hilbert and his followers would emphasize the latter standpoint and thus shift substantially the conception of the subject. Meanwhile, Peirce and Schröder lacked a modern conception of formal

---

[1] This is said explicitly in [Carnap 1931, 46].

[2] There are indications that Hilbert abandoned that view in favor of logicism during the War (see [1918, 153; Hilbert & Ackermann 1928]), but only to return to it in the early 1920s.

[3] [Skolem, 1928, 517]. This was essentially Hilbert's viewpoint in [1904].

systems and formal proof, but they systematically studied the question of possible interpretations of logical expressions on given domains. Thus, Frege and Russell prepared the syntactic ingredient of modern logic, refined and transformed by Hilbert, while Schröder and Löwenheim prepared its semantic ingredient. Only the convergence of the three traditions, effected gradually in the work of Skolem, Hilbert, Gödel and others, could give rise to the distinctive 20th century approach to logic, the metalogical approach [Goldfarb 1979, 356].

Hilbert and his followers became the foremost proponents of the formal, finite standpoint; they were in a privileged position to effect the above-mentioned convergence of traditions. In their work, a most careful attention to formal systems was joined by free consideration of different possible interpretations, in the tradition of algebra and axiomatics.[1] Insistence on purely formal systems would be a guiding principle leading to first-order logic. But it is interesting to note that the earliest proponents of first-order logic as the basic system were constructivists (§5). The whole development of logic in the 1920s and 30s, particularly the increasing concentration on formal systems, owed much to the context of the foundational debate and, specifically, to the need to confront constructivist criticism.

The most interesting aspect of this period, for our purposes, is that there were extreme divergences regarding the problems of set theory and the related issue of higher-order logic. Of course, the extent to which set theory was acceptable was very much in discussion: whether the Axiom of Choice, or even the Axiom of Infinity could be accepted, was a matter of opinion in which different authors diverged greatly. In the tradition of *Principia*, the problem of the paradoxes was regarded as a logical problem, that called for a 'natural' logical solution – Russell's type theory, which left no need for an independent theory of sets. The followers of Schröder had no common position. As I have said, this was a weak tradition, much weaker by 1920 than that of *Principia*. An orthodox position in line with Schröder's would take set theory to be a part of logic, since his calculus of relatives involved sets and relations over a domain of individuals. This seems to have actually been the viewpoint of Löwenheim, but he was never influential as long as basic foundational questions go. Much more important were the opinions of Skolem, who developed a powerful critique of the axiomatic movement and the 'classical' mentality. He argued forcefully for the exclusive use of first-order logic in axiomatics, using this as a weapon against formalism (§5.2).

Meanwhile, the formalists sided with Zermelo in treating set theory as a mathematical theory that called for axiomatization. Hilbert and his followers stressed the need for an underlying logic in axiomatics and chose a strong higher-order logic, in fact a version of type theory (see §4). At the same time, they were among the first to study first-order logic as an autonomous system, which they regarded as an important *subsystem* of logic [Moore 1988; 1997]. It was only gradually that younger members of the Hilbertian school became partisans of first-order logic, particularly

---

[1] A similar tendency can be found among North-American authors such as Post, influenced by the axiomatic movement, and also, though less clearly, in Skolem (see [Dreben & van Heijenoort 1986, 44–48]).

after Gödel's results. More details on the evolution of logic during the interwar period will be found in the following sections, particularly §§2 and 3. But from the foregoing overview it should be clear that the scope of logic and the problems of set theory were topics of heated debate within the context of the 'foundational crisis.' As a natural result, insecurity reigned and caution was the keyword.

3. The Road to the Simple Theory of Types

Until about 1930, *Principia Mathematica* left an unmistakable mark on the conception of mathematical logic. True, Russell's system faced important problems when presented as the foundation for a purely logical development of modern mathematics. But one should differentiate the issue of the appropriate basic system of logic, from that of type theory as a foundational system for mathematics. Regarding the Axioms of Infinity and Choice (Multiplicative axiom) as properly mathematical ones, type theory became a purely logical system that seemed quite unobjectionable at the time. We shall still call this higher-order logical system 'type theory,' because it includes axioms of comprehension and extensionality, that is, a general theory of classes. But the reader should keep in mind that no assumption is being made that the system affords a sufficient foundation for mathematics.

The most influential simplification of TT was suggested by Chwistek and Ramsey. Ramsey was able to avoid the ramification introduced by Russell's theory of orders (§IX.5) and get along with the hierarchy of types alone. In the process, the system was completely extensionalized and he had to adopt a Platonistic viewpoint on the existence of classes that are satisfactory as a foundation for classical analysis. TT became an impredicative system, which implied the need to find a motivation for it quite different from Russell's. The work of Ramsey is extremely interesting as the first indication of a convergence between both systems, type theory and set theory, that began to develop. But before coming to him, we shall begin discussing what was chronologically the first great simplification of the system.

3.1. Ordered pairs.   Classes and relations are treated separately in *Principia*, so that at each level of the hierarchy of types we have to consider different types for classes, for binary relations, for *n*-ary relations. Likewise, Whitehead & Russell had to formulate two axioms of Reducibility, one for classes and one for binary relations. Already in 1895 Schröder had treated binary relations as classes of ordered couples, and Peano did so too, but the notion of couple or ordered pair had not been satisfactorily analyzed. This situation affected set theory as much as type theory: a set-theoretical definition of ordered pair would greatly simplify the picture, by allowing a reduction of relations and functions to sets. In 1914, Norbert Wiener presented such a definition within the system of *Principia*. Simultaneously, Hausdorff looked for one within set theory.

Wiener obtained a Ph.D. at Harvard in 1913, with a thesis comparing Schröder's treatment of relations with that of Whitehead and Russell; this may have given the initial motivation for his improvement. While living in England on a fellowship, and attending a course on *Principia* by Russell, he published a short paper with his definition. Suffice it to say that Wiener employed the null set for differentiating the members of the ordered pair.[1] This had the pleasant effect of dispensing with one of the two forms that Whitehead & Russell had to give the Axiom of Reducibility [see Wiener 1914, 224–25].

Notably, Russell did not regard this contribution as an important step forward for his theory. In the introduction to the second edition of *Principia* (1925) he stressed the significance of Sheffer's stroke for simplifying propositional logic, but he did not even mention Wiener's new definition of ordered pairs. The reason seems to be the following. While it was a simple step technically, conceptually it involved a wide change, since Wiener is treating relations extensionally. For philosophical reasons, Russell (like Frege) always preferred an intensional treatment of propositional functions – i.e., of concepts and relations. The situation would change after the adoption of simple TT in the form proposed by Ramsey, which was extensional from the beginning.

Simultaneously with Wiener, Hausdorff [1914, 32] employed two distinguished objects, called 1 and 2, to define the ordered pair $<a,b>$ as $\{\{a,1\},\{b,2\}\}$. This was acceptable in the informal context in which he developed his work, but cumbersome within axiomatic set theory. As is well known, the most economical definition came seven years later, proposed by Kasimierz Kuratowski [1921]: $<a,b>$ is defined as the set $\{\{a\},\{a,b\}\}$. It is noteworthy that this definition was a natural byproduct of Kuratowski's investigation of the theory of ordered sets [Hallett 1984]. With these definitions, axiomatic set theory and extensional type theory have no need for a separate notion of function or of relation, in addition to the basic one of set. The reader will recall that Dedekind treated both notions as primitive (§VII.2), and the same has happened occasionally in modern treatments of set theory, such as Bourbaki's.

3.2. Toward simple type theory. The *Principia* version of type theory turned out to be highly controversial because of the Axiom of Reducibility. Most critics, even those who formulated systems very close to TT, regarded it as unacceptable – examples are Weyl, Wittgenstein, Hilbert, Ramsey, Gödel, Waismann, and Quine. The motivation behind the vicious circle principle could hardly be other than that properties are not given, existing in some Platonistic realm, but are constructed. If properties existed independently of our definitions or constructions, there would be no reason to prohibit impredicative definitions.[2] The Axiom of Reducibility decrees

---

[1] See [Wiener 1914, 225]. Translated into modern notation, $< x,y > =_{Df.} \{ \{\{x\},\varnothing\}, \{\{y\}\} \}$; this complies with type restrictions, since there is a null set for each type.

[2] One may conjecture that part of the difficulty during this period was that many people lacked a clear understanding of the background motivations for the different viewpoints ad-

that, for any property of any order, there is a property of order 1 coextensive with it. But, if properties are constructed, one should be able to prove that proposition, otherwise it establishes an arbitrary decree. One can hardly avoid the conclusion that the axiom was in effect a renunciation from the vicious circle principle.

In [1903], Russell characterized logical propositions as those which are true and absolutely general, but this was still somewhat vague. His student Wittgenstein, in the *Tractatus* finished in 1918, pointed out that accidental general validity must be distinguished from logical validity. One can perfectly well imagine a world in which the Axiom of Reducibility is invalid, therefore it is not a logical proposition but – if true – a happy accident [Wittgenstein 1921, §§6.1232, 6.1233]. Ramsey was also of the opinion that the only proposition in Russell's system that is not tautological, "the blemish," is actually Reducibility [1926, 162–3, 208].[1] Similarly, Weyl [1918, 36] regarded the Axiom as a clear indication of the "abyss" that separated him from Russell. As he later said, it was a symptom of Russell's "complete volteface" and abandonment of the road of purely logical analysis [Weyl 1946, 272]. Russell himself acknowledged in 1925 that it was highly desirable to avoid the axiom, since its only justification was "purely pragmatic" [Whitehead & Russell 1910/13, vol.1, xiv]. From about 1920, several authors worked on this project.

Prominent among them was the Polish logician Leon Chwistek, who in 1921 published the first proposal of simple type theory:

For the elimination of this antinomy there suffices the simple theory of types, depending on distinction of individuals, functions of individuals, functions of these functions, and so forth. Distinction of orders of functions of a given argument, and introduction thereby of predicative functions, and in further consequence appealing to the principle of reducibility is from this point of view a superfluous complication of the system. It should be noted that removal of the above elements from the theory of types of Whitehead & Russell would render this theory extraordinarily simple and perspicuous. If therefore Whitehead & Russell could not make up their minds to the simplification, then they undoubtedly did that as a result of the conviction that a system of logic admitting the antinomy of Richard cannot be regarded as a final expression of that which it is possible to attain in the given sphere. Leaving this matter aside, we restrict ourselves to the assertion that the [ramified] theory of types together with the principle of reducibility cannot be maintained, because either it is false or else it represents in intricate form that which fundamentally is simple.[2]

---

vanced, in the present case, of the relationship between platonistic assumptions and impredicativity.

[1] But in fact Ramsey lacked a clear notion of tautology; by 'tautological' he means 'true for an intended interpretation' (see, e.g., [1926, 209–10]). Strictly speaking, a proposition is a tautology iff it is true for any combination of truth-values of its constituent elementary propositions [Wittgenstein 1921, §4.46]. The doctrine of Wittgenstein and the Vienna Circle is flawed, since the notion of tautology only seems to make sense within the realm of propositional logic, as Tarski pointed out in 1934 [Tarski 1986, vol. 4, 694; 1956, 419–20].

[2] [Chwistek 1921], as translated by Church [1937a, 169].

Thus, Chwistek distinguished two kinds of type theory, 'primitive' or 'simple' TT, and 'pure,' branched' or 'constructive' TT. But he never accepted simple type theory as other than a provisional system, because it presupposes a realistic viewpoint that he could not accept. In 1924–25, he presented a system of 'constructive' TT without the Axiom of Reducibility, in which he accepted only finite types and those propositional functions that are constructible according to predetermined rules.[1] Russell pointed out in the second edition of *Principia* [1910/13, vol.1, xiv] that Chwistek's was a "heroic" decision, which led to sacrificing much of classical mathematics. But, as Church remarks [1937a], the introduction of arithmetic by means of special axioms makes his proposal more reasonable, in fact somewhat close to Weyl's system.

At any rate, the work of Chwistek remained little known because he published mostly in Polish, used a strange symbolism, and liked to enter into many obscure philosophical asides. The most influential modification of TT came in the mid-20s with Frank P. Ramsey, a student of Russell and friend of Wittgenstein, inspired to a large extent by the sophisticated understanding of logic proposed in the *Tractatus*. In 1926 he published a paper on 'The Foundations of Mathematics' which presented a version of simple type theory and attempted to justify it. He explicitly advanced the mathematical motivation of offering a sound foundation for classical analysis. In reference to ramified TT, he wrote:

as I can neither accept the Axiom of Reducibility nor reject ordinary analysis, I cannot believe in a theory which presents me with no third possibility [Ramsey 1926, 180].

Ramsey defended forcefully the idea that mathematics is "a calculus of extensions" [1926, 165], which had great implications for his proposed modifications of TT. He saw three main defects in the system of *Principia*. 1) The introduction of orders and ramification with the purpose of solving some of the paradoxes, which creates the need for the Axiom of Reducibility. 2) The treatment of identity, which creates problems in the interpretation of the Axiom of Infinity, making it look far from being a tautology. And 3) not admitting infinite "indefinable classes," which leads to problems with the Multiplicative Axiom [*op.cit.*, 173–181]. Such 'indefinable' classes cannot be mentioned by themselves, but they could be under the scope of quantifiers, and Ramsey thinks their assumption is essential for mathematics.

Concerning problem 1), Ramsey proposed to distinguish two radically different kinds of paradoxes, as Peano had already suggested in 1906. Those which depend only on basic logical or mathematical notions, like Russell's or Burali-Forti's paradox, are called "logical paradoxes" [Ramsey 1926, 171–72]. These are the main problem, while the second group of paradoxes cannot affect a neatly specified logical system, for they depend on "some psychological term, such as meaning, defining, naming or asserting" [Ramsey 1978, 227–28].[2] The second kind of paradoxes

---

[1] See [Grattan-Guinness 1979, 74], [Fraenkel, Bar-Hillel & Levy 1973, 200–05] and further references given here.

[2] Nowadays, Ramsey's 'logical' paradoxes are called set-theoretical; his terminology reflects

are solved by applying the vicious circle principle to the logical symbolism, but not to the propositional functions that are supposedly referred by it [Ramsey 1926, 192–200]. Subsequent authors would simply point out that paradoxes of the second kind cannot be formulated in the formal languages of logical systems. In general, the proposals of Ramsey have the typical flavor of the pre-metatheoretical period, since he does not differentiate (and interrelate) clearly the spheres of syntax and semantics. But, although he failed to present the idea in its full simplicity, his differentiation made it possible to justify reliance on simple TT. The need for ramification and Reducibility disappeared, while the principle of the types of arguments was enough to stop the logical paradoxes.

Trying to solve problems 2) and 3), Ramsey experimented with a system based on what he called "predicative propositional functions" [Ramsey 1926, 190]. This was somewhat similar to the predicative system of Weyl (see §5.1), but Ramsey found it unsatisfactory, "every bit as inadequate as *PM* to provide an extensional logic," i.e., a foundation for mathematics [*op.cit.*, 201]. For those reasons, he came to the conclusion that the only way to obtain a complete theory of classes was to introduce non-predicative functions, which forced him "to treat propositional functions like mathematical functions, that is, to extensionalize them completely" [*op.cit.*, 203]. In complete rupture with Russell and the tradition of Frege, he was abandoning the idea of the primacy of intensions and joining the proponents of set theory in their preference for extensions. Variables of the first type were to range over "functions in extension," while variables of the second type would range over "predicative functions" [*op.cit.*, 207]. This is not yet the simple TT that would be used by later authors, but it indicates the way.

Ramsey had found a way to intuitively motivate a system that is formally very similar to that of *Principia*, but in spirit and interpretation much closer to set theory. He extensionalized completely, adopting a Platonistic viewpoint on the existence of "functions in extension" (classes) that provide a foundation for classical analysis. He found ways of dispensing with the Axiom of Reducibility, but there still were many problems with the formal structure of the new system, in particular with how to delimit the realm of "functions in extension." Ramsey lacked an independent characterization of what these functions are. All we seem to know is that they are the counterparts of arbitrary classes, but without a more detailed specification we seem to relapse into a pre-axiomatic viewpoint again, into naive set theory. Later authors would characterize the available propositional functions syntactically, but Ramsey was still far from this formal viewpoint.

Ramsey's class realism made understandable the Axiom of Choice (Multiplicative Axiom), and he also tried to find arguments showing the Axiom of Infinity to be tautological. He toyed with the idea that we do not need infinitely many basic objects, as Russell thought, but only "some infinite type" with infinitely many elements, which we can then take as the type of individuals. But the admission of

---

the fact that he viewed the type-theoretic theory of classes as a part of logic. After Tarski's work on formal semantics, the other kind of paradoxes were called 'semantic.'

types of level ω and greater also created huge difficulties, contradicting the initial spirit of TT. Since for these reasons he was not able to rescue Russell's logicist project, Ramsey concluded his article on a pessimistic note [1926, 210–212].[1] Nevertheless, early in the 1930s Gödel and Tarski would formulate versions of simple TT that are quite satisfactory from a formal standpoint, and, though comparable to set theory in being Platonistic, much weaker than this theory in their existential assumptions.

4. Type Theory at its Zenith

The success of type theory as a logical system can be substantiated through writings of the most important logicians of the period. From about 1928, *Principia* ceased to be the main reference work in logic, since books such as Hilbert & Ackermann's *Grundzüge der theoretischen Logik* [1928] became available. This introduced a more advanced standpoint: the different formal systems are presented with greater clarity and precision, and metatheoretical questions are addressed. The work of Hilbert & Ackermann can actually be regarded as the first textbook that is paradigmatic of the new period in the development of logic, the metatheoretical period. It included a detailed study of advanced logical systems that were called 'functional calculi,' the so-called "restricted functional calculus" [engerer Funktionenkalkül] – a system of first-order logic, and the "expanded functional calculus" [erweiterter Funktionenkalkül] – a peculiar version of type theory. The restricted calculus was regarded as an interesting subsystem worthy of being studied independently, but for the purposes of an axiomatic development of mathematics the expanded calculus was thought indispensable. Set theory and arithmetic cannot be adequately treated in the restricted system, since the principle of induction, and notions such as number, membership, and cardinality, have to be formulated in higher-order logic in order to capture their intuitive meaning. Hilbert never abandoned this appraisal of the situation (see [Hilbert & Bernays 1934/39, vol. 2]), which was also shared by other logicians, notably by Church [1956].

Similarly, though less surprisingly, Rudolf Carnap presented versions of type theory in his different books on logic, beginning with [Carnap 1929]. Carnap was strongly influenced by Russell in matters of logic and philosophy, being the most prominent defender of logicism at the time in question. His [1929] was a summary and popularization of the logical system of *Principia*, but developed along the lines of Ramsey's simple TT. Type theory is, again, the system proposed in Carnap's defense of logicism [1931] at the famous 1930 Conference on epistemology of the exact sciences at Königsberg, where Gödel announced his first incompleteness theorem. In his most important work, *The Logical Syntax of Language*, Carnap

---

[1] Ramsey then tended to formalism with a 1928 contribution to the decision problem, establishing the theorem that bears his name, and a 1929 article for the *Britannica*; and finally he tended to intuitionism, according to a review by Russell in *Mind* **40** (1931).

studied two formal systems: a form of first-order logic, called "language I," which included primitive recursive arithmetic and was inspired by intuitionist ideas (see [Carnap 1937, §16]); and a form of type theory built on top of the previous system, called "language II."[1]

We could go on citing further examples, for instance from articles by Quine and Church. The latter in [Church 1940] presented a noteworthy and influential reformulation of simple type theory based on his lambda calculus, which produced a very elegant result; he presented axioms of infinity and choice because he wanted to preserve TT as a foundational system alternative to set theory.[2] But if one had to choose the two most important authors as of 1930, most logicians and historians would select Gödel and Tarski. Both of them seem to have regarded type theory as the natural (or at least the customary) system of logic at the time. TT is at the basis of their renowned contributions, Gödel's paper on undecidable propositions and Tarski's on the concept of truth. Though published in different years, both papers were actually written in 1930–31, and the systems presented in them are coincidental. Quine [1986, 11] has written that TT received its "classical thumbnail formulation" with Tarski and Gödel.

The simplest version is Gödel's [1931, 150–55], while Tarski's has the peculiarity of being formulated in his metatheoretical symbolism, which makes it slightly more difficult to interpret. Besides, the only difference is that Tarski presents an Axiom of Infinity, while Gödel has no need of it since his purpose is to give a type-theoretic formulation of Peano arithmetic.[3] Gödel [1931, 150–55] employs a formal language with indexed variables 'x_i', 'y_i', 'z_i', ... for all natural numbers i, where i is the type of the objects that the variable refers to. Elementary formulas are of the form '$x_{i+1}(x_i)$' and we find the usual recursive definitions for well-formed formulas. He lists five groups of axioms, the first group being that of the Peano axioms, while the rest synthesizes the logic of *Principia*. The system looks very much like current ones for first-order logic, except for the last two groups of axioms, devoted to the characteristic principles of type theory [1931, 154–55]. The *Axiom of Comprehension*, group IV, is any instance of

$$\exists u_{n+1} \, \forall v_n \, (u_{n+1}(v_n) \leftrightarrow \alpha)$$

---

[1] See [Carnap 1937] and [Sarkar 1992]. Of the intuitionist authors which Carnap quotes, the one who uses a system closer to language I is Weyl (§5.1). Type theory was still the preferred logical system in Carnap's *Introduction to Symbolic Logic*, a handbook of 1954 in the original German.

[2] On Church's life and work, including a description of his system for TT and comments on the implications it had for Henkin, see [Manzano 1997] (particularly [227–29]).

[3] See [Tarski 1933, 241–43] and for motivation of the system [Tarski 1931, 213–17]. Infinity was formulated by asserting the existence of a class of type 3 whose properties make necessary the existence of infinitely many individuals. Formulating the Axiom of Choice does not present any particular problem either.

where 'u_{n+1}' is not free in the arbitrary formula 'α.'[1] The *Axiom of Extensionality*, group V, is any instance of

$$\forall x_{i-1}\ [x_i(x_{i-1}) \leftrightarrow y_i(x_{i-1})] \rightarrow x_i = y_i$$

for any natural number i.[2] The reader should keep in mind that *predication* plays here the role that membership plays in set theory. Otherwise said, instead of '$u_{n+1}(v_n)$' we might as well write '$v_n \in u_{n+1}$', instead of '$x_i(x_{i-1}) \leftrightarrow y_i(x_{i-1})$' we might also write '$x_{i-1} \in x_i \leftrightarrow x_{i-1} \in y_i$'.[3] These purely formal changes should help the reader grasp the relation between this version of TT and set theory.

Gödel's system is formally very simple. It keeps the Axiom of Comprehension, which was responsible for the emergence of paradoxes in Frege's system, but avoids these by means of the type restrictions, imposed by allowing only elementary formulas of the form '$x_i \in x_{i+1}$'. The intuitive idea is that a class of type $i+1$ can only have classes of type i as its elements. There is an important difference in the formulation of this axiom, compared with Russell's or Ramsey's. Russell did not specify clearly what the domain of propositional functions is, corresponding to 'α' in the Axiom of Comprehension above. He seemed to assume the existence of a realm of propositional functions (concepts or attributes) given as a Platonistic domain. Gödel and Tarski, on the other hand, are under the powerful influence of Hilbert's axiomatics and formalism; in their system, 'α' represents an arbitrary well-formed formula of the language. The difference is huge, since a vague intuitive notion has been replaced by a precise formal one. With this last change, simple type theory receives its definitive formulation.

Tarski's paper on the concept of truth in formalized languages is notable because it gives the ultimate rationale of his preference for TT as of 1931 and 1933. The main reason was his adherence to the theory of semantical categories of his teacher, the philosopher and logician Stanisław Leśniewski. Tarski started his career, early in the 1920s, devoting much time to set theory under the guidance of Sierpiński, and to the logic of *Principia Mathematica* as the subject of his doctoral thesis. In the context of set theory he used to employ Zermelo's axiom system,[4] and he was well aware that this system was more convenient for mathematical purposes, and more powerful too [Tarski 1986, vol.1, 186]. Nevertheless, in his joint paper with Lindenbaum recording their set-theoretical results, published in 1926, a choice was left open between the systems of Zermelo, Whitehead & Russell, and Leśniewski (see [*op.cit.*, 173]). We may conjecture that the influence of Leśniewski was

---

[1] Tarski, following his teacher Leśniewski, called instances of this axiom "pseudo-definitions."

[2] I have not formulated the axioms exactly as Gödel did, but introduced slight simplifications that involve no conceptual change. He wrote: $(Eu)(v\prod(u(v) \equiv a))$, making explicit the type restrictions, and $x_1\prod[x_2(x_1) \equiv y_2(x_1)] \supset x_2 = y_2$, indicating that we may elevate the types.

[3] This is actually what Tarski did. He remarks on the relation between predication and membership in [Tarski 1931, 214n].

[4] See, e.g., his papers of 1924 in [Tarski 1986, vol.1, 41, 67].

getting stronger or becoming more conscious by this time. Be that as it may, in 1930 Tarski regarded himself a follower of the "intuitionistic formalism" of Leśniewski [Tarski 1956, 62]. When he started publishing on metamathematics and semantics, in 1930, he needed to make explicit a basic system of logic, and chose a Russell- or Leśniewski-style one [1986, vol. 1, 313ff].[1]

According to Tarski, Leśniewski's notion of semantical category is formally analogous to that of type, and the theory of semantical categories can be seen as an extension of simple type theory. Semantical categories are regarded as essential for a sound explanation of the meaningfulness of expressions: grammatical or syntactical rules are not enough to guarantee meaningfulness.[2] Every linguistic expression belongs to exactly one category, and the theory thus makes it necessary to differentiate the categories of individuals, classes of individuals, binary, ternary, etc. relations of individuals, classes of classes of individuals, binary relations of classes of individuals, and so on [Tarski 1933, 215–219]. In Tarski's view,

the theory of semantical categories penetrates so deeply into our fundamental intuitions regarding the meaningfulness of expressions, that it is scarcely possible to imagine a scientific language in which the sentences have a clear intuitive meaning but the structure of which cannot be brought into harmony with the above theory. [Tarski 1933, 215]

As a result, Tarski avoided any theoretical means going beyond TT, finding that the notion of truth – which he showed how to define for languages of finite order – was not definable for what he called "languages of infinite order," such as the full theory of types [Tarski 1935, 241ff].[3]

Type theory kept being extremely influential throughout the 1930s. Logicians published important contributions to simple TT, among which we may mention Church's formulation [1940] employing an original notation for types. Tarski, Gentzen and Beth gave proofs of the consistency of simple TT without Infinity and Choice.[4] Fitch [1938; 1939] proved the consistency of *ramified* TT, thus shown to be quite a weak system. Some authors proposed cumulative versions of simple TT, where a class of type n can have members of any type lower than n [Church 1939, 69]; this brings TT even closer to set theory. Whenever the foundations or the formalization of mathematics were being discussed during this period, TT was mentioned as one of the two main alternative systems, the other being axiomatic set theory.[5] Nevertheless, in the course of the 1930s some key figures distanced themselves from TT; most notably Tarski and Gödel did so, as we shall see (§XI.5).

---

[1] This is the period reflected in the logical papers collected for [Tarski 1956].

[2] See [Fraenkel, Bar-Hillel & Levy 1973, 188–190]. Leśniewski's theory has an important precedent in Husserl, whose work affected Weyl in a similar way.

[3] He also thought it impossible to give a formal definition of the concept of logical consequence, a view that he would correct two or three years later (see [Tarski 1956, 293–295, 413 note 2]).

[4] See *Journal of Symbolic Logic* **2** (1937), 44. The question, however, is quite elementary, and there was corresponding work on ZF [Ackermann 1937].

[5] See, e.g., [Gödel 1931, 144–45], [Tarski 1986, vol. 1, 236 and *passim*], [Church 1939, 69–70], or the retrospective comments in [Carnap 1963, 33].

5. A Radical Proposal: Weyl and Skolem on First-Order Logic

During the 1920s, only authors of a constructivist tendency, particularly Weyl and Skolem, regarded first-order logic as a natural underlying logic for mathematics. This is acknowledged by Carnap in his well-known 1931 presentation of the logicist viewpoint. After stating that he shared with the intuitionists the view that only those expressions which are constructed in finitely many steps should be recognized as properties, he went on:

> The difference between us lies in the fact that we recognize as valid not only the rules of construction which the intuitionists use (the rules of the so-called "restricted functional calculus"), but in addition, permit the use of the expression 'for all properties' (the operations of the so-called "expanded functional calculus"). [Carnap 1931, 52]

Carnap here followed the logical terminology of Hilbert & Ackermann (see §4). What he says is obviously wrong for Brouwer and his strict followers, since these authors do not use a classical logic like the first-order 'restricted functional calculus.' The only authors of a clear constructivist tendency (not exactly intuitionists) that Carnap could have in mind are Weyl in his [1918] and Skolem.

Within the context of the 1920s, a decade marked by *Principia*, the proposal that logical theory had to be restricted to the extent of abandoning quantification over properties could only appear excessively radical. But in the course of time this minority position came to be generally accepted. Von Neumann seems to have been one of the earliest authors to join Weyl and Skolem, the third important proponent of first-order logic in the 1920s; but he did not argue explicitly for this move (§5.2). In the course of the 1930s most came to agree that, if one is looking for a *strictly formal* system, first-order logic is the only reasonable choice. Thus, the situation was finally ripe for the emergence of the modern first-order axiomatization of set theory. The reasons why first-order logic won the day are related to the needs of axiomatics, the goals of Hilbert's *Beweistheorie*, and the basic metatheoretical results gathered around 1930. Reviewing the work of Weyl and Skolem will help us understand the more subtle and less well-known of these motives.

5.1. Weyl's proposals.   As we have seen (§1), Weyl developed a system of predicative set theory and analysis, based on a detailed analysis of logic and certain convictions regarding the notion of infinite set. His reflections on the scope of logic led him to the conclusion that there are basically two possibilities for a mathematician. In mathematics as in other fields, one deals with previously given objects, properties and relations. We are never given 'all' possible properties or relations, only a few, and we need to determine what the admissible properties are, as is shown by the fact that a naive notion of property, like naive set theory, leads to contradictions.[1]

---

[1] One just has to consider the property version of Russell's paradox, or the paradox of 'heterological' [Weyl 1918, 2].

One must either abandon the notion of 'all properties' or give it a clear sense. The latter happens when we consider those properties and relations that can be constructed from the primitive by means of logical operations. This road leads to a predicative logic: one cannot accept impredicative concept formation since we are taking properties and relations to be constructed (see also §3.2). Weyl's full predicative logic was essentially equivalent to Russell's ramified type theory, but of course without the Axiom of Reducibility. On the other hand, a restricted form of logic, corresponding to our first-order, would correspond to the abandonment of 'all properties.' Trivially, it complies with type restrictions, since it only has variables for individuals. Thus, Weyl saw two possible logical frames in which to develop pure mathematics: a "strict procedure" [engeres Verfahren] and a "broad procedure" [weiteres Verfahren] employing predicative higher-order logic. But when analysis is developed predicatively, we must differentiate real numbers of different levels, and it becomes artificial and impracticable. This is why Weyl opted for the "strict procedure" [Weyl 1918, 21–23]. We shall call Weyl's preferred logical frame 'restricted logic,' in adaptation of his words, in order to avoid begging the question of interpreting its relation to first-order logic.[1]

One should mention that Weyl's way of dealing with logical matters is somewhat peculiar, and does not measure up to the highest standards employed by authors of the two following decades. For instance, he does not clearly differentiate propositions from inference rules, nor does he make a difference between variables and schematic letters. Perhaps he intentionally avoided some of these ideas (particularly the first difference) because he wished to stay far from formalistic mathematics and logic, and closer to intuitive conceptions and intensional logic. Most importantly, the logical principles that he presents are given under the form of principles for a logic of relations, although he suggests that parallel to these principles are some corresponding forms of inference [Weyl 1918, 9–10, 29]. His emphasis on relations is easy to understand considering that his purpose was to establish set theory as a theory of relation- and concept-extensions.

Under the heading 'Principles of the combination of judgements,' Weyl [1918, 4–6] presents a logic of relations based on six operations. The principles are clearly established, reflect a careful analysis of logical inference, and can easily be seen to correspond to a first-order logic. We find principles allowing the negation of a relation, the conjunction and disjunction of two relations (principles 1, 3, 4). Principle no. 2 allows the "identification" of free variables in relations, that is, the step from $R(x,y)$ to $R(x,x)$. And two further principles allow the "*filling-out*" [*Ausfüllung*] of a free variable in a relation. One (no. 5) is a principle of instantiation, i.e., it allows the construction of $U(x,y,a)$ from $U(x,y,z)$, a being one of the given objects (in modern terminology, a constant for individuals). Principle no. 6 allows the existential filling-out of free variables, i.e., the

---

[1] Another possible denomination is "finite logic" [*op.cit.*, 21 note, 32], responding to the idea that higher-order predicative logic is "transfinite" because it requires us to overview all the derived properties and relations, not simply the initially given individuals as in restricted logic.

step from $U(x,y,z)$ to $U(x,y,*)$, which is read: "*there is* an object z (of our category) such that the relation $U(x,y,z)$ obtains" [1918, 5].[1]

Weyl was well aware of the role played by existential propositions in mathematics, and, of course, of the relation between general and existential propositions [1918, 4, 7]. In the "strict procedure" we employ the first-order existential quantifier, for we are only allowed to quantify over individuals, objects of the basic given category, such as natural numbers in the case of arithmetic [*op.cit.*, 21, 31]. In contrast with Russell, Weyl was not guided by the goal of reducing mathematics to logic – on the contrary, he was of the opinion that the basic idea of iteration is a first foundation, and that reducing it to set theory, or even to axiomatic mathematics, is circular and superfluous [*op.cit.*, 12, 37–38]. This made it more natural for him to renounce the expanded predicative logic and adopt first-order logic as a frame. To remedy the weakness of this logic, he attempted to find additional principles, of a specifically mathematical kind, that would be sufficient to establish an alternative edifice of analysis [1946, 274].[2]

At the time when Weyl published *Das Kontinuum*, Hilbert was also considering first-order logic in his lectures of 1917/18.[3] From Hilbert's standpoint, this restricted logic was a subsystem of type theory, which he carefully singled out because he found it worthy of independent metatheoretical study. The main difference between both men is that Hilbert never accepted the radical idea of restricting logic to first-order (§4). It may have been that both came to first-order logic independently, but one should also consider the possibility of direct influence. For Weyl started to consider that system while working on his attempt to improve Zermelo's axioms for set theory, and this happened in the early 1910s, while he was still a *Privatdozent* at Göttingen.[4]

Once Weyl's viewpoint is understood, one can easily grasp a problem that logicians in the 1920s and 1930s must have considered. When we speak of higher-order logic, does it mean a Platonistic logic, treating the domain of properties as a pre-existing realm, or predicative logic? This raised doubts on the admissibility of even second-order logic, since it happens to be intimately entangled with precisely those questions regarding the meaning of existence in mathematics that were in discussion among the different parties in the foundational debate.

---

[1] In modern notation $\exists z\, U(x,y,z)$. Weyl realized that his notation was inconvenient, but he was not interested in formalizing mathematics. Similar principles, and in particular the analogue of first-order existential quantication, were already presented in [Weyl 1910].

[2] Thus, to the above logical principles of construction, he added further mathematical principles – of substitution and above all of iteration (principles 7 and 8 [*op.cit.*, 24–28]). We shall not discuss them in detail; suffice it to say that they establish the possibility of recursive definitions, what Weyl calls "iteration." For a detailed analysis see [Feferman 1988].

[3] For detailed analysis of these lectures, see [Moore 1997]. They were the basis for the later book [Hilbert & Ackermann 1928].

[4] See [Weyl 1910] and [1918, 35–36]. Weyl moved to Zürich in 1913.

5.2. Skolemism.   In [1932], Zermelo criticized harshly what he called *"Skolemism"* for its negative effects on axiomatic mathematics in general, and set theory in particular. The referent of his exacerbation, Albert Thoralf Skolem, was a Norwegian mathematician who had studied under Sylow, among others, and taught at Oslo from 1918 [Nagell 1963; Skolem 1970]. Skolem worked on number theory, group theory, and logic and foundations, among other topics. His algebraic work shows the influence of Kronecker, so it is not surprising that he had a strong interest in constructive proofs of known theorems. Skolem started the study of primitive recursive arithmetic in a paper that was published in 1923, but written in 1919. This constituted, for him, the soundest foundation on which to base mathematics, a position he had in common with Poincaré and Weyl [Skolem 1923, 299–300]. Since the consistency of Zermelo's axiom system can only be established by metamathematics, which in turn needs recursive definitions and inductive inferences, it would be circular to insist on reducing the notion 'finite' to set theory.

Skolem is mainly remembered for his decisive contributions to mathematical logic, above all the first important metalogical result ever proved, the Löwenheim–Skolem theorem. Already in the winter of 1915–16, during a post-doctoral stay at Göttingen, he communicated to Bernstein the surprising and radically new result that the notions of axiomatic set theory are unavoidably relative [Skolem 1923, 300]. This was simply a consequence of the (downward) Löwenheim–Skolem theorem, and the anecdote implies that by 1916 he had already studied Löwenheim's work [1915]. Skolem's streamlined proof, and extensions, of Löwenheim's main result appeared in [1920], but he did not address its implications for axiomatic set theory until 1922. The reason was that he believed it was "so clear" that axiomatic set theory "was not a satisfactory ultimate foundation of mathematics;" but having seen, to his surprise, that many mathematicians thought otherwise, he decided to present a critique. The occasion came in August 1922, at a congress of Scandinavian mathematicians [Skolem 1923, 300–01].

Skolem's papers [1920; 1923] reveal detailed knowledge of the work of Cantor, Dedekind, Zermelo, and, among logicians, Schröder, Löwenheim, and Whitehead & Russell.[1] His 1922 address is a masterpiece. Sharply conceived and clearly written, the author deals with foundational matters in a way that anticipates by a decade the level of precision that would become customary among mathematical logicians. Nevertheless, it was probably difficult to understand at the time, since most mathematicians lacked even an elementary knowledge of logical theory. The talk dealt with several questions surrounding Zermelo's axiomatization. It showed some of its deficiencies and explained how to amend them; it offered a new, simpler proof of the Löwenheim–Skolem theorem and discussed Skolem's paradox in an authoritative way; it elaborated on the issues of the consistency and non-categoricity of the axiom system.

---

[1] It was after studying *Principia Mathematica* in 1919 that he wrote his piece on "the recursive mode of thought," i.e., primitive recursive arithmetic.

The only point that is not sufficiently explained in the address, and a very important one for us, is Skolem's insight that axiomatizing set theory means giving a first-order formalization of it [1923, 292, 295–96, 300]. Some historians have interpreted Skolem's proposals as implying that he did not accept the possibility of any logic going beyond first-order, but this cannot be maintained in the face of later papers [1928, 516–17]. It has to be said that Skolem does not speak of first-order logic; rather, he talks of 'Zählaussagen,' which can be translated as "number-statements."[1] The intended meaning of this expression is that the propositions we employ refer only to individuals, to basic objects; we quantify only over individuals (in Skolem's eyes, numbers are the foremost representatives of simple individuals). Skolem does not deny the possibility of what we call second-order logic but, like Weyl, he sees problems in it, because the question: what is the totality of all predicates?, is a difficult one [Skolem 1928, 516].[2]

But Skolem's basic insight lies elsewhere. The spirit of the axiomatic method implies the use of 'number-statements,' because in axiomatics one proceeds *as if* one knew nothing about the objects of the axiomatization, except what is explicitly formulated in the axioms. Since the 19th century it had been stressed that any objects whatsoever may constitute a model of the axiom system, provided we can interpret the basic relations of the system appropriately. This is why "the objects of the axiomatization (in set theory the 'sets') will assume the role of individuals" [Skolem 1928, 517]. Should we act as if the expression 'all predicates of individuals' had a clear sense, for individuals of any possible model, we would be contradicting the spirit of axiomatic mathematics – or else assuming an important part of set theory. Think of predicates extensionally, as sets: 'all predicates of individuals' would then mean 'all subsets of the domain.' Using second-order logic presupposes that the meaning of 'all subsets' is fixed beforehand, independently of axiomatic set theory. In the context of the foundational debate, that was certainly an untenable position.[3] Thus, axiomatic set theory, and in fact any formal axiom system, has to be formulated by means of 'number-statements.'

Skolem believed, and he was probably right, that most mathematicians do not conceive of set theory axiomatically, but think of sets as given by specification of arbitrary collections. As proof of this fact, he pointed to the polemics surrounding the Axiom of Choice, which would be pointless with regard to a strictly axiomatic theory [1923, 300]. Since the notion of set cannot be employed naively, one proceeds to axiomatize; but "when founded in such an axiomatic way, set theory can-

---

[1] Löwenheim [1915] talks about "number-expressions."

[2] He proposes two ways of solving the problem. One is essentially the same as Weyl's predicative (expanded) logic, discussed above. The other would be to axiomatize the notion of predicate, but the result would be essentially equivalent to axiomatic set theory. Thus, non-predicative higher-order logic is certainly not a good candidate for a logical basis on which to axiomatize set theory.

[3] If one has goals different from providing a *foundation* for a given theory, for instance semantic goals, it may be sound to rely on higher-order logic [Shapiro 1991]. This makes some sense of Zermelo's work around 1930 (see [Moore 1980]).

not remain a privileged logical theory" [*op.cit.*, 292]. The basic relation of membership ceases to have a fixed meaning, it can be interpreted at will in models of axiomatic set theory.

As for deficiencies in Zermelo's system, Skolem points above all to the imprecise notion of "definite statement," which he regards as a very deficient point. His
proposal is the now standard one: a definite proposition is a first-order formula, a
"finite expression constructed from elementary propositions of the form $a \in b$ or
$a = b$ by means of the five operations mentioned," conjunction, disjunction, negation, universal and existential quantification [1923, 292–93]. This he regards as a
very natural way of explaining Zermelo's obscure notion, that immediately suggests itself. In fact, on the face of Skolem's proposal it seems easy to read it back in
Zermelo's original explanation.[1] But Zermelo disagreed [1929, 342], preferring a
strong rendering of the Axiom of Separation in higher-order logic, which avoids the
"finitistic prejudice" of "Skolemism" [1932] with its unfortunate consequence: the
Skolem paradox.

With the notion of definite proposition amended, Zermelo's system is turned
into a clearly specified set of 'Zählaussagen' or first-order propositions. Actually it
becomes an infinite sequence of such propositions, because the Axiom of Separation has to be regarded as an axiom-schema, to be replaced by its denumerably
many instances [Skolem 1923, 294–95]. Skolem goes on to offer a rather weak
proof of the Löwenheim–Skolem theorem, which relies only on the recursive mode
of thought [*op.cit.*, 293–94].[2] It thus becomes clear that the theorem does not presuppose anything from set theory. The result reads as follows:

Let there be given an infinite sequence U_1, U_2, ... of number-statements numbered with the
integers; if, now, it is consistent to assume that all these propositions hold simultaneously,
they can all be simultaneously satisfied in the infinite sequence of the positive integers, 1, 2,
3, ... by a suitable determination of the class and relation symbols occurring in the propositions.[3]

Since this immediately applies to Zermelo's system, it is clear that among the domains which satisfy the axioms one can find denumerable domains. This brings
with it the unavoidable relativity of the notions of axiomatic set theory, particularly
the notion of cardinality [1923, 292, 296].

On the basis of the axioms one can prove that there exists a non-denumerable
set M, although from the outside it is clear that no such sets can be found in a denumerable domain. This is paradoxical but, as Skolem explains, there is no contra-

---

[1] That a definite proposition is one for which "the fundamental relations of the domain [of the
form $a \in b$], by means of the axioms and the universaly valid laws of logic, determine without
arbitrariness whether it holds or not" [Zermelo 1908, 201].

[2] Previously Skolem [1920] had offered a different proof using the Axiom of Choice and
Dedekind's theory of chains, with the stronger result that every model of the axiom system has a
denumerable submodel.

[3] [Skolem 1923], as translated in [van Heijenoort 1967, 293].

diction. It just means that, *within* the model or domain, there is no one-to-one mapping of *M* onto the set representing the number sequence [1923, 295]. "In the axiomatization, 'set' does not mean an arbitrarily defined collection," but simply an object, an individual which is connected with some others by means of certain relations [*ibid.*]. And even if the domain consists of sets, the model's power set of *M* may not contain all 'possible' subsets of *M*, i.e., it may not contain some subsets which we might be able to determine from the outside. The resulting relativity of the notion of cardinality is, according to Skolem, the most important point in his address [*op.cit.*, 293].[1] He also speculated with the possibility that, within the domain, some sets may be declared infinite which are from without finite; and that in different domains one may even find representatives of Z_0 which are different from each other.[2]

Skolem went on to indicate that Zermelo's axiom system had to be supplemented in order to be sufficient for Cantorian set theory, and finally he dealt with questions of consistency and categoricity (see §§XI.1 and 2). He showed that the axiomatization is not categorical, for which reason Skolem thought that the system may not decide the Continuum Problem [1923, 299 note]. The richness of the questions addressed by Skolem is evident, but his paper was probably little read, although it was cited by both Fraenkel [1925, 250–51] and von Neumann [1925, 405], and the latter presented a detailed discussion of the Skolem paradox. At any rate, Skolem's viewpoint that axiomatic set theory is, or has to be, formulated in first-order logic was very slow in being accepted (see chapter XI).

Von Neumann seems to have accepted Skolem's viewpoint that formal axiom systems must be framed within first-order logic. His systems of the 1920s accord with that tendency, as does his emphasis on the distance between naive set theory and its formal counterpart [1925, 395–97, 404–05; 1928a, 344]. But he did not emphasize that feature explicitly, nor did he argue for that view, at least in print. It actually seems difficult to ascertain what his opinion was on the issue around 1925, because most authors lacked a precise notion of the distinction between first- and higher-order logic at the time. At any rate, von Neumann's axiom system reveals good knowledge of recent logical work, and in [1925] he discussed the Skolem paradox in detail.[3]

Later on, authors like Gödel, Bernays and Quine would gradually adopt that conception, particularly after Gödel [1931] proved that formal systems going beyond first-order logic are incomplete, i.e., not completely formalizable (§XI.5). By "*Skolemism*", Zermelo understood the doctrine that "*every* mathematical theory, including set theory, is realizable in a *countable model*," which is a result of the presupposition that "all mathematical concepts and propositions must be repre-

---

[1] As we have seen, the result was known to him as early as 1916. He came back to the topic later, in a long paper where he also formulated explicitly a first-order version of ZF [Skolem 1929].

[2] Following this trend of thought, Skolem was the first to study non-standard models of arithmetic in the 1930s.

[3] Further research, particularly on unpublished material, might settle that point.

sentable by a *fixed finite sign-system.*"[1] Unfortunately for him, the 1930s was a period in which mathematicians working on foundational questions were increasingly infected by the virus of Skolemism. It is quite telling that Zermelo's antidote, the radical abandonment of that presupposition and the use of a powerful form of infinitary logic, found no echo during that period.[2] Logic was increasingly coming to mean a theory based on a finitary formal system.

---

[1] [Zermelo 1932, 85]: "Von der Voraussetzung ausgehend, dass alle mathematischen Begriffe und Sätze durch ein *festes endliches Zeichensystem* darstellbar sein müssten, gerät man ... in die bekannte '*Richardsche Antinomie*', wie sie neuerdings, nachdem sie schon lange erledigt und begraben schien, im *Skolemismus*, der Lehre, dass *jede* mathematische Theorie, auch die Mengenlehre, in einem *abzählbaren Modell* realisierbar sei, ihre fröhliche Auferstehung gefunden hat."

[2] A detailed discussion of Zermelo's proposal and his discussion with Skolem can be found in [Moore 1980].

XI · Consolidation of Axiomatic Set Theory

> We shall carefully investigate those ways of forming notions and those modes of inference that are fruitful; we shall nurse them, support them, and make them usable, whenever there is the slightest promise of success. No one shall be able to drive us from the paradise that Cantor created for us.[1]

After 1918, most important contributions to the foundations of abstract set theory relied on modern axiom systems. But until the 1920s few authors adopted Zermelo's axiom system explicitly.[2] As we saw in the preceding chapter, many favored the theory of types because it seemed to offer a safer framework, and at the same time it was sufficient for the limited amount of set theory that is necessary in so-called classical mathematics. As late as 1939 Alonzo Church was writing that the simplified theory of types and Zermelo's set theory were essentially similar, and the "safest cities of refuge" for classicist mathematicians at the time.[3] But the Zermelo system had to compete with another alternative, the system of von Neumann, presented in 1925 and developed later by Bernays and Gödel (see §3).

Furthermore, between 1910 and 1940 a good number of treatises on set theory were still developed in naive style. An example is Sierpiński's *Leçons sur les nombres transfinis* [1928], where the author does not care to clarify the notion of set, nor to formulate a single axiom except AC – which is, to be sure, very carefully investigated.[4] Similarly, Luzin in his *Leçons sur les ensembles analytiques et leurs applications* [1930] works informally and even regards as the goal of "set theory" the following open question: whether it is acceptable to understand the continuum atomistically, as a set of points [*op.cit.*, 2]. One should also mention that there were very interesting contributions to the foundations of set theory developed within a naive, non-axiomatic context. This was the case with Mirimanoff, who advanced a great many new conceptions that would be taken up by other authors we shall review. For expository reasons, his work will be mentioned briefly in §2, although it was published before that of Fraenkel (§1).

---

[1] [Hilbert 1926] as translated in [van Heijenoort 1967, 375–76].

[2] Among the few examples are Hessenberg [1909] and Hartogs [1915]. The latter proved that, in Zermelo's system without AC, the comparability of cardinals implies Well-Ordering.

[3] See [Church 1937, 95; 1939, 69–70].

[4] Sierpiński was interested in open problems, like those related to AC and CH, their consequences and equivalences. See [Moore 1982; 1989].

Nevertheless, the situation was rapidly changing, mostly under the impact of the foundational debate. As we have seen, simple type theory was the result of an evolution in which Russell's system was transformed to become closer to axiomatic set theory in its setup and spirit. At the same time, the Zermelo axioms were complemented and simplified, coming to be called Zermelo–Fraenkel system. In the hands of Skolem, von Neumann and Gödel it was formulated as a "logistic" system, namely within the framework of formal logic, and in this way it came to resemble more closely (while superseding it) the work of Whitehead & Russell. Finally, the Zermelo–Fraenkel system came to incorporate traits of simple type theory, and in the process it gained a satisfactory backing in the idea of the cumulative hierarchy (§2.3).

The present chapter focuses upon contributions to the foundations of set theory, with an eye to innovations at the basic level of the axiom system(s), and to some crucial metatheoretical results (§4 above all). We shall also finish reviewing the above-mentioned convergence of viewpoints, and reconstruct the emergence of the modern first-order axiomatization (§§3–5). Finally, we shall glance ahead with the aim to provide a glimpse of the manifold attitudes toward foundations and set theory after the Second World War (§6).

It is convenient to clarify some abbreviations we shall employ, even though they are customary in works on set theory. ZF refers to the Zermelo–Fraenkel system without the Axiom of Choice, ZFC refers to the system with Choice. NBG refers to the system of von Neumann–Bernays–Gödel without Choice. By the end of the period we shall consider, after World War II, the choice between the systems ZF and NBG was still open among experts in mathematical logic. The following abbreviations will continue to be used as we have done in preceding chapters: TT (type theory), AC (Axiom of Choice), CH (Continuum Hypothesis), GCH (Generalized Continuum Hypothesis).

1. The Contributions of Fraenkel

The name of Abraham Fraenkel has come to be closely associated with axiomatic set theory, since the most common axiom system was and is called the Zermelo–Fraenkel system. The name was first used by Zermelo himself [1930, 29], and the rationale for its adoption was that Fraenkel suggested in 1922 that the system might be complemented by the Axiom of Replacement – but so did Mirimanoff in 1917 and Skolem in 1922. Historically considered, the name given to the ZF system reflects Fraenkel's role in the diffusion of set theory through his textbooks [1923; 1927; 1928], his consistent reliance on Zermelo's axiom system, and his refinements and investigations of that system. Neither Mirimanoff nor Skolem regarded Zermelo's axioms as a convincing foundation.

The name Zermelo–Fraenkel does not represent faithfully the actual importance of Fraenkel's contributions to set theory as contrasted with those of other authors. He worked on improving the notion of 'definite proposition,' but his proposal adolesced of some defects, as von Neumann showed [1928], and in the end it was not so influential as the contemporary one of Skolem's. It may come as a surprise for the reader that, for years, Fraenkel himself refused to adopt the Axiom of Replacement in general set theory.[1] And he did not foster the emergence of modern versions of set theory, based on a refined conception of logic, in the same measure as Skolem and von Neumann did. None of these remarks, however, is meant to deny the importance of his contributions in the 1920s.

Fraenkel devoted the early part of his career to algebraic work, which he started under the guidance of Hensel, giving axioms for *p*-adic systems and attempting to elaborate an abstract theory of rings [Corry 1996, 190–215]. He had also received the strong influence of an uncle who acquainted him with the North-American trend of postulate theory, an outgrowth of Hilbert's early axiomatics fostered by E.H. Moore, Veblen and Huntington, among others [*op.cit.*, 198–202]. This would become an important background for Fraenkel's set-theoretical work, since his investigations of the Zermelo axiom system focused on typical questions of postulate theory – independence and categoricity. The first, 1919 edition of his famous introduction to set theory emerged from lessons he gave to prisoner friends during World War I. The work, addressed to mathematicians and philosophers, was substantially improved in subsequent versions during the 1920s [1923; 1928], becoming a widely used reference in questions of general and axiomatic set theory.[2]

Fraenkel's earliest publications on set theory, 'On the Foundations of the Cantor–Zermelo Set Theory' [1922] and 'On the Notion 'Definite' and the Independence of the Axiom of Choice' [1922a], were the outcome of research on the independence of Zermelo's postulates. In the course of it, Fraenkel noticed that some of the axioms could be given simpler formulations,[3] and that there was room for adding a new axiom of Replacement [1922, 231, 234]. He also noticed that the system, as given by Zermelo, was far from categorically characterizing a realm of sets, and he introduced a new postulate trying to remedy that shortcoming (see below). Fraenkel's most important result was that the Axiom of Choice is independent from the other postulates, which he was able to establish by considering a model $\mathfrak{M}$ of set theory with urelements, in which there are denumerably many urelements [1922a, 287]. This problem was also particularly fruitful because it led him to realize the need to clarify Zermelo's notion of 'definite property' [1922a, 286]. By the very nature of his problem, Fraenkel had to show that all of Zermelo's axioms, except Choice, are valid in the model $\mathfrak{M}$ in question, and this was impossible for Separation until the notion 'definite' used in it had been sharpened.

Fraenkel's solution was different from the now standard one, which Skolem was simultaneously suggesting (§X.5.2). He introduced a new notion of 'function' by

---

[1] The first to argue in earnest for Replacement, and to work out its consequences, was von Neumann in 1923. The history of Replacement has been carefully studied by Hallett [1984].

[2] Fraenkel, who was a zionist, moved to Palestina and became a professor at the Hebrew University in Jerusalem in 1929. See his autobiography [Fraenkel 1967].

[3] The most important instance was Separation, apart from which only the axiom of Elementary Sets was reduced to postulating pairs $\{a,b\}$ (whence the name in use today, Axiom of Pairs).

means of a recursive definition [1922a, 286; 1925, 254; 1926, 132–33]. Such 'functions' are not general mappings, and Fraenkel did not even regard the notion as a new primitive idea of the axiom system [1922a, 286]. It might be found adequate to name them 'set-conditions,' but we shall use the name 'Fraenkel function,' or 'F-function,' to remind the reader that we are not dealing with functions in the usual sense of mappings. Intuitively, Fraenkel's idea was the following [1922a, 286]: given any element x of a set, and possibly other objects, we may form a new set by means of a *finite* process consisting in prescribed applications of Axioms II–VI of Zermelo (§IX.4.2). The outcome is, for each particular x, a certain well-determined set, but we can also leave x indeterminate and conceive of the process as defining a 'Fraenkel function' $\varphi(x)$. It turned out to be possible to define F-functions independently and to link them to the axiom of Separation without a vicious circle.

Fraenkel kept refining his definition of F-functions in subsequent papers. The final formulation [1926, 132–33] was as follows. We define recursively the 'F-functions of x:'

1. x and any constant c are F-functions of x; and so are the power set $\wp(x)$ and the union set $\cup(x)$.
2. Let $m(x)$ be an F-function, $\varphi(x,y)$ and $\psi(x,y)$ F-functions of y, which may also depend on x, and let $\circ$ be one of the relations $=, \neq, \in, \notin$. Then the subset of $m(x)$ determined by $\varphi(x,y) \circ \psi(x,y)$ – i.e., the set of all elements y in $m(x)$ such that $\varphi(x,y) \circ \psi(x,y)$, guaranteed to exist by Separation – , is also an F-function of x.
3. Let $\varphi(x)$ and $\psi(x)$ be two F-functions, then so are $\{\varphi(x),\psi(x)\}$ (an unordered pair) and $\varphi(\psi(x))$.

The Axiom of Separation was now given the following formulation [1925, 254]: To every set m and F-functions $\varphi(y)$, $\psi(y)$, there is a set $m_{\varphi(y) \circ \psi(y)}$ that contains all and only the elements y of m for which $\varphi(y) \circ \psi(y)$ holds (with $\circ$ defined as above).

Fraenkel was aware that his proposal was restrictive. By so sharpening the axiom of Separation, he might be narrowing it too much, so it was doubtful whether the resulting system was equivalent to the original one of Zermelo [1922a, 286; 1925, 251]. For this reason, he devoted another paper [1925] to formulating his system carefully and developing the theory of cardinal equivalence, just as Zermelo had done (the paper bore the same title as Zermelo's [1908]). This was followed in [1926] by a detailed axiomatic development of the theory of ordered sets, and six years later by a paper on the theory of well-ordering. Actually, F-functions were too narrowly defined, but this only became apparent in the context of the Axiom of Replacement (§2.2 and [von Neumann 1928, 322–24]). Shortly afterwards, Skolem showed that the Fraenkel functions are all equivalent to conditions (propositional functions) in the language of first-order logic with the basic set-theoretical relation $\in$ [Skolem 1929, 231–33]. Considering his rendering of the F-functions, one can see that there are first-order conditions that correspond to no Fraenkel function. For this reason, the proposals of Skolem and Fraenkel for sharpening the Zermelo system were not equivalent. It was von Neumann [1928] who refined Fraenkel's and turned it into equivalent with Skolem's.

Already in [1922], simultaneously with Skolem, Fraenkel had noticed that certain sets which Cantor regarded as existent could not be obtained from Zermelo's axioms. Calling Z_0 the set of natural numbers, Z_1 the power set $\wp(Z_0)$, Z_2 the set $\wp(\wp(Z_0))$, and so on, it was not possible to show that $\{Z_0, Z_1, Z_2, ...\}$ is a set. This means, assuming the Generalized Continuum Hypothesis, that $\aleph_\omega$ cannot be shown to exist [1922, 230–31; Skolem 1922, 296–97]. Reflecting on this issue, both mathematicians were led to considering the Axiom of Replacement. Skolem gave it a precise formulation, but Fraenkel only gave a hint at a formulation by saying that, given a set M, if we replace each of its elements by a "thing of the domain" we obtain a new set. The axiom still lacked a precise statement in which the intuitive idea of replacing was sharpened. Fraenkel would suggest one formulation later [1925, 260, 271], but he did not develop the idea because he felt that Replacement would be too powerful a postulate to assume it in general set theory [1922, 231, 233; 1925, 252, 271]. In his view, the axiom was apt for "special" topics in set theory, such as the existence of particular cardinalities. Von Neumann pressed for its acceptance, in connection with his proposal of a new axiomatic definition of the ordinal numbers.

As I have said, Fraenkel was also interested in the question of categoricity of Zermelo's system [Fraenkel 1922, 233–34]. He remarked that, as formulated by Zermelo, the system made room for non-mathematical objects (called urelements), and that there was also room for 'extraordinary sets' in the sense of Mirimanoff. All of this means that the system is not categorical, e.g., that there must be non-isomorphic models. To remedy this shortcoming, he proposed to adopt a new 'axiom' analogous (but inverse) to Hilbert's axiom of completeness (see §IX.1.1). Fraenkel's "axiom of restriction" required that the "domain" or model of axiomatic set theory be restricted to the smallest compatible with the remaining axioms [1922, 234; 1923, 219].[1] That is, there would only be those sets whose existence is strictly required by the other axioms. Formulated as above, the 'axiom' is unacceptable – it is no condition on sets but on models of set theory, i.e., it is not an axiom but a meta-axiom. As von Neumann remarked [1925, 404–05], a precise formulation would require notions of naive set theory or reliance on a 'higher set theory' at the metatheoretical level. And, in the second case, the 'restriction' would not be absolute but relative to the particular model of the 'higher set theory' being used.

The issue of Fraenkel's axiom of restriction is a good example of the difficulties that most mathematicians must have had with formalized axiomatics during this early period. It was easy to relapse into naive set theory one way or another. At any rate, his reflections led Fraenkel to propose restricting axiomatic set theory to pure sets, that is, avoiding urelements other than $\varnothing$ [1922, 234], as is normally done today. Slightly later, von Neumann [1925, 412; 1929, 498] found the way to exclude Mirimanoff's extraordinary sets by means of a new axiom, Foundation, which was independently found and adopted by Zermelo (see §2.3).

---

[1] The idea of such an axiom had already been considered by Weyl in his work of the early 1910s; see [Weyl 1910, 304; 1918, 36].

2. Toward the Modern Axiom System: von Neumann and Zermelo

The contributions of von Neumann advanced substantially in the direction of modern axiomatic set theory. But the fact that he established an important alternative to the ZFC system has led many authors to forget that he was also responsible for innovations in ZFC that have shaped its modern versions (see [Hallett 1984]). Some of von Neumann's proposals had been anticipated in unpublished work of Zermelo, some were taken up and investigated further by him in the important paper [Zermelo 1930]. This presented a clear theory of models of second-order ZF suggesting the idea of the cumulative hierarchy, which would later be emphasized by Gödel. To complement our review of their work, we shall consider the earlier contributions of Mirimanoff, who advanced in this same direction, but non-axiomatically.

2.1. Dimitri Mirimanoff.   A professor at Geneva, Mirimanoff did important work in number theory, and worked on issues in set theory during the War. Here he attempted to explore the universe of sets as freely as possible, without committing himself to arbitrary restrictions beforehand. Thus he was led to distinguish between "ordinary" and "extraordinary sets" [1917, 42]. The latter are nowadays called non-well-founded sets, and they are characterized by giving rise to infinite "descents" $... \in x_2 \in x_1 \in x$. Examples would be any set that is an element of itself, $x \in x$, or a couple of sets such that $x \in y$ and $y \in x$. These sets are 'circular' in some sense, or 'ungrounded' in that we do not reach their 'roots' after finitely many steps. They played an important role in investigations of models of set theory, and they led to the Axiom of Foundation (see §2.3). Mirimanoff was also the first to employ the von Neumann ordinals (§2.2), under the guise of ordered sets called "S sets" [1917, 44–48; 1917a, 213–17]. He showed how the S sets can be defined independently of the notion of well-ordering, proved their basic properties, and presented clearly the essential relation between transfinite ordinals and S sets.[1]

Mirimanoff's main purpose was to find a solution to the "fundamental problem" of set theory, posed by the paradoxes: "what are the necessary and sufficient conditions for a set of individuals to exist?"[2] He showed that every 'ordinary set' has a given "rank," which is the fundamental idea behind the cumulative hierarchy (§2.3), and solved the 'fundamental problem' of set theory (for the case of ordinary sets) as follows. A set of ordinary sets exists if and only if the ranks of its elements have a Cantorian bound, i.e., if there is an ordinal greater than all of those ranks [1917, 51]. Trying to make his presuppositions explicit, Mirimanoff [*op.cit.*, 49]

---

[1] But, unlike von Neumann, he could not see whether this detoured way of dealing with ordinals "presents real advantages," although it throws new light on Cantor's theory [1917a, 217]. Mirimanoff [1917, 45] considered also a criterion of set existence that would be characteristic of von Neumann's axiomatization of set theory.

[2] [Mirimanoff 1917, 38]: "Quelles sont les conditions nécessaires et suffisantes pour qu'un ensemble d'individus existe?"

was the first after Cantor to state naively, by reference to an arbitrary law, a principle that would later be included among the ZF axioms as the Axiom of Replacement (see §1).

As the reader can see, Mirimanoff's work was extremely rich and introduced a great many of the important novelties of the 1920s in the foundations of set theory. But he worked naively, regarding Zermelo's approach as foreign to his own.[1] His three papers were published in *L'Enseignement Mathématique*, which was certainly not a research journal, but still a widely read one. Nevertheless, all one can find is two citations of his work in Fraenkel and von Neumann, which acknowledge only his notions of 'extraordinary sets' and 'descent.' One is left wondering whether the fact that Mirimanoff wrote in French in a Swiss journal, combined with the tense post-War atmosphere, may have been the reason why German-speaking authors did not cite him. If so, political reasons would have caused his work to receive much less credit than it deserves. But let us proceed to those authors who have certainly been influential in the history of axiomatic set theory.

2.2. John von Neumann.   Von Neumann is a legendary figure of 20th century mathematics. In 1926 he received a diploma in chemical engineering from Zürich and a Ph.D. from the University of Budapest with a dissertation on set theory – his first field of research in pure mathematics. By the age of 30, when he was appointed professor at the Institute for Advanced Study in Princeton, he had made fundamental contributions to set theory, Hilbert's program, the foundations of quantum mechanics, and had started work in game theory.

First of all, one should mention that von Neumann presented an original axiom system that was substantially different from Zermelo's (see §3). His contributions advanced substantially in the direction of modern axiomatic set theory, and he was responsible for innovations in ZFC that have shaped its modern versions.[2] In this connection, one should mention the Axioms of Replacement and Foundation, the modern axiomatic theory of ordinal and cardinal numbers, and the theorem that establishes the possibility and univocalness of transfinite recursive definitions. These are the aspects of his work that are of our interest at this point.

Those innovations are the topic of two of his set-theoretical papers, 'On the Introduction of Transfinite Numbers' [1923] and 'On Definitions by Transfinite Induction and Related Questions in General Set Theory' [1928]. Both papers contained a definition of the ordinals that was apt to "give unequivocal and concrete form to Cantor's notion of ordinal number" in the context of axiomatized set theories [1923, 347]. The well-known von Neumann ordinals are, to put it in Cantor's terminology, representatives of the order types of well-ordered sets:

---

[1] See [Mirimanoff 1919, 35]. In this later paper his lack of understanding for axiomatic issues comes out clearly.

[2] This point was emphasized by Hallett [1984].

What we really wish to do is to take as the basis of our considerations the proposition: 'Every ordinal is the type of the set of all ordinals that precede it.' But, in order to avoid the vague notion 'type,' we express it in this form: 'Every ordinal is the set of the ordinals that precede it.' This is not a proposition proved about ordinals; rather, it would be a definition of them if transfinite induction had already been established. [von Neumann 1923, 347]

We thus obtain the series: $\emptyset$, $\{\emptyset\}$, $\{\emptyset,\{\emptyset\}\}$, $\{\emptyset,\{\emptyset\},\{\emptyset,\{\emptyset\}\}\}$, ..., $\{\emptyset, \{\emptyset\}, \{\emptyset,\{\emptyset\}\}, \{\emptyset,\{\emptyset\},\{\emptyset,\{\emptyset\}\}\}, ... \}$, ... Such representatives turn out to be particularly convenient in an axiomatic context, since they can be defined on the basis of the membership relation alone, being well-ordered by strict inclusion or else by membership. Moreover, since each ordinal has the appropriate cardinality, they are also particularly convenient for establishing the connection between ordinals and cardinals. In [1928, 325], von Neumann defined a well-ordered set M to be an ordinal number if and only if, for all $x \in M$, x is equal to the initial section of M determined by x itself (as he wrote, $x=A(x;M)$). The elements of an ordinal number are also ordinal numbers. An ordinal is called a cardinal number if it is not equipollent to any of its elements [1928, 332–33].

In his paper [1923], von Neumann assumed the notions of well-ordered set and similarity and went on to prove that, to each well-ordered set, there is a unique corresponding ordinal. He presented the theory naively, but indicated clearly that it could be incorporated in an axiomatic framework, as he himself did later on [1928; 1928a]. The only restriction is that, in order to do so, one needs the Axiom of Replacement, "Fraenkel's axiom" as he said at the time [1923, 347]. From this point on, von Neumann became the most consistent advocate of Replacement, which contrasts with Fraenkel's doubts regarding the convenience of assuming such a powerful axiom in general set theory. As we have seen, Fraenkel had made precise Zermelo's notion of a 'definite' condition by axiomatizing a certain notion of 'function,' which was then used in the rigorous formulation of Separation and Replacement. In [1928, 322–24], von Neumann showed that Fraenkel's definition of 'function' was insufficient for a satisfactory version of Replacement.[1] Indeed, he proved that, when formulated on the basis of 'Fraenkel functions,' the axiom was superfluous, being provable from the rest of the system. Von Neumann went on to amend the ZF system, presenting a strengthened definition of 'F-functions' that was sufficient for his purposes and turned Replacement into an axiom that expanded Zermelo's system essentially. He also indicated that the Axiom of Separation is a consequence of Replacement [1929, 497].

The last part of [von Neumann 1928] was devoted to establish that it is always possible to satisfy a definition by transfinite recursion based on the ordinals, and that such definitions are univocal. Von Neumann proved that, given any condition φ in two variables, there is one and only one 'F-function' f, with domain the ordinals, such that, for each ordinal α, we have $f(\alpha) = \varphi(f(\alpha),\alpha)$.[2]

---

[1] This had become clear to him in correspondence with Fraenkel himself [1928, 323].

[2] [von Neumann 1928, 334]. I have introduced a slight change to simplify the original notation.

For many years, since Zermelo's axiomatization, Cantor's ordinal and cardinal numbers lacked formalized counterparts and had been avoided in axiomatic set theory. It had even been customary to look for ways to avoid relying on transfinite numbers, or transfinite induction, in mathematical reasoning (see, e.g., [Kuratowski 1921; 1922]). With von Neumann's work, the use of ordinals and transfinite recursion became customary again, and axiomatic set theory began to look as it nowadays does. In this work, von Neumann was apparently led by a desire to rescue as much as possible from Cantor's approach to set theory, and to secure the widest possible scope for the theory. This was taken further in his original axiom system, giving his work a peculiar flavor, since he made bold proposals that must have looked quite risky at the time (§3).

Von Neumann was also one of the first authors to investigate the metatheory of axiomatic set theory. This he did in connection with his own axiom system, analyzing questions of categoricity [1925] and of relative consistency [1929]. He was probably the first author to call attention to the Skolem paradox as a serious result which stamps axiomatic set theory "with the mark of unreality" and gives reasons to "entertain reservations" about it [1925, 405–09]. As the reader will recall, the Skolem paradox is a consequence of the Löwenheim–Skolem theorem that applies to first-order formulations of set theory. Although von Neumann was not sufficiently explicit regarding this point, it seems that he accepted Skolem's proposal that axiom systems ought to use first-order logic. His systems of the 1920s [von Neumann 1925; 1927; 1928; 1928a; 1929] seem to be intended as first-order, and certainly are formalizable within that frame. If that was his intention, von Neumann was the first mathematician to accept Skolem's (and Weyl's) views.

For the purpose of his metatheoretical investigations, von Neumann introduced further axioms that served to make his axiom system more restrictive [1925, 411–12; 1929, 498]. He restricted the objects that the theory deals with to pure sets, i.e., he avoided urelements, following an idea of Fraenkel, and he formulated the Axiom of Foundation. Ideas behind this axiom had first been discussed by Mirimanoff and Skolem, and the issue was subsequently taken up by Fraenkel and Zermelo. Foundation is a very interesting axiom from our viewpoint, since it makes axiomatic set theory look much more similar to type theory. Actually, it is likely that some of the authors who worked on related ideas were looking for differences between type theory and set theory, as was probably the case with Skolem. With Foundation, ZF can be regarded as an extension of (cumulative) type theory to transfinite types, described in a logical language that is simpler than Russell's (§§5 and X.4). This rapprochement of TT and ZF suggested an intuitive picture justifying the latter system, and it seems to have given reasons to feel more convinced of the consistency of ZF.

On the basis of the Axiom of Foundation, von Neumann developed in [1929, 503–05] the cumulative hierarchy in detail. Starting from the assumption that there is a domain satisfying an analogue of ZF, he employed the ordinals and Foundation to define a cumulative subdomain. Such a subdomain can be decomposed into 'sections' that are currently called the *ranks*, indexed by transfinite ordinals. To every set in the model there is a corresponding rank, namely the rank of least ordinal at which that particular set can be found [1929, 505]. The ranks are cumulative, i.e., every rank contains all of the sets that appear in previous ranks – therefore the name *cumulative hierarchy* (see §2.3). Thus, the technical details of the definition of the cumulative hierarchy go back to von Neumann. Nevertheless, it should be added that von Neumann considered Foundation merely as a tool for investigations in the metatheory of his axiom system: he used the cumulative hierarchy to show that, if a certain system resembling ZF is consistent (has a model), then NBG is also consistent (see §3). But, apparently, von Neumann did not entertain adoption of the Axiom of Foundation seriously.

2.3. Zermelo's cumulative hierarchy.   After a long period of silence, motivated by health problems, Zermelo published on set theory again in the late 1920s. Most important is his paper 'On Boundary Numbers and Set-Domains' [1930] where he investigated models of set theory – what he called 'domains.' On the basis of the "Zermelo–Fraenkel axioms" [1930, 29] supplemented with the "Axiom of Foundation" [*op.cit.*, 31], Zermelo was able to produce a greatly illuminating picture of ZF-models.

Zermelo was the first author who explicitly included Foundation among the axioms of ZF, and one of the first to accept Replacement wholeheartedly. His 1930 version of the system was thus quite close to modern versions, but it differed by not being formulated in first-order logic. The system he outlined can be interpreted as second-order, as comes out particularly clearly in his formulation of Separation and Replacement. He explicitly emphasizes that the propositional functions (conditions or predicates) used for separating-off subsets, as well as the replacement functions, can be "entirely *arbitrary*" [ganz *beliebig*; 1930, 30]. This means that they may include higher-order quantification and that the Skolem paradox does not apply (see [Zermelo 1929; 1931]).[1] By giving a second-order formulation of ZF, Zermelo obtained a system that is very powerful in the way of characterizing models, although quite weak from a foundational standpoint.

The new Axiom of Foundation had been considered previously by von Neumann, who gave two different formulations in the context of his own axiom system [1925, 404, 411–12; 1929, 494–508]. Zermelo is said to have introduced the axiom independently [Bernays 1941, 6], and gave it the following form:

---

[1] Zermelo did not spell out the details of his logical standpoint, particularly the effect of second-order logic for the Skolem paradox. But see [Zermelo 1931] and [Grattan-Guinness 1979; Moore 1980; Dawson 1985a]. A clear discussion of the logical aspects can be found in [Shapiro 1991] or [Lavine 1994].

Axiom of Foundation: Every (inverse) chain of elements, each member of which is an element of the previous one, breaks up with finite index into a urelement. Or, what is equivalent: Every subdomain T [of a ZF-model] contains at least one element t_0, that has no element t in T.[1]

The first form of the axiom prohibits infinite $\in$-descending chains, linking with the ideas of Mirimanoff and Skolem; as we can see, Zermelo kept allowing urelements. The second, alternative form had been given by von Neumann [1929, 498]; Zermelo feels free to speak of arbitrary submodels, which reflects his reliance on second-order logic.[2] He freely applies the set-theoretical terminology to models because, as we shall see, in his picture the domain of every model becomes a set in a higher model. Zermelo goes on to say that Foundation excludes all kinds of 'circular' sets, in particular sets that 'contain themselves,' and in general any 'ungrounded' sets. The new axiom, he argued, was actually valid in all previous applications of set theory and therefore, "provisionally" [vorläufig], brought no essential restriction to the theory [1930, 31].

The Axiom of Foundation was employed to great effect. Restricting sets in models of ZF to those that are well-founded, Foundation makes possible a decomposition of the model into what Zermelo called "layers" [Schichten] and "sections" [Abschnitte], presently called 'ranks.'[3] Every set is associated to a given rank, so that its elements belong to previous ranks, and the set itself serves as material for sets in the next ranks [1930, 29–30]. One may describe the ranks as follows (see [*op.cit.*, 36]):

1. Rank zero, V_0, is the collection of all urelements (one of which Zermelo identified with the empty set);

2. $V_{\alpha+1}$ is the union of V_α and a new 'layer;' it coincides with the power set $\wp(V_\alpha)$;

3. if α is a limit ordinal, V_α is the union $\bigcup_{\beta<\alpha} V_\beta$ of all previous ranks.

The model itself is then identical with $V_\pi = \bigcup_{\beta<\pi} V_\beta$ where π is the 'characteristic' of the model, defined below.

---

[1] [Zermelo 1930, 31]: "*Axiom der Fundierung*: Jede (rückschreitende) Kette von Elementen, in welcher jedes Glied Element des vorangehenden ist, bricht mit endlichem Index ab bei einem Urelement. Oder, was gleichbedeutend ist: Jeder Teilbereich T enthält wenigstens ein Element t_0, das kein Element t in T hat."

[2] Today one would require (with first-order quantification) that every non-empty set s in the domain have at least one element t, such that no element of t is also an element of s.

[3] The 'layers' Q_α of Zermelo are not cumulative, in contrast to his 'sections' P_α [1930, 36]. Therefore it is the second which correspond to the ranks of the usual cumulative hierarchy; instead of P_α we shall write V_α for the ranks, as has become customary.

In this work, Zermelo relied on what he called "*basic sequences*" [Grundfolgen; 1930, 31], which are essentially the same as the von Neumann ordinals. According to Bernays, who worked with him at Zürich during the early 1910s, Zermelo had come to that idea independently around 1915, but had not published it (see [Bernays 1941, 6, 10; von Neumann 1928, 321]).

Second-order ZF may be said to be quasi-categorical. The main result in Zermelo's paper is that every model of his version of ZF is fully characterized by two numbers: the cardinality of the "basis," i.e., of the totality of its urelements, and the "characteristic" of the model, i.e., the least "basic sequence" or ordinal number not in the model [1930, 29, 40–42]. One can prove that the 'characteristic' must be what Zermelo called a "boundary number" [Grenzzahl]: not any ordinal, but a strongly inaccessible initial ordinal (here the fact that we are dealing with second-order ZF is essential). Given those two numbers – cardinality of the 'basis' and 'characteristic' – two models of second-order ZF are isomorphic, and this is why one may call the theory 'quasi-categorical.'[1] If we disregard urelements, as is usual, models are categorically determined by the 'characteristic,' and any two models stack in a neat way, since one must be isomorphic to a section of the other [*op.cit.*, 41].[2]

Zermelo thus provided a general analysis of the possible "structures" of models of ZF, a theory of the "model-types" [Modelltypen; 1930, 42]. Each model-type is determined by two numbers, 'basis'-number and 'characteristic,' which Zermelo also calls, metaphorically, the "breadth" and the "height" of the model. He was working by analogy with Steinitz's pathbreaking research on fields, as he acknowledged in the context of an automorphism theorem which he proved [*op.cit.*, 42–43]. The fact that the axiom system is not categorical, and we encounter an unlimited series of essentially different models, was no hindrance in his eyes. On the contrary, it was an "*advantage*" [Vorzug; *op.cit.*, 45] because it afforded a satisfactory explanation of the "ultrafinite antinomies" and enriched the field of application of set theory [*op.cit.*, 29, 45]. As regards the paradoxes, one can see that the "ultrafinite non-sets" [ultrafinite Un- oder Übermengen] of one model become legitimate sets in the next model. Only a confusion of the non-categorical theory of sets itself with a particular one of its models could give rise to the impression that the theory entails contradiction [*op.cit.*, 47]. One also gets a picture suggesting why 'all-embracing' sets are never reached.[3]

As regards motivation, it would seem that Zermelo, a strict defender of actual infinity, was trying to fix the Cantorian paradise. It was to this end that he kept emphasizing the need of a higher-order logic against Skolem [Zermelo 1929; 1931], and it was to this end that he developed his theory of the different models of

---

[1] This term is taken from [Hellman 1989] and [Lavine 1994].

[2] The same applies to different models with a single, common 'basis' of urelements (that differ in their 'characteristic'). Likewise, of two different models with the same 'characteristic,' one is always isomorphic to a subdomain of the other [1930, 42].

[3] For a careful analysis of this paper, see Hallett's introduction to his translation of the paper in [Ewald 1996, vol.2].

ZF. As he emphasized, his work presupposed the existence of such models, which can be taken to be a basic article of faith for the modern mathematician. But he also argued that, assuming the consistency of the Zermelo–Fraenkel axioms, "the (mathematical, i.e., ideal) *existence*" of the models can be proved [1930, 43].

The Axiom of Foundation is quite different from the other axioms in that it has no single known consequence for actual mathematical work outside set theory. But it clarified immensely the possibilities for models of ZFC, particularly in the second-order version of Zermelo. And, by forbidding 'circular' and 'ungrounded' sets, it incorporated one of the crucial motivations of TT – the principle of the types of arguments. As a result of Zermelo's clear analysis of the cumulative hierarchy of sets within a model, there emerged a Zermelo–Fraenkel system that was closer to type theory. We have seen that during the 1920s TT was transformed into a system much closer in spirit to set theory, a system dealing with extensional classes or sets that worked on an impredicative basis, sharing to that extent the Platonism of set theory (§X.3.2). Still, TT seemed to be safer than set theory, since it was more restrictive and its guiding principles seemed clearer than those of ZF (see, e.g., [von Neumann 1929, 495 note 7]). TT was well suited for developing analysis, but it was not powerful enough to develop set theory in all its extension. On the other hand, the way in which the paradoxes and any possible contradictions were avoided appeared clearer in the case of TT.

Zermelo's 'layers' are essentially the same as the types in the contemporary versions of simple TT offered by Gödel and Tarski (§X.4). One can describe the cumulative hierarchy into which Zermelo developed his models as the universe of a cumulative TT in which transfinite types are allowed. (Once we have adopted an impredicative standpoint, abandoning the idea that classes are constructed, it is not unnatural to accept transfinite types.) Thus, simple TT and ZF could now be regarded as systems that 'talk' essentially about the same intended objects. The main difference is that TT relies on a strong higher-order logic, while Zermelo employed second-order logic, and ZF can also be given a first-order formulation. The first-order 'description' of the cumulative hierarchy is much weaker, as is shown by the existence of denumerable models (Skolem paradox), but it enjoys some important advantages (§§5 and X.5).

Zermelo did not yet present the idea of a single 'universe' of sets. His was a dynamic conception, essentially based on the idea of an open-ended sequence of bigger and bigger models, each of which can be identified with a set in the next model.[1] He obtained the cumulative hierarchies *from* axiomatic set theory, never trying to *justify* the latter by the notion of a cumulative hierarchy. Nevertheless, his work suggested precisely this, that one may give an intuitive argument for the ZFC system on the basis of the so-called iterative conception. Gödel spelled out what the iterative conception comes to almost two decades later:[2]

---

[1] Indeed, a double sequence of models, since we also have to take into account the many possible 'bases' of urelements [Zermelo 1930, 42, 47].

[2] Gödel had already presented this conception in conferences during the 1930s; see volume 3 of his *Collected Works*.

As far as sets occur and are necessary in mathematics (at least in the mathematics of today, including all of Cantor's set theory), they are sets of integers, or of rational numbers (i.e., of pairs of integers), or of real numbers (i.e., of sets of rational numbers), or of functions of real numbers (i.e., of sets of pairs of real numbers), etc. ... This concept of set, however, according to which a set is anything obtainable from the integers (or some other well-defined objects) by iterated application of the operation 'set of,' and not something obtained by dividing the totality of all existing things into two categories, has never led to any antinomy whatsoever; that is, the perfectly 'naïve' and uncritical working with this concept of set has so far proved completely self-consistent. [Gödel 1947, 180]

As he explained in a footnote, Gödel did not regard this as the general concept of set, only as a specification (that is aptly expressed in the Axiom of Foundation); he also made clear that one must include transfinite iterations of the operation 'set of.'[1] As one can see, he was the first to establish a radical contrast between this viewpoint and the old one, embodied in the work of Frege and Russell (but also Riemann and Dedekind) and based on the principle of comprehension.[2]

Gödel went on to say that the iterative conception of 'set' explains at once "that a set of all sets or other sets of similar extension cannot exist, since every set obtained in this way immediately gives rise to further application of the operation 'set of' and, therefore, to the existence of larger sets" [1947, 180]. This makes even clearer that he had in mind the picture of the cumulative hierarchy as suggested by Zermelo's work. Below (§4) we shall see that he had used that picture in his crucial work of the late 1930s. It is noteworthy that, as Gödel presented it, the picture links directly to the old tradition of a step-by-step 'construction' or definition of the number system from the integers (§IV.1, §§VII.1 and 3). It should also be clear that set theory, as conceived by Zermelo and Gödel in these works, is not motivated by a principle of limitation of size.

3. The System von Neumann–Bernays–Gödel

Zermelo's axiom system had been preferred by those who worked not so much on foundational questions, but on advanced issues in set theory and related areas. With the work of von Neumann, and particularly with the later developments and simplifications introduced by Bernays and Gödel, the system NBG became a workable alternative. Von Neumann was in the possession of the essential ideas for his new axiomatization already in 1923, at age twenty. He described them in a letter to Zermelo, where he wrote that he owed the stimulus for this work exclusively to

---

[1] Strictly speaking, it is wrong to say that Gödel offered the iterative conception as justification of the ZFC axioms. This was done by later authors. See [Klaua 1964], [Kreisel 1965], [Shoenfield 1967], [Boolos 1971], [Wang 1974], [Parsons 1977].

[2] Comprehension obtains sets by dividing the universal class into two categories – objects that comply or do not comply with a given condition.

Zermelo's paper of 1908, but that he had deviated from his approach at a few essential points:

1. The notion of 'definite property' had been avoided, presenting instead the "acceptable schemas" for the construction of functions and sets.
2. The Axiom of Replacement had been assumed, since it was necessary for the theory of ordinal numbers. (Later he also emphasized, like Fraenkel and Skolem, that it is needed in order to establish the whole series of cardinalities [von Neumann 1928a, 347].)
3. Sets that are "too big" (e.g., the set of all sets) had been admitted, which he regarded as necessary to formulate Replacement. But "too big" sets were taken to be inadmissible as *elements* of sets, which sufficed to avoid the paradoxes.[1]

Von Neumann later distinguished [1925, 403; 1928a, 348] between "domains" [Bereiche] and "sets" [Mengen]; it has become customary to call the first 'classes.' A class or 'domain' is defined, essentially, by means of the principle of comprehension; von Neumann seems to have regarded this principle as the quintessence of what he called "naive set theory" [1923, 348; 1928, 325; 1929, 496].

Von Neumann's approach to axiomatic set theory was strongly based on the idea of limitation of size:[2] a class is a set if and only if it is not "too big." For this notion, he established a strong criterion by means of his axiom IV.2, which we shall call von Neumann's Axiom:

A [class or domain] is "too big" if and only if it is equivalent to the [class or domain] of all things.[3]

That is, a class c is not a set (i.e., is a *proper* class) if and only if there is a function g, such that to every thing x there is an element $y \in c$ for which $g(y) = x$ [1928a, 345, axiom IV.2]. Of course, the function g is not a set, but a proper class.

In von Neumann's original presentation, the axiom system looked strongly deviant from Zermelo's because he employed the notion of function, not that of set (resp. set-membership), as the primitive notion. But, as he himself remarked, this difference was only important from a technical point of view.[4] He emphasized that the notions of set and function can easily be reduced to one another – a set can be regarded as a function that takes only one of two values (intuitively: being and not being an element) – and a function can be regarded as a set of pairs. Much more important is the fact that von Neumann's Axiom turned out to imply the axioms of

---

[1] Von Neumann to Zermelo, August 1923, partly reproduced in [Meschkowski 1967, 289–91].

[2] The latter had already been stressed by Fraenkel, whose work seems to have guided many of von Neumann's reflections.

[3] [Meschkowski 1967, 290]: "Eine Menge ist dann und nur dann 'zu gross', wenn sie der Menge aller Dinge aequivalent ist." In this letter, von Neumann employed naive terminology, as he would keep doing later, but it is clear that he was aware of the dangers.

[4] In his opinion the development of the system became much simpler this way [1928a, 346; 1929, 494].

Separation, Replacement, and Choice. Actually it yields Global Choice: there is a single relation (a class, not a set) that simultaneously selects an element from each set of the universe. This is because one can use the axiom to derive the existence of a well-ordering of the universal class. Intuitively, the proof goes as follows: the class of all ordinals leads to the Burali-Forti paradox, therefore it is "too big" and, by axiom IV.2, equivalent to the class of all things. Since the class of ordinals is well-ordered, we obtain a well-ordering of the universal class, by which every set or class (i.e., every subclass of the universe) is well-ordered [1925, 398; 1929, 496].[1] Moreover, von Neumann's Axiom goes beyond what is strictly required for an axiomatic reconstruction of Cantorian set theory, since it warrants that all classes whose cardinality is smaller than that of the universal class are sets [1929, 496].

Von Neumann regarded these results as clear indication that his Axiom looked dangerous, and was thus led to prove the consistency of his axiom system *relative to* a simpler system, in which von Neumann's Axiom is substituted by Replacement and Choice [von Neumann 1929]. Call the simpler system *S**, von Neumann's original one *S*. His strategy was, first, to prove that *S** retains its consistency when Foundation is added and urelements are not allowed; and second, to show that *S* follows from the resulting augmented system [*op.cit.*, 499]. This was the first noteworthy result in the metatheory of axiomatic set theory, and, as one can see, it included a proof of the consistency of Foundation relative to von Neumann's system [*op.cit.*, 498–506].

The system was later transformed and simplified by several authors, above all Bernays and Gödel, for which reason it was given the name of NBG (von Neumann–Bernays–Gödel). Bernays simplified it by bringing it closer to the traditions of logic and set theory:

The purpose of modifying the von Neumann system is to remain nearer to the structure of the original Zermelo system and to utilize at the same time some of the set-theoretic concepts of the Schröder logic and of *Principia Mathematica* which have become familiar to logicians. As will be seen, a considerable simplification results from this arrangement. [Bernays 1937, 65]

By formulating the system directly in terms of the primitive ideas of set and class, Bernays avoided the foreign appearance of the original system. His classes behaved in much the same way as those of the logical tradition, since they complied with the laws of Boolean algebra: there is a complement to any given class, etc.

Establishing some rather natural "axioms for construction of classes" [1937, 69], Bernays obtained a powerful device which allowed him to recover an analogue of the principle of comprehension. This was already a feature of von Neumann's original system [1925, 400], but it became clearer in Bernays' version. Here, one postulates class-existence axioms like the following: there exists a class which is the graph of the $\in$-relation; to any given class there is a complementary class; for

---

[1] The formal derivation of the result can be found in [1928a, 396–99].

any two classes there exists another which is their intersection; for any class A there exists the class of pairs $A \times V$, where V is the universal class. As one can see, the class axioms do not include a comprehension principle directly, but they suffice to prove a certain comprehension principle. Bernays [1937, 76–77] proved a meta-theoretical result, establishing that there is a class that corresponds to any condition in the language of the system, where one quantifies *only* over sets, not over classes.[1]

Meanwhile, Bernays' set-existence axioms were very similar to those of Zer-melo–Fraenkel. This applies in particular to his Axioms of Infinity, Separation, Replacement, Union, and Power Set [1941, 2–5]. Shortly afterward, Gödel simpli-fied a bit more the class- and set-existence axioms of Bernays [Gödel 1940, 37]. Bernays had not identified co-extensional sets and classes [1937, 67]. Gödel found it simpler to identify them, postulating that every set is a class, and that, if a class X is a member of another, then X is a set [1940, 35]. As regards sets, Gödel simplified a bit by merging Separation and Replacement [1940, 38]. The Axiom of Choice was given in a stronger form by Gödel, bringing the system closer to the original very strong system of von Neumann. He postulated Global Choice, so that a single (class-)relation selects, simultaneously, an element from each non-empty set of the universe [1940, 39]. Finally, Bernays included a "restrictive axiom," Foundation, in the second formulation given to it by Zermelo (§2.3). As Gödel wrote [1940, 38], Foundation is not indispensable, but it simplifies considerably the later work, and von Neumann [1929] had proved its consistency with the rest of the NBG system.

NBG enjoyed wide acceptance from the late 1930s. A noteworthy contrast to ZFC is that the new system was finitely axiomatizable. In first-order ZFC we must use axiom-schemas of Separation and Replacement, which means that there are denumerably many axioms generated from each schema. For this reason, the sys-tem cannot be axiomatized by finitely many axioms. Meanwhile, NBG proceeds with finitely many axioms because classes play the role of 'definite properties' or first-order conditions in ZFC, and there are finitely many axioms for 'construction' of classes. Once again, this feature was already visible in von Neumann's original presentation, but it became clearer with Bernays and Gödel.

In principle, the systems NBG and ZFC might have been essentially different, and this must have been the impression of some authors in the 1920s. However, the reformulation of Bernays, and his 'class theorem,' intimated the opposite. The situation was finally clarified in 1950, when several authors proved the metatheo-retical result that NBG (without Choice) is a conservative extension of ZF. This means that, if a theorem about sets can be proved in NBG, a corresponding propo-sition can also be proved in ZF. The result was established on the basis of the com-pleteness theorem of first-order logic by Rosser & Wang, Novak, and Mostowski, independently of each other.[2] This showed that NBG is not stronger as an axiomatic

---

[1] This is the "*class theorem*" proved in the first installment of Bernays' series of papers, nowadays called theorem of predicative existence of classes.

[2] In the 60s, Kripke, Cohen and Solovay, working independently, established that NBG with Global Choice is also a conservative extension of ZFC. The result was published later by Felgner [1971].

set theory than ZFC is. From then on, it became increasingly common to employ the axioms of Zermelo–Fraenkel in advanced work on the metatheory of set theory.[1]

4. Gödel's Relative Consistency Results

Systems like type theory – not to mention weaker systems, due to the Hilbertian metamathematical school or to the constructivists – had been seriously entertained mostly by authors working on logic and foundations, not by practicing mathematicians. Reduction of the distance between TT and ZF eliminated the impression that TT was much safer than set theory; ZF gained a greater intuitive plausibility and doubts regarding its consistency diminished. The process continued during the 1930s. Stricter formalization of ZF within the background of first-order logic, promoted (after Skolem) by authors like Gödel, Bernays, Tarski and Quine, was also a source of renewed confidence (§5). But the most important change happened in 1938, when Gödel showed that AC and GCH are consistent relative to the Zermelo–Fraenkel system. These consistency results undermined any remaining cautionary reason to adopt TT. They showed that, if one is determined to adopt some version of 'classical' mathematics, axiomatic set theory (ZF or NBG) is a safe framework for the work, and perhaps the most natural one.

After establishing his famous incompleteness theorems, Gödel turned to set theory with the aim of settling fundamental questions on AC and CH. Quite early he came to consider the so-called 'constructible' sets as a model for axiomatic set theory without Choice (see Feferman in [Gödel 1986, 9, 21]). In 1935 he wrote to von Neumann that he had proven AC to be valid for the constructible sets, i.e., he had established its consistency relative to the other axioms. Two years later he proved the relative consistency of GCH.[2] The results were announced in the *Proceedings of the National Academy of Sciences* of the U.S.A. for 1938, with a detailed outline of the proof following the year after. In 1940, a monograph based on lectures given at Princeton in the fall of 1938 was published, with full proofs that are, however, less perspicuous than those of 1939. Interestingly, each time Gödel stated the result for a different axiom system: in the announcement [1938] von Neumann's system, in [1939] ZF, in [1940] his own (slight) modification of the system of Bernays. He also indicated that a corresponding theorem can be proven for the system of *Principia*, namely type theory [1938, 26].

---

[1] We shall not enter into more details regarding NBG here. Readers interested in a detailed analysis may turn, e.g., to [Fraenkel, Bar-Hillel & Levy 1973]. I would also like to emphasize that the historical interplay between ZFC and NBG ought to be the subject of more detailed research.

[2] He refrained from publishing immediately, since he hoped to get further results regarding the independence of AC and CH (see Moore in [Gödel 1990, 158]).

Gödel's method was to define on the basis of axiomatic set theory what would later be called an 'inner model.'[1] From our standpoint, it is noteworthy that type theory seems to have provided essential background for the work. Regarding the constructible universe, Gödel wrote:

This model, roughly speaking, consists of all 'mathematically constructible' sets, where the term 'constructible' is to be understood in the semi-intuitionistic sense which excludes impredicative procedures. This means 'constructible' sets are defined to be those sets which can be obtained by Russell's ramified hierarchy of types, if extended to include transfinite orders. The extension to transfinite orders has the consequence that the model satisfies the impredicative axioms of set theory, because the axiom of reducibility can be proved for sufficiently high orders. [Gödel 1938, 26–27; see also 1944, 147]

Much later, Gödel would refer to this idea as one of the most fruitful outcomes of his Platonistic attitudes: a constructivist would never have considered going beyond finite orders.[2] By viewing the theory of orders within the framework of ordinary (impredicative) mathematics, he was able to extend it to transfinite orders [Gödel 1944, 136]. Now, it became possible to prove that every propositional function is extensionally equivalent to one of order α – where the ordinal α is so great that it presupposes impredicative set-formation. But "all impredicativities are reduced to one special kind, namely the existence of certain large ordinal numbers (or well-ordered sets) [e.g., ω_1] and the validity of recursive reasoning for them" [*ibid.*].

One can thus describe the universe of constructible sets directly, within the formalism of set theory, by transfinite recursion on the ordinals. This definition makes it appear similar to Zermelo's cumulative hierarchy, but also to simple type theory, which is how Gödel presented it [1938, 31 note]. Take

1. $L_0 = \varnothing$;

2. $L_{\alpha+1}$ as the set of all subsets of L_α which can be defined by first-order conditions restricted to L_α, i.e., conditions containing only the following notions: negation $\neg$, disjunction $\vee$, the $\in$-relation, elements of L_α as parameters, and quantifiers $\forall, \exists$ for variables with range L_α;

3. $L_\beta = \bigcup L_\alpha$, for all $\alpha < \beta$, if β is a limit ordinal.[3]

A set s is called "constructible" if there exists an ordinal α such that $s \in L_\alpha$. The essential difference between this and the usual cumulative hierarchy is the restriction on parameters and quantifiers imposed in 2. At stage $\alpha+1$, one does not have the collection of all subsets of L_α, but only those subsets which can be defined by reference to 'all sets of order α;' here lies the predicative element. As a conse-

---

[1] Shepherdson [1951/53]. It was a very special inner model: the minimal one that contains all the ordinals and is transitive, i.e., such that, whenever x is in the model, so are elements of x.

[2] Letter to Wang, 1968, in [Wang 1974, 10].

[3] See the 1939 abstract in [Gödel 1990, 27]. Gödel writes M_α instead of the usual L_α, and I have modernized his logical symbolism slightly.

quence, new subsets of L_α may appear at stages later than $\alpha+1$, as happens for all infinite stages.

Gödel was well acquainted with previous work of Russell, Skolem, and Zermelo, including his [1930] treatment of the cumulative hierarchy. He did not just combine the cumulative hierarchy of Zermelo with the first-order perspective of Skolem in order to obtain the constructible universe.[1] The predicative element taken from Russell's ramified type theory was also essential. As he consistently presented the matter [1938; 1939, 31; 1944, 136], Gödel was led to the constructible sets by the comparison between axiomatic set theory and the two versions of type theory, ramified TT and simple TT. Taking into account the development of logical theory from type theory to first-order logic, and his earlier work within simple TT (§X.4), this seems rather natural.

Gödel was able to prove that the axioms of ZF hold in the universe of constructible sets, which is thus a model of ZF. He established that L_{ω_ω} is a model of Zermelo's axioms, and that the Axiom of Replacement is satisfied in L_Ω, Ω being the first inaccessible number [1938, 31]. Moreover, he was able to prove that AC and GCH also hold in the model. The proof was in two steps, which we shall present in reverse order. One part is to prove that the statement 'every set is constructible' is true in the model in question, L_{ω_ω} or L_Ω. To this end, one has to show that the statement is *absolute*, in the sense that it has the same meaning within the model as when regarded from without, from any 'bigger' model. Gödel establishes that the operation of forming the set of all first-order definable subsets of a given set is absolute: the outcome is the same when the operation is carried in the cumulative hierarchy, or relativized to a given model. As a result, the statement 'every set is constructible,' the *axiom of constructibility*, is absolute and holds in each of the models. Within the context of the NBG system it was natural to formulate the axiom by writing $V = L$ – the universal class is the class of constructible sets [Gödel 1940, 81]. This is the form in which it is now customarily written.[2]

The other step consisted in showing that AC and GCH are consequences of $V = L$. The first was, according to Gödel [1990, 27] an "incidental result" of his work on the Continuum Hypothesis. One can associate with each constructible set a unique first-order condition (having ordinals as parameters) as its 'definition,' and use these first-order expressions to establish a well-ordering of all constructible sets [1938, 29]. Since $V = L$ is absolute, the well-ordering is also absolute in the above sense. Establishing the result on GCH took Gödel two more years. It depends on a key lemma: any subset of L_{ω_α} which is an element of some L_β is already an element of $L_{\omega_{\alpha+1}}$. The proof employs "a generalization of Skolem's method for constructing enumerable models" (Gödel in [1990, 27]); it uses an argument analogous to the Löwenheim–Skolem theorem, and what is nowadays called the 'Mostowski

---

[1] As Solovay writes in [Gödel 1990, 8].

[2] In ZF one writes: $\forall x \exists \alpha \ (x \in L_\alpha)$. Gödel's axiom has been extensively studied by logicians, specially since the 60s; see Solovay in [Gödel 1990, 14–25].

collapse' to a transitive set.[1] That suffices to establish the Generalized Continuum Hypothesis, in virtue of the fact that the cardinalities of L_{ω_α} and $L_{\omega_{\alpha+1}}$ within the constructible universe are $\aleph_\alpha$ and $\aleph_{\alpha+1}$.

It is crucial that the whole reasoning outlined above – defining the models and proving that AC and GCH are valid in them – can be carried through within axiomatic set theory without Choice, for instance within ZF. This means that a contradiction derived from ZFC+GCH could be transformed into a contradiction derived from ZF alone [Gödel 1938, 32]. If ZF is consistent, so is ZFC+GCH; in that sense, the Axiom of Choice and the Generalized Continuum Hypothesis are safe assumptions.[2]

Once these profound metatheoretical results had been established, there was no further reason to stick to simple TT or to regard type theory as a safer system. Zermelo's system (without Replacement but with Foundation) and simple type theory with Infinity can be regarded as alternative descriptions of one and the same domain – a cumulative universe of sets (§2.3). Gödel's results showed that introducing the Axiom of Choice, and even the Generalized Continuum Hypothesis, does not imperil the consistency of the system. Therefore, there is no reason to think that ZFC+GCH is more dangerous than TT with Infinity. And the fact that ZFC can be formulated in first-order logic became an argument for it (§5), reinforcing even more the position of axiomatic set theory.

Gödel's relative consistency results stood as a rather isolated landmark in mathematical logic for a decade. After 1950, however, there was renewed activity in the study of models of set theory, beginning with inner models *à la* Gödel (see, e.g., [Shepherdson 1951/53]), and continuing with the powerful method of forcing introduced by Paul Cohen in the 1960s. As is well known, Cohen showed that CH is independent from the usual ZFC axioms of set theory (Gödel had tried to establish this last result in the early 1940s, but unsuccessfully).[3] Cohen's work was regarded by Gödel as the most important development in set theory since its axiomatization; it opened up a new era of intense activity in the metatheory of the system. The study of constructibility, together with the topics of large cardinals (§IX.6) and forcing, has formed one of the major areas of study in modern axiomatic set theory.

---

[1] Theorem 2 in [1938, 29]. This is the version of the axiom of reducibility that Gödel mentioned (see above). For further details, see Solovay in [Gödel 1990, 8–12] and [Gödel 1938].

[2] In [1940], working within NBG, Gödel was able to give detailed proofs of his results without having to deal with metatheoretical notions within axiomatic set theory. The result, however, was a much less intuitive proof than the one offered for ZF in [1938]. Gödel himself admitted that the first exposition exhibited more clearly the basic idea of the proof in a note added in 1965 [Gödel 1940, 97] (see also Solovay in [Gödel 1990, 12–13]).

[3] See [Moore 1988a], [Kanamori 1996] and Moore in [Godel 1990, 158–59]. Forcing is a powerful method, based on first-order logic, for defining models with prescribed properties.

Figure 10. *Alfred Tarski (left) and Kurt Gödel in 1935.*
Courtesy Bancroft Library, Berkeley.

5. First-Order Axiomatic Set Theory

Skolem had started to argue for a first-order formulation of axiom systems, particularly Zermelo set theory, as early as 1922. As we have seen (§X.5), this was a radical proposal, shared only by authors of constructivist tendencies like Weyl. Von Neumann seems to have been the first to join Skolem; his new system was formulated in such a way that it was translatable into first-order logic, and he seems to have shared Skolem's attitude toward axiomatics. But the fact that his axiom system was "elementary," in the sense of being formalizable in first-order logic, was only made explicit by Bernays [1937, 65]. According to Bernays [*ibid.*], von Neumann's was the first example of a system both adequate to arithmetic and elementary.

Since the early century, formalists had emphasized the metatheoretical questions of independence, categoricity, and consistency of axiom systems.[1] In 1928, Hilbert posed the problem of completeness for first-order logic [Hilbert & Acker-

---

[1] The Hilbertians laid much emphasis, too, on the decision problem [Entscheidungsproblem].

mann 1928], a problem solved in the affirmative by Gödel in his 1929 Ph.D. thesis and [1930]. But, as a result of his first incompleteness theorem [1931], it turned out that second-order logic is incomplete. In the atmosphere of foundational skepticism reigning during the interwar period, that afforded a new argument for first-order logic, to be added to those used previously by Skolem and others (§X.5). Since Leibniz, and particularly since Boole, it had been customary to conceive of logic as a calculus. Gödel's striking results implied that second-order inference cannot be completely formalized – it cannot be codified by means of a satisfactory calculus. Thus, first-order logic seemed much better as a purely formal system of the kind that foundational studies in the 1930s required. And, of course, Skolem's argument that axiomatic set theory can only rely on first-order logic retained its force. From about 1935, a number of authors emphasized that ZFC and NBG can be formalized within this elementary logic.

That was the case with Tarski, Quine, Gödel and Bernays. Up to 1933, or at least 1931, Tarski had been of the opinion that one *must* incorporate some kind of type restriction into logical systems (§X.4). But the 1935 postscript to his famous paper on the notion of truth records his abandonment of Leśniewski's theory of semantic types [Tarski 1935, 268]. Now, Tarski acknowledged that first-order logic is an acceptable system that suffices for codifying set-theoretical proofs. In contrast to type theory, first-order set theory is a "much more convenient and actually much more frequently applied apparatus" [*op.cit.*, 271n]. Similarly, while presenting a modification of ZF, Quine [1936] emphasized that, by formulating the system in first-order logic, one obtains the Skolem interpretation of 'definite property' automatically in the axiom-schema of Separation.

Three years later, Gödel's first published proof of the relative consistency of AC and GCH was established for the ZF system, using Skolem's interpretation of 'definite property.' A 'definite property' (or relation) was understood as a "propositional function over the class of all sets," that is, as a condition with one (or more) free variable(s) in the language of first-order set theory [Gödel 1939, 28, 31]. The quantifiers were clearly restricted to first-order quantification over the intended domain of sets. In the booklet that he published the year after, Gödel [1940] presented a modification of the von Neumann–Bernays system and emphasized that, to everyone familiar with mathematical logic, it should be clear that the proofs could be formalized using Hilbert's 'engerer Funktionenkalkul,' i.e., first-order logic [Gödel 1940, 34]. As we have seen, the fact that NBG is an elementary system, formalizable in first-order, had already been indicated by Bernays [1937, 65–66]. Bernays and Gödel were definitely no relativists or constructivists in the style of Skolem, but they regarded first-order logic as the adequate underlying system for metatheoretical analysis of mathematical axiom systems.[1]

Thus, it was in the context of axiom systems for set theory, in the second half of the 1930s, that first-order logic came to be regarded as a distinguished logical system. From the standpoint of the foundational debate and metamathematical studies,

---

[1] Gödel even suggested that set and concept (*viz* class) are 'higher' logical notions (see [Gödel 1944]). This makes clear that he was no first-order reductionist.

the proof-theoretical properties of that logic were perceived as highly desired ones, although the reasons remained unmentioned. First-order theories of arithmetic, the real numbers, or sets do not characterize categorically the intended objects, but they have the advantage that the basic logic system is completely formal. Second-order logic buys categoricity at the price of not being completely formalizable [Gödel 1931]. In an atmosphere of foundational skepticism and strict requirements concerning axiomatization, authors required tacitly that a logic ought to be proof-theoretically well-behaved, even if it did not work well from a semantic standpoint. Therefore, one may conclude that the main reasons for choosing first-order logic were its simplicity and completeness, its perfect agreement with the spirit of axiomatics and basic foundational work (particularly proof theory), and the fact that it did not pre-judge questions like the meaning of 'all subsets.'[1]

After World War II, mathematical logic enjoyed a new status at University departments, particularly in the United States. Tarski's Berkeley and Church's Princeton became leading centers for research in mathematical logic, and a new generation trained in metatheoretical work emerged. In line with the preferences of Bernays and Gödel, the new generation of logicians increasingly acknowledged first-order logic as the standard logic system. Moreover, with the rise of model theory around 1950, first-order logic became a most interesting system. Its weakness turned out to afford a powerful method for transferring results from one model to another. From this point onwards, most mathematicians and experts in foundational studies have agreed that first-order axiomatic set theory is a very satisfactory framework. This came to mean that ZFC is the natural axiomatic framework for most of mathematics.[2]

6. A Glance Ahead: Mathematicians and Foundations after World War II

After the end of the War, the situation was ripe for axiomatic set theory to finally consolidate its key role within modern mathematics. During the 1930s it had been satisfactorily reformulated, and its logical basis had been deeply investigated. The heated foundational controversies had been left behind, and almost all mathematicians were ready to continue working in line with the abstract tradition, which found its interpreter, encyclopedic codifier, and public figure in Nicolas Bourbaki. But, of course, not all mathematicians emerged from the interwar period with equal confidence in the classical tradition. To conclude, we shall briefly review the views expressed in the late 1940s by some notable figures, including von Neumann,

---

[1] As second-order logic can be seen to do, when the system is interpreted extensionally (§X.5).

[2] I skip here over the issue of choosing between ZFC and NBG (see §3).

Gödel, Weyl, and Bourbaki.[1] Needless to say, I shall make no attempt to deal with the topic comprehensively; my aim is just to sample a few relevant views, in order to convey the reader a feeling for the manifold possible standpoints.

Von Neumann started to doubt the soundness of the abstract tradition very early on. Even in his contributions to axiomatic set theory, he expressed caution and skepticism regarding the validity of the systems he examined. Much later, in a paper for the general reader, he mentioned that his own views regarding mathematical truth had changed substantially, and humiliatingly easily, three different times in quick succession [1947, 6]. Although he did not clarify what these changes actually were, it would seem that he was referring to the following periods. Up to [1923] he seems to have fully accepted Cantorian set theory and so he became interested in taking axiomatic set theory to the limits. But two years later he sounded a note of caution as a result of the constructivists' critique, the work of Skolem, and his own investigations [1925, 395–96, 408–09, 412–13]. He discussed in detail the Skolem paradox, indicating that axiomatic set theory leaves room for relativism, and he presented his reflections on the lack of categoricity of systems of set theory. Von Neumann thought that these might be arguments for intuitionism [1925, 412]. He also stressed the distance between naive and formalized set theory, and the arbitrariness of the restrictions introduced in axiomatic set theory [1925, 396; 1928a, 347; 1929, 495]. But shortly afterward, working on Hilbert's program, he came quite close to proving the consistency of Peano arithmetic [von Neumann 1927]. By this time he must have believed, like many others, that Hilbert would soon win the battle,[2] but Gödel's incompleteness theorems shattered his beliefs and convinced him that Hilbert's program was hopeless [1947, 6]. After 1931 he ceased publishing on foundational issues. It seems that the blow dealt to Hilbert's program by Gödel's results finally convinced him that mathematicians should devote themselves to topics in areas of mathematics closer to applications.

Von Neumann's paper ended with the suggestion that, although mathematical ideas have empirical origins, they are creatively developed in a theoretical way governed mainly by aesthetic motives. And there is danger of degeneration:

As a mathematical discipline travels far from its empirical source, or still more, if it is a second or third generation only indirectly inspired by ideas coming from 'reality,' it is beset with very grave dangers. It becomes more and more purely aestheticizing, more and more purely *l'art pour l'art*. ... In other words, at a great distance from its empirical source, or after much 'abstract' inbreeding, a mathematical subject is in danger of degeneration. [von Neumann 1947, 9]

---

[1] Another relevant but slightly earlier contribution is [Bernays 1935], which deals with Platonism in mathematics.

[2] By the long sought proof of consistency of the theory of real numbers. As regards axiomatic set theory, von Neumann thought that a consistency proof was beyond reach [1929, 495].

Such a discipline risks being developed along lines of least resistance, it may even end up as a disorganized mass of details and complexities. It is difficult not to read those lines as a reflection on von Neumann's long experience with set theory. His last word was to caution that, when that stage is reached, the only remedy is the rejuvenating return to the source, the reinjection of more or less directly empirical ideas [*ibid.*]. It seems this is the remedy he applied to himself back in the early 1930s.

The skepticism of von Neumann contrasts sharply with the confident Platonism expressed by Gödel by the end of the War. This came out for the first time in his paper on Russell's logic, where he wrote that classes and concepts may be conceived as real objects, existing independently of our definitions and constructions [1944, 128]. The assumption of such objects is quite as legitimate as the assumption of physical bodies – classes and concepts are necessary to obtain a satisfactory system of mathematics, "in the same sense" as bodies are necessary for a satisfactory theory of our sense perceptions [*ibid.*; see also 131]. Furthermore, nothing expresses better the meaning of the term 'class' than the axioms of Separation and Choice, which had thus come to seem obvious to Gödel [1944, 139]. The only difficulty is that we do not perceive the notions of 'concept' and 'class' with sufficient distinctness, as the paradoxes show. This means that the axioms of set theory may have to be supplemented by new basic propositions, a position that Gödel developed in detail three years later, in his only expository article [1947].[1]

In connection with a review of the volume on Russell's philosophy which included Gödel's contribution, Weyl [1946] summarized his own perceptions of the question 'Mathematics and Logic' and how it had evolved in the last half century. His aim was at least twofold: to show that the whole drift of research had been away from the Frege–Russell thesis that pure mathematics is logic [1946a, 601], and to emphasize again the importance of the debate between Brouwer and Hilbert, from which Gödel seemed to be retiring [1946a, 603]. He stressed once again the "transcendental character" and "high degree of arbitrariness" involved in systems like simple type theory and axiomatic set theory [1946, 278; 1946a, 603]. He wrote that Russell had not founded mathematics on logic, but on "a sort of logician's paradise," an axiomatic world system, and that "belief in this transcendental world taxes the strength of our faith hardly less than the doctrines of the early Fathers of the Church" [1946, 272]. The situation was essentially the same with the world of sets postulated in Zermelo-style axiomatics: there is no assurance of its consistency except the empirical support of having not yet led to any contradiction. Still, Weyl acknowledged that the ZF system is essentially simpler than Russell's and "seems to be the most adequate basis for what is actually done in present-day mathematics," i.e., for the familiar "existential" or abstract mathematics [1946, 276–77, 278–79].

---

[1] Here the realistic view is again insinuated [1947, 179–81], but it was only in the second, 1964 edition of the paper on Cantor's continuum problem that Gödel took it even further.

Weyl emphasized that we are less certain than ever about the ultimate foundations of mathematics. There is a choice between several different possible systems. Some emphasize the constructive tendency and so remain within the bounds of what may be legitimately called 'evident;' this is done most clearly and deeply in Brouwer's intuitionism, less so, but remaining closer to customary mathematics, in Weyl's system. Some systems tend to postulate a world of mathematical objects by means of axiomatic systems; this is done austerely by Hilbert and his followers, much less so by Russell, and with the greatest freedom in Zermelo's axiomatic set theory. Weyl summed up the situation in the following diagram:

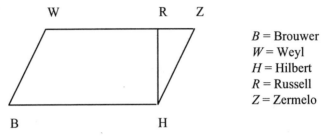

B = Brouwer
W = Weyl
H = Hilbert
R = Russell
Z = Zermelo

Systems located toward the bottom are taken to be deeper, more basic foundations; those located toward the left are more of a constructive tendency, toward the right more of an axiomatic tendency. Needless to say, Weyl's sympathies fell on the left side, that of B and W.

In spite of all their differences, there is one feature that von Neumann, Weyl and Gödel shared. They no longer believed in the centuries-old tradition that mathematics plainly consists of true, evident statements. In their view, many mathematical propositions have a hypothetical character; they are introduced as hypotheses that serve to explain and unify more concrete material. This is precisely the relation between the axioms of set theory and the theorems of arithmetic and analysis – the former are strong explanatory hypotheses, the latter are much more convincing, true-like, or evident. This interesting conclusion – diametrically opposed to the simplistic idea that mathematics is tautological or purely formal – seems to have been one of the profoundest outcomes, among deep thinkers, of the foundational debates. Russell began to express this new 'hypothetical' mood in *Principia Mathematica* [1910], Hilbert and Weyl went on in the 1920s comparing the real number axioms with the hypotheses of physics. Bernays, Gödel, von Neumann and many others followed.[1]

To some extent, the hypothetical conception was shared by a keynote speaker at the 1948 meeting of the Association for Symbolic Logic, a professor named Bourbaki from the University of "Nancago" who would become extremely influential among practicing mathematicians.[2] His talk [1949] was devoted to laying out the

---

[1] See [Lakatos 1967], who calls this 'quasi-empiricism,' while I would prefer to speak of a *hypothetical* conception of (large parts of) mathematics. The most important popularizer of this view among philosophers has been Quine; see his [1953].

[2] For Bourbaki's hypothetical conception, see [1949, 3]. Perhaps I should mention that Bour-

foundations of mathematics in a way that he regarded as sound, relatively safe, and flexible enough for the working mathematician. As a peculiar, many-headed follower of Hilbert, Bourbaki discussed his conception of logic as an integral part of mathematics and presented his version of first-order logic, the basic sign-language and grammar of the discipline. Having explained the meaning of the words 'axiom,' 'proof,' and 'theory,' he proceeded to list his basic axioms for mathematics. These are simply a version of first-order ZFC, for

as every one knows, all mathematical theories can be considered extensions of the general theory of sets. [Bourbaki 1949, 7]

The system had some peculiarities. It employed as primitives the notions of equality, ordered pair, and membership. It lacked axioms of Union and Replacement, but included an axiom postulating the existence of the Cartesian Product of any two given sets.

On these foundations, I state that I can build up the whole of the mathematics of the present day; and, if there is anything original in my procedure, it lies solely in the fact that, instead of being content with such a statement, I proceed to prove it in the same way as Diogenes proved the existence of motion; and my proof will become more and more complete as my treatise grows. [Bourbaki 1949, 8]

Motion is proved walking. Bourbaki went on to present within the set-theoretical framework his interesting idea of the 'mother structures' [1950], and to systematize large parts of the mathematical tradition. But in the course of walking, the French group also came to experience some limitations to the confident belief they had expressed in 1949.[1] Today we do not feel so sure that axiomatic set theory is the ultimate foundation, but certainly all mathematicians are accustomed to using it as the basic language to teach and learn in order to become a mathematician. To this extent, Bourbaki was right and simply expressed the working convictions of the 20th century mathematician.

---

baki was no real person, but a fictitious character invented by a group of French mathematicians, who published their joint work under that name.

[1] When the first volume of his treatise was published, Bourbaki [1954] had changed his axiom system slightly. Some indication of the difficulties can be found in [Corry 1996].

Bibliographical References

Abbreviations employed for frequently mentioned journals.

AHES Archive for History of Exact Sciences
AM   Acta Mathematica
AMM   American Mathematical Monthly
AS   Annals of Science
DMV   Jahresbericht der Deutschen Mathematiker-Vereinigung
FM   Fundamenta Mathematica
HM   Historia Mathematica
HPL   History and Philosophy of Logic
JrM   Journal für die reine und angewandte Mathematik (Crelle)
JSL   Journal of Symbolic Logic
LMS   Proceedings of the London Mathematical Society
MA   Mathematische Annalen
NAS   Proceedings of the National Academy of Sciences, U.S.A.
NG   Nachrichten der Königliche Gesellschaft der Wissenschaften zu Göttingen
NTM   International Journal of History and Ethics of Natural Sciences, Technology and Medicine
RHS   Revue d'histoire des sciences et de leurs applications

ACKERMANN, W.
   1937 Die Widerspruchsfreiheit der allgemeinen Mengenlehre, *MA* **114**, 305–15.
   1938 Mengentheoretische Begründung der Logik, *MA* **115**, 1–22.

ALEKSANDROV, P.S.
   1916 Sur la puissance des ensembles mesurables B, *Comptes Rendus, Académie des Sciences de Paris* **162**, 323–25.

ARISTOTLE
   Organon, in J. Barnes, ed., *The Complete Works of Aristotle*, 2 vols., Princeton University Press, 1984.

ARNAULD, A. & P. NICOLE
   1662 *La Logique ou l'Art de Penser*, Paris, Flammarion, 1970.

AUSLEIHREGISTER
   Handwritten record of books and journals loaned to professors and students of the University of Göttingen, kept at the Niedersächsische Staats- und Universitätsbibliothek (Handschriftenabteilung).

BEKEMEIER, B.
   1987 *Martin Ohm (1792–1872): Universitätsmathematik und Schulmathematik in der neuhumanistischen Bildungsreform*, Göttingen, Vandenhoeck & Ruprecht.

BENACERRAF, P. & H. PUTNAM, eds.

1983 *Philosophy of Mathematics: selected readings*, Cambridge University Press.

BENDIXSON, I.

1883 Quelques théorèmes de la théorie des ensembles, *AM* **2**, 415–429.

BENSAUDE-VINCENT, B.

1983 A Founder Myth in the History of Science? The Lavoisier Case, in L. Graham, W. Lepenies & P. Weingart, eds. *Functions and Uses of Disciplinary Histories*, Dordrecht, Reidel, 53–78.

BERNAYS, P.

1930 Die Philosophie der Mathematik und die Hilbertsche Beweistheorie, *Blätter für Deutsche Philosophie* **4**, 326–67. References to [Bernays 1976], 17–61.

1935 Hilberts Untersuchungen über die Grundlagen der Arithmetik, in [Hilbert 1932/35], vol. 3, 196–216.

1935a Sur le platonisme dans les mathématiques, *L'Enseignement Mathématique* **34**, 52–69. German trans. in [Bernays 1976], English in [Benacerraf & Putnam 1983], 258–71.

1937/42 A System of Axiomatic Set Theory, Parts I–III, *JSL* **2**, 65–77, **6**, 1–17, **7**, 65–89, 133–45.

1958 *Axiomatic Set Theory*, Amsterdam, North-Holland.

1976 *Abhandlungen zur Philosophie der Mathematik*, Darmstadt, Wissenschaftliche Buchgesellschaft.

BERNSTEIN, F.

1905 Untersuchungen aus der Mengenlehre, *MA* **61**, 117–155.

BIERMANN, K.R.

1959 *Johann Peter Gustav Lejeune Dirichlet. Dokumente für sein Leben und Wirken*, Berlin, Akademie-Verlag.

1966 Karl Weierstrass. Ausgewählte Aspekte seiner Biographie, *JrM* **223**, 191–220.

1966a Richard Dedekind im Urteil der Berliner Akademie, *Forschungen und Fortschritte* **40**, 301–302.

BOCHEŃSKI, I.M.

1956 *Formale Logik*, München, Alber.

DU BOIS-REYMOND, P.

1875 Versuch einer Classification der willkürlichen Functionen reeller Argumente nach ihren Aenderungen in den kleinsten Intervallen, *JrM* **79**, 21–37.

1880 Der Beweis des Fundamentalsatzes der Integralrechnung, *MA* **16**, 115–128.

BÖLLING, R.

1997 Georg Cantor – Ausgewählte Aspekte seiner Biographie, *DMV* **99**, 49–82.

BOLLINGER, M.

1972 Geschichtliche Entwicklung des Homologiebegriffs, *AHES* **9**, 94–170.

BOLZANO, B.

1817 *Rein analytischer Beweis des Lehrsatzes dass zwischen je zwei Werten die ein entgegengesetztes Resultat gewähren, wenigstens eine reelle Wurzel der Gleichung liege*, Prag, Haase. Reprint in Ostwald's Klassiker no. 153. English trans. in [Ewald 1996, vol.1].

1837 *Wissenschaftslehre*, 4 vols., Seidel, Sulzbach. Reprinted in 1937. Selections translated as *Theory of Science*, Oxford, Blackwell, 1972.

1851 *Paradoxien des Unedlichen*, Leipzig, Reclam. Reprint in Leipzig, Meiner, 1920, and Hamburg, 1975. English trans. London, Routledge, 1950.

BOOLE, G.

1847 *The mathematical analysis of Logic*, Cambridge, Macmillan. References to the reprint Oxford, Basil Blackwell, 1951. Reprinted in [Ewald 1996, vol.1].

BOOLOS, G.

1971 The Iterative Concept of Set, *Journal of Philosophy* **68**, 215–31. Reprint in [Benacerraf & Putnam 1983].

BOREL, E.

1898 *Leçons sur la theorie des fonctions*, Paris, Gauthier-Villars.

BORGA, M., P. FREGUGLIA & D. PALLADINO

1985 *I contributi fondazionali della scuola di Peano*, Milano, Angeli.

BOS, H., H. MEHRTENS & I. SCHNEIDER, eds.

1981 *Social history of nineteenth century mathematics*, Boston, Birkhäuser.

BOTTAZZINI, U.

1977 Riemanns Einfluss auf Betti und Casorati, *AHES* **18**, 27–37.

1985 Dall'Analisi Matematica al Calcolo Geometrico: Origini Delle Prime Ricerche di Logica di Peano, *HPL* **6**, 25–52.

1986 *The* Higher Calculus*: A History of Real and Complex Analysis from Euler to Weierstrass*, New York, Springer.

BOURBAKI, N.

1949 Foundations of Mathematics for the Working Mathematician, *JSL* **14**, 1–8.

1950 The Architecture of Mathematics, *AMM* **67**, 221–32. Reprinted in [Ewald 1996], vol.2.

1954 *Théorie des ensembles*, Paris, Hermann.

1994 *Elements of the history of mathematics*, Berlin, Springer (original edition 1969).

BROUWER, L.E.J.

1910 Zur Analysis Situs, *MA* **68**, 422–34. English trans. in [Brouwer 1975/76].

1911 Beweis der Invarianz der Dimensionszahl, *MA* **70**, 161–65. English trans. in [Brouwer 1975/76].

1918 Begründung der Mengenlehre unabhängig vom logischen Satz vom ausgeschlossenen Dritten, I, *Verhandelingen der Koninklijke Nederlandse Akademie van Wetenschappen te Amsterdam* **12**. English trans. in [Brouwer 1975/76].

1925 Zur Begründung der intuitionistischen Mathematik, I, *MA* **93**, 244–57. English trans. in [Brouwer 1975/76].

1975/76 *Collected Works*, A. Heyting and H. Fredenthal, eds., 2 vols., Amsterdam, North-Holland.

BUNN, R.

1980 Developments in the Foundations of Mathematics from 1870 to 1910, in [Grattan-Guinness 1980].

BURALI-FORTI, C.

1896 Le classi finite, *Atti della Academia delle Scienze di Torino* **32**, 34–52.

1897 Una questione sui numeri transfiniti, *Rendiconti del Circolo Matematico di Palermo* **11**, 154–64. References to the English trans. in [van Heijenoort 1967], 104–11.

BUTZER, P.L.

   1987 Dirichlet and His Role in the Founding of Mathematical Physics, *Archives internationales d'histoire des sciences* **37**, 49–82.

CANEVA, K.L.

   1978 From Galvanism to Electrodynamics: The Transformation of German Physics and Its Social Context, *Historical Studies in the Physical Sciences* **9**, 63–159.

CANTOR, G.

   1872 Über die Ausdehnung eines Satzes aus der Theorie der trigonometrischen Reihen, *MA* **5**, 123–32. References to [Cantor 1932], 92–101.

   1874 Über eine Eigenschaft des Inbegriffes aller reellen algebraischen Zahlen, *JrM* **77**, 258–62. References to [Cantor 1932], 115–118. English trans. in [Ewald 1996], vol.2.

   1878 Ein Beitrag zur Mannigfaltigkeitslehre, *JrM* **84**, 242–58. References to [Cantor 1932], 119–133.

  1879/84 Über unendliche, lineare Punktmannichfaltigkeiten, *MA* **15**, 1–7, **17**, 355–58, **20**, 113–21, **21**, 51–58, 545–86, **23**, 453–88. References to [Cantor 1932], 139–244.

   1883 *Grundlagen einer allgemeinen Mannigfaltigkeitslehre*, Leipzig, 1883 (separate edition of [Cantor 1879/84], part 5). References to [Cantor 1932], 165–208. English trans. in [Ewald 1996], vol.2.

   1886 Über die verschiedenen Standpunkte in bezug auf das aktuelle Unendliche, *Zeitschrift für Philosophie und philosophische Kritik* **88**, 224–33. References to [Cantor 1932], 370–76.

  1887/88 Mitteilungen zur Lehre vom Transfiniten, *Zeitschrift für Philosophie und philosophische Kritik* **91**, 81–125, 252–70, **92**, 250–65. References to [Cantor 1932], 378–439.

   1892 Über eine elementare Frage der Mannigfaltigkeitslehre, *DMV* **1**, 75–78. References to [Cantor 1932], 278–280. English trans. in [Ewald 1996], vol.2.

  1895/97 Beiträge zur Begründung der transfiniten Mengenlehre, *MA* **46**, 481–512, **49**, 207–46. References to [Cantor 1932], 282–351.

   1915 *Contributions to the founding of the Theory of Transfinite Numbers*, Chicago, Open Court. Reprint New York, Dover, 1955.

   1932 *Gesammelte Abhandlungen mathematischen und philosophischen Inhalts*, ed. E. Zermelo, Berlin, Springer. Reprint Hildesheim, G. Olms, 1966.

   1970 Principien einer Theorie der Ordnungstypen, in [Grattan-Guinness 1970], 83–101.

CANTOR, G. & R. DEDEKIND

   1932 Aus dem Briefwechsel zwischen Cantor und Dedekind (1899 letters), in [Cantor 1932], 443–450. English trans. in [Ewald 1996], vol.2.

   1937 *Cantor–Dedekind Briefwechsel*, ed. E. Noether and J. Cavaillès, Paris, Hermann. French translation including the 1899 letters in [Cavaillès 1962], 187–249. English trans. in [Ewald 1996], vol.2.

   1976 Unveröffentlichter Briefwechsel, in [Dugac 1976], 223–262.

CARNAP, R.

   1929 *Abriss der Logistik, mit besonderer Berücksichtigung der Relationstheorie und ihrer Anwendungen*, Wien, Springer, 1929.

   1931 Die logizistische Grundlegung der Mathematik, *Erkenntnis* **2**, 91–105. References to the English translation in [Benacerraf & Putnam 1983, 41–52].

   1937 *The Logical Syntax of Language*, New York, Harcourt, Brace.

1963 Intellectual Autobiography, in P.A. Schilpp, ed., *The Philosophy of R. Carnap*, La Salle, Ill., Open Court, 1963.

CAUCHY, A.L.

1821 *Cours d'analyse algébrique*, Paris. In *Oeuvres complètes*, ser. 2, Paris, Gauthier-Villars, 1897–99, vol. 3.

CAVAILLÈS, J.

1932 Sur la deuxième definition des ensembles finis donnée par Dedekind, *FM* **19**, 142–148.

1962 *Philosophie mathématique*, Paris, Hermann.

CHURCH, A.

1937 Review of E.T. Bell, *Men of Mathematics*, *JSL* **2**, 95.

1937a Review of L. Chwistek, Überwindung des Begriffsrelativismus, *JSL* **2**, 169? (this includes an interesting short analysis of Chwistek's logical work).

1939 The present situation in the foundations of mathematics, in F. Gonseth, *Philosophie mathématique*, Paris, Hermann, 1979.

1940 A Formulation of the Simple Theory of Types, *JSL* **5**, 56–68.

1956 *Introduction to Mathematical Logic*, Princeton University Press.

CHWISTEK, L.

1921 Über die Antinomien der Principien der Mathematik, *Mathematische Zeitschrift* **14**.

CLIFFORD, W.K.

1882 *Mathematical Papers*, New York, Chelsea, 1968.

COFFA, J.A.

1979 The Humble Origins of Russell's Paradox, *Russell* **33–34**, 31–37.

COOKE, R.

1991 Uniqueness of Trigonometric Series and Descriptive Set Theory, 1870–1985, draft of communication to the *Second Annual Göttingen Workshop on the History of Modern Mathematics*.

CORRY, L.

1996 *Modern Algebra and the Rise of Mathematical Structures*, Basel, Birkhäuser.

COURANT, R.

1926 Bernhard Riemann und die Mathematik der letzten hundert Jahre, *Naturwissenschaften* **14**, 813–818, 1265–1277.

COUTURAT, L.

1896 *De l'infinie mathématique*, Paris, Alcan.

1901 *La logique de Leibniz, d'apres des documents inédits*, Paris, Alcan.

CREATH, R., ed.

1990 *Dear Carnap, Dear Van: The Quine–Carnap correspondence and related work*, University of California Press.

CURRY, H.B.

1939 Remarks on the Definition and Nature of Mathematics, *Dialectica* **8** (1954), 228–233. References to reprint in [Benacerraf & Putnam 1983].

VAN DALEN, D.

1995 The War of the Frog and Mice, *The Mathematical Intelligencer* .

DAUBEN, J.W.

1979 *Georg Cantor. His Mathematics and Philosophy of the Infinite*, Cambridge, Harvard University Press.

1980   The development of Cantorian set theory, in [Grattan-Guinness 1980], 235–282.

1981   ed. *Mathematical Perspectives. Essays on mathematics and its historical. development*, New York, Academic Press.

DAWSON, J.W., Jr.

1984   Discussion on the Foundations of Mathematics, *HPL* 5, 111–129.

1985   The Reception of Gödel's Incompleteness Theorems, in *Philosophy of Science Association 1984*, vol. 2, p. 253–271.

1985a Completing the Gödel–Zermelo Correspondence, *HM* 12, 66–70.

DEAÑO, A.

1980   *Las concepciones de la Lógica*, Madrid, Taurus.

DEDEKIND, R.

Cod. Ms. Dedekind. Scientific legacy or *Nachlass* kept at the Niedersächsische Staats- und Universitätsbibliothek (Handschriftenabteilung) in Göttingen.

1854   Über die Einführung neuer Funktionen in der Mathematik, [Dedekind 1930/32], vol.3, 428–438. English trans. in [Ewald 1996], vol.2.

1871   Über die Komposition der binären quadratischen Formen, Supplement X to 2nd edn. of [Dirichlet 1894]. References to the partial reprint [Dedekind 1930/32], vol.3, 223–261.

1872   *Stetigkeit und irrationale Zahlen*, Braunschweig, Vieweg. References to [Dedekind 1930/32], vol.3, 315–334. English trans. in [Ewald 1996], vol.2.

1872/78 Gedanken über die Zahlen (first draft for [Dedekind 1888]), in [Dugac 1976], 293–309.

1876   Bernhard Riemann's Lebenslauf, in [Riemann 1892], 541–58.

1877   Sur la théorie des nombres entiers algébriques, *Bulletin des Sciences mathématiques et astronomiques*, 11 (1876), 1 (1877). Separate edn. Paris, Gauthier-Villars, 1977. References to the partial reprint [Dedekind 1930/32], vol.3, 262–296.

1879   Über die Theorie der ganzen algebraischen Zahlen, Supplement XI to 3rd edn. of [Dirichlet 1894]. Partial reprint in [Dedekind 1930/32], vol.3, 297–314.

1888   *Was sind und was sollen die Zahlen?*, Braunschweig, Vieweg. References to [Dedekind 1930/32], vol.3, 335–390. English trans. in [Ewald 1996], vol.2.

1894   Über die Theorie der ganzen algebraischen Zahlen, Supplement XI to [Dirichlet 1894].

1895   Über die Begründung der Idealtheorie, *NG*, 106–13. [Dedekind 1930/32], vol. 2, 50–58.

1901   *Essays on the Theory of Numbers*, Chicago, Open Court. English translation of [Dedekind 1872] and [1888]. Reprint New York, Dover, 1963.

1930   *Gesammelte mathematische Werke*, ed. R. Fricke, E. Noether and Ö. Ore, Braunschweig, 3 vols. Reprint in 2 vols. New York, Chelsea, 1969.

1981   Eine Vorlesung über Algebra, in [Scharlau 1981], 59–100.

DEDEKIND, R. & H. WEBER

1882   Theorie der algebraischen Funktionen einer Veränderlichen, *JrM* 92, 181–290. References to [Dedekind 1930/32, vol. 1, 238–350].

DE MORGAN, A.

1858   On the Syllogism, III, *Transactions of the Cambridge Philosophical Society* 10, 173–230. References to [De Morgan 1966].

1966   *On the Syllogism and other logical writings*, ed. P. Heath, London, Routledge

DETLEFSEN, M.
  1986 *Hilbert's Program. An essay on mathematical instrumentalism*, Dordrecht, Reidel.
DHOMBRES, J.
  1978 *Nombre, Mesure et Continue*, Paris, Cedic-Nathan.
DIEUDONNÉ, J., ed.
  1978 *Abrégé d'histoire des mathématiques, 1700–1900*, 2 vols., Paris, Hermann.
DINI, U.
  1878 *Fondamenti per la teorica delle funzioni di variabili reali*, Pisa, Nistri.
DIRICHLET, P.G. Lejeune
  1829 Sur la convergence des séries trigonométriques qui servent à représenter une fonc-
        tion arbitraire entre des limites données, *JrM* **4**, 157–69. References to [Dirichlet
        1889/97], vol.1, 283–306.
  1837 Über die Darstellung ganz willkürlicher Functionen durch Sinus- und Cosinus-
        reihen, *Repertorium der Physik* **1**, 152–74. References to [Dirichlet 1889/97],
        vol.1, 133–160.
  1889/97 *G. Lejeune Dirichlet's Werke*, Berlin, Reimer. Reimpreso en New York, Chel-
        sea, 1969.
  1894 *Vorlesungen über Zahlentheorie*, ed. R. Dedekind, 4th edn., Braunschweig,
        Vieweg (first 1863, second 1871, third 1879). Reprint New York, Chelsea, 1968.
DROBISCH, M.W.
  1836 *Neue Darstellung der Logik nach ihren einfachen Verhältnissen*, Leipzig, Voss.
DUGAC, P.
  1973 Elèments d'analyse de Karl Weierstrass, *AHES* **10**, 41–176.
  1976 *Richard Dedekind et les fondements des mathématiques (avec de nombreux textes
        inédits)*, Paris, Vrin.
  1981 Richard Dedekind et l'application comme fondement des mathématiques, in
        [Scharlau 1981], 134–144.
  1981a Des fonctions comme expresions analytiques aux fonctions représentables ana-
        lytiquement, in [Dauben 1981], 13–36.
  1984 Georg Cantor et Henri Poincaré, *Bolletino di Storia delle Scienze Mathematiche* **4**,
        65–96.
ECKERMANN, J.P.
  1909 *Gespräche mit Goethe in den letzten Jahren seines Lebens*, Leipzig, Brockhaus.
EDWARDS, H.M.
  1980 The Genesis of Ideal Theory, *AHES* **23**, 321–378.
  1981 Kronecker's Place in History, in [Dauben 1981], 139–144.
  1983 Dedekind's Invention of Ideals, *Bulletin of the London Mathematical Society* **15**,
        8–17. Reprint in [Phillips 1987], 8–20.
  1989 Kronecker's Views on the Foundations of Mathematics, in [Rowe & McCleary
        1989], 67–77.
EDWARDS, H.M., O. NEUMANN & W. PURKERT
  1982 Dedekinds «Bunte Bemerkungen» zu Kroneckers «Grundzüge», *AHES* **27**, 49–85.
EPPLE, M.
  1995 Branch Points of Algebraic Functions and the Beginnings of Modern Knot Theory,
        HM **22**, 371–401.

EYNDE, R. vanden
1992 Historical Evolution of the Concept of Homotopic Paths, *AHES* **45**, 127–88.

EULER, L.
1796 *Vollständige Anleitung zur niedern und höhern Algebra*, Berlin. References to *Opera Omnia*, vol. 1, Leipzig, Teubner, 1911.

EWALD, W.B.
1996 *From Kant to Hilbert: A source book in the foundations of mathematics*, 2 vols., Oxford University Press.

FECHNER, G.T.
1864 *Über die physikalische und philosophische Atomenlehre*, Leipzig, Mendelssohn.

FEFERMAN, S.
1988 Weyl vindicated:· "Das Kontinuum" 70 Years Later, in *Atti del Congreso Temi e propettive della logica e della filosofia della scienza contemporanea*, Bologna, CLUEB, vol. 1, 59–93

FELGNER, U.
1971 Comparison of the Axioms of Local and Universal Choice, *FM* **71**, 43–62.

FERREIRÓS, J.
1990 ¿Por qué el álgebra simbólica británica no fue un álgebra estructural?, in A. Díaz et al., eds., *Structures in Mathematical Theories*, Universidad del País Vasco, 241–244.
1991 La confluencia de álgebra y lógica en la obra de Augustus De Morgan, in *Actas del V Congreso de la S.E.H.C.yT.*, Universidad de Murcia, vol. 3, 2072–86.
1992 Sobre los orígenes de la matemática abstracta: Richard Dedekind y Bernhard Riemann, *Theoria* **16–17–18**, tomo A, 473–498. Short version in S. Garma, D. Flament & V. Navarro, eds., *Contre les titanes de la routine*, Madrid, CSIC, 1994, 301–18.
1993 On the Relations between Cantor and Dedekind, *HM* **20**, 343–363.
1993a *El nacimiento de la teoría de conjuntos, 1854–1908*, Madrid, Publicaciones de la Universidad Autónoma.
1995 «What Fermented in Me for Years»: Cantor's Discovery of Transfinite Numbers, *HM* **22**, 33–42.
1996 Traditional Logic and the Early History of Sets, 1854–1908, *AHES* **50**, 5–71.
1997 Notes on Types, Sets, and Logicism, 1930–1950, *Theoria* **12**, 91–124.

FITCH, F.B.
1938 The Consistency of the Ramified *Principia*, *JSL* **3**, 140–149.

FORSGREN, K.A.
1992 *Satz, Satzarten, Satzglieder. Zur Gestaltung der traditionellen Grammatik von K.F. Becker bis K. Duden 1830–1880*, Münster, Nodus.

FOURIER, J.
1822 *Théorie analytique de la chaleur*, Paris, Gabay, 1988.

FRAENKEL, A.
1922 Über den Begriff "definit" und die Unabhängigkeit des Auswahlaxioms, *Sitzungsberichte der Preussischen Akademie der Wissenschaften*, 253–57. References to the English trans. in [van Heijenoort 1967], 285–89.
1922a Zu den Grundlagen der Cantor–Zermeloschen Mengenlehre, *MA* **86**, 230–37.
1923 *Einleitung in die Mengenlehre. Eine elementare Einführung in das Reich des Unendlichgrossen*, 2nd augmented edn., Berlin, Springer (first published in 1919).

1925 Untersuchungen über die Grundlagen der Mengenlehre, *Mathematische Zeitschrift* **22**, 250–73.

1926 Axiomatische Theorie der geordneten Mengen, *JrM* **155**, 129–58.

1928 *Einleitung in die Mengenlehre*, 3rd augmented edn., Berlin, Springer.

1930 Georg Cantor, *DMV* **39**, 189–266. Short version in [Cantor 1932], 452–483.

1932 Axiomatische Theorie der Wohlordnung, *JrM* **167**, 1–11.

1967 *Lebenskreise. Aus den Erinnerungen eines Jüdischen Mathematikers*, Stuttgart, Deutsche Verlags-Anstalt.

FRAENKEL, A. & Y. BAR-HILLEL

1958 *Foundations of set theory*, Amsterdam, North-Holland.

FRAENKEL, A., Y. BAR-HILLEL & A. LEVY

1973 *Foundations of set theory*, Amsterdam, North-Holland.

FRAPOLLI, M.J.

1992 The Status of Cantorian Numbers, *Modern Logic* **2**, 365–382.

FREGE, G.

1879 *Begriffschrift*, Halle, Nebert. Reprint in Hildesheim, Olms, 1964.

1884 *Die Grundlagen der Arithmetik*, Breslau, Koebner.

1893 *Grundgesetze der Arithmetik*, vol.1, Jena, Pohle. Reprint Hildesheim, Olms, 1966.

1895 A Critical Elucidation of some Points in E. Schröder, *Vorlesungen über die Algebra der Logik*, in [Frege 1984], 210–228.

1903 *Grundgesetze der Arithmetik*, vol.2 of [Frege 1893].

1969 *Nachgelassene Schriften*, H. Hermes, F. Kambartel, F. Kaulbach eds., Hamburg, Meiner.

1980 *Philosophical and mathematical correspondence*, Oxford, Basil Blackwell.

1984 *Collected papers on mathematics, logic and philosophy*, Oxford, Basil Blackwell.

FREI, G. & U. STAMMBACH

1994 *Die Mathematiker and den Zürcher Hochschulen*, Basel, Birkhäuser.

FREUDENTHAL, H.

1962 The Main Trends in the Foundations of Geometry in the 19th Century, in *Logic, Methodology and Philosophy of Science*, Stanford Univ. Press, 613–21.

1981 Riemann, in [Gillispie 1981], vol. 10, 447–56.

FRISCH, J.C.

1969 *Extension and Comprehension in Logic*, New York, Philosophical Library.

FROBENIUS, G.

1893 Gedächtnisrede auf Leopold Kronecker, in *Gesammelte Abhandlungen* vol.3, Berlin, Springer, 1968, 705–724.

GANA, F.

1985 Peirce e Dedekind: La Definizione di Insieme Finito, *HM* **12**, 203–218.

GARCIADIEGO, A.R.

1985 The Emergence of Some of the Non-Logical Paradoxes of the Theory of Sets, *HM* **12**, 337–351.

1986 On Rewriting the History of the Foundations of Mathematics at the Turn of the Century, *HM* **13**, 39–41.

1992 *Bertrand Russell and the origins of the set theoretical paradoxes*, Basel, Birkhäuser.

1994 The Set-theoretic Paradoxes, in I. Grattan-Guinness, ed. , 629–34.

GAUSS, C.F.
  1801 *Disquisitiones arithmeticae*, Leipzig, Fleischer. References to [Gauss 1863/1929], vol. 1. English translation New Haven, 1966.
  1827 Disquisitiones generales circa superficies curvas, [Gauss 1863/1929], vol.4, 217–258. German translation in *Gaussche Flächentheorie, Riemannsche Räume und Minkowski-Welt*, Leipzig, Teubner, 1984.
  1831 Anzeige der [Gauss 1832], *Göttingische gelehrte Anzeigen*. References to [Gauss 1863/1929], vol.2, 169–178. English trans. in [Ewald 1996], vol.1.
  1832 Theoria residuorum biquadraticorum, commentatio secunda, [Gauss 1863/1929], vol.2, 93–148.
  1849 Beiträge zur Theorie der algebraischen Gleichungen, [Gauss 1863/1929], vol.3, 71–102.
  1863/1929 *Werke*, 12 vols., Göttingen, Dieterich. Reprint Hildesheim, Olms, 1973.

GEISON, G.L.
  1981 Scientific Change, Emerging Specialties, and Research Schools, *History of Science* **19**, 20–40.
  1993 Research Schools and New Directions in the Historiography of Science, *Osiris* **8**, 227–38.

GERICKE, H.
  1970 *Geschichte des Zahlbegriffs*, Mannheim, B.I. Hochschultaschenbuch.
  1971 Zur Geschichte des Zahlbegriffs, *Math. Semesterberichte* **18**, 161–173.
  1973 Vorgeschichte der Mengenlehre, *Math. Semesterberichte* **20**, 151–170.

GILLISPIE, C.C.
  1981 *Dictionary of scientific biography*, 16 vols. en 8, New York, Scribner's Sons.

GÖDEL, K.
  1931 Über formal unentscheidbare Sätze der *Principia mathematica* und verwandter Systeme I, *Monatshefte für Mathematik und Physik* **38**, 173–198. References to [Gödel 1986/90], vol. 1, 144–95.
  1938 The Consistency of the Axiom of Choice and of the Generalized Continuum Hypothesis, *NAS* **24**, 556–57. References to [Gödel 1986/90], vol. 2, 26–27.
  1939 Consistency Proof for the Generalized Continuum Hypothesis, *NAS* **25**, 220–24. References to [Gödel 1986/90], vol. 2, 28–32.
  1940 *The Consistency of the Axiom of Choice and of the Generalized Continuum Hypothesis with the Axioms of Set Theory*, Princeton University Press; reprinted with additional notes in 1951 and 1966. References to [Gödel 1986/90], vol. 2, 33–101.
  1944 Russell's Mathematical Logic, in P.A. Schilpp, ed., *The philosophy of Bertrand Russell*, La Salle, Ill., Open Court. References to [Gödel 1986/90], vol. 2, 119–41.
  1947 What is Cantor's continuum problem?, *AMM* **54**. References to [Godel 1986/90], vol. 2, 176–87. Revised and expanded version of 1964 in [Benacerraf & Putnam 1983].
  1986/90 *Collected Works*, ed. S. Feferman et al., Oxford University Press, 3 vols.

GOLDFARB, W.
  1979 Logic in the Twenties: The Nature of the Quantifier, *JSL* **44**, 351–368.
  1988 Poincaré Against the Logicists, in W. Aspray & Ph. Kitcher, eds., *History and Philosophy of Modern Mathematics*, Minneapolis, University Press.

GRABINER, J.
  1981 *The Origins of Cauchy's Rigorous Calculus*, Cambridge, Mass.

GRASSMANN, H.
1844 *Die Lineale Ausdehnungslehre*, Leipzig, Teubner. In [Grassmann 1894], vol.1, 1–319.
1861 *Lehrbuch der Arithmetik für höhere Lehranstalten*, Berlin, Enslin.
1894 *Gesammelte mathematische und physikalische Schriften*, 3 vols., Leipzig, Teubner, 1894–1911. Reprint New York, Chelsea, 1969.

GRATTAN-GUINNESS, I.
1970 An unpublished paper by Georg Cantor: Principien einer Theorie der Ordnungstypen. Erste Mitteilung, *AM* **124**, 65–107.
1971 Towards a Biography of Georg Cantor, *AS* **27**, 345–391.
1971a The correspondence between Georg Cantor and Philip Jourdain, *DMV* **73**, 111–130.
1972 Bertrand Russell on his Paradox and the Multiplicative Axiom: An unpublished letter to P. Jourdain, *Journal of Philosophical Logic* **1**, 103–10.
1974 The rediscovery of the Cantor–Dedekind correspondence, *DMV* **76**, 104–139.
1978 How B. Russell Discovered his Paradox, *HM* **5**, 127–37.
1979 In Memoriam Kurt Gödel: His 1931 correspondence with Zermelo on his incompletability theorem, *HM* **6**, 294–304.
1980 ed. *From the Calculus to Set Theory, 1630–1910*, Londres, Duckworth, 1980.
1980a The Emergence of Mathematical Analysis and the Progress in its Foundations from 1780 to 1880, in [Grattan-Guinness 1980], 125–193.
1981 On the Development of Logics Between the Two World Wars, *AMM* **88**, 495–509.
1984 Notes on the Fate of Logicism from *Principia Mathematica* to Gödel's Incompletability Theorem, *HPL* **5**, 67–78.

GRAY, J.
1989 *Ideas of Space. Euclidean, Non-Euclidean and Relativistic*, Oxford, Clarendon.

HADAMARD, J., R. BAIRE, E. BOREL & H. LEBESGUE
1905 Cinq lettres sur la théorie des ensembles, *Bulletin de la Société Mathématique de France* **33**, 261–273. English translation in [Moore 1982].

HALLETT, M.
1984 *Cantorian Set Theory and Limitation of Size*, Oxford, Clarendon.
1991 Hilbert's Axiomatic Method and the Laws of Thought, in A. George, ed., *Mathematics and Mind*, Oxford University Press, 1994, 158–200.

HAMILTON, W.R.
1837 Theory of Conjugate Functions, or Algebraic Couples; with a preliminary and elementary Essay on Algebra as the Science of Pure Time, *Transactions of the Royal Irish Academy* **17**, 293–422. References to [Hamilton 1967], 3–96.
1853 Preface to *Lectures on Quaternions*, Dublin, Hodges & Smith. References to [Hamilton 1967], 117–155. Reprint in [Ewald 1996], vol.1.
1967 *The Mathematical Papers*, vol.3, Cambridge, University Press.

HANKEL, H.
1867 *Vorlesungen über komplexe Zahlen und Funktionen. I. Theil: Theorie der komplexen Zahlensysteme*, Leipzig, Voss.
1870 *Untersuchungen über die unendlich oft oszillierenden und unstetigen Funktionen* (Gratulationsprogramm, Tübingen, 1870), *MA* **20** (1882), 63–112.

HANKINS, T.L.
   1976 Triplets and Triads: Sir William Rowan Hamilton on the Metaphysics of Mathematics, *Isis* **68**, 175–193.
   1980 *Sir William Rowan Hamilton*, Baltimore.

HARNACK, A.
   1882 Vereinfachung der Beweise in der Theorie der Fourier'schen Reihe, *MA* **19**, 235–279, 524–528.
   1885 Über den Inhalt von Punktmengen, *MA* **25**, 241–250.

HARTOGS, F.
   1915 Über das Problem der Wohlordnung, *MA* **76**, 436–43.

HAUBRICH, R.
   1993 Die Herausbildung der algebraischen Zahlentheorie als wissenschaftliche Disziplin, unpublished draft of conference for the 4th Annual Göttingen Workshop on the History of Modern Mathematics.
   1999 *The Genesis of Richard Dedekind's Algebraic Number Theory,* (unpublished manuscript).

HAUSDORFF, F.
   1904 Der Potenzbegriff in der Mengenlehre, *DMV* **13**, 569–71.
   1908 Grundzüge einer Theorie der geordneten Mengen, *MA* **65**, 435–505.
   1914 *Grundzüge der Mengenlehre*, Leipzig, Veit. Reprint New York, Chelsea, 1949.
   1916 Die Mächtigkeit der Borelschen Mengen, *MA* **77**, 430–37.
   1927 *Mengenlehre*, Berlin, W. de Gruyter. 2nd, shortened edn. of [Hausdorff 1914].
   1937 3rd, revised edn. of [Hausdorff 1914]. Translated as *Set Theory*, New York, Chelsea, 1957.

HAWKINS, T.W.
   1970 *Lebesgue's theory of integration. Its origins and development*, New York, Chelsea, 1975.
   1980 The origins of Modern Theories of Integration, in [Grattan-Guinness 1980], 194–234.
   1981 The Berlin School of Mathematics, in [Bos, Mehrtens & Schneider 1981], 233–245.
   1984 The *Erlanger Programm* of Felix Klein: Reflections on its place in the history of mathematics, *HM* **11**, 442–470.

VAN HEIJENOORT, J.
   1967 *From Frege to Gödel. A source book in mathematical logic, 1879–1931*, Cambridge/London, Harvard Univesity Press.

HEINE, E.
   1870 Über trigonometrische Reihen, *JrM* **71**, 353–365.
   1872 Die Elemente der Funktionenlehre, *JrM* **74**, 172–188.

HELLMAN, G.
   1981 How to Gödel a Frege–Russell: Gödel's incompleteness theorems and logicism, *Noûs* **4**, 451–68
   1989 *Mathematics Without Numbers*, Oxford University Press.

HELMHOLTZ, H.
   1868 Über die Tatsachen, die der Geometrie zu Grunde liegen, *NG*, 193–221. Reprinted in [Helmholtz 1921].

1887 Zählen und Messen erkenntnistheoretisch betrachtet, in *Philosophische Aufsätze für E. Zeller*, Leipzig, Fues. References to [Helmholtz 1921], 70–108. English trans. in [Ewald 1996], vol.2.

1921 *Schriften zur Erkenntnistheorie*, ed. P. Hertz and M. Schlick, Berlin, Springer.

HEMPEL, C.G.

1945 On the Nature of Mathematical Truth, *AMM* **52**, 543–556. References to reprint in [Benacerraf & Putnam 1983, 377–393].

HENDRY, J.

1984 The Evolution of William Rowan Hamilton's View of Algebra as the Science of Pure Time, *Studies in the History and Philosophy of Science* **15**, 63–81.

HERBART, J.F.

1808 Hauptpunkte der Logik, in [Herbart 1964], vol. 2, 217–26.

1824 *Die Psychologie als Wissenschaft, neu gegründet auf Erfahrung, Metaphysik und Mathematik*, vol. 1, in [Herbart 1964], vol. 5, 177–402.

1825 *Die Psychologie als Wissenschaft, neu gegründet auf Erfahrung, Metaphysik und Mathematik*, vol. 2, in [Herbart 1964], vol. 6, 1–339.

1837 *Lehrbuch zur Einleitung in die Philosophie*. Zweyter Abschnitt: Die Logik, in [Herbart 1964], vol. 4, 67–104.

1964 *Sämtliche Werke in chronologischer Reihenfolge*, 19 vols., Langensalza, 1879–1912. Reprint Aalen, Scientia, 1964.

HESSENBERG, G.

1906 *Grundbegriffe der Mengenlehre*, Göttingen, Vandenhoeck & Ruprecht.

1909 Kettentheorie und Wohlordnung, *JrM* **135**, 81–133

HEYTING, A.

1956 *Intuitionism: An introduction*, Amsterdam, North-Holland.

HILBERT, D.

1897 Die Theorie der algebraischen Zahlkörper, *DMV* **4**, 175–546. References to [Hilbert 1932/35], vol.1, 63–363.

1900 Mathematische Probleme. Vortrag, gehalten auf dem internationalen Mathematiker-Congress zu Paris, *NG*, 253–97. References to [Hilbert 1932/35], vol. 3, 290–339. English translation in *Bulletin of AMS* **8** (1902), 437–79. Partial translation in [Ewald 1996], vol. 2.

1900a Über den Zahlbegriff, *DMV* **8**, 180–94. References to [Hilbert 1930], 241–46. English trans. in [Ewald 1996], vol.2.

1904 Über die Grundlagen der Logik und Arithmetik, in *Verhandlungen des dritten internationalen Mathematiker-Kongresses in Heidelberg*, Leipzig, Teubner, 1905. Reprint in [Hilbert 1930], 247–61. References to English translation in [van Heijenoort 1967], 129–38.

1918 Axiomatisches Denken, *MA* **78**, 405–15. References to [Hilbert 1932/35], vol. 3, 146–56. English trans. in [Ewald 1996], vol. 2.

1922 Neubegründung der Mathematik, *Abhandlungen mathematischen Seminar Universität Hamburg* **1**, 157–77. References to to [Hilbert 1932/35], vol. 3, 157–77. English trans. in [Ewald 1996], vol. 2.

1923 Die logischen Grundlagen der Mathematik, *MA* **88**, 151–65. References to [Hilbert 1932/35], vol. 3, 178–91. English trans. in [Ewald 1996], vol. 2.

1926 Über das Unendliche, *MA* **95**, 161–90. Short version in [Hilbert 1930]. References to the English translation in [van Heijenoort 1967], 367–92.

1928 Die Grundlagen der Mathematik, *Abhandlungen mathematischen Seminar Universität Hamburg* **6**, 65–85. English trans. in [van Heijenoort 1967].
1929 Probleme der Grundlegung der Mathematik, *MA* **102**, 1–9.
1930 *Grundlagen der Geometrie*, 7th edn. (1st 1899, 2nd 1903), Leipzig, Teubner.
1931 Die Grundlegung der elementaren Zahlenlehre, *MA* **104**, 485–94. Partial reprint in [Hilbert 1932/35], vol. 3, 192–95. References to the English translation in [Ewald 1996], vol. 2.
1932/35 *Gesammelte Abhandlungen*, 3 vols., Berlin, Springer.

HILBERT, D. & W. ACKERMANN
1928 *Grundzüge der theoretischen Logik*, Berlin, Springer. 2nd edn. 1938.

HILBERT, D. & P. BERNAYS
1934/39 *Grundlagen der Mathematik*, 2 vols., berlin, Springer.

HOFFMANN, L.
1858/67 *Mathematisches Wörterbuch*, Berlin, Wiegandt und Hempel.

HURWITZ, A.
1894 Über die Theorie der Ideale, *NG*, 291–98. Reprint in *Werke*, vol. 2, 191–97.
1895 Über einen Fundamentalsatz deer arithmetischen Theorie der algebraischen Grössen, *NG*, 230–40. Reprint in *Werke*, vol. 2, 198–207.

JAHNKE, H.N.
1987 Motive und Probleme der Arithmetisierung der Mathematik in der ersten Hälfte des 19. Jahrhunderts – Cauchys Analysis in der Sicht des Mathematikers Martin Ohm, *AHES* **37**, 101–182.
1990 *Mathematik und Bildung in der Humboldtschen Reform*, Göttingen, Vandenhoeck & Ruprecht.

JAHNKE, H.N. & M. OTTE
1981 Origins of the Programm of «Arithmetization of Mathematics», in [Bos, Mehrtens & Schneider 1981], 21–49.

JANÉ, I.
1995 The Role of the Absolute Infinite in Cantor's Conception of Set, *Erkenntnis* **42**, 375–402.

JOHNSON, D.M.
1979/81 The Problem of the Invariance of Dimension in the Growth of Modern Topology, *AHES* **20**, 97–188, **25**, 85–266.

JORDAN, C.
1870 *Traité des substitutions et des équations algébriques*. Paris, Gauthier-Villars. Reprint 1957.
1893 *Cours d'analyse de l'Ecole Polytechnique*, 2nd edn., Paris, Gauthiers-Villars.

JOURDAIN, P.E.B.
1906/14 The Development of the Theory of Transfinite Numbers, *Archiv für Mathematik und Physik* **10**, 254–281, **14**, 287–311, **16**, 21–43, **22**, 1–21.
1910/13 The Development of Theories of Mathematical Logic and the Principles of Mathematics, *Quarterly Journal for pure and applied Mathematics* **41**, 324–352, **43**, 219–314, **44**, 113–128.

JUNGNICKEL, C. & R. McCORMACH
1986 *Intelectual Mastery of Nature. Theoretical Physics from Ohm to Einstein. Vol. I. The Torch of Mathematics, 1800–1870*, University of Chicago Press.

KANAMORI, A.
 1994 *The Higher Infinite*, Berlin, Springer.
 1995 The Emergence of Descriptive Set Theory, in J. Hintikka, ed., *From Dedekind to Gödel*, Dordrecht, Kluwer.
 1996 The Mathematical Development of Set Theory from Cantor to Cohen, *BSL* 2.

KANT, I.
 1787 *Kritik der reinen Vernunft*, Riga, Hartknoch. In *Kants Werke*, vol. 3, Berlin, Akademie, 1911, reprint 1968. All references to edition B. English translation London, Macmillan, 1933.

KEFERSTEIN, H.
 1890 Über den Begriff der Zahl, *Mitteilungen der Mathematische Gesellschaft zu Hamburg* 2, 119–125.

KEMENY, J.
 1950 Type Theory vs. Set Theory, *JSL* 15, 78.

KENNEDY, H.C.
 1974 *Giuseppe Peano*, Basel, Birkhäuser.

KIERNAN, M.
 1971 The Development of Galois Theory from Lagrange to Artin, *AHES* 8, 40–154.

KLAUA, D.
 1964 *Allgemeine Mengenlehre*, Berlin, Springer.

KLEENE, S.C. & J.B. ROSSER
 1935 The Inconsistency of Certain Formal Logics, *Annals of Mathematics* 36, 630–36.

KLEIN, F.
 1893 Vergleichende Betrachtungen über neuere geometrische Forschungen (Universitätsprogramm, Erlangen, 1872), *MA* 43, 63–100. References to [Klein 1921/23], vol. 1, 460–97.
 1895 Über Arithmetisierung der Mathematik, *NG*, Geschäftliche Mitteilungen. References to [Klein 1921/23], vol. 2, 232–40. English trans. in *Bulletin of AMS* 2, 241–249, reprinted in [Ewald 1996], vol. 2.
 1897 Riemann und seine Bedeutung für die Entwicklung der modernen Mathematik, *DMV* 4, 71–87.
 1921/23 *Gesammelte mathematische Abhandlungen*, 3 vol., Berlin, Springer. Reprinted in 1973.
 1926 *Vorlesungen über die Entwicklung der Mathematik im 19. Jahrhundert*, 2 vols., Berlin, Springer, 1926, 1927. Reprint Berlin/New York, Springer, 1979.

KLINE, M.
 1972 *Mathematical Thought from Ancient to Modern Times*, New York, Oxford University Press.
 1980 *Mathematics: The loss of certainty*, Oxford University Press.

KLÜGEL, G.S.
 1803/08 *Mathematisches Wörterbuch*, 3 vol. (A-P), Leipzig, Schwickert.

KNEALE, W. & M. KNEALE
 1962 *The Development of Logic*, Oxford, Clarendon.

KNOBLOCH, E.
 1981 Symbolik und Formalismus im mathematischen Denken des 19. und beginnenden 20. Jahrhunderts, in [Dauben 1981], 139–165.

KOCH, H.

1982 *Über das Leben und Werk Johann Peter Gustav Lejeune Dirichlets: zu seinem 175 Geburtstag*, Berlin, Akademie Verlag.

KÖNIG, G. ed.

1990 *Konzepte des mathematisch Unendlichen im 19. Jahrhundert*, Göttingen, Vandenhoeck & Ruprecht.

KÖNIG, J.

1905 Über die Grundlagen der Mengenlehre und das Kontinuumproblem, *MA* **63**, 217–21. References to English trans. in [van Heijenoort 1967], 145–49.

1914 *Neue Grundlagen der Logik, Arithmetik und Mengenlehre*, ed. D. König, Leipzig, Veit.

KOPPELMAN, E.

1971 The Calculus of Operations and the Rise of Abstract Algebra, *AHES* **8**, 155–242.

KOSSAK, E.

1872 *Die Elemente der Arithmetik*, Schulprogrammschrift des Friedrich-Werderschen Gymnasiums, Berlin.

KOWALEWSKI, G.

1950 *Bestand und Wandel*, München, Oldenbourg

KREISEL, G.

1965 Mathematical Logic, in T.L. Saaty, ed., *Lectures on Modern Mathematics*, New York, Wiley, vol.3, 95–195.

1978 Hilbert's Programme, 2nd revised version, in [Benacerraf & Putnam 1983].

KRONECKER, L.

1882 Grundzüge einer arithmetischen Theorie der algebraischen Grössen, *JrM* **92**, 1–122. Reprint in [Kronecker 1895/1930], vol.2, 237–387.

1886 Über einige Anwendungen der Modulsysteme auf elementare algebraische Fragen, *JrM* **99**, 329–71. Reprint in [Kronecker 1895/1930], vol.3, 145–208.

1887 Über den Zahlbegriff, in *Philosophische Aufsätze für E. Zeller*, Leipzig, Fues. also in *JrM* **101**, 337–55. References to [Kronecker 1895/1930], vol.3/1, 249–274. English trans. in [Ewald 1996], vol.2.

1895/1930 *Werke*, 5 vols., Leipzig, Teubner. Reimpreso en New York, Chelsea, 1968.

KUHN, T.S.

1962 *The Structure of Scientific Revolutions*, Chicago Univ. Press (2nd edn. 1970).

KUMMER, E.

1860 Gedächtnisrede auf Gustav Peter Lejeune Dirichlet, in [Dirichlet 1889/97], vol.2, 311–344.

1975 *Collected Papers*, vol. 1. Berlin, Springer.

KURATOWSKI, K.

1921 Sur la notion de l'ordre dans la théorie des ensembles, *FM* **2**, 161–71.

1922 Une méthode d'elimination des nombres transfinis des raisonnements mathématiques, *FM* **3**, 76–18.

1924 Sur l'etat actuel de l'axiomatique de la théorie des ensembles, *Annales de la Société Polonaise de Mathématiques* **3**, 146–47.

KUSHNER, D.S.

1993 Sir George Darwin and a British School of Geophysics, *Osiris* **8**, 196–223.

LAKATOS, I.
1967 A Renaissance of Empiricism in the Recent Philosophy of Mathematics?, in *Mathematics, Science and Epistemology. Philosophical Papers Vol. 2*, eds. J. Worrall and G. Currie, Cambridge University Press, 1978, 24–42.
1970 Falsification and the Methodology of Research Prgrammes, in *The Methodology of Scientific Research Programmes. Philosophical Papers Vol. 1*, Cambridge University Press, 1978.

LANDAU, E.
1917 Richard Dedekind – Gedächtnisrede, *NG*, 50–70.

LAUGWITZ, D.
1996 *Bernhard Riemann 1826–1866. Wendepunkte in der Auffassung der Mathematik*, Basel, Birkhäuser.

LAVINE, S.
1994 *Understanding the Infinite*, Harvard university Press.

LEIBNIZ, G.W.
1704 *Nouveaux essais sur l'entendement humain*, in *Die philosophischen Schriften*, vol. 5, 39–509. Ed. C.J. Gerhardt, 7 vols., Berlin, Weidmann, 1875/90. Reprint Hildesheim, G. Olms, 1960/61.
1714 *Monadologie*, in *Die philosophischen Schriften*, vol. 6, 607–623.
1966 *Logical Papers*, Oxford, Clarendon Press.
1976 *Die mathematischen Studien zur Kombinatorik*, Wiesbaden.

LEIBNIZ, G.W. & S. CLARKE
1717 Correspondence, in [Leibniz 1875/90, vol.7].

LEWIS, A.C.
1977 H. Grassmann's 1844 Ausdehnungslehre and Schleiermacher's Dialektik, *AS* **34**, 103–162.

LIPSCHITZ, R.
1864 De explicatione per series trigonometricas instituenda functionum unius variabilis arbitrariarum, et praecipue earum, quae per variabilis spatium finitum valorum maximorum et minimorum numerum habent infinitum, disquisitio, *JrM* **63**, 296–308. French translation in *AM* **36**, 281–295.
1877 *Grundlagen der Analysis*, Bonn, Cohen & Sohn.
1986 *Briefwechsel mit Cantor, Dedekind, Helmholtz, Kronecker, Weierstrass und anderen*, Braunschweig/Wiesbaden, Vieweg.

LOREY, W.
1916 *Das Studium der Mathematik an den Deutschen Universitäten seit Anfang des 19. Jahrhunderts*, Leipzig/Berlin, Teubner.

LOTZE, H.
1856/64 *Mikrokosmus, Ideen zur naturgeschichte und Geschichte der Menschheit, Versuch einer Antropologie*, 3 vols., Leipzig, Hirzel. English translation in 2 vols., Edinburgh, Clark, 1885.

ŁUKASIEWICZ, J.
1957 *Aristotle's Syllogistic*, Oxford, Clarendon Press.

LUZIN, N.
1930 *Leçons sur les ensembles analytiques et leurs applications*, Paris, Gauthier-Villars.

MANHEIM, J.R.

1964 *The Genesis of Point Set Topology*, New York, Macmillan.

MANNING, K.R.

1975 The emergence of the Weierstrassian approach to complex analysis, *AHES* **14**, 297–383

MANZANO, M.

1997 Alonzo Church: His life, his work, and some of his miracles, *HPL* **18**, 211–32.

MATHEWS, J.

1978 William Rowan Hamilton's Paper of 1837 on the Arithmetization of Analysis, *AHES* **19**, 177–200.

MAY, K.O.

1969 Review of F.A. Medvedev: *The development of the theory of sets in the nineteenth century* [1965], *Mathematical Reviews* **37**, 4–5.

McCARTY, D.C.

1995 The Mysteries of Richard Dedekind, in J. Hintikka, ed., *From Dedekind to Gödel*, Dordrecht, Kluwer.

McCLELLAND, Ch.

1980 *State, Society, and University in Germany, 1700–1914*, Cambridge Univ. Press.

MEDVEDEV, F.A.

1965 *Razvitie teorii mnozhestv v XIX veke*, Moscow, Nauka.

1984 Über die Abstrakten Mengenlehren von Cantor und Dedekind, *Berichte zur Wissenschaftsgeschichte* **7**, 195–200.

MEHRTENS, H.

1979 *Die Entstehung der Verbandstheorie*, Hildesheim, Gerstenberg.

1979a Das Skelett der modernen Algebra. Zur Bildung mathematischer Begriffe bei Richard Dedekind, in *Disciplinae Novae*, Göttingen, Vandenhoeck & Ruprecht, 25–43.

1982 Richard Dedekind Der Mensch und die Zahlen, *Abhandlungen der Braunschweigischen Wissenschaftlichen Gesellschaft* **33**, 19–33.

1990 *Moderne – Sprache – Mathematik. Eine Geschichte des Streits um die Grundlagen der Disziplin und des Subjekts formaler Systeme*, Frankfurt, Suhrkamp.

MESCHKOWSKI, H.

1965 Aus den Briefbüchern Georg Cantors, *AHES* **2**, 503–519.

1967 *Probleme des Unendlichen. Werk und Leben Georg Cantors*, Braunschweig, Vieweg. Reprinted with additions as *Georg Cantor. Leben, Werk und Wirkung*, Mannheim, Bibliographisches Institut, 1983.

MESCHKOWSKI, H. & W. NILSON

1991 *Georg Cantor: Briefe*, Berlin, Springer.

MIRIMANOFF, D.

1917 Les antinomies de Russell et de Burali-Forti et le probléme fondamental de la théorie des ensembles, *L'Enseignement Mathématique* **19**, 37–52.

1917a Remarques sur la théorie des ensembles et les antinomies cantoriennes, I, *L'Enseignement Mathématique* **19**, 209–17.

1919 Remarques sur la théorie des ensembles et les antinomies cantoriennes, II, *L'Enseignement Mathématique* **21**, 29–52.

MITTAG-LEFFLER, G.
    1884 Sur la représentation analytique des fonctions d'une variable indépendante, *AM* **4**, 1–79.
    1923 Weierstrass et Sonja Kowalewsky, *AM* **39**, 133–198.
    1927 Zusätzliche Bemerkungen to [Schoenflies 1927], *AM* **50**, 25–26.

MONNA, A.F.
    1975 *Dirichlet's Principle, a mathematical comedy of errors and its influence on the development of analysis*, Utrecht.

MOORE, G.H.
    1978 The Origins of Zermelo's Axiomatization of Set Theory, *Journal of Philosophical Logic* **7**, 307–329.
    1980 Beyond First-order Logic: The Historical Interplay between Mathematical Logic and Axiomatic Set Theory, *HPL* **1**, 95–137.
    1982 *Zermelo's Axiom of Choice. Its Origins, Development and Influence*, Berlin, Springer.
    1987 A House Divided Against Itself: The emergence of first-order logic as the basis for mathematics, in [Phillips 1987], 98–136.
    1988 The emergence of First-Order Logic, in W. Aspray & Ph. Kitcher, eds., *History and Philosophy of Modern Mathematics*, Minneapolis, University Press, 95–135
    1988a The Origins of Forcing, in F.R. Drake and J.K. Truss, eds., *Logic Colloquium '86*, Amsterdam, North-Holland, 143–73.
    1989 Towards a History of Cantor's Continuum Problem, in [Rowe & McCleary 1989], 79–121.
    1993 Introduction to vol. 3 of *The Collected Papers of Bertrand Russell*, ed. G.H. Moore (general ed. J. Passmore), London, Routledge.
    1995 The Prehistory of Infinitary Logic: 1885–1955, preprint, to appear in *Proceedings of LMPS 95*.
    1997 Hilbert and the Emergence of Modern Mathematical Logic, *Theoria* **12**, 65–90.
    forthcoming. Logic, Early Twentieth Century, to appear in E. Craig, ed., *Routledge Encyclopedia of Philosophy*, London, Routledge.
    forthcoming. Paradoxes of Set and Property, to appear in E. Craig, ed., *Routledge Encyclopedia of Philosophy*, London, Routledge.

MOORE, G.H. & A. GARCIADIEGO
    1981 Burali-Forti's Paradox: A reappraisal of its origins, *HM* **8**, 319–50.

MOSTOWSKI, A.
    1965 Thirty Years of Foundational Studies ... 1930–1964, *Acta Philosophica Fennica* **17**, 1–180. References to *Foundational Studies. Selected works*, vol. 1, Amsterdam, North-Holland, 1979.

NAGEL, E.
    1935 «Impossible Numbers»: A Chapter in the History of Modern Logic, in *Teleology Revisited and Other Essays*, Columbia University Press, 1979, 166–194.
    1939 The Formation of Modern Conceptions of Formal Logic in the Development of Geometry, in *Teleology Revisited*, Columbia University Press, 1979, 195–259.

NAGELL, T.
    1963 Thoralf Skolem in Memoriam, *AM* **110**, i–xi.

NAUMANN, B.
    1986 *Grammatik der deutschen Sprache zwischen 1781 und 1856*, Berlin, E. Schmidt.

NETTO, E.
1878 Beitrag zur Mannigfaltigkeitslehre, *JrM* **86**, 263–268.
1908 Kombinatorik, in M. Cantor, ed., *Vorlesungen über Geschichte der Mathematik*, vol. 4, Leipzig, Teubner.

NEUENSCHWANDER, E.
1981 Lettres de Bernhard Riemann à sa famille, *Cahiers du Séminaire d'Histoire des Mathématiques* **2**, 85–131.
1981a Über die Wechselwirkung zwischen der französischen Schule, Riemann und Weierstrass. Eine Übersicht mit zwei Quellenstudien, *AHES* **24**, 221–255.
1988 A brief report on a number of recently discovered sets of notes on Riemann's lectures and on the transmission of the Riemann's «Nachlass», *HM* **15**, 101–113.

VON NEUMANN, J.
1923 Zur Einführung der transfiniten Zahlen, *ALS* **1**, 199–208. References to English trans. in [van Heijenoort 1967], 347–54.
1925 Eine Axiomatisierung der Mengenlehre, *JrM* **154**, 219–40. Reprint in [von Neumann 1961]. References to English trans. in [van Heijenoort 1967], 394–413.
1927 Zur Hilbertschen Beweistheorie, *Mathematische Zeitschrift* **26**, 1–46. References to [von Neumann 1961], 256–300.
1928 Über die Definition durch transfinite Induktion und verwandte Fragen der allgemeinen Mengenlehre, *MA* **99**, 373–91. References to [von Neumann 1961], 320–38.
1928a Die Axiomatisierung der Mengenlehre, *MZ* **27**, 669–752. References to [von Neumann 1961], 339–422.
1929 Über eine Widerspruchsfreiheitsfrage in der axiomatischen Mengenlehre, *JrM* **160**, 227–41. References to [von Neumann 1961], 494–508.
1947 The Mathematician, in R.B. Heywood, ed., *The Works of the Mind*, University of Chicago Press. References to [von Neumann 1961].
1961 *Collected Works, Vol. I: Logic, theory of sets, and quantum mechanics*, ed. A.H. Taub, Oxford, Pergamon.

NEUMANN, O.
1980 Zur Genesis der algebraischen Zahlentheorie, *NTM* **17**, (1) 32–48, (2) 38–58.
1981 Über die Anstösse zu Kummers Schöpfung der «Idealen Complexen Zahlen», in [Dauben 1981], 179–199.

NEUMANN, O. & W. PURKERT
1981 Richard Dedekind – zum 150. Geburtstag, *Mitteilungen der Mathematische Gesellschaft der DDR* **2**, 84–110.

NIDDITCH, P.H.
1987 *The Development of Mathematical Logic*, Londres, Routledge & Kegan.

NOVY, L.
1973 *Origins of modern algebra*, Leyden, Noordhoff.

NOWAK, G.
1989 Riemann's *Habilitationsvortrag* and the Synthetic *A Priori* Status of Geometry, in [Rowe & McCleary 1989], 17–46.

OHM, M.
1853 *Versuch eines vollkommen consequenten Systems der Mathematik. Lehrbuch der Arithmetik, Algebra und Analysis. Vol. 1*, Berlin, T.H. Riemann.
1855 *Versuch eines vollkommen consequenten Systems der Mathematik, Vol. 2*, Berlin, T.H. Riemann.

OTTE, M.
1989 The Ideas of Hermann Grassmann in the Context of the Mathematical and Philosophical Tradition since Leibniz, *HM* **16**, 1–35.

PARSONS, C.
1977 What is the Iterative Concept of Set?, in R. Butts & J. Hintikka, eds., *Logic, Foundations of Mathematics and computability theory*, Dordrecht, Reidel, 335–67. Reprint in [Benacerraf & Putnam 1983].
1987 Developing Arithmetic in Set Theory without Infinity: Some Historical Remarks, *HPL* **8**, 201–213.
1990 The Structuralist View of Mathematical Objects, *Synthese* **84**, 303–346.

PAULSEN, F.
1896/97 *Geschichte des gelehrten Unterrichts auf den deutschen Schulen und Universitäten*, 3 vols., Leipzig, Teubner.

PEANO, G.
1889 *Arithmetices principia, nova methodo exposita*, Torino, Bocca. Partial English trans. in [van Heijenoort 1967], 83–97.
1891 Sul concetto di numero, *Rivista di matematica* **1**, 87–102, 256–67.
1906 Additione, *Rivista di matematica* **8**, 143–57.
1908 *Formulaire de mathématiques*, 5th edn. Torino, Bocca. Reprint Roma, Cremonese.
1957/59 *Opere Scelte*, 3 vols., Roma, Cremonese.

PECKHAUS, V.
1990 «Ich habe mich wohl gehütet, alle Patronen auf einmal zu verschiessen». Ernst Zermelo in Göttingen, *HPL* **11**, 19–58.
1991 *Hilbertprogramm und kritische Philosophie*, Göttingen, Vandenhoeck & Ruprecht.
1993 Ernst Schröder und der Logizismus, in W. Stelzner, ed., *Philosophie und Logik*, Berlin, de Gruyter, 108–19.
1994 Wozu Algebra der Logik? Ersnt Schröders Suche nach einer universalen Theorie der Verknüpfungen, *Modern Logic* **4**, 356–81.

PEIRCE, C.S.
1931/60 *Collected Papers*, 7 vols., Harvard University Press.

PHILLIPS, E.R.
1987 *Studies in the History of Mathematics*, Mathematical Association of America.

PIEPER, H., ed.
1897 *Briefwechsel zwischen A. von Humboldt und C.G.J. Jacobi*, Berlin, Akademie Verlag.

PINCHERLE, S.
1880 Saggio di una introduzione alla teoria delle funzioni analitiche secondo i principii del prof. C. Weierstrass, *Giornale di Mathematiche* **18**, 178–254, 314–357.

PLÜCKER, J.
1828 *Analytisch-geometrische Entwicklungen*, vol. 1, in *Gesammelte wissenschaftliche Abhandlungen*, Leipzig, Teubner, vol. 1, 1895.

POINCARÉ, H.
1894 Sur la nature du raisonnement mathématique, *Revue de métaphysique et de morale* **2**. Reprinted with modifications in *La science et l'hypothèse*, Paris, Flammarion, 1902. References to the English translation in [Benacerraf & Putnam 1983], 394–402.

414 *Bibliographical References*

1905/06 Les mathématiques et la logique, *Revue de métaphysique et de morale* **13**, **14**. Short version in *Science et méthode*, Paris, Flammarion, 1908. Translation of original version in [Ewald 1996], vol. 2.

PONT, J.C.
1974 *La topologie algébrique des origines à Poincaré*, Paris, P.U.F..

PRINGSHEIM, A.
1898 Irrationalzahlen und Konvergenz unendlicher Prozesse, in W.F. Meyer, ed., *Enzyklopädie der mathematischen Wissenschaften*, Leipzig, Teubner, vol. 1, 47–146.

PURKERT, W.
1973 Zur Genesis des abstrakten Körperbegriffs, *NTM* **10**, (1) 23–37, (2) 8–20.
1977 Ein Manuskript Dedekinds über Galois-Theorie, *NTM* **13**, (2) 1–16.
1989 Cantor's Views on the Foundations of Mathematics, in [Rowe & McCleary 1989], 49–65.

PURKERT, W. & H.J. ILGAUDS
1987 *Georg Cantor 1845–1918*, Basel/Boston/Stuttgart, Birkhäuser.

PUTNAM, H.
1967 The Thesis that Mathematics Is Logic, in *B. Russell, Philosopher of the Century*, ed. R. Schoenman, London, Allen & Unwin. Reprinted in *Mathematics, Matter, and Method*, Cambridge University Press, 1975.

PYCIOR, H.
1983 De Morgan's Algebraic Work: The Three Stages, *Isis* **74**, 211–26.
1987 British Abstract Algebra: Development and Early Reception, in Grattan-Guinness, ed., *History in Mathematics Education*, Paris, 1987, 152–168.

PYENSON, L.
1983 *Neohumanism and the Persistence of Pure Mathematics in Wilhelmian Germany*, Philadelphia, American Philosophical Society.

QUINE, W.V.
1934 *A System of Logistic*, Cambridge, Mass., Harvard University Press.
1936 Set-theoretic foundations for logic, *JSL* **1**, 45–57.
1937 New Foundations for Mathematical Logic, *AMM* **44**. Revised version in [Quine 1953].
1940 *Mathematical Logic*, New York, Norton. 2nd revised edn. 1951, based on a modification proposed in [Wang 1950]. Reprint New York, Harper & Row, 1962.
1941 Whitehead and the Rise of Modern Logic, in P.A. Schilpp, *The Philosophy of A.N. Whitehead*, La Salle, Ill., Open Court, 127–63.
1953 *From a Logical Point of View*, Cambridge, Mass., Harvard University Press.
1963 *Set Theory and its Logic*, Harvard University Press. 2nd. edn. 1969.
1986 Autobiographical Notes, in L.E. Hahn & P.A. Schilpp, eds., *The Philosophy of W.V. Quine*, La Salle, Ill., Open Court, p. 1–46.

RAMSEY, F.P.
1926 The Foundations of Mathematics, *LMS* **25**. References to *Foundations. Essays in philosophy, logic, mathematics, and economics*, London, Routledge & Kegan Paul, 1978.

RANG, B. & W. THOMAS
1981 Zermelo's Discovery of the 'Russell Paradox,' *HM* **8**, 15–22.

REID, C.
1970 *Hilbert*. Berlin, Springer.

RICHARD, J.
1905 Lettre à Monsieur le rédacteur de la Revue générale des Sciences, *AM* **30**, 295–96. References to English trans. in [van Heijenoort 1967], 143–44.

RIEMANN, B.
Cod. Ms. Riemann. Scientific legacy or *Nachlass* kept at the Niedersächsische Staats- und Universitätsbibliothek (Handschriftenabteilung), Göttingen.
1851 Grundlagen für eine allgemeine Theorie der Functionen einer veränderlichen complexen Grösse (Inauguraldissertation). References to [Riemann 1892], 3–45.
1854 Über die Hypothesen, welche der Geometrie zu Grunde liegen (Habilitationsvotrag), *Abhandlungen der Königlichen Gesellschaft der Wissenschaften zu Göttingen* **13** (1868). References to [Riemann 1892], 272–287. English translation in [Clifford 1882], reprinted in [Ewald 1996], vol. 2.
1854a Über die Darstellbarkeit einer Function durch eine trigonometrische Reihe (Habilitationsschrift), *Abhandlungen der Königlichen Gesellschaft der Wissenschaften zu Göttingen* **13** (1868). References to [Riemann 1892], 227–265.
1857 Theorie der Abel'schen Functionen, *JrM* **54**, 115–55. References to [Riemann 1892], 88–142.
1876 Fragmente philosophischen Inhalts, in [Riemann 1892], 509–538.
1892 *Gesammelte mathematische Werke und wissenschaftlicher Nachlass*, Leipzig, Teubner. Ed. H. Weber in collaboration with R. Dedekind (first edn. 1876, with different pagination). Reprinted with *Nachträge*, ed. M. Noether and W. Wirtinger (1902), in New York, Dover, 1953. All references to this reprint. There is a new edition with up-to-date bibliography in Berlin, Springer/Teubner, 1990.

RINGER, F.K.
1969 *The Decline of the German Mandarins. The German academic community, 1890–1933*, Harvard University Press.

ROBINSON, R.M.
1937 The Theory of Classes. A modification of von Neumann's system, *JSL* **2**, 29–36.

RODRIGUEZ CONSUEGRA, F.
1987 Rusell's Logicist Definitions of Numbers, 1898–1913: Chronology and Significance, *HPL* **8**, 141–169.
1991 *The Mathematical Philosophy of Bertrand Russell: Origins and Development*, Basel, Birkhäuser.

ROSSER, J.B.
1939 On the Consistency of Quine's "New Foundations for Mathematical Logic", *JSL* **4**, 15–24.
1942 The Burali-Forti Paradox, *JSL* **7**, 1–17.
1953 *Logic for Mathematicians*, New York, McGraw.

ROSSER, J.B. & H. WANG.
1950 Non-standard Models for Formal Logics, *JSL* **15**, 113–129. (Errata p. iv.)

ROWE, D.
1989 Klein, Hilbert, and the Göttingen Mathematical Tradition, *Osiris* **5**, 186–213.
1995 David Hilbert (1862–1943), in D. Rauschning & D. v. Nerée, eds., *Die Albertus-Universität zu Königsberg und ihre Professoren*, Berlin, Duncker & Humblot, 543–52.

1998 Mathematics in Berlin, 1810–1933, in: H.G.W. Begehr *et al.* eds., *Mathematics in Berlin*, Basel, Birkhäuser, p. 9–26.

2002 Mathematical Schools, Communities, and Networks, in Mary Jo Nye, ed., The Cambridge History of Science, vol. 5, Modern Physical and Mathematical Sciences, Cambridge University Press, to appear in 2002.

ROWE, D. & J. McCLEARY, eds.

1989 *The History of Modern Mathematics. Vol. I: Ideas and their reception*, Boston/London, Academic Press.

RUSSELL, B.

1903 *The principles of mathematics*, Cambridge, University Press. 2nd edn. 1937. Reprint London, Allen & Unwin, 1948.

1906 On Some Difficulties in the Theory of Transfinite Numbers and Order Types, *LMS* **4**, 29–53. Reprint in *Essays in Analysis*, ed. D. Lackey, London, Allen & Unwin.

1906a Les paradoxes de la logique, *Revue de métaphysique et de morale* **14**, 627–50. English trans. in *Essays in Analysis*, ed. D. Lackey, London, Allen & Unwin.

1908 Mathematical Logic as Based on the Theory of Types, *American Journal of Mathematics* **30**, 222–262. References to reprint in [van Heijenoort 1967], 150–182.

1919 *Introduction to mathematical philosophy*, London, Allen & Unwin.

SARKAR, S.

1992 "The Boundless Ocean of Unlimited Possibilities": Logic in Carnap's *Logical Syntax of Language*, *Synthese* **93**, 191–237.

SCHARLAU, W.

1981 ed. *Richard Dedekind 1831/1981. Eine Würdigung zu seinem 150. Geburtstag*, Braunschweig/Wiesbaden, Vieweg.

1981a Erläuterungen zu Dedekinds Manuskript über Algebra, in [Scharlau 1981], 101–108.

1982 Unveröffentliche algebraische Arbeiten Richard Dedekinds aus seiner Göttinger Zeit 1855–1858, *AHES* **27**, 335–367.

SCHOENFLIES, A.M.

1900/08 Die Entwicklung der Lehre von den Punktmannigfaltigkeiten, *DMV* **8**, 1–251, *Ergänzungsband* **2**, 1–331.

1922 Zur Erinnerung an Georg Cantor, *DMV* **31**, 97–106.

1927 Die Krisis in Cantor's mathematischem Schaffen, *AM* **50**, 1–23.

SCHOLZ, E.

1980 *Geschichte des Mannigfaltigkeitsbegriffs vom Riemann bis Poincaré*, Stuttgart, Birkhäuser.

1982 Riemanns frühe Notizen zum Mannigfaltigkeitsbegriff und zu den Grundlagen der Geometrie, *AHES* **27**, 213–232.

1982a Herbart's influence on Bernhard Riemann, *HM* **9**, 413–440.

1990 ed. *Geschichte der Algebra. Eine Einführung*, Mannheim/Wien/Zürich, BI-Wissenschaftsverlag.

1990a Riemann's Vision of a New Geometry, draft for International Coloquium *1830–1930: Un siècle de géométrie, de C. F. Gauss et B. Riemann a H. Poincaré et E. Cartan. Épistémologie, histoire et mathématiques*, Paris, 1989.

SCHRÖDER, E.
  1873 *Lehrbuch der Arithmetik und Algebra für Lehrer und Studirende. Erster Band: Die sieben algebraischen Operationen*, Leipzig, Teubner.
  1877 *Der Operationskreis des Logikkalküls*, Leipzig, Teubner. Reimpreso en Stuttgart, Teubner, 1966.
  1890/95 *Vorlesungen über die Algebra der Logik*, 3 vols., Leipzig, Teubner. Fourth vol. posthumously edited in 1905. Reprint New York, Chelsea, 1966.
  1898 Über Pasigraphie, ihren gegenwärtigen Stand und die pasigraphische Bewegung in Italien, in F. Rudio, ed., *Verhandlungen des ersten Internationalen Mathematiker-Kongresses*, Leipzig, Teubner, 147–62. English trans. in *The Monist* **9** (1899).
  1898a Über zwei Definitionen der Endlichkeit und G. Cantorsche Sätze, *Nova Acta Leopoldina* **71**, 303–62.

SCHUBRING, G.
  1982 Ansätze zur Begründung theoretischer Terme in der Mathematik – Die Theorie des Unendlichen bei Johann Schultz, *HM* **9**, 441–84.
  1983 *Die Entstehung des Mathematiklehrerberufs im 19. Jahrhundert*, Weinheim/Basel, Beltz.
  1984 Die Promotion von P.G. Lejeune Dirichlet, *NTM* **21**, 45–65.

SERVOS, J.W.
  1993 Research Schools and Their Histories, *Osiris* **8**, 3–15.

SHAPIRO, S.
  1991 *Foundations without Foundationalism: A case for second-order logic*, Oxford, Clarendon Press.

SHEPHERDSON, J.C.
  1951/53 Inner Models for Set Theory, *JSL* **16**, 161–90, **17**, 225–37, **18**, 145–67.

SHOENFIELD, J.R.
  1967 *Mathematical Logic*, Reading, Mass., Addison-Wesley.

SIERPIŃSKI, W.
  1928 *Leçons sur les nombres transfinis*, Pairs, Gauthier-Villars.
  1975 *Oeuvres choisies, tome II: Théorie des ensembles et ses applications (1908–1929)*, Warszawa, PWN – Éditions scientifiques de Pologne.

SIMONS, P.M.
  1987 Frege's Theory of Real Numbers, *HPL* **8**, 25–44.

SINACEUR, M.A.
  1971 Appartenance et inclusion: Un inedit de Richard Dedekind, *RHS* **24**, 247–254.
  1974 L'infini et les nombres. Commentaires de R. Dedekind a «Zahlen». La correspondance avec Keferstein, *RHS* **27**, 251–278.
  1979 La méthode mathématique de Dedekind, *RHS* **32**, 107–142.
  1990 Dedekind et le programme de Riemann, *RHS* **11**, 221–296.

SKOLEM, T.
  1920 Logisch-kombinatorische Untersuchungen über die Erfüllbarkeit oder Beweisbarkeit mathematischer Sätze nebst einem Theoreme über dichte Mengen, *Videnskaps-selskapets Skrifter*, 1–36. References to partial English trans. in [van Heijenoort 1967], 252–63.

1922 Einige Bemerkungen zur axiomatischen Begründung der Mengenlehre, in *Dem femte skandinaviska mathematikerkongressen*, Helsinki, Akademiska Bokhandeln. Also in [Skolem 1970]. References to English translation in [van Heijenoort 1967], 290–301.

1928 Über die mathematische Logik, *Norsk mat. tids.* **10**, 125–42. References to English translation in [van Heijenoort 1967], 512–24.

1929 Über einige Grundlagenfragen der Mathematik, *Videnskaps-selskapets Skrifter*, 1–49. References to [Skolem 1970].

1930 Einige Bemerkungen zu der Abhanlung von E. Zermelo [1929], *FM* **15**, 337–41.

1970 *Selected Works in Logic*, ed. J.E. Fenstad, Oslo, Universitetsforlaget.

SMITH, H.J.S.

1875 On the Integration of Discontinuous Functions, *LMS* **6**, 140–153.

SPALT, D.

1990. Die Unendlichkeiten bei Bernard Bolzano, in [König 1990], 189–218.

SPECKER, E.

1953 The Axiom of Choice in Quine's "New Foundations", *NAS* **39**, 972–975.

STEIN, H.

1981 *Logos*, Logic, and *Logistiké*: Some Philosophical Remarks on Nineteenth-Century Transformation of Mathematics, in [Dauben 1981], 238–259.

STEINER, J.

1832 *Systematische Entwickelung der Abhängigkeit geometrischer Gestalten von einander*, Berlin, Fincke.

STEINITZ, E.

1910 Algebraische Theorie der Körper, *JrM* **137**, 167–309.

STOLZ, O.

1884 Über einen zu einer unendlichen Punktmenge gehörigen Grenzwerth, *MA* **23**, 152–156.

STYAZHKIN, N.I.

1969 *History of mathematical logic from Leibniz to Peano*, Cambridge/London, MIT Press.

TARSKI, A.

1924 Sur les ensembles finis, *FM* **6**, 45–95. Reprint in [Tarski 1986], vol. 1.

1933 Poj cie prawdy w j zykach nauk dedukcyjnych, *Prace Towarzystwa Naukowego Warszawskiego* **34** (written 1931). German translation, with postscript, as 'Der Wahrheitsbegriff in den formalisierten Sprachen,' *Studia philosophica* **1** (1935), 261–405. Reprint in [Tarski 1986], vol. 2. References to English translation in [Tarski 1956], 152–278.

1938 Über unerreichbare Kardinalzahlen, *FM* **30**, 68–89. Reprint in [Tarski 1986], vol. 3.

1941 *Introduction to Logic and the Methodology of Deductive Sciences*, Harvard University Press.

1956 *Logic, Semantics, Metamathematics: Papers from 1923 to 1948*, Oxford, Clarendon. Reprint in Hackett, 1983, ed. J. Corcoran.

1964 2nd edn. of [Tarski 1941].

1986 *Collected Papers*, ed. S.R. Givant & R.N. McKenzie, Basel, Birkhäuser, 4 vols.

TORRETTI, R.
  1984 *Philosophy of Geometry from Riemann to Poincaré*, Dordrecht, Reidel

TROELSTRA, A.S. & D. VAN DALEN
  1988 *Constructivism in Mathematics: An introduction*, Amsterdam, North-Holland.

UEBERWEG, F.
  1882 *System der Logik und Geschichte der logischen Lehren*, 5th edn., Bonn, A. Mar-
        cus.

ULLIAN, J.S.
  1986 Quine and the Field of Mathematical Logic, in L.E. Hahn & P.A. Schilpp, eds., *The
        Philosophy of W.V. Quine*, La Salle, Ill., Open Court, 569–593.

ULLRICH, P.
  1989 Weierstrass' Vorlesung zur «Einleitung in die Theorie der analytischen Funk-
        tionen», *AHES* **40**, 143–172.

VAUGHT, R.
  1974 Model Theory before 1945, in *Proceedings of the Tarski Symposium on Pure
        Mathematics*, Providence, AMS, 153–72.

VOSS, A.
  1914 Heinrich Weber, *DMV* **23**, 431–444.

VAN DER WAERDEN, B.L.
  1930 *Moderne Algebra*, Berlin, Springer.

WALTHER-KLAUS, E.
  1987 *Inhalt und Umfang. Untersuchungen zur Geltung und zur Geschichte der Rezi-
        prozität von Extension und Intension*, Hildesheim, Olms.

WANG, H.
  1950 A Formal System of Logic, *JSL* **15**, 25–32.
  1957 The axiomatisation of arithmetic, *JSL* **22**, 145–158. Reprint in *A survey of mathe-
        matical logic*, Amsterdam, North Holland, 1964, 68–81.
  1974 The concept of set, in *From Mathematics to Philosophy*, London, Routledge. Ref-
        erences to reprint in [Benacerraf & Putnam 1983], 530–570.
  1986 Quine's Logical Ideas in Historical Perspective, in L.E. Hahn & P.A. Schilpp, eds.,
        The Philosophy of W.V. Quine, La Salle, Ill., Open Court, 623–648.

WATERHOUSE, W.C.
  1979 Gauss on Infinity, *HM* **6**, 430–436.

WEBER, H.
  1893 Die allgemeinen Grundlagen der Galoisschen Gleichungstheorie, *MA* **43**, 521–49.
  1893a Leopold Kronecker, *MA* **43**, 430–36.
  1895/96 *Lehrbuch der Algebra*, 2 vols., Braunschweig, Vieweg. References to the re-
        print New York, Chelsea, 1961.
  1906 Elementare Mengenlehre, *DMV* **15**, 173–184.

WEIERSTRASS, K.
  unp. *Einleitung in die Theorieen* [sic] *der analytischen Functionen* (Mitschrift Hettner,
        1874), .
  1894/1927 *Mathematische Werke*, 7 vols., Berlin, Mayer & Müller.
  1986 Einführung in die Theorie der analytischen Funktionen (Mitschrift Killing, 1868),
        Schriftenreihe des Math. Instituts der Univ. Münster, 2nd series, **38**.

1988 *Einleitung in die Theorie der analytischen Funktionen* (Mitschrift Hurwitz, 1878), Brauschweig, Vieweg.

WEIL, A.

1979 Riemann, Betti and the birth of topology, *AHES* **20**, 91–96.

WEYL, H.

1910 Über die Definitionen der mathematischen Grundbegriffe, *Mathematisch-naturwissenschaftliche Blätter* **7**. References to [Weyl 1968], vol. 1, 298–304.

1913 *Die Idee der Riemannschen Fläche*, Leipzig: Teubner.

1918 *Das Kontinuum: Kritische Untersuchungen über die Grundlagen der Analysis*, Leipzig, Veit. References to the reprint New York, Chelsea.

1921 Über die neue Grundlagenkrise der Mathematik, *Mathematische Zeitschrift* **10**, 39–79. References to [Weyl 1968], vol. 2, 143–180.

1927 Comments on Hilbert's Second Lecture on the Foundations of Mathematics, in [van Heijenoort 1967, 482–84].

1944 David Hilbert and his Mathematical Work, *Bulletin of the AMS* **50**, 612–54. References to [Weyl 1968], vol. 4, 130–72. Also in [Reid 1970].

1946 Mathematics and Logic. A brief survey serving as a preface to a review of "The Philosophy of Bertrand Russell," *AMM* **53**, 2–13. References to [Weyl 1968], vol. 4, 268–279.

1946a Review: The Philosophy of Bertrand Russell, *AMM* **53**. References to [Weyl 1968], vol. 4, 599–605.

1954 Erkenntnis und Besinnung, in [Weyl 1968], vol. 4, 631–49.

1968 *Gesammelte Abhandlungen*, ed. K. Chandrasekharan, 4 vols., Berlin, Springer.

WHITEHEAD, A.N. & B. RUSSELL

1910 Vol. 1 of [Whitehead & Russell 1910/13].

1910/13 *Principia Mathematica*, Cambridge University Press. 2nd edn. 1925/27. References to the 1978 reprint.

WIENER, N.

1914 A Simplification of the Logic of Relations, *Proceedings of the Cambridge Philosophical Society* **17**, 387–90. Reprint in [van Heijenoort 1967].

WITTGENSTEIN, L.

1921 *Tractatus Logico-Philosophicus*, London, Routledge & Kegan Paul, 1961. First edn. in *Annalen der Naturphilosophie* **14**, 185–262.

WUSSING, H.

1969 *The Genesis of the Abstract Group Concept*, Cambridge/London, MIT Press, 1984.

YOUNG, W.H. & M. YOUNG

1906 *The Theory of Sets of Points*, Cambridge, University Press.

YOUSCHKEVITCH, A.P.

1976 The Concept of Function up to the Middle of the 19th Century, *AHES* **16**, 37–85.

ZERMELO, E.

1904 Beweis, dass jede Menge wohlgeordnet werden kann (Aus einem an Herrn Hilbert gerichteten Briefe), *MA* **59**, 514–516. References to English translation in [van Heijenoort 1967], 139–141.

1908 Untersuchungen über die Grundlagen der Mengenlehre, I, *MA* **65**, 261–281. References to English translation in [van Heijenoort 1967], 199–215.

1908a Neuer Beweis für die Möglichkeit einer Wohlordnung, *MA* **65**, 107–128. References to English translation in [van Heijenoort 1967], 183–198.

1909 Sur les ensembles finis et le principe de l'induction complète, *AM* **32**, 185–193.

1929 Über den Begriff der Definitheit in der Axiomatik, *FM* **14**, 339–44.

1930 Über Grenzzahlen und Mengenbereiche. Neue Untersuchungen über die Grundlagen der Mengenlehre, *FM* **16**, 29–47. English trans. in [Ewald 1996], vol.2.

1932 Über Stufen der Quantifikation und die Logik des Unendlichen, *DMV* **41**, 85–88.

Index of Illustrations

Figure 1. Gustav Lejeune Dirichlet (1805–1859). 2

Figure 2. Bernhard Riemann (1826–1866) in 1863. 40

Figure 3. Doubly and triply connected surfaces, from [Riemann 1857].
Riemann explains the behavior of transversal cuts and closed curves. 56

Figure 4. Richard Dedekind (1831–1916) in 1868. 80

Figure 5. Georg Cantor (1845–1918) around 1870. 146

Figure 6. Curve showing [0, 1] and (0, 1] to be equipollent, from [Cantor 1878].
The curve consists of the infinitely many segments ab, $a'b'$, ... and the isolated
point c; the points b, b', ... do not belong to the curve.
($op = pc = 1$; points a_i and b_i are obtained by halving intervals.) 193

Figure 7. Title page of Dedekind's *What are numbers and what could they be?*
[also: *..and what are they for?*] [1888].
Notice the Greek motto: "man eternally arithmetizes." 216

Figure 8. Title page of Cantor's *Foundations of a general theory of manifolds*
[1883]. The subtitle indicates that it is a "mathematico-philosophical attempt"
to contribute to the "theory of infinity." 258

Figure 9. The Göttingen Mathematics Society in 1902. Sitting at the table we find
Klein and Hilbert, at the extreme right Zermelo; standing in the second row are
E. Schmidt (behind Klein) and Bernstein (behind Zermelo).
Courtesy Niedersächsiche Staats- und Universitätsbibliothek Göttingen. 298

Figure 10. Alfred Tarski (left) and Kurt Gödel in 1935. Courtesy Bancroft Library,
Berkeley. 386

Name Index

A

Abel, Niels Henrik (1802–1829), 8; 12; 32–33; 53; 70; 83; 91; 149

Ackermann, Wilhelm (1896–1962), 344; 346; 353; 357; 359; 387

Aleksandrov, Pavel S. (1896–1985), 212; 333; 336

Ampère, André-Marie (1775–1836), 19

Apelt, Ernst Friedrich (1812–1859), 14

Appell, Paul Emile (1855–1930), 283

Archimedes (c. 287–212 b.C.), 6

Aristotle (384–322 b.C.), 18; 22; 41; 48–51; 64; 70; 209; 260

Arnauld, Antoine (1612–1694), 49

Artin, Emil (1898–1962), 115

B

Bacon, Francis (1561–1626), 201; 266; 285

Bachmann, Paul (1837–1920), 114

Baire, René L. (1874–1932), 315; 323; 333; 338

Banach, Stefan (1892–1945), 334

Bar-Hillel, Yehoshua (1915 1975), 326; 351; 356; 382

Beltrami, Eugenio (1835–1900), 72; 78–79; 123

Bendixson, Ivar (1861–1935), 172; 205–06; 208; 210–14; 267; 268; 272; 282

Berkeley, George (1685–1753), 220; 334; 388

Berlin, xvii–xx; 3; 6–9; 12; 21–24; 26; 29; 32–38; 53; 111; 115; 119; 122; 124; 127; 140; 142; 149; 154–55; 158–59; 177; 180; 183; 185; 191–92; 197–201; 222; 238; 261; 277; 284–85; 317

Bernays, Paul (1888–1978), 326; 344–45; 353; 363; 365–66; 375–76; 378; 380–82; 386–88; 391

Bernstein, Felix (1878–1956), 185; 191; 211; 230; 239–40; 248; 289–90; 292; 300; 302; 309; 312; 314; 324; 333; 360

Bessel, Friedrich Wilhelm (1784–1846), 15

Beth, Evert Willem (1908–1964), 356

Bettazzi, Rodolfo (1861–1941), 251; 305; 313

Betti, Enrico (1823–1892), 60

Biermann, Kurt-R. (b. 1919), 8–9; 32–33; 35; 111; 186; 199

Bjerknes, Carl Anton (1825–1903), 77

Bolyai, Janos (1802–1860), 61

Bolzano, Bernard (1781–1848), xvi; 10; 18–21; 36–37; 42; 52; 64; 75–76; 88; 109; 118; 128; 139–44; 158–59; 181–82; 189; 210; 234; 244–46; 259

Boole, George (1815–1864), 50; 52; 69; 122; 244; 304; 387

Borchardt, Carl Wilhelm (1817–1880), 141; 197

Borel, Emile (1871–1956), 167; 196; 212; 239; 283; 300; 313–15; 333; 338

Borewicz, Senon I., 112

Bottazzini, Umberto, 26; 36; 54; 118; 147; 152

Bourbaki, Nicolas, xiii; 3; 63; 69; 81; 94; 349; 388; 391–92

Braunschweig, 31; 81; 116; 176; 200

Brouwer, Luitzen E. Jan (1881–1966), 196; 264; 302; 335; 338–44; 346; 357; 390–91

Burali-Forti, Cesare (1861 1931), 251; 305–07; 313; 327; 380

C

Cantor, Georg (1845–1918), xi–xvi; xix; 1;
  3–4; 8; 14; 18–24; 31; 34; 37–39; 44;
  62; 65; 68; 70–76; 81–82; 87; 90; 93;
  109; 111; 117–20; 122; 124–31; 133–46;
  153; 155–67; 171–213; 221–22; 225;
  229; 233–37; 239–42; 244; 248–52; 254;
  258–96; 299–310; 312–14; 317; 319–24;
  326–27; 333–34; 337–39; 341; 343; 360;
  363; 365; 367; 369–73; 376; 378; 380;
  389–90
Carnap, Rudolf (1891–1970), 330; 346;
  353–54; 356–57
Cauchy, Augustin-Louis (1789–1857), 10;
  12; 15; 18; 27–31; 33; 36; 45; 54–55;
  59; 117–19; 121; 124; 127–28; 133; 135;
  141; 147–51; 233; 280; 341
Cavaillès, Jean (1903–1944), 124; 175;
  225; 239; 270
Cayley, Arthur (1821–1895), 45; 86
Chasles, Michel (1793–1880), 24
Christoffel, Elwin Bruno (1829–1900), 9;
  78–79
Church, Alonzo (1903–1995), 337–38;
  350–51; 353–54; 356; 365; 388; 390
Chwistek, Leon (1884–1944), 348–51
Clarke, Samuel (1675–1729), 46
Clebsch, Rudolf F. A. (1833–1872), xviii;
  26; 31; 56; 84; 90; 127; 159
Clifford, William Kingdon (1845–1879),
  47; 61; 63; 79
Cohen, Paul Joseph (b. 1934), 382; 385
Corry, Leo, xx; 75; 81; 84; 86; 93; 95; 113;
  115–16; 367; 392
Couturat, Louis (1868–1915), 253; 308;
  326
Crelle, August Leopold (1780–1855), 8; 32;
  159; 180; 182; 199

D

d'Alembert, Jean le Rond (1717–1783),
  147
Darboux, Jean-Gaston (1842–1917), 146
Dauben, Joseph W., 20; 35; 147; 154; 175–
  77; 191; 196; 205–06; 261; 285

De Morgan, Augustus (1806–1871), 123;
  244
Dedekind, J. W. Richard (1831–1916), xii–
  xvi; xviii–xx; 1; 3–4; 9; 12–14; 17; 19–
  20; 22–31; 35–39; 41; 48; 52–53; 62; 68;
  69; 71; 73–79; 81–116; 117–39; 142;
  146; 148–49; 151; 161–62; 167; 171–92;
  194–202; 204; 209; 214; 216–55; 258;
  263–64; 266; 268–70; 274–75; 282; 284;
  289–96; 299; 301; 304–06; 310; 313;
  317–22; 324; 334–35; 337–41; 343; 349;
  360; 362; 378
Descartes, René (1596–1650), 3; 229; 260
Dieudonné, Jean (1906–1994), xii; 114;
  116; 124; 339
Dini, Ulisse (1845–1918), 1; 146; 161;
  165–66; 282; 300
Diogenes (404–c. 323 b.C.), 392
Dirichlet, Gustav P. Lejeune (1805–1859),
  xviii; 1; 7–9; 12; 21; 26–28; 31–36; 38;
  53–54; 59; 76; 78–79; 83; 87; 91; 97;
  99–100; 111–13; 137–39; 141; 145;
  147–52; 154–57; 161–65; 173; 221; 229;
  238
Dreben, Burton (b. 1937), 347
Drobisch, Moritz Wilhelm (1802–1896),
  48; 50; 52
du Bois-Reymond, Paul (1831–1889), 1;
  72; 145–46; 149; 153; 161–62; 164–66;
  185; 202–03; 210; 249; 282–84; 287
Dugac, Pierre, xii; 15; 28; 35–36; 77–78;
  84; 92–93; 102; 107; 114; 121; 124–26;
  138; 140; 147; 149; 175; 184–86; 198;
  200; 216; 224; 228; 233; 242; 244; 249;
  254; 282–83; 318

E

Edwards, Harold M., 37–38; 90; 96–97;
  99–100; 102; 109–12; 116; 185; 198;
  245; 260
Einstein, Albert (1879–1955), 339
Eisenstein, F. Gotthold (1823–1852), 9–10;
  28; 83; 96; 98
Epicurus (341–270 b.C.), 209
Epimenides (6th cent. b.C.?), 310

Euclid (4th–3rd cent. b.C.), 15–16; 20–21; 58; 61–62; 68; 72; 75; 95; 119; 123; 131–32; 136; 189; 233; 247

Euler, Leonhard (1707–1783), 27; 41–42; 52; 56; 95; 147

F

Faraday, Michael (1791–1867), 19

Fechner, Gustav Theodor (1801–1887), 19

Fermat, Pierre de (1601–1665), 9

Fitch, Frederic B. (1908-1987), 356

Fourier, J.B. Joseph (1768–1830), 9–10; 27; 147–52; 157

Fraenkel, Adolf A.H. (1891–1965), 174–77; 294; 309; 324; 326; 338; 342; 351; 356; 363; 366–74; 377; 379; 381

Franzelin, Johannes B. (1816–1886), 20

Fréchet, Maurice (1878–1973), 335

Frege, F.L. Gottlob (1848–1925), 16–17; 51–52; 68; 123; 137; 227; 229–30; 234; 237; 243–250; 252–53; 255; 264; 279; 290–92; 303–06; 308–09; 317; 325; 330–32; 337; 340; 343; 345–46; 349; 352; 355; 378; 390

Fries, Jakob F. (1773–1843), 7; 14; 17–18

Frobenius, Ferdinand Georg (1849–1917), 31; 35; 93; 102; 176; 185; 242

Fuchs, I. Lazarus (1833–1902), 31

G

Galilei, Galileo (1564–1642), 189; 233

Galois, Evariste (1811–1832), 75; 81; 83–84; 87–91; 95; 97–98; 110; 113–15; 238; 242

Gauss, Carl Friedrich (1777–1855), xiv; 1; 10; 12; 15–16; 20–21; 24–26; 28–29; 31–32; 40–45; 47; 53–55; 57–58; 60–61; 65; 69–70; 73–74; 79; 83–84; 87–88; 90–91; 95–97; 99; 110; 137; 149; 194; 216; 244

Gentzen, Gerhard (1909–1945), 356

Gerbaldi, Francesco (1858–1934), 305

Gergonne, Joseph Diaz (1771–1859), 51

Gillispie, Charles Coulston, 33–34; 339

Gödel, Kurt (1906–1978), xiii; 227; 296; 309; 330–32; 338; 344–45; 347–49; 353–56; 363; 365–66; 370; 377–78; 380–91

Goethe, Johann Wolfgang von (1749–1832), 5

Grassmann, Hermann G. (1809–1877), 8; 13; 15; 42; 45; 64; 69; 120; 123; 222; 248

Grattan-Guinness, Ivor, xiii; 90; 118; 123; 147; 149; 167; 172; 174–75; 186; 197–01; 270; 275; 278–81; 284–85; 306; 318; 351; 374; 395–97; 401; 403–04; 414

Gudermann, Christoph (1798–1852), 13; 33

Gutberlet, Constantin (1837–1928), 20

H

Hadamard, Jacques (1865–1963), 296; 300; 315–18; 323

Halle, 5; 32; 42; 111; 157–58; 164; 173; 176; 200–01; 239; 285

Hallett, Michael, 210; 212; 260–61; 265; 292; 345; 349; 367; 370–71; 376

Hamel, Georg (1877–1954), 314; 320

Hamilton, William Rowan (1805–1865), 14; 16; 43; 220–21; 242–44

Hankel, Hermann (1839–1873), 1; 13; 72; 120–21; 123; 145; 149; 154–57; 159; 160–65; 219

Harnack, C.G. Axel (1851–1888), 146; 161–62; 165–67; 185; 202; 210; 282–84

Hartogs, Friedrich (1874–1943), 365

Haubrich, Ralf, xx; 28; 33; 78; 82–84; 86; 90–91; 95–101; 105–07; 112

Hausdorff, Felix (1868–1942), xi; 212; 299; 302; 312; 320; 333–35; 348–49

Hawkins, Thomas W., 3; 11; 35; 72; 139; 146–48; 151; 154; 156–57; 161–62; 165–67; 196; 283; 300

Hegel, G.W. Friedrich (1770–1831), 7; 18

Heine, Heinrich Eduard (1821–1881), 1; 9; 28; 72; 121; 127; 133; 135; 138; 141–42; 145; 153; 155; 157–60; 196; 200; 313

Helmholtz, Hermann von (1821–1894), 14;
    71–72; 79; 137; 194; 221; 242; 249
Henle, Jakob (1809–1885), 99
Hensel, Kurt (1861–1941), 367
Herbart, Johann Friedrich (1776–1841), 7;
    16–19; 40–41; 43; 45–51; 59; 63–66;
    241
Hermite, Charles (1822–1901), 38; 266;
    282–83
Hessenberg, Gerhard (1874–1925), 217;
    319; 322–24; 365
Hettner, Georg (1854–1914), 118; 124;
    140–41; 184
Hilbert, David (1862–1943), xi–xii; 24; 31;
    33; 52; 54; 95; 114–16; 119; 123; 132;
    186; 232; 238; 243; 246; 253–55; 263;
    278; 290–92; 295; 299; 301–02; 305;
    307; 309; 311–13; 315–18; 321; 330;
    332; 335; 337–39; 342–49; 353; 355;
    357; 359; 365; 367; 369; 371; 382; 386–
    87; 389–92
Hindenburg, C.F. (1739–1808), 11; 15
Hoffmann, L., 6; 42
Humboldt, Alexander von (1769–1859), 6;
    8–10; 32
Hume, David (1711–1776), 13
Huntington, E. (1847–1952), 236; 367
Hurwitz, Adolf (1859–1919), 28; 102; 115–
    16; 124–26; 300; 318
Husserl, Edmund (1859–1938), 264; 309;
    339; 356

I

Ilgauds, Hans Joachim, 20; 38; 111; 127;
    158; 166; 172–73; 175; 177; 186; 199–
    201; 213; 260–62; 282; 285; 291–92;
    300; 311

J

Jacobi, Carl Gustav J. (1804–1851), xviii;
    6–8; 10; 21; 26; 28; 32–34; 45; 53; 70;
    96; 216
Johnson, D.M., 72–73; 168; 196; 335
Jordan, Camille (1838–1922), 139; 146;
    166; 283; 300; 315

Jourdain, Philipp E.B. (1879–1919), 124;
    147; 253; 314; 326
Jürgens, Enno (1849–1907), 196; 282

K

Kanamori, Akihiro, 168; 212; 333–34; 385
Kant, Immanuel (1724–1804), 7; 10–18;
    21; 45–46; 48–49; 51; 67; 137; 220;
    241–44; 247; 295; 309; 326
Keferstein, Hans (1857–?), 93; 223–24;
    230–31; 246
Killing, Wilhelm K.J. (1847–1923), 35; 124
Kleene, Stephen C. (1909–1994), 337
Klein, Felix (1849–1925), xviii; 7–8; 23–
    24; 26; 31; 54–55; 71–72; 122–23; 160;
    166; 173; 193; 249; 267; 282; 285
Klügel, Georg Simon (1739–1812), 42
Kneser, Adolf (1862–1930), 124
König, Julius (1849–1913), 7; 19; 21; 299;
    309–10; 312; 314; 324; 327
Kossak, Ernst (1839–1892), 119; 121; 124
Kripke, Saul A. (b. 1940), 382
Kronecker, Leopold (1823–1891), xiii;
    xviii–xix; 4; 6; 9; 12; 15; 28; 31; 33–35;
    38; 96–102; 111; 115–16; 141; 185–86;
    198–200; 213; 216–17; 227; 244–45;
    249; 259–60; 262; 268–69; 276; 283;
    311; 315; 338; 341–43; 360
Krull, Wolfgang (1899–1970), 93
Kummer, Ernst Eduard (1810–1893), xviii;
    8; 10; 18; 32–34; 83; 95–101; 103–05;
    141; 157; 173; 185–86
Kuratowski, Kasimierz (1896–1980), 217;
    336; 349; 373

L

Lagrange, Joseph Louis (1736–1813), 11;
    45; 70; 96; 141
Lakatos, Imre (1922–1974), xix; 391
Landau, Edmund J. (1877–1938), 176; 234
Langbehn, Julius (1851–1907), 285
Lavine, Shaugan, 265; 374; 376
Lebesgue, Henri Léon (1875–1941), 158;
    313; 315–16; 333; 338
Legendre, A.-M. (1752–1833), 9; 141

Leibniz, Gottfried Wilhelm (1646–1716), 3; 11; 13; 18–20; 45–46; 57; 64–66; 117–18; 223; 226; 229; 241; 243–45; 248; 260; 266; 307; 387

Leśniewski, Stanislaw (1886–1939), 355–56; 387

Levy, Azriel (b. 1934), 326; 351; 356; 382

Lie, M. Sophus (1842–1899), 86

Lindemann, Ferdinand (1852–1939), 173; 194

Liouville, Joseph (1809–1882), 177; 184

Lipschitz, Rudolf O.S. (1832–1903), 9; 28–29; 38; 78–79; 93–94; 98; 103–04; 110–11; 113; 119; 132; 134–36; 145; 150; 154–56; 157; 159–60; 176; 222

Lobachevsky, Nikolai I. (1792–1856), 61

Locke, John (1632–1704), 260

Lorey, Wilhelm (1873–1955), 24–26

Lotze, Rudolf Hermann (1817–1881), 19

Löwenheim, Leopold (1878–1957), 304; 345–47; 360–62; 373; 384

Lüroth, Jakob (1844–1910), 196; 282

Luzin, Nikolai N. (1883–1950), 333; 365

M

Mahlo, Paul (1883–1971), 334

May, Kenneth O. (1915 1977), xiv–xv; 176; 186–87; 200; 284

Medvedev, Fedor Andreevich (1923–1993), xiv–xv; 174; 232

Mehrtens, Herbert, xx; 100; 113; 304; 316; 339

Mèray, H. Charles (1835–1911), 118–19

Meschkowski, Herbert, xiii; 20; 36–37; 118; 141; 158; 175; 177; 185–86; 194; 199–200; 203; 205–06; 241; 261; 266; 269; 276–77; 279; 282–83; 285; 291; 293; 295; 379

Minkowski, Hermann (1864–1909), 62; 255

Mirimanoff, Dimitry (1861–1945), 365–66; 369–71; 373; 375

Mittag-Leffler, Gösta, 15; 141; 160; 176; 185; 198–201; 203; 206; 213; 266; 268–69; 278; 281–85; 300

Möbius, A. F. (1790–1868), 15; 23–24; 88

Montel, Paul (1876–1975), 155

Moore, Gregory H., xx; 48; 52; 73; 140; 190; 210–12; 251; 255; 283; 302; 304; 306–18; 320; 323; 345; 347; 359; 367; 374; 382; 385

Mostowski, Andrzej (1913–1975), 381; 384

N

Nagel, Ernest (1901–1985), 13; 43

Netto, Eugen (1846–1919), 11; 196; 282

Neumann, Carl ( ), 33

Newton, Isaac (1643–1727), 119

Nicole, Pierre (1625–1695), 49

Noether, A. Emmy (1882–1935), 81; 116; 138; 175; 254–55

Novak, I., 381

O

Ohm, Martin (1792–1872), 12–13; 15; 32; 35; 120–23; 125; 219; 221–23

Oken, Lorenz (1779–1851), 8

Oresme, Nicole (*c.* 1323–1382), 70

P

Paris, 9; 26; 198; 201; 284

Pasch, Moritz (1843–1930), 243–44; 305

Paulsen, Friedrich (1846–1908), 5; 7

Peacock, George (1791–1857), 121–23; 219

Peano, Giuseppe (1858–1932), 50–51; 139; 194; 228; 235; 250–51; 300; 304–05; 308; 310; 313; 321; 332; 344–46; 348; 351; 354; 389

Peirce, Charles S. (1839–1914), 244; 250–51; 253; 346

Pestalozzi, J. Heinrich (1746–1827), 22

Pfaff, Johann Friedrich (1765–1825), 70

Phragmén, Lars Edvard (1863–1937), 282

Picard, Emile (1856–1941), 283

Pieri, Mario (1860–1913), 305

Pincherle, Salvatore (1853–1936), 36; 73; 124; 146

Plato (427–347 b.C.), 3; 51; 216; 265–66

Plücker, Julius (1801–1868), 23–24
Plutarch (3rd cent.), 216
Poincaré, Jules Henri (1854–1912), 282; 284; 309; 316; 318; 324; 326–27; 335; 338–39; 342; 360
Porphyry (3rd cent.), 49; 51
Post, Emil (1897–1954), 347
Pringsheim, Alfred (1850–1941), 124
Proclus (410–485), 70
Puiseux, Victor (1820–1883), 55
Purkert, Walter, 20; 38; 82–83; 90; 111; 127; 158; 166; 172–73; 175; 177; 186; 199–201; 213; 260–62; 282; 285; 291–92; 300; 311

Q

Quine, Willard Van Orman (b. 1908), 51; 309; 325–26; 329–32; 337; 349; 354; 363; 382; 387; 391

R

Ramsey, Frank P. (1903–1930), 310; 329; 332; 338; 348–53; 355
Richard, Jules-Antoine (1862–1956), 310; 327; 350
Riemann, G.F. Bernhard (1826–1866), xv–xvi; xviii–xix; 1; 3–4; 7–9; 17–31; 36; 38–48; 51–79; 81–83; 85–88; 90–91; 94; 99–100; 110; 112; 114; 131; 133–39; 142–55; 157–59; 162; 165–67; 172–73; 187; 193–94; 196; 202–03; 218–21; 233; 238; 241–43; 281; 292; 316; 378
Roch, Gustav (1839–1866), 56; 114
Rosser, J. Barkley (1907–1989), 337; 381
Rowe, David, xviii; xx; 7–8; 24; 31; 33; 199
Russell, Bertrand A.W. (1872–1970), xi; xiii; xix; 17; 45; 123; 227; 229; 234; 237; 246; 249–53; 286; 292; 299; 303–10; 317–18; 320–34; 340; 342–52; 355; 358–60; 366; 373; 378; 383–84; 390–91

S

Safarevic, Igor R., 112
Salmon, George (1819–1904), 45
Scharlau, Winfried, 6; 26–28; 76; 81–86; 88; 90; 98
Schelling, Friedrich Wilhelm Joseph (1775–1854), 7
Schering, Ernst C.J. (1833–1897), 77
Schiller, Friedrich von (1759–1805), 6
Schleiermacher, Friedrich Ernst Daniel (1768–1834), 13
Schmidt, Erhard (1876–1959), 313; 318
Schoenflies, Arthur Moritz (1853–1928), 20; 147; 168; 198–201; 213; 260; 281–82; 300; 302; 312; 314; 335
Scholz, Erhard, xx; 43–47; 53–61; 63; 69–70; 72; 83; 241
Schröder, F.W.K. Ernst (1841–1902), 50; 137; 230; 239; 244; 247; 250–53; 300; 304–05; 332; 345–49; 360; 380
Schultz, Johann (1739–1805), 21
Schumacher, Heinrich Christian (1780–1850), 20
Schwarz, Hermann Amandus (1843–1921), 31; 37; 118; 121; 124; 141; 145; 158; 176; 184; 200; 233; 283
Shakespeare, William (1564–1616), 201; 285
Sheffer, Henry (1882–1964), 349
Shepherdson, John C. (b. 1926), 383; 385
Sierpiński, Wacław (1882–1969), 320; 334; 355; 365
Skolem, Thoralf (1887–1963), 304; 330; 332; 345–47; 357; 360–63; 366–69; 373–77; 379; 382; 384–87; 389
Smith, Henry John Stephen (1826–1883), 1; 162–65; 203
Solovay, Robert M., 382; 384–85
Spinoza, Baruch (1632–1677), 130; 177; 260; 265–66
Steiner, Jacob (1796–1863), 1; 8; 18; 21–24; 32; 88; 188
Steinitz, Ernst (1871–1928), 320; 376
Stern, Moritz A. (1807–1894), 13; 25; 35
Stevin, Simon (1548–1620), 41; 119
Stolz, Otto (1842–1905), 146; 166–67
Suslin, Mikhail (1894–1919), 333

Sylow, Peter Ludvig M. (1832–1918), 360
Sylvester, James Joseph (1814–1897), 45

T

Tannery, Jules (1848–1910), 281
Tarski, Alfred (1901–1983), 233; 330; 334; 338; 350–56; 377; 382; 387–88
Thieme, K. Gustav Hermann (1852–1926), 124
Thomae, Johannes Karl (1840–1921), 196; 282

U

Ueberweg, Friedrich (1826–1871), 48
Ulrich, Georg K.J. (1798–1879), 25

V

van der Waerden, Bartel Leendert (1903–1996), 81; 84; 116; 337
van Heijenoort, Jean (1912–1986), 48; 223–24; 231–32; 234; 249; 255; 299; 308–09; 320–23; 332; 342; 345; 347; 365
Veblen, Oswald (1880–1960), 236; 367
Vitali, Giuseppe (1875–1932), 320; 333
Vivanti, Giulio (1859–1949), 211; 240; 266; 305
Volterra, Vito (1860–1940), 162; 165
von Neumann, John (1903–1957), 293–94; 324; 337; 342; 344–45; 357; 363; 365–82; 386–91
von Staudt, Karl Georg C. (1798–1867), 238; 305

W

Waismann, F. (1896–1959), 349
Wang, Hao (b. 1921), 223; 227; 231; 287; 296; 326; 337; 378; 381–83
Wangerin, Albert (1844–1944), 201
Waring, Edward (1736–1798), 28–29

Weber, E. Heinrich (1842–1913), 19–20; 31; 59; 78; 84; 86; 88; 93; 97; 102–03; 111; 114–16; 134; 176; 189; 198; 217; 219; 228; 233; 235
Weber, Wilhelm (1804–1891), 19; 25–26; 53
Weierstrass, Karl T.W. (1815–1897), xviii–xix; 1; 12–13; 15; 30; 33–38; 42; 53–54; 64; 73–74; 78; 117–27; 131; 133; 135; 137; 139–45; 149; 153; 155; 157–59; 171; 173; 177; 181–86; 194; 198; 202; 206; 209–10; 222; 243; 259; 266; 268; 281; 283; 305; 311; 338
Weyl, Hermann (1885–1955), 54–55; 116; 132; 252; 254–55; 264; 328; 330; 332; 335; 338–44; 346; 349–52; 354; 356–61; 369; 373; 386; 389–91
Whitehead, Alfred North (1861–1947), 253; 325; 327–33; 344; 349–50; 355; 360; 366
Wiener, Norbert (1894–1964), 348–49
Wilson, John (1741–1793), 28
Wittgenstein, Ludwig (1889–1951), 332; 349–51
Wussing, Hans, 10; 23; 86

Z

Zeno of Elea (5th cent. b.C.), 141
Zermelo, Ernst F.F. (1871–1953), xii–xiii; xix; 143; 174–75; 188; 202–03; 217; 225; 230; 234; 237; 239; 244; 246; 255; 270–71; 278; 280; 286; 289; 294–96; 299; 302; 304; 309–24; 326–27; 330; 334; 337–39; 343; 347; 355; 359–86; 391
Zorn, Max (1906–1993), 333

Subject Index

A

A priori, 3; 11; 13–17; 137; 165; 216; 241–44; 247

Abbildung, see mapping, 55; 89–90; 108; 224; 228–29; 236; 275

Absolute, 18; 261; 291

Abstraction, 57; 74; 93; 235; 243–44; 279; 311; 314–15

Acta Mathematica, 155; 198; 201; 213; 282; 284

Aleph numbers, 197; 272; 289–95; 301; 307; 310; 312; 314; 318

Algebra, xiii; 1; 11–14; 16; 25; 27; 31; 37; 43; 54; 69; 79; 81–84; 86; 90–92; 95; 97; 99; 103; 107; 108; 111; 113–17; 120–24; 139; 173; 218–21; 225; 237–38; 243–44; 249–51; 253; 260; 265; 301–02; 320; 325; 347; 380
  modern, xiii; 81; 84; 95; 109; 320
  symbolical, British school, 11; 13; 69; 120; 123–24

Algebraic integer, 30; 94–98; 100–01; 104–06; 112–13; 116

Algebraic number theory, 28; 38; 75; 79; 82–83; 90; 93–100; 107–08; 114–16; 139; 173; 185; 204; 217; 253; 262; 301; 313

Analysis, xiii; 1; 4; 9–16; 22; 27–28; 33; 35–38; 46; 49–51; 57; 59–61; 69; 71; 73–75; 82; 90; 100; 103; 107; 109; 113; 115–17; 119–22; 124; 127; 130–33; 139; 145; 149–53; 159; 167–68; 173–77; 182; 184; 190; 193–94; 202; 210; 212; 218; 221; 223; 231; 234; 237–38; 243; 246; 254; 258; 260; 279; 281; 283; 295; 300–01; 305; 308; 313; 320–24; 328; 331; 339; 341; 343–45; 348; 350–52; 357–59; 376–77; 382; 387; 391

Analysis situs, *see* topology, 57; 59; 69

Anschaulichkeit, see intuitiveness, 14; 16; 55; 241–43

Anschauung, see intuition, 14; 59; 224; 292

Antinomies, *see* paradoxes, 67; 309; 316; 321; 323; 327; 350; 376; 378

Arithmetic, xiii; 1; 12; 14; 35–36; 41; 43; 64–65; 68–73; 82; 83; 85; 88; 90–91; 95–97; 100; 102–03; 106; 118; 120–26; 128; 131; 135; 137; 141; 173–74; 181; 190–93; 197; 202; 211; 216–24; 231; 233; 235–38; 242–43; 247–49; 251–53; 264–66; 269; 271; 281; 293; 295; 303; 305; 308–12; 321; 324; 331; 344; 346; 351–54; 359–60; 363; 388–91
  transfinite, 211; 269

Arithmetization, 3; 35; 37; 100; 102; 260

Automorphism, 115; 238; 280; 376

Axiomatization, xii–xiii; 85–86; 119–24; 132; 135; 222; 237–38; 243; 246–48; 251; 254; 296; 299; 301; 305; 316–18; 320; 324; 330; 334–38; 342–47; 349; 352–53; 355; 357; 359–63; 365–73; 377; 378–83; 385–92

Axioms, 58; 60; 74; 86; 120–23; 132–33; 135–37; 222; 230; 233–38; 243–44; 247; 255; 263; 293–95; 301–02; 305; 311–18; 320–24; 329–30; 332; 334–35; 337; 339; 344; 348; 350–51; 354–55; 359–63; 365–87; 390–92
  Axiom of Completeness, *see* continuity, 91; 118; 120; 137; 141; 222–23; 302; 369; 381; 386; 388
  Axiom of Choice, 189; 237; 248; 278; 295; 299; 311–15; 318; 320; 322; 324; 333; 339; 347; 354; 361–62; 365–67; 381–85; 387
  Axiom of Foundation, 216; 370; 373–77; 380–81; 385
  Axiom of Infinity, 295; 322; 331; 347; 351; 354

Axiom of Pairs, 322; 367
Axiom of Power Set, 286; 306; 322–23; 381
Axiom of Reducibility, 328–32; 349–52; 358
Axiom of Replacement, 291; 294–95; 366–69; 371–72; 374; 379–81; 384–85; 392
Axiom of Separation, 294; 322; 362; 367–68; 373–74; 380–81; 387; 390
Axiom of Union, 291; 294; 322; 381; 392
Principle or Axiom of Comprehension, 48–49; 52; 104–05; 108; 113; 226–27; 250–53; 290; 299; 304–05; 308; 321–22; 325; 329; 338; 348; 378–80
Principle or Axiom of Extensionality, 108; 226; 252; 264; 304; 311; 322; 348; 355
von Neumann's Axiom, 294; 363; 370; 379

B

Begriff, see concept, 50; 63–64; 68; 90; 143; 151; 216; 224; 246; 263; 277; 308
Beiträge [Cantor 1895/97], 69; 175; 240; 259; 265; 270; 275–82; 287–90; 292; 294; 300; 317
Belegung, see covering, 288; 314
Berlin school, xvii–xix; 32; 34–36; 38; 122; 158–59; 185–86; 191–92; 199–200; 261; 283
Beweistheorie, see proof theory, 344–46; 357
Bildung [education or formation], 5; 7
Bolzano–Weierstrass principle, 37; 182
Bolzano–Weierstrass theorem, 37; 140

C

Cantor's Theorem, 259; 264; 286; 288; 291; 306–07; 324; 334
Cantor–Bendixson theorem, 172; 191; 205–06; 210–12; 239–40; 267–68; 272; 289
Cardinality, 24; 73; 75–76; 171–72; 177; 188; 190; 192–94; 202–03; 206–07; 210;

213; 229; 239; 258; 264–65; 268; 270–75; 277–78; 283; 286; 288–91; 294–95; 306–07; 310; 312; 314; 318; 320; 324; 353; 363; 369; 372; 376; 380; 385
Cardinals, *see* power, *Mächtigkeit*, 8; 134; 171–72; 185; 188–89; 192–94; 197; 202–03; 210–13; 218; 239; 258; 265; 268; 270–74; 277–78; 282; 286–90; 307; 312; 317–18; 320; 323–24; 331; 334; 365; 369; 372; 385
comparability of, 188; 277; 365
inaccessible, 334
large, 323; 334; 385
Chain theory, 224–25; 230–32; 235; 239–40; 244; 249; 251; 255; 269; 319; 339; 362; 375
Classes, *see Klasse*, sets, 1; 36; 50–53; 64; 71; 73; 86–88; 96; 104; 106; 110; 116; 128; 130; 133; 135; 149; 151; 153; 155; 157; 161; 189–90; 207; 212–13; 221; 226; 229; 237; 243; 249–52; 263–64; 268; 272–77; 279; 281–83; 286; 289; 293; 300–08; 313–14; 319; 321–23; 325–27; 329–31; 337; 342; 345; 348; 351–56; 362; 365; 377–81; 384; 387; 390
logical, 50–53; 104; 301–08; 325–27
von Neumann's, 379–81; 384
Collections, *see Inbegriff*, sets, 68; 92; 107; 110; 178; 180–82; 202; 206; 227; 258; 262–65; 291–95; 321; 327; 363; 375; 383
consistent and inconsistent, 295
Combinatorial tradition, xviii; 11–12; 33; 35
Completeness, 4; 20; 36; 56; 91–92; 113; 118; 120; 124; 141; 177; 195; 203; 222–23; 232; 252; 283; 288; 302; 310; 330; 350; 352; 369; 381; 386; 388; 392
Concepts, *see Begriff*, xvi; 18; 21; 27–28; 31; 40; 42; 43; 46–53; 59–61; 63; 66–71; 73; 77; 85–86; 90–91; 100–02; 104–05; 107–08; 111; 116; 125; 127; 143; 148; 151–52; 156; 205; 208; 216; 221; 224; 226–28; 230; 232; 234; 237; 242; 245; 250; 252–53; 261–63; 265; 267; 276–79; 281; 290; 303; 307–08; 321; 339–40; 349; 354–56; 358; 378; 380; 387; 390

extension of, *see Umfang*, xvi; 49–50;
    252–53; 296
intension of, 49–50; 52
Conceptual approach, 3; 10; 27–28; 31; 74;
    100; 107; 121–23; 142; 148
    abstract, 3; 31; 36; 100; 107; 112; 142;
        148; 221
    formal, 35; 126; 142; 159
Congruence, 23; 79; 87; 95–96; 100–01;
    105; 109–10; 112
Consistency, 111; 122–23; 234; 244–46;
    293; 301; 338; 343–44; 356; 360; 363;
    373–74; 377; 380–82; 385–87; 389
    relative, 122; 373; 382; 385; 387
Constructivism, 36; 101; 126; 134; 174;
    184–85; 260; 271; 313–16; 339–41; 351;
    360; 391
Content (outer), 37; 44–45; 49; 51; 55; 87;
    91; 102; 106; 114; 130; 146; 156; 161–
    68; 171; 174; 177; 180; 197; 203; 211;
    235; 288; 300; 392
Continuity, *see* Axiom of Completeness,
    27; 46; 47; 67; 74; 114; 118–20; 131–33;
    135–37; 141; 148; 151; 153; 159; 182;
    194; 196; 222; 243–44; 313
Continuum, xi; xiii; 47; 63; 142; 171–72;
    178; 187; 190–95; 203; 208–13; 220;
    254; 268; 270; 273; 278–79; 287; 310;
    312; 315; 341; 365; 390
Continuum Hypothesis, 171–72; 177; 194;
    199; 202; 205; 210–13; 258; 268; 273;
    280–81; 285–86; 289; 301–02; 312; 333;
    365; 369; 382–85
Contradictions, *see* paradoxes, antinomies,
    inconsistencies, 21; 66; 75; 110; 123;
    180; 189; 227–28; 234; 236; 260; 262;
    271; 287; 291–94; 295; 305–10; 312;
    317; 321; 325; 327; 329; 332; 337; 357;
    363; 376–77
Convergence, xiii; 9; 36; 118; 126–28; 133;
    142; 145; 147; 149; 152–53; 157–59;
    211; 261; 266; 338; 341; 347–48; 366
    uniform, 145; 153; 157
Correspondence, xiii; xvii; xix; 14; 19; 31;
    119; 131–32; 135–36; 158; 171–72;
    174–80; 183; 185–86; 194–97; 201; 206;
    211–13; 225; 261; 283–85; 292; 301
    [Korrespondenz], *see* mapping, 23: 75;

89; 148; 177; 179; 190–91; 194–96;
    211; 237; 239; 275; 278–79; 283; 372
Covering, *see Belegung*, mapping, 166–67;
    288–89; 312; 314; 318
Cumulative hierarchy, 366; 370; 373–75;
    377–78; 383–84

D

Dedekind cut, 56; 85; 102; 103; 124; 131–
    35; 340
Definition, 13; 27–29; 30–31; 36–38; 40–
    42; 47–48; 51; 53; 57; 62–66; 68; 70–74;
    84–85; 87; 89; 91; 94; 97–99; 103; 105–
    09; 112–14; 116; 118–19; 121; 125–28;
    131–33; 135; 137; 141; 146; 148–49;
    151–52; 155; 166–67; 172–73; 179; 184;
    188–89; 195; 203; 208–09; 214; 219–23;
    226; 228–38; 241; 243–45; 250; 253–54;
    263–65; 268; 274; 276; 279–80; 283;
    286; 288–90; 292; 294; 300–01; 303;
    306–08; 315; 318–19; 321–24; 327; 330;
    332; 335; 339; 348–49; 356; 368–69;
    371–74; 378; 383–84
    impredicative, *see* predicativism, 327;
        332; 349
Denumerability, xii; 109; 136; 166; 167;
    171; 174; 177–80; 183–85; 187; 189–90;
    194; 196; 200; 203; 206–12; 263; 265;
    267; 272–73; 279–80; 284; 286–87; 289;
    300–01; 313–14; 333; 335; 341; 362;
    377
Deutsche Mathematiker Vereinigung, 8;
    201; 306
Diagonalization, 181; 286–88; 308
Dimension, 44; 58; 66; 70; 171; 187–88;
    194–96; 209; 283; 335
Ding [thing], 89; 108; 226; 228; 293

E

Empiricism, 10; 391
Epistemology, 13; 15; 17; 45–46; 48; 241;
    243–45; 247; 263–64; 353
Equations, theory of, 83–84; 115
Equipollence, 73; 75–76; 171; 188; 192–
    93; 210–11; 229–30; 233; 236–37; 239–
    40; 273; 274; 279; 282; 287–89; 294–95;

273; 274; 279; 282; 287–89; 294–95; 301; 303; 372

Equivalence class, 87; 128; 130; 237; 264

Existence, mathematical, 4; 12–13; 16; 38; 50; 54; 57; 66; 78–79; 86; 96; 111; 119; 123; 132; 138; 150; 157; 164; 177; 180–82; 186; 220; 222; 227; 233–34; 244; 246; 262–63; 268; 271; 273; 276; 292; 294–95; 301; 308–09; 314–16; 318; 321; 325; 332–34; 340; 342; 348; 352; 354–55; 359; 369–70; 377–78; 380; 383; 392

Extensional, xvi; 38; 50–52; 104; 174; 227–28; 252–53; 303; 306; 325; 349; 352; 361; 381; 383; 388

F

Factorization, 1; 96–101; 104; 111–13

Fields, *see Körper*, 10; 24; 27–30; 37–38; 43; 71–72; 78; 81–84; 86–95; 97–98; 100–02; 105–06; 109–11; 114–15; 139; 153; 173–74; 179; 182; 186; 189; 216–17; 242–43; 253; 283; 300–02; 311; 316–17; 320; 340; 357; 371; 376

Finitism, 260

Formal system, xiii; 337; 344–45; 347; 353–54; 357; 363; 387

Formalism, 10–13; 15; 123–24; 331; 337; 342; 345; 347; 353; 356; 358; 383

Forms, 12–14; 29–31; 38; 46–47; 64; 77; 87–88; 90–92; 94; 96–98; 100–02; 110; 112–13; 115; 121; 125; 128; 130; 162; 181; 190; 211–12; 245; 249; 276; 288; 290; 327; 329; 354–55; 362

    quadratic, 90–92; 96–97; 100

Foundational debate, xiii; 299; 309; 311; 314; 316–17; 338; 344–48; 359–61; 366; 387; 391

Foundations of mathematics, xiii; xvii; 1; 4; 11; 20; 28–30; 43; 58; 65; 68; 71–74; 78; 81–82; 85; 97; 102; 107; 113–15; 120; 123; 137; 145; 167; 173–75; 194; 217–19; 221–22; 238; 242; 246; 250; 252; 255; 289; 296; 299; 305; 307; 309–10; 318; 321; 331; 337–38; 342–44; 346; 356; 360; 365–66; 371; 382; 391–92

Function theory, xiii; 1; 21; 27–29; 31; 53–54; 57; 59; 70–71; 77; 85–87; 88; 90–91; 114; 117; 137; 139; 145; 155; 168; 194; 196; 198–200; 206; 208; 210; 262; 264–65; 281–82; 284; 300

Functions, *see* mapping, xiii; 1; 12; 15–16; 21; 25; 27–31; 33–36; 42–43; 45; 53–57; 59–60; 66; 69–73; 75; 77–78; 85–88; 90–91; 97; 102; 110; 114; 116–18; 124; 131; 137–39; 142; 144–65; 167; 183–84; 190; 194–96; 198–200; 206; 208; 210; 218; 223; 229; 238; 262; 264–66; 281–82; 284; 286–88; 289; 300; 303–04; 306; 313–16; 318; 321–22; 325–29; 333–35; 340; 348–52; 355; 367; 372–74; 378–79; 383; 387

    Abelian, 33–34; 53; 60; 70; 77

    abstract notion of, 27; 151

    algebraic, 34; 53; 88; 97; 102; 114; 116; 198

    analytic, 30; 34; 36; 54–55; 69; 91; 124; 140; 142; 206; 282; 300

    arbitrary, 148; 150; 153–54; 157

    continuous, 27; 35–36; 78; 117–18; 138; 140; 148–49; 151; 153; 184; 195

    differentiable, 36

    discontinuous, 1; 72; 148; 150; 152–56; 159; 162

    elliptic, 25; 33; 53; 77; 218

G

Galois theory, 75; 83; 87–89; 91; 97–98; 110; 113–15; 238; 242

Gebiet [domain], *see* sets, 40; 66; 73; 86–87; 90; 129; 140; 142; 160; 218

Genetic approach, 119–20; 122; 218–22

Geometry, xiii; 1; 13–16; 21–23; 25; 33–34; 37; 41–45; 53–55; 57–62; 65; 70–76; 78–79; 82–83; 87–88; 90–91; 114; 117; 120; 123; 131–32; 135–39; 173; 194; 196; 210; 216; 220; 238; 242; 247; 264–65; 305; 308; 343

    algebraic, 114

    differential, 21; 42–43; 53; 58; 60–62; 70–72; 79; 173

    Euclidean, 15–16; 21; 58; 72; 75; 123; 132; 136

projective, 21; 75; 83; 238
Gesamtheit [totality], *see* sets, 226–27; 234
Gesellschaft Deutscher Naturforscher und Ärtzte, 8
God, 5; 67; 216–17; 245; 258; 261–62; 268; 291; 312
Gödel's incompleteness theorems, 345; 353; 382; 387; 389
Göttingen, xviii–xx; 3; 5; 13; 24–26; 28–29; 31–32; 36; 39; 41; 45; 53; 71; 99–100; 115; 127; 138; 149; 176; 201; 218; 220; 238; 254–55; 285; 302; 316–17; 339; 344; 359–60
Göttingen group, xix; 24; 36; 100
Grösse, see magnitude, *Zahlgrösse*, 20; 30; 42; 64; 70; 140
Groups, group theory, xix; 5; 11; 32; 81; 84–86; 89; 95; 99; 102; 115; 228; 238; 309; 354; 360
Grundlagen [Cantor 1883], 30; 113; 160; 208; 211; 222; 253; 259–61; 267; 269; 271; 274–75; 277; 280–81; 283; 286; 289; 292; 300; 321
Gymnasium, 6–9; 12; 25; 32; 52; 223; 243

H

Hilbert's program, 345; 371; 389
Homomorphism, 84; 89; 91; 93; 102; 108

I

Ideal, xii; xvi; 1; 5–7; 29–30; 33; 38; 78–79; 81–82; 89; 91; 93–94; 96–97; 99–107; 109–16; 131; 134; 138; 187; 204; 238; 242; 245–46; 253; 296; 377
— theory, xii; 1; 29; 79; 82; 89; 91; 93–94; 96–97; 102–03; 105–07; 109–11; 113; 116; 138; 187; 238; 242; 245
Idealism, 7; 10; 18; 45; 48; 266; 307
Inbegriff, see collections, sets, 92; 108; 110; 178; 180; 184; 226–27; 263–65; 293
Incompleteness, 65; 138; 290; 329; 344–45; 353; 363; 382; 387; 389
Inconsistencies, *see* contradictions, 118; 293–94; 307; 319; 321
Independence, 224; 343; 367; 382; 386

Induction, 25; 160; 217; 222–24; 227; 230; 232; 235–36; 324; 328–29; 353; 372–73
principle of mathematical, 217; 222–24; 227; 230; 232; 236; 324; 328–29
transfinite, 160; 372–73
Infinitesimal, 21; 150; 163
Infinitism, 38; 82; 107; 109–10; 120; 131; 177
Infinity, xi–xiv; 4; 18–24; 38; 56; 65–68; 88; 109–10; 125–26; 130; 144–45; 160; 183; 193; 203; 205; 207; 224; 232–34; 241; 244–46; 250; 254–55; 259–61; 267–70; 291–92; 305; 308; 315–16; 334; 340–41; 354; 362; 376
actual or proper, xii; xiv; 4; 18; 20–21; 23; 37; 53; 65–66; 68; 88; 111; 203; 241; 245–46; 260–61; 341; 376
Dedekind infinite, his definition of, 68; 107; 109; 173; 189; 230; 233–34; 244–45; 251
Dedekind's theorem of, 241; 244; 292; 305
potential or improper, 18; 21–22; 38; 65; 260
simply infinite, 102; 234–36; 246
symbols of infinity, 144; 160; 203; 205; 207; 267–70
Integral, 5; 27; 52; 71–72; 95; 99; 113–14; 145–46; 147; 149–54; 157; 161; 163–64; 166; 178; 190; 251; 267–68; 304; 311; 391
Riemann, 27; 145–46; 150; 153–54
Integration theory, 26; 137; 144; 146–47; 150; 152–54; 156–57; 161–62; 165–67; 177; 210; 218; 283; 300; 333
Intellectualism, 10; 15; 134
Intensional, 51–52; 228–29; 252–53; 299; 303; 349; 358
Intuition, *see Anschauung*, 13–16; 55; 58–59; 74; 131; 137–38; 220–21; 224; 242–44; 249; 255; 292; 295; 310; 316; 326
intuitiveness, *see Anschaulichkeit*, 14; 16; 55; 242; 244
Intuitionism, 10–11; 337–39; 341–43; 353; 356; 383; 389–91
Isomorphism, 89; 225; 278; 320
Iterative conception of sets, 296; 377–78

J

Journal für die reine und angewandte Mathematik, 8; 56; 98; 127; 159; 180; 182–83; 192; 197–99; 285; 356

K

Kantianism, 10–17; 45–46; 241; 243–44; 248; 295
Klasse, see classes, sets, 87; 133; 190; 308
Körper, see fields, 30; 81; 91–92; 94; 139; 179; 189

L

Law of thought, 226; 244; 277
Limit point, 129; 139–42; 155; 159; 160; 162; 165; 205–06; 208; 210–12; 271; 280
Limitation of size, 326; 378; 379
Limits, 19–21; 34; 38; 65; 67; 73; 118–19; 122; 125; 131; 140–41; 149; 202; 216; 235; 261; 335; 389
Logic, *see* law of thought, xiii; xix–xx; 16–17; 40; 47–52; 62–64; 67; 88; 106; 120; 122; 137; 151; 217; 226; 230; 234; 241–44; 246–53; 263–64; 266; 292; 296; 299–304; 308–10; 316; 323; 326; 330–32; 337–38; 340; 344–63; 366–68; 374–77; 380–81; 384–88; 390–92
   classical, 346; 357
   first-order, xiii; 17; 297; 301; 332; 338; 345; 347; 353–54; 357–59; 361–63; 366; 368; 373–75; 377; 381–88; 392
   formal, 16–17; 47–52; 225; 252; 311; 346; 366
   higher-order, 330; 346–48; 353; 358–59; 361–63; 374; 376–77
   modern, 304; 330; 346
   propositional, 49; 249; 349–50
   second-order, 301–03; 359; 361; 370; 374–77; 387–88
   traditional, 40; 47–52; 63; 88; 106; 151; 243; 246; 292; 345
Logicism, xiii; 15; 17; 20; 48; 137; 217; 241–54; 266; 305; 316; 325–26; 331; 346; 353; 357

M

Mächtigkeit, see power, cardinals, 24; 171; 187–88; 212–13; 272; 274; 277; 287
Magnitude, *see Grösse, Zahlgrösse*, 15; 20; 37–38; 40–42; 44; 47; 51; 53; 57–60; 63–71; 91; 103; 117; 119; 122; 125–29; 131; 135; 140–41; 151; 184; 190; 202; 222; 242; 275; 281; 294
   continuous, 37; 57; 65; 69; 243
   discrete, 68–69
   variable, 131; 140
Manifolds, *see Mannigfaltigkeit*, 1; 4; 39–48; 51; 53; 57–79; 87; 91; 94; 108; 127; 131; 134; 136; 138; 142; 162–63; 166; 175; 187–90; 194–96; 202–03; 205; 207; 209; 212; 226; 258; 263–65; 267; 270; 274; 277; 281; 287; 304; 335; 341; 366; 389
   continuous, 57–58; 60–61; 63; 66; 68–69; 72–73; 134; 187; 195; 203
   discrete, 47; 61; 63; 66; 68–69; 74; 136
Mannigfaltigkeit, see manifolds, sets, xvi; 40; 44; 47; 58; 63; 66; 68–70; 72–73; 83; 87; 107; 134; 140; 187; 195; 226; 287; 304
Mapping, *see Abbildung*, correspondence, covering, function, 23–24; 55; 88–94; 107–08; 173; 191; 195–96; 206; 217; 223–32; 234–38; 240; 243; 246–48; 250; 253; 275; 281; 288; 294–95; 313; 318–19; 324; 363
   bijective, one-to-one, 23–24; 71–72; 75; 89; 148; 177–80; 187–88; 190–92; 194; 206; 211; 225; 229; 239; 275; 278; 282; 283; 294–95; 303; 319–20; 363
   injective, 89; 93; 108; 229–32; 235; 240; 246; 313; 318
Mathematics, *passim*
   modern, xi–xii; xiv; xix; 31; 253; 255; 259; 337; 340; 342–44; 346; 348; 388
   pure, 1; 6; 11–13; 26; 33; 42; 53; 65; 71–72; 81–82; 94; 117; 167; 173–74;

216–17; 232; 238; 244; 250; 260; 262; 265–66; 281; 341; 358; 371; 390

Mathematische Annalen, 114; 127; 156; 159; 162–63; 202; 267; 282; 285; 291; 312; 314

Measure theory, 146; 167; 334–35

Menge, see sets, xvi; 21–22; 50; 58; 65; 79; 126; 142–43; 160; 208; 245; 263–65; 275; 277; 279; 292–93; 341; 379

Metamathematics, metatheory, xiii; 302; 311; 345; 352–56; 357; 359–60; 366; 369; 373–74; 380–82; 385–88

Metaphysics, 19; 48; 66; 262; 266; 273; 332

Methodology, xviii; 25; 27–28; 31; 34; 42; 45; 49; 77; 82; 84; 97; 99–100; 103–05; 111; 123; 131; 133–34; 141; 149; 199; 221; 227; 259–60; 262; 264; 270; 283; 288–89; 321

Model theory, 230; 304; 388

Modules, 81; 93–94; 99; 113; 313

N

Naturforscherversammlung, 176; 276; 285

Naturphilosophie, 7; 8; 266

NBG (von Neumann–Gödel–Bernays set theory), 366; 374; 378; 380–88

Neighborhood, 140; 142; 206; 209; 213; 334–35

Neohumanism, 5; 7–9; 33

Non-denumerability, xiii; 109; 136; 167; 171; 177; 183–85; 196; 207–08; 212; 263; 268; 284; 286–87; 362

Number theory, xi; 1; 9–10; 16; 25; 27–28; 30; 32; 34; 38; 43; 75; 77–79; 82–83; 86–87; 90–102; 107–08; 111–16; 139; 141; 150; 157; 162; 173; 185; 204; 217; 221–22; 225; 242–43; 249; 254; 262; 276–77; 280; 301; 313; 360; 370

Number-classes, 207; 212; 268–69; 272–77; 281–83; 290

Numbers, *passim*
  algebraic, 1; 16; 28; 38; 75; 79; 82–83; 90; 93–100; 107–08; 111; 114–16; 132; 136; 139; 173; 177–81; 183; 185–86; 189; 200; 203–04; 217; 254;

262; 301; 313
  cardinal, 224; 229; 234; 236–37; 242; 269–70; 288–89; 291; 295; 301; 303; 324; 372–73
  complex, 14; 28; 38; 43; 54–55; 92; 95–96; 98; 119; 125; 219–21; 237
  ideal, 33; 78; 96; 100–01; 106; 109
  integer, 30; 38; 43; 94–95; 97–106; 109–10; 111–13; 115–17; 121; 128; 163; 178–80; 189–90; 217–19; 221; 224; 267; 277; 331; 362; 378
  irrational, 37; 64; 77; 100; 109–10; 125; 127; 129; 131–32; 134; 183; 187; 191–94; 219–20; 222; 274; 278; 343
  natural, 12; 14; 20; 35; 37; 50; 64; 82; 85; 88; 103; 107; 110; 121; 125; 137; 177; 179; 214; 217–19; 222–25; 230–31; 232–35; 237–38; 241; 243–47; 249; 253–54; 260; 268–69; 279; 301; 305; 320; 339; 341; 354; 359; 369
  ordinal, 21; 205; 213; 217; 236; 258; 266; 269–72; 275; 289; 294; 306; 323–24; 369; 372; 379; 383
  rational, 92; 119; 122; 125–30; 132; 134; 178; 184; 192; 203; 209; 211; 219; 221; 238; 278–79; 331; 340; 378
  real, xii; 1; 14; 37; 54; 58; 74; 81; 85; 100; 103; 109–10; 117–44; 146; 159; 171; 173; 177; 180–85; 196; 203; 211; 220–22; 243–47; 254; 260; 263–64; 279; 286; 287; 290; 301–02; 328–29; 331; 333; 340–42; 344; 358; 378; 388–89
  transcendental, 177; 180; 186
  transfinite, xiii; 4; 130; 144; 160; 201–03; 207; 210; 214; 239; 259–62; 267–68; 271–78; 282–84; 289; 291; 305; 318; 373

O

Order, xii–xiv; xix; 3–4; 8; 14; 17–19; 21; 23; 26; 38; 40; 44–46; 51; 53; 56–57; 65; 75; 77–78; 82–83; 87; 95; 98–99; 102–03; 105; 108; 115; 117; 120–21; 125–30; 133; 135–36; 138; 140; 142–43; 153; 160; 162–66; 171; 177; 179; 183–84; 189–92; 200; 203–05; 208–09; 211–

4; 189–92; 200; 203–05; 208–09; 211–
  13; 219–20; 222; 224–25; 230–32; 235–
  37; 242; 245–46; 251; 253–55; 258;
  264–66; 269–70; 273–80; 289; 291–92;
  294–95; 300–04; 308; 311; 314–16;
  318–22; 327–30; 333; 335; 338; 341–48;
  350; 353–63; 366; 368; 370; 371–77;
  380–89; 392
Order types, 209; 211; 258; 265; 274; 278–
  80; 289; 302; 320; 333; 335; 371
Ordinals, 21; 205; 207; 210–13; 217–18;
  225; 236; 240; 258–59; 262; 265–66;
  268–76; 278; 280; 284; 286; 289–90;
  294–95; 306; 310; 314; 319; 323–24;
  326; 331; 369–70; 371–74; 376; 379–80;
  383–84
  von Neumann, 370; 371; 376

P

Paradoxes, *see* antinomies, contradictions,
  17; 21; 48; 63; 67; 104; 167; 175–76;
  186; 227; 234; 237; 252; 254; 259; 262;
  266; 290–95; 299; 304–11; 314; 316–17;
  322–23; 325–27; 332; 334; 337; 340;
  342–43; 345; 347; 351–52; 355; 358;
  360; 362–63; 373–74; 376–77; 379–80;
  389–90
  Burali-Forti paradox, 314; 351
  Russell paradox, 304; 309; 317; 323
Philosophy, xviii; xx; 3–16; 18; 20; 22; 32;
  36; 41–47; 52; 59–60; 62; 66; 77; 88;
  90; 120; 130; 134; 177; 195; 201; 205;
  209; 217; 220; 228; 234–35; 241; 245;
  250; 259–62; 266–67; 281; 283; 285;
  288; 292; 294; 305; 307; 311; 326; 332;
  338–39; 342; 346; 349; 351; 353; 390–
  91
Philosophy of mathematics, 10–18; 41–47;
  90; 99–111; 119–24; 134–37; 241–46;
  259–66; 311–16; 326; 338–45; 388–92
Physics, 9–11; 19; 25–27; 33; 43–44; 46–
  47; 53; 58; 60–62; 74; 77; 119; 136;
  147; 196; 245; 266; 281; 317; 390–91
  experimental, 25
  mathematical, 9–10; 25; 27; 281
Platonism, 241; 245; 262; 310; 388; 390

Polynomials, 36–38; 54; 83; 90; 98–103;
  115; 118; 179; 205
Power, *see Mächtigkeit*, cardinals, 8; 12;
  24; 34; 36; 51; 78; 103–04; 134; 171;
  185; 187–94; 197; 202–03; 205; 210–13;
  218; 225; 229; 235; 239; 254; 258; 264–
  65; 268; 270–79; 282–84; 286–88; 294;
  296; 312; 314–15; 322; 324; 334; 343;
  363; 368–69; 375
Predicativism, *see* definition, impredicative,
  254; 264; 326–29; 339–40; 349; 352;
  357–59; 361; 381; 383
Principia Mathematica [Whitehead &
  Russell 1910/13], 253; 325; 329–32;
  352; 355; 360; 380; 391
Proof, 9–10; 20; 28; 37; 43; 54–55; 83–84;
  86; 92; 95–96; 100–07; 111–14; 116;
  118; 123; 131–33; 135; 137–38; 141–42;
  149; 151; 156; 158; 160; 162–63; 171;
  174–75; 177; 179–97; 203; 206–07;
  210–13; 216–17; 232–34; 236; 239–40;
  244–47; 249; 254–55; 259; 263; 267;
  272; 274; 277–80; 282–87; 290–91;
  294–97; 300–01; 307–08; 311–14; 317–
  20; 323–24; 327; 335; 338; 341; 343–46;
  360–62; 380; 382; 384; 388–89; 392
Proof theory, *see Beweistheorie*, 344; 388
Proposition, 49–50; 64; 133; 135–36; 182;
  194; 195; 217; 233–34; 240; 244–46;
  312; 325; 350; 362; 366; 372; 381
Propositional functions, 322; 325–29; 349;
  351–52; 355; 368; 374; 383; 387
Psychology, 19; 45–46; 48; 120; 122; 129;
  233–34; 242; 250; 351

Q

Quantifiers, 126; 253; 331; 351; 359; 383;
  387

R

Rank of sets, 277–78; 280; 370; 373–75
Recursion, 236; 290; 372; 383
  recursive definition, 222–23; 236; 354;
  359–60; 362; 368; 371; 383

theorem of, 236
transfinite, 236; 290; 373; 383
Relations, xix; 3; 16; 20; 23; 43–44; 49; 51;
　57; 59; 65; 69; 83; 91; 97; 105; 112;
　115; 118; 122; 128; 135; 158; 171–72;
　176; 185; 200; 204; 210; 220; 232; 238–
　39; 250; 252; 262; 269; 278–79; 281–82;
　289; 293; 299; 301; 305; 322–23; 325–
　26; 329; 331; 340; 343; 346–49; 356–58;
　361–63; 368
　equality, 6; 49; 64; 121; 126; 128–29;
　　　184; 226; 241; 304; 392
　identity, 51; 113; 129; 204; 228–29; 351
　logical, 250–51;262; 269; 278–79;
　　　281–82; 289; 293; 299; 301; 305;
　　　322–23; 325–26; 329; 331; 340; 343;
　　　346–49; 356–58; 361–63; 368
　membership, 33; 51; 228; 305; 307; 321;
　　　330; 353; 355; 362; 372; 379; 392
　similarity, 22–23; 55; 229; 335; 338; 372
　successor, 32; 223; 275
Riemann surface, 30; 54–60; 72; 77; 114
Rigor, 10; 12; 14; 28; 34; 36; 38; 54; 59–
　60; 69; 74; 76–78; 83–84; 96; 117; 119;
　121–22; 124; 126; 134; 138; 144; 149;
　162–63; 174; 188; 190–91; 202; 207;
　213; 216–17; 219; 222; 230; 242–43;
　247; 252; 265; 270; 289; 295; 304; 309;
　331
Rings, 15; 32; 38; 81; 95; 98–99; 101–102;
　105; 109–110; 114; 116; 367

S

Science, xvii; xix–xx; 4–9; 11; 14–15; 17;
　19; 21; 25; 28; 32–33; 41–42; 46; 64;
　74; 81; 84; 89; 91; 107; 114; 117; 122;
　131; 158; 216–17; 220; 226; 232; 242–
　44; 247; 261–62; 266; 284; 299–300;
　308; 317; 330; 356
School of research, xvii–xix; 3; 8–9; 11;
　31–32; 34–36; 38; 48; 120; 122; 158–59;
　185–86; 191–92; 198–200; 245; 250;
　253–54; 261; 284; 300; 320; 333–34;
　336; 338; 348; 382
Schröder–Bernstein theorem, *see* Cantor–
　Bernstein, 239

Sequences, 37; 66; 82; 85; 118; 124; 127–
　31; 133–35; 141; 143; 165; 179; 181;
　189–90; 192; 209–11; 213; 223–24; 228;
　230–31; 234–36; 255; 258; 261; 263;
　271–72; 275–76; 278–80; 286–87; 289;
　291; 294; 313; 324; 335; 339; 341; 362–
　63; 376–77
　fundamental, 127–31; 134; 278; 280
Series, 9; 11–12; 27; 32–36; 44; 65; 71–72;
　74; 77; 87; 93; 104; 107; 124–25; 127–
　29; 131; 136–37; 141–43; 145; 147–54;
　157–61; 163; 165–66; 168; 173; 189;
　196; 203; 207; 212; 218; 224; 242; 249;
　258–59; 261; 263; 282; 291; 326; 330;
　338; 372; 376; 381
　convergent, 36; 127
　divergent, 12; 149
　Fourier, 9; 27; 147; 149; 152; 157
　power, 12; 36
　trigonometric, 34; 71–72; 127; 129;
　　　136–37; 142–43; 145; 147; 150–54;
　　　157–58; 160; 167; 173; 258
Sets, *see* classes, collections, manifolds,
　system; *see* iterative conception, rank;
　　*see Gebiet, Gesamtheit, Inbegriff,
　　Klasse, Mannigfaltigkeit, Menge, Um-
　　fang, Vielheit,* 1; 3–4; 10; 16–17; 20–24;
　29–30; 35; 38–39; and *passim*
　Borel sets, 212; 333
　bounded, 140; 166; 340
　Cantor sets, 163; 206
　Cartesian product of, 225; 288; 313
　closed, 139; 167; 212; 300–02; 333
　complement of, 72; 138; 199; 211; 240;
　　　266; 277–79; 289; 312; 316; 325–26;
　　　333; 370; 380
　concept of, xiii–xvi; 1; 17; 21–22; 38–
　　　40; 42; 48; 64; 77; 85; 92; 104; 106;
　　　107; 117–18; 124–25; 127; 142; 173;
　　　197; 220–21; 226; 228; 250–52; 254;
　　　259; 263–65; 292; 306–07; 310; 342;
　　　361; 365; 378
　constructible, 382–84
　definition of, 107–08; 263–66; 292;
　　　321–22
　dense, 152; 154–56; 160–62; 164; 196;
　　　203; 211
　denumerable, 167; 189; 190; 207; 211–
　　　13; 268; 273; 313; 335

3; 268; 273; 313; 335
derived, 129; 137; 141–46; 153–56;
159–60; 163–67; 172; 189; 195; 202–
08; 210; 212; 258; 263; 267; 270–71;
282; 319
empty set, 159–60; 189; 207; 227–28;
313–14; 318; 321–22; 330; 375; 381
finite, 38; 87; 209; 224; 236–37; 277;
289; 293; 320; 324; 346
first species, 143; 156–57; 159–62; 165;
177; 189; 193; 207; 282
infinite, xiv; 21; 38; 68; 75; 87; 107;
109–11; 127; 171; 173; 183–84; 188–
89; 206; 229–30; 233–36; 244–46;
260; 275; 286; 289; 292; 313; 323–
24; 340; 342; 346; 357
intersection of, 51; 93–94; 113; 204–05;
227; 231–32; 247; 305; 319; 381
nowhere dense, 150; 154–56; 160–63;
165–66; 203; 209–11; 282
open, 74; 139
perfect, 167; 207–11; 278–79; 282
point-sets, xii; xiv; 1; 62; 70–74; 93;
117; 129; 137; 139; 142–47; 150;
154–57; 159–63; 165–69; 172–73;
189; 193–95; 202–14; 232; 238; 258–
59; 263; 267; 269–70; 276; 280–82;
284; 290; 300; 302; 315; 334–35; 341
power set of, 171; 225; 264; 286; 288;
314; 322; 324; 334; 363; 368–69; 375
second species, 143; 156; 159–65; 282
simply infinite, 102; 234–36; 246
transfinite, xii; 171–72; 185; 187–88;
193; 214; 225; 232; 259; 262; 265;
267; 272; 277; 285; 288–91; 294;
307; 319–20
union of, 51; 93–94; 107–08; 113; 163;
165; 190; 192; 204; 207; 211; 225;
227; 232; 240; 247; 268; 275; 288;
294; 313–14; 322; 333; 368; 375
universal set, xv; 27; 49; 126; 227; 234;
293; 307; 309; 323; 325–26; 330;
339; 362; 378; 380–81; 384
well-founded, 370; 375
well-ordered, 5; 205; 210; 225; 236;
264–65; 269–70; 273–78; 280–81;
289; 294; 306; 310; 312; 314; 318;
320; 324; 371; 380; 383

Set Theory, xi–xvii; xix–xx; 1; 3–4; 17; 20;
38–39; 48; 50–52; 62; 71; 74–76; 81–82;
88–89; 94; 103; 109; 117–18; 123; 142–
43; 145; 147; 156; 162; 165; 167–68;
171–74; 184–85; 187–88; 191–93; 197–
200; 202; 204; 205; 210–12; 214; 217;
221–22; 224–26; 230; 233–35; 237–39;
243–44; 246; 250–55; 258–59; 261;
263–68; 270; 281; 283; 286; 288; 290;
292; 295–96; 299–311; 313; 315–17;
319–21; 324–26; 328; 331–35; 337–49;
352–57; 359–63; 365–67; 369–74; 376–
80; 382–83; 385–92
abstract, xiii–xv; xvii; 1; 74; 76; 172;
184; 204; 225; 230; 255; 258; 290;
300–01; 324; 336; 365
axiomatic, 237; 296; 317; 334; 338; 349;
360–63; 366–67; 369–71; 373; 377;
379–83; 385–92
Cantorian, 81; 225; 306; 317; 341; 363;
380; 389
descriptive, 212; 333
general, 94; 147; 172; 217; 239; 244;
281; 307; 335; 337; 342; 367; 369;
372
naive, 352; 357; 363; 369; 379
point-set theory, xii; xiv; 1; 72; 137;
139; 144; 154; 160; 162; 165; 167–
69; 202; 208; 258–59; 281–82; 284;
300; 335
Set-theoretical approach, xi; xiii–xv; xvii;
1; 82; 106; 139; 174; 255–56; 297; 301
Similarity, xiv; xviii; 9; 19; 22–23; 29; 42;
44–45; 47; 55; 73–75; 86; 91; 94; 98;
102; 110–11; 119; 123; 125; 152; 164;
185; 196; 198; 218; 221; 229; 233; 236;
245; 252; 263; 275; 278; 280; 283; 301;
329; 335; 338; 340; 347; 352; 356; 365;
372–73; 378; 383
Space, 14–16; 18–19; 22–23; 46–47; 53;
58–62; 66–67; 69; 72–74; 79; 117; 132;
135–40; 142; 196; 209; 216; 220–21;
238; 242–44; 250; 335
Structures, xiv; xvi; 30; 34; 38; 51; 75; 81;
84–86; 90; 93–94; 103; 112–15; 120;
123; 167; 225; 230; 234–35; 238; 241;
246; 302; 305; 330; 337; 343; 352; 356;
376; 380; 392

System, *see* sets, xvi; 3; 5; 13; 17; 22; 24;
    44; 47; 54; 67; 81–82; 87; 89; 91–92;
    102–03; 105–10; 113–15; 122; 131–35;
    139–40; 190–91; 195; 217; 219; 221–22;
    224; 226–37; 241; 243; 247; 253; 263;
    267; 281; 292–96; 301–02; 307
    axiom, xiii; 122–23; 300–05; 307; 309–
        11; 316–17; 320–21; 323–24; 326–
        27; 330–32; 334–35; 337–39; 342;
        344; 346–57; 359–63; 365–82; 384–
        90; 392
    number, 69; 74–75; 81–82; 100; 102–03;
        110; 114–15; 117–18; 119–24; 135;
        140; 169; 173–74; 217; 219; 221–22;
        246

T

Theology, 6; 45; 130; 201; 259; 261; 266;
    283; 285; 291
Topology, xii–xiv; xvi; 41–45; 54–58; 60–
    63; 65; 68–76; 78–79; 81; 91; 114; 119;
    121; 137–39; 144; 146; 150; 161; 167;
    172–73; 181; 195; 196; 205; 208–09;
    214; 258; 274; 280; 282; 300–02; 334–
    35; 341
    topological space, 69; 209; 335
Tradition of research, xv; xviii–xix; 7; 9;
    11–12; 31; 33; 36; 59; 74; 121–23; 225;
    250; 254; 304; 345–47; 352; 378; 380;
    388–89; 391–92
Transfinite, xii–xiv; 4; 21; 75; 130; 144;
    160; 171–73; 185; 187–88; 192; 197;
    201–03; 205–07; 210–11; 213; 225; 233;
    236; 239–40; 255; 258–62; 265–69;
    271–80; 282–85; 286–91; 294; 303;
    305–07; 312; 317–20; 333; 358; 371–73;
    377–78; 383
    — ordinal, 207; 218; 259; 262; 265; 268;
        270–76; 278; 280; 286; 289–90; 294–
        95; 310; 314; 319; 326; 331; 370–74;
        376; 380; 383–84
TT (type theory), xiii; 310; 325–26; 332–
    33; 338; 346–59; 366; 373; 377; 382–85;
    387; 390
    ramified, 326; 332; 333; 351; 356; 358
    simple, 325; 338; 346; 349; 350; 351;

352; 353; 354; 355; 356; 366; 377;
    383; 384; 385; 390

U

Umfang, see concepts, extension of, 49;
    151; 303; 308
Universities, xviii; 3–9; 25–26; 31–34; 52;
    83; 115; 186; 200

V

Variable, 30; 38; 44; 57; 61; 72; 79; 87;
    110; 131; 140; 145; 147; 154–55; 190–
    92; 194; 202; 300; 322–23; 326–28; 352;
    354; 358; 372; 383; 387
Vicious circle principle, 326–28; 332; 340;
    350; 352
Vielheit [multitude], *see* sets, 76; 293–94

W

Well-order, 169; 189; 254–55; 264;
    269–70; 274; 277; 301; 311; 314–15;
    318–20; 333; 368; 370; 380; 384
    Well-Ordering theorem, 255; 277; 289;
        291; 293–95; 297; 299; 301; 312;
        317–18; 333; 341

Z

Zahlen [Dedekind 1888], 50; 79; 81; 87;
    90; 92; 98; 101; 103; 107; 110; 113;
    128; 132; 134; 173; 179–80; 183–84;
    190; 216–18; 222–25; 227–40; 243–47;
    249; 250–51; 253–54; 261; 272; 274;
    277; 296; 308; 313; 335
Zahlgrösse, see magnitude, *Grösse*, 42; 65;
    125–29; 142; 159; 178; 184
Zeitschrift für Philosophie, 285
ZF (Zermelo–Fraenkel set theory), 326;
    338; 356; 363; 366; 370–77; 381–85;
    387; 390
    with Choice, ZFC, 342; 366; 370; 377–
        78; 381–82; 385; 387–88; 392